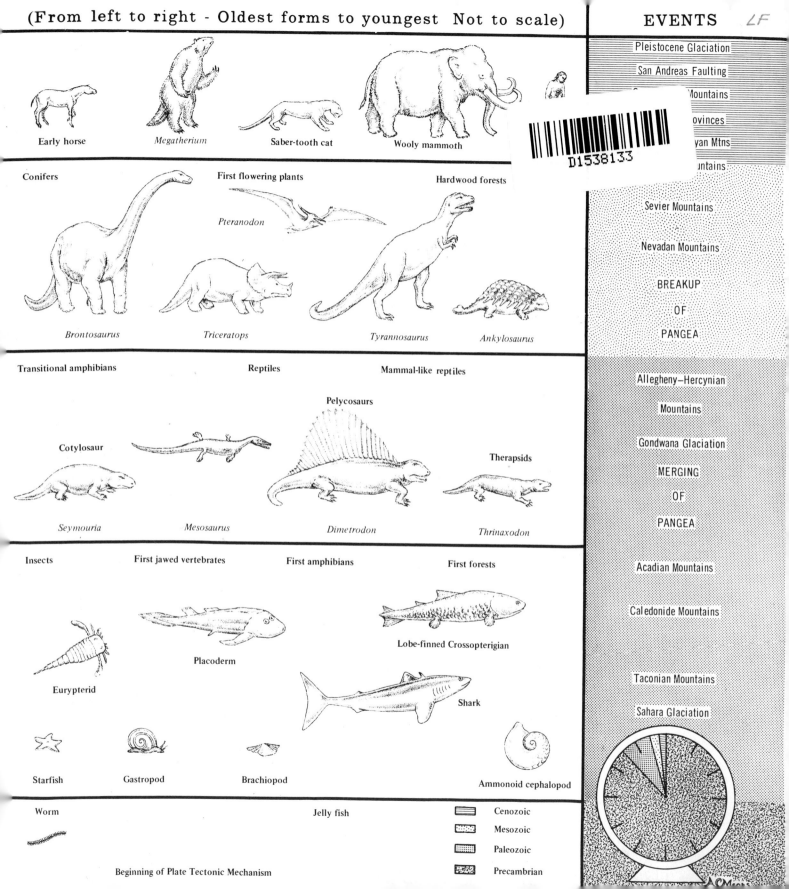

(From left to right - Oldest forms to youngest Not to scale)

Early horse

Megatherium

Saber-tooth cat

Wooly mammoth

Conifers

First flowering plants

Hardwood forests

Pteranodon

Brontosaurus

Triceratops

Tyrannosaurus

Ankylosaurus

Transitional amphibians

Reptiles

Mammal-like reptiles

Pelycosaurs

Cotylosaur

Therapsids

Seymouria

Mesosaurus

Dimetrodon

Thrinaxodon

Insects

First jawed vertebrates

First amphibians

First forests

Lobe-finned Crossopterigian

Eurypterid

Placoderm

Shark

Starfish

Gastropod

Brachiopod

Ammonoid cephalopod

Worm

Jelly fish

Beginning of Plate Tectonic Mechanism

	Cenozoic
	Mesozoic
	Paleozoic
	Precambrian

EVENTS *LF*

Pleistocene Glaciation

San Andreas Faulting

Mountains

ovinces

yan Mtns

ntains

Sevier Mountains

Nevadan Mountains

BREAKUP

OF

PANGEA

Allegheny–Hercynian

Mountains

Gondwana Glaciation

MERGING

OF

PANGEA

Acadian Mountains

Caledonide Mountains

Taconian Mountains

Sahara Glaciation

THE CHANGING EARTH, Introduction to Geology

SECOND EDITION

THE CHANGING

Brainerd Mears, Jr.

University of Wyoming

EARTH

INTRODUCTION TO GEOLOGY

SECOND EDITION

D. Van Nostrand Company

New York Cincinnati Toronto London Melbourne

To Don Blackstone and Dave Love

Two old pros who have forgotten more geology than most of us will ever know.

**Graphic illustrations
designed by Anne Carter Mears.**

D. Van Nostrand Company Regional Offices:
New York   Cincinnati

D. Van Nostrand Company International Offices:
London   Toronto   Melbourne

Library of Congress Catalog Card Number: 76-50261
ISBN: 0-442-25310-9

Published by D. Van Nostrand Company
450 West 33rd Street, New York, N.Y. 10001

10 9 8 7 6 5 4 3 2

Preface

The Second Edition of THE CHANGING EARTH is a major revision of a text designed for a one-term introductory course that discusses both the physical and historical aspects of geology. The book assumes no prior training in science and recognizes that for many students the course for which this book is written will be their sole exposure to a science at the college level.

My rationale for a second edition was to add a background theme—the changing nature of those human schemes whose interrelated patterns make up the science of geology. It is introduced in a new first chapter and occurs throughout the text in historical bits of data. Other revisions include the early presentation of plate tectonics and its integration throughout the fabric of the book; a section on astrogeology in the treatment of physical evolution; a discussion, in the first part of the book, of geological time and principles of historical reconstruction; and a presentation of the possible methods of predicting earthquakes. Measurements are given in metric units with the English equivalents following in parentheses. The appendix contains a metric conversion table for easy reference.

In light of recent developments in the teaching of geology and in the relevant applications of geological principles to our lives, I have added several new chapters. Chapter 1 introduces the nature and history of geology. Chapter 16 discusses new and changing concepts and possibilities for future work in regard to the world's mountains, continents, and ocean basins. Chapter 17, a new concluding chapter, presents the relevance of geology to population explosion, agricultural lands, climatic changes, geological resources, and the energy crisis.

Major new sections have also been added to existing chapters. An introduction to the Earth's internal dynamics and architecture as related to the plate tectonics theme has been added to Chapter 2. Chapter 4 now has a section on the historical interpretation of major rock groups based on sedimentary rocks and a discussion integrating the major rock groups into mountain building and the new world view of plate tectonics. Chapter 6 presents an updated discussion of earthquake prediction. In Chapter 7 the treatment of volcanism now includes some descriptive geochemistry of magmas and the resulting rocks. In Chapter 12 Precambrian Earth history is followed by an updated section on the Moon and a new section on Mars, which are used as indications of the Earth's earliest history.

From the first edition the text incorporates the treatments of the Earth's planetary aspects, materials, external and internal architecture, and dynamics. Topics including important fossil groups, the evolving forms of life, and the Earth's changing panoramas are also covered.

Acknowledgments

For helpful criticisms, I thank D. L. Blackstone, D. W. Boyd, E. R. Decker, J. Eaton, R. W. Fairbridge, W. E. Frerichs, W. D. Gunter, R. S. Houston, J. A. Lillegraven, J. D. Love, P. O. McGrew, H. T. Ore, R. B. Parker, S. B. Smithson, and H. D. Thomas. The influence of the late S. H. Knight on the diagrams will be recognized by many. V. J. Anderson, J. Berdan, E. A. Carter, W. B. Hall, C. MacClintock, T. Nichols, H. Pownall, L. J. Prater, J. Shelton, W. M. Sutherland, and many others helped greatly in the quest for photographs. In revising presentations and culling slips that pass in the type, comments by users of the first edition were most helpful. Special thanks go to the following individuals who read the manuscript and made valuable suggestions: Dr. Donald Owen and Dr. Les Walters, Bowling Green State University, and Dr. Monte Wilson, Boise State University. Most of all I am indebted to my wife, Anne Carter Mears, for her encouragement and her countless hours of work on illustrations, including many new ones especially designed for this edition, and for her assistance in writing and proofreading.

Contents

ON THE HISTORY OF LIFE   359

13 "DRAMATIS PERSONAE" IN THE HISTORY OF LIFE   361

14 A GREATER GENEALOGY   403

PALEOZOIC AND LATER PHYSICAL HISTORY   433

15 NORTH AMERICA AFTER THE PRECAMBRIAN   435

GEOLOGIC PERSPECTIVES AND SOME HUMAN CONCERNS 523

Introduction

Let's begin by tracing through time the much-debated and ever-changing ideas that concern the planet we live on. In Chapter One, we will examine the historical development of geology as a science. Chapter Two will present an overview of the reasoning leading to the model of the globe and a preliminary description of the Earth's internal structure.

The Andromeda Galaxy. Our Galaxy of the Milky Way is thought to resemble this spiral galaxy. Courtesy of Lick Observatory.

1

The Science of the Earth

Somewhere in the immensity of astronomical space, there is a speck of matter—we call it the planet Earth. It is one of the smallest of nine planets wheeling around the Sun, which is itself an average star among hundreds of thousands of stars in a cluster called the Galaxy of the Milky Way. This galaxy is only one of many millions of such star groups that lie in every direction as far as modern telescopes can detect.

ON THE NATURE OF GEOLOGY

And yet—the cosmic speck called Earth is important because we live here. For those of us curious about this globe we live on, geology holds an endless fascination.

Images of the Earth

Geology is the study of the Earth: its composition, structure, and history. It is the study of a volcano erupting in glowing gas, clouds of ash, and streams of molten rock. It is a mental picture of the Earth's structure from the outer surface to the center 6400 kilometers (about 4000 miles)[1] inward; or of the neat arrangement of the invisible particles that give a mineral crystal its outer form. It is the reconstruction of prehistoric plants and animals on a landscape drained by rivers flowing into long-gone seas—a scene that disappeared hundreds of millions of years ago. It is a concept of lands rising from the oceans to form lofty mountains, which are later worn away to form low rolling plains—all during an immensity of time. It is an image of the Earth revealing gigantic crustal blocks that split, drift apart to make ocean basins, and collide with each other to make mountains.

1 The kilometer is a unit of length in the metric system. One kilometer is 1000 meters and equals approximately 6/10 of a mile. In this book, a quantity will normally be given in metric units, with the equivalent in common (English) units following in parentheses. Conversion factors between the two sets of units will be found in Appendix B.

3

Geology is a collection of imaginative human schemes that help us visualize the Earth's architecture, mechanical workings, and changes through time. As with any science, the schemes vary in reliability. Some of the explanations, those called theories, are quite well established and have stood much testing. Others, called hypotheses, are reasonable explanations, although they may later be modified or possibly rejected. Some, the working hypotheses, are interesting speculations that need much more investigation; but they are important because they provide a direction for future work and thinking, even though many of them are later proved wrong. Many of these concepts, of varying reliability, are considered models. That is, they are simplified representations, such as diagrams and mathematical equations, that leave out many details of the real world but are helpful in scientific thinking and prediction. From the vast store of facts and interrelated schemes about the Earth, those that are presently accepted or considered helpful make up the subject matter of modern geology.

A few of our present-day concepts (such as the shape and size of our planet) were known to the ancient Greeks, and some basic ideas (i.e., the way rock layers are deposited by seas and rivers) date from the Renaissance. Most of our modern geologic knowledge, however, comes from the last 250 years—when geology became an organized study that attracted more and more hard-working, thoughtful, and often argumentative people to its particular way of life.

Modern Geology

Since its beginnings in the mid-1700s, the notable progress of geology has hinged on several major developments.

Instruments   Most spectacular has been the spurt of growth during the last few decades—an explosion of new geological discoveries based on twentieth-century technology. Sophisticated instruments have tremendously expanded mankind's range of perception, which in earlier times was limited to the world we could see with our unaided eyes. Today, such inventions as X-ray machines have opened up the incredibly minute world, the microcosm, of a mineral's atomic structure. In the larger world around us, shipboard echo sounders continuously record the invisible ocean-floor topography of the abyssal depths. Seismographs record information from the Earth's completely inaccessible interior. Manned and robotlike space vehicles now survey the Earth's surface and extend our geologic observations to other bodies in the solar system.

Space technology has added an ultramodern dimension to our investigations. Rocks brought back from the Moon by astronauts are now routine laboratory samples. They supplement a flood of information from space vehicles whose sophisticated "hardware" flashes back impulses through hundreds of thousands of kilometers of space. These messages, enhanced and translated, give geological data such as temperatures, magnetic fields, and chemical compositions, as well as remarkable photographic images of mountains, craters, and other features on the surface of the Moon. Similar information is becoming available from unmanned space vehicles that have visited the neighboring planets Mercury, Venus, Mars, and Jupiter.

Allied Sciences   Twentieth-century geology has become immersed in physics, chemistry, and mathematics. This development goes hand-in-glove with the technological revolution. In studying rocks and minerals, geologists now apply chemical-physical concepts such as atoms, crystals, and chemical equilibria. In determining the age of rocks, they use findings on radioactive isotopes from atomic physics. The use of statistical techniques and mathematical models—aided by the invention of the digital computer—has greatly expanded. Biology is yet another allied science. Its detailed anatomical information and evolutionary theories—which geologists were using in the nineteenth century—are now supplemented by findings in molecular biology, as tools for paleontologists. They study the geologic record of "ancient life" recorded in the rocks.

The modern geologic emphasis on technology and allied sciences disturbs some reflective people.[2] Many investigations by geologists are now based on readings from dials, flashing lights, and squiggley lines on graph paper provided by sophisticated instruments. The physical things they record are completely inaccessible to unaided human senses, and much of these "black box" data can only be presented in highly abstract mathematical equations. So, will geology become so entangled with physical and chemical laboratory manipulations that new discoveries become meaningless to intelligent nonprofessionals—totally divorced from ordinary human reality? I don't think so.

The Nature of Geologic Concepts It was a liberal injection of technology and the allied sciences that recently led geologists to accept a brand new view of the Earth—that which involves the global drifting of continental and oceanic plates. The essence of this broad concept can be represented in highly pictorial diagrams (see p. 40). Most of our concepts, even the abstract idea of geologic time—for which drawings provide only poor models—can be appreciated, if not always fully grasped, without reliance on a sophisticated background in the allied sciences. Moreover, it takes no fancy equipment to make instructive observations concerning minerals, rocks, fossils, maps—and also the great outdoors.

Spectacular mountains, magnificent canyons, vast plains, and rugged coasts can be enjoyed simply for their beauty, which is enough for some people. But such scenic features have always been grist for the mills of geologic speculations. What are the forces that make mountains? When was that canyon created? Were those rolling plains formerly overwhelmed by glacial ice? Why the rocky headlands here, and broad sandy beaches along other coasts? Answers to such questions come from the schemes of geology. Even though professional investigations may be technical, the final explanations are usually simple, comprehensible, and tangible. For the nonspecialist, they can add the interesting new dimension to life that is enjoyed by professional geologists.

THE DEVELOPMENT OF GEOLOGIC PATTERNS OF THOUGHT

The flurry of discoveries in latter-day geology, based on technology, should not blind us to the past. Present-day geologists are neither more nor less intelligent than their predecessors. The modern accomplishments stem from a long history of the development of ideas, of scientific methods, and of techniques.

Opposing Views of the Earth's Physical History

The early history of geology, when it was becoming a science, was marked by controversies which sometimes reached bitter proportions. Geologists with a conservative frame of mind were strongly opposed to what then seemed like radical new ideas.

The Uniformity of Nature Such founders of modern geology as James Hutton (1726–1797) in the late 1700s and later his disciple Charles Lyell (1797–1875) in the 1800s forced a necessary viewpoint into studies of the Earth. This fundamental creed, now known as the doctrine of *uniformitarianism*, requires continuous and gradual change of the Earth's surface over an immensity of geologic time. Hutton believed that the visible workings of

---

2 Also college freshmen taking geology as the only science they think they can pass.

Fig. 1-2 James Hutton (1726–1797). Hutton demonstrated the plutonic origin of granite and recognized the significance of angular unconformities. A few scientists before him had reasoned in uniformitarian terms, but it was Hutton and his followers at Edinburgh—though vehemently attacked for decades—who introduced the necessary concept that "the present is the key to the past" into the mainstream of geologic thinking.

streams, waves, volcanic eruptions, and other natural processes created the Earth's observable features. He rejected unnatural and extraordinary events, the likes of which were not operating at present, as geologic processes. In essence he accepted the scientific faith in an orderly universe, but he went a step further by extending the idea of the uniformity of nature back through the Earth's long history. Although he had no way of measuring geologic time, Hutton sensed its enormity, saying: "I see no vestiges of a beginning, nor prospects of an end."

Hutton's ideas were presented in his two-volume book *Theory of the Earth* in 1795. Although he was a brilliant conversationalist, Hutton's style of writing was difficult to read and obscured his meaning. Fortunately, John Playfair (1748–1819), a mathematician in Hutton's circle of friends, stated the views clearly in his book, *Illustrations of the Huttonian Theory*, published in 1802, and contributed important ideas of his own. Playfair's law states that streams have cut the valleys in which they flow. This was so, he believed, because streams are proportionate in size to their valleys, and where two streams join their surfaces are ac-

Fig. 1-3 John Playfair (1748–1819). Playfair, mathematician and philosopher at Edinburgh, demonstrated the erosional origin of valleys by streams in his Law of Accordant Junctions. An even greater contribution was his *Illustrations of the Huttonian Theory of the Earth,* in which he clearly restated and explained the brilliant ideas that might well have been lost in James Hutton's confusing and difficult style of writing.

Fig. 1-1 "And Noah removed the covering of the ark and looked, and behold, the face of the ground was dried" (Genesis 8:13). *The dove sent forth from the ark,* a print by the French artist Gustave Doré. By permission, William H. Wise & Co.

Fig. 1-4 Abraham G. Werner (1749–1817). Although he published little, Werner's teaching attracted many able students. He contributed greatly to the systematic organization of mineral studies. A firm believer in natural causes for geologic phenomena, he attempted to establish the major succession of geologic events which led to his Neptunist theory, for which he is most often remembered.

cordant (not separated by waterfalls). This seemingly simple concept, that ordinary stream erosion over a long period of time made valleys, was considered nonsense by most of the other early-day geologists. They thought large valleys were ready-made by cataclysmic rendings of the ground or devastating worldwide floods—and thereafter streams spilled into them.

Neptunists and Vulcanists   The Neptunists of the late eighteenth and early nineteenth centuries were led by Professor Abraham G. Werner (1749–1817), an excellent mineralogist and imaginative teacher whose lectures inspired a generation of his students to go forth and apply his grand scheme. Werner taught that *all* the Earth's surface rocks had been precipitated in successive layers from water in a former universal sea. The scheme was based on his observations of rocks in the countryside around Freiburg, Germany. Since the concept involved a flood and required no great age for the Earth, it fitted the biblical account of the Earth's origin as described in the Book of Genesis. Theologically it satisfied the many geologists who were shocked at Hutton's concepts of gradual change and the great antiquity of the Earth.

Opposition to Werner's scheme came quickly from France. Field observations by Jean Guettard (1715–1786) and Nicolas Desmarest (1725–1804) clearly showed that many dark-colored rocks (called basalts) in the Auvergne district of central France had been consolidated from molten streams of lava erupted by former volcanos.

Hutton's group in Edinburgh, Scotland, became the leaders of the Vulcanists, who opposed the Neptunists by stressing the work of the Earth's internal heat. Field observations by the Vulcanists indicated that besides the dark volcanic-type rocks, a light-colored and more coarsely crystalline rock (called granite) had also originated from a formerly hot molten material. The granites seemed to represent melted material that was injected into older rocks inside the Earth, but never broke out and erupted onto the surface. Because the deep-seated melt was insulated by surrounding rocks, it cooled slowly. Slow cooling allowed larger crystals to form than in volcanic fluids that were quick-chilled by air when erupting on the Earth's surface. From the existence of volcanos and his reasoning on the origin of granites, Hutton concluded that beneath an outer solid crust the Earth had a completely molten interior.

Furthermore, Hutton speculated that the Earth's internal heat solidified the loose deposits on the Earth's surface into solid rock; and that the expansive force of the internal heat and upwelling of molten masses from the depths lifted parts of the Earth's surface—contorting overlying rock layers and creating mountains. Although Hutton's speculations on the completely molten interior and the origin of mountains have not stood the test of time, his direct observations and reasoning about granites are incorporated in modern geology. As to the Neptunist-Vulcanist controversy, which had become quite bitter, Werner's grand scheme collapsed in the early 1800s when some of his more widely traveled and open-minded followers studied the rocks around volcanos.

Catastrophist Geology Not so easily dismissed were the Catastrophists, who included many of the most able nineteenth-century geologists. They considered geology as different from the other physical sciences, where natural laws remain the same. The past history of the Earth, they believed, demanded violent forces unlike any now in operation. Although completely rejecting Hutton's uniformitarian ideas, they readily absorbed his vulcanist views into their belief in worldwide cataclysms. The violent forces of the Earth's central inferno had raised whole continents and suddenly created whole mountain ranges in catastrophic volcanic blasts. The Catastrophists were experienced field geologists who smiled at Werner's naive scheme, but they too were impressed with floods.

The basic physical mechanism in their scheme for the Earth's past development came from the great authority of Georges Cuvier (1769–1832), an ardent believer in floods and the leading French scientist of the time. Cuvier made major contributions to the founding of geology. He recognized large bones, then being excavated, as those of extinct monsters, dinosaurs and mastodon elephants. He also recognized that different rock layers, lying one upon the other, had contrasting groups of fossils (remains of "pre-historic" life). Thus the Earth had been inhabited by different plants and animals in the geologic past. He also noticed that rock layers may be separated by sharp boundaries, as is sometimes the case, and that some contain rounded cobbles and boulders, and that some of the older rocks are severely contorted.

Fig. 1-5 Baron Georges Cuvier (1769–1832). One of the founders of vertebrate paleontology, Cuvier recognized that large bones discovered in France were those of extinct animals rather than living types. He and Alexandre Brongniart gave geology an essential technique for unraveling Earth history in their discovery of faunal succession. Despite his recognition that fossil groups differed, Cuvier denied the possibility of organic evolution and died a confirmed Catastrophist.

Based on his observations, Cuvier adopted a dramatic explanation: the Earth had undergone a succession of catastrophic deluges. Violent, rushing flood waters had deformed older layers, planed them off, and deposited new layers on top of them. In the process, all living things on land were exterminated. After the deluges receded, the lands were repopulated by new forms of life. Each of the successive populations had been a new and special creation; once living things appeared they never changed. In the far north, elephants (now known to be woolly mammoths) with hair, hides, and flesh undecomposed had been found encased in ice. This, Cuvier said, proved the frightful nature of the devastations, because such animals must have lived in a far milder climate that ended abruptly when they were entombed in ice.

Thus the catastrophist geologists envisioned a past history of cataclysmic volcanic explosions followed by annihilating deluges that deposited new rocks. The violent surges of torrential waters also scoured deep valleys into the Earth's surface. Once created, the valleys existed with little or no change.

Geology, A Natural Theology Professor William Buckland (1784–1856) of Oxford University, a dean of Westminster Abbey, was a leading British geologist whose widely read and influential writings represent the thinking of many prominent geologists until late in the 1800s. Buckland enthusiastically adopted Cuvier's catastrophic mechanism and added his own deeply held conviction: the ultimate purpose of geology is to give evidence for the biblical account of the Creation and the Mosaic Deluge. A liberal Protestant, Buckland believed that the Scriptures do require interpretation (for which he was attacked by fundamentalists); hence the six days of the Creation were figurative. Thousands of ages could have separated the Creation and the final, or Mosaic, Deluge.

Adam Sedgewick (1785–1873) and Sir Roderick Murchison (1792–1871), who set up much of the geologic time scale we use today, as well as Louis Agassiz (1807–1873), who gained acceptance for the concept of an "Ice Age," were among the many prominent geologists of the nineteenth century who advocated *catastrophism* and agreed with Buckland's theological views. Thus much of the field work and many concepts developed by such men have survived the former connections with their speculative philosophic mechanism and are incorporated in modern geologic thinking. Moreover, catastrophic views on such matters as the excavation of valleys were logical in the nineteenth century because in those days the great weight of geologic opinion favored an age for the Earth of only a few thousands or tens of thousands of years. Thus, they reasoned, there simply was not time to carve out major river valleys by the observably slow action of existing streams. At that time, Hutton's concept of the great length of geologic time was not proven—final confirmation awaited twentieth-century techniques of measuring the age of radioactive minerals in rocks. The gradual decline of catastrophist geology, however, accompanied the growing influence of Buckland's former student, Charles Lyell.

Establishing Geologic Uniformity Charles Lyell, who coined the word "uniformitarianism" for Hutton's doctrine of gradual change, invented no grand new theory for the Earth's geologic features. Instead, he made an airtight case (Lyell had legal training) for Hutton's rejection of extraordinary

Fig. 1-6 Charles Lyell (1797–1875). This Englishman's patient collection of evidence during the nineteenth century finally settled the violent debate among geologists as to whether slow, observable, natural processes acting through a long time or unobserved catastrophic changes accounted for most of the Earth's geologic surface features. Although by modern standards he overemphasized uniformitarianism, his work established a sound theoretical basis for geology.

geologic events in the Earth's history. Lyell traveled widely and read extensively in amassing the evidence.

In challenging the Catastrophists' version of volcanic workings, Lyell adopted the thinking of George P. Scrope (1797–1876), who had suggested that observable eruptions created volcanos. Scrope rejected the catastrophist view that volcanos appeared full-grown in single cataclysmic blasts; rather they grew by sporadic eruptions over a long period of time. Lyell visited many active volcanos, including Mount Vesuvius and Mount Etna to establish that ongoing eruptions could indeed gradually build up the large volcanos. From a human point of view, some eruptions are devastating, but they are catastrophies of a sort that are observable or well recorded in written human history.

Lyell gradually demolished the Catastrophists' most fundamental mechanism: Cuvier's notion of cataclysmic, worldwide deluges. The vast quantities of sands and muds presently being carried along and deposited at the mouths of large rivers could—given time—account for the thicknesses of all the Earth's surface rock layers. The large boulders scattered over many places in northern Europe and North America were not all blasted from incredible volcanic eruptions, nor were they relics of global torrents. Former glaciers of the Ice Age, landslides, and devastating but local floods scatter very large boulders on the landscape. Modern textbooks present much of Lyell's reasoning.

Lyell developed his viewpoint in *Principles of Geology,* a textbook in three volumes (the third later became the separate *Elements of Geology*) that went through 12 constantly updated and revised editions from 1830 until his death 45 years later. The clear, nontechnical, and unemotional style of his presentation made his books enormously popular and greatly influenced nineteenth-century thinking. His uniformitarian concepts are not complex, abstract, or even original ideas, but Lyell's brilliance lay in his presentation of massive evidence, clear reasoning, and critical evaluation that made them seem almost self-evident. It was his work that led other geologists to a proper understanding of their field observations.

In his role as the "high priest of uniformitarianism," Lyell was perhaps overzealous. His insistence that geology admit only the everyday, observable processes is too rigid by modern standards. Lyell recognized that the Earth has had different climates in the past; in fact he attributed animal extinctions to local changes in environmental conditions rather than global deluges. But it now seems clear that along with climatic changes, the intensity of geologic processes has differed markedly at some times in the past—during Ice Ages, for example. Far greater floods Ch. 16 than any remembered or seen in written human history—although clearly not global deluges—now seem possible from geologic evidence. In its primeval geologic development, and occasionally in later times, the Earth may well have had catastrophic encounters with great meteoroids from the solar system—a possibility based on modern studies of space-vehicle images recently sent back from the little-changed faces of the Moon, Mars, Venus, and Mercury. Moreover, although the existing volcanos clearly were built by long-continued eruptions, some of them have been virtually obliterated by geologically sudden, extremely violent eruptions followed by cataclysmic collapse.

But if Lyell was overly enthusiastic, a failing of many evangelists, his work was absolutely essential in establishing a basis for worthwhile geologic thinking. No modern geologist is about to espouse the mechanisms of nineteenth-century catastrophism. And, in proving that volcanic cataclysms and annihilating deluges had not shaped the Earth's surface, he undermined the physical mechanisms for global exterminations of past life. Gradualism in the Earth's past physical history led inevitably to a similar view of slow and continuous change in the development of life. The implication was clear to religiously devout and perceptive catastrophist geologists such as Adam Sedgewick, a professor at Cambridge University, who considered uniformitarianism repugnant. To such thoughtful geologists as Sedgewick and Buckland, uniformitarianism was unthinkable because it also meant slow gradual change in the history of life. Lyell himself originally considered life forms as unchanging—the lack of some later plants and animals in older rock layers he attributed to the spotty nature of fossil preservation and gaps in the record. Lyell, however, as his more mature catastrophist confreres quickly noted, was a quick convert to organic evolution when Charles Darwin (1809–1882) presented his concept.

Organic Evolution

"There is nothing new under the Sun" (Ecclesiastes 1:9), and in a way this is true of the scientific theory of evolution.

Early Schemes In Greece, six centuries before Christ, Thales of Miletus speculated that the teeming waters of the seas were the mother of all living things. Aristotle in the fourth century B.C. proposed a progression of life by gradual stages from plants to plant-animals to animals to man. In later times, because of the authority of Scripture, such ideas lay dormant—but by the late 1700s they had surfaced again.

Erasmus Darwin (1731–1802) was a competent medical doctor, something of a dreamer, and an unconventional man. He forecast the steamship, automobile, and airplane long before their times. Tied to his medical practice, he could not engage in world travel; so he substituted the study of biological specimens from all over the world which were available in London's museums. He did not memorize impressive scientific labels of plants and animals, but rather wondered about their methods of defense and concealment and many of their other adaptations to their environments. An om-

nivorous reader, he digested the works of David Hume, the philosopher who rejected transcendental causes and said that all knowledge comes from observation and practical experience; that similar causes produce similar effects; and that people and their institutions gradually evolve. Erasmus also read the economist Adam Smith, who fathered the "free enterprise" system and believed that human development resulted from competition.

From such unlikely sources, Erasmus Darwin concocted the idea that in the world of living things free competition among plants and animals involving purely natural causes would lead—if there was no outside interference—to the progressive improvement of living things and the evolution of more advanced forms. In a British society distressed at the excesses of the French Revolution, Erasmus's truly radical ideas caused not a ripple of dissent—for he presented his views in epic poems. One especially long one was entitled *Zoonomia, or the Laws of Life*. His works were accepted as "good literature," but scientists—perhaps because the ideas were in poetry—seemed unaware of them.[3] Erasmus Darwin's most lasting contribution to the theory of evolution was a grandson, named Charles.

At about the same time, in early-nineteenth-century France, Jean Baptiste de Monet Lamarck (1774–1829) proposed an evolutionary scheme. He set up a still-valid classification of the invertebrates—animals without backbones, such as clams, crabs, insects, and worms. In studying their hearts, digestive systems, and other organs, Lamarck noticed a continuous gradation in anatomy from simple, minute, jellylike polyps through increasingly more complicated invertebrates, thence from fish, reptiles, and mammals to man. Thus he suggested that all animal species had gradually evolved, over a long period of time, from the primitive jellylike globules through a continuous chain of life to man.

3 C. P. Snow's modern essay *Two Cultures* is relevant to the situation.

Lamarck also had an explanation. All living things, he said, have an "inner force" tending to improve them and convert them into more perfect, higher forms. The mystical idea of inner forces (which goes back to Aristotle) has appealed to some philosophers and attracted scientists from time to time. Most modern biologists and geologists, however, prefer scientifically demonstrable explanations; and here Lamarck proposed a rational mechanism. Characteristics acquired during an organism's lifetime, or over several generations, would eventually become hereditary and be passed along to offspring. For example, if generations of short-necked giraffes kept stretching to eat the leaves on trees, the exercise would lengthen their necks (just as an athlete can build up muscles); the longer necks would eventually be inherited by newborn offspring.

In his day, Lamarck's ideas fell into obscurity from the ridicule of his former protégé, the brilliant and articulate Georges Cuvier, author of the catastrophist-creationist theory of unchanging organic species. Some 50 years later, Lamarck's ideas were rediscovered and restored to grandeur by the French as an alternative evolutionary mechanism to the one proposed by Charles Darwin.

The Origin of Species Charles Darwin's greatness lay in forcing the theory of organic evolution into the mainstream of scientific thinking. He gained acceptance for a previously offbeat—and unpopular—speculation; and, most important, he discovered a reasonable natural mechanism. Darwin succeeded where others had failed because he gathered a mass of evidence and spent 20 years thinking, observing, experimenting, and thinking again—before publishing, in 1859, his major work: *On the Origin of Species by Means of Natural Selection, or the Preservation of Favored Races in the Struggle for Life.*

Even if Darwin had never written his monumental biological works, he would still be remembered as a geologist. His first exposure to the

Fig. 1-7 Charles Darwin (1809–1882). Darwin ultimately gained scientific acceptance for the concept of organic evolution and suggested a reasonable mechanism—a major intellectual achievement of the nineteenth century. The opposition generated in his time (and in some circles today) was violent, but today the scientific evidence, much from paleontology, overwhelmingly supports the concept, although the mechanism is still being investigated.

subject was a bit unfortunate. At medical school in Edinburgh (he dropped out because he couldn't stand the sight of blood), Darwin attended Professor Robert Jameson's geology lectures, which were so incredibly dull that he vowed never to study the subject again.[4] His interest revived, however, on a long field trip with Professor Adam Sedgewick. During the excursion, he learned that science involved grouping facts as a basis for drawing general conclusions. Darwin's future was determined

4 But then, Jameson was a mineralogist (the leading British follower of Abraham G. Werner).

by his selection as naturalist for a five-year world voyage, starting in 1831, on H.M.S. *Beagle*. During the trip he enthusiastically absorbed Charles Lyell's *Principles of Geology;* studied uplifted wave-cut terraces, fossils, and volcanos; and developed a theory (confirmed by modern work) for the origin of ring-shaped coral islands, called atolls, of tropical seas. Perhaps most important, Darwin credited the influence of Charles Lyell and his uniformitarian views as leading to his own ultimate recognition of the fact of evolution.

In South America, Darwin saw piles of bones lying on the ground, from animals killed in a drought. He also found many fossil bones of large extinct animals in gravel deposits. The extinct animals had still-living close relatives, and bones of these modern types were also in the gravel deposits. He began to wonder if the ancient animals whose bones were in the gravels had also died in a drought or by other ordinary causes. It led him to question the catastrophist theory (Cuvier's idea of deluges and special creations). "The present is the key to the past" might apply to the world of life, just as Lyell believed it did to the Earth's physical history.

Darwin also studied living plants and animals in South America. Later the *Beagle* sailed to the Galapagos Islands. There the birds, lizards, plants, and other living organisms were South American types—but they differed slightly from island to island, and were notably different from comparable forms on the South American mainland. On mulling over this new discovery after returning to England, Darwin concluded that South American animals and plants had become isolated on the Galapagos Islands, and then gradually changed— the uniformitarian view in geology had led him to the same idea for the organic world—in short the concept of organic evolution.

The mechanism for evolution, however, was suggested to Darwin by another source—Thomas Robert Malthus's *Essay on Population*. The Reverend Malthus (1766–1834), an early advocate of family planning, predicted that mankind's unrestrained reproduction would outstrip food supplies and cause widespread famine.[5] In nature, Malthus said, the struggle for limited food supplies checked overpopulation. In thinking this over, Darwin noted two obvious facts: many organisms produce a tremendous number of offspring; but populations remain relatively constant. For, in the "balance of nature," big fish eat many smaller fish, who eat many smaller animals and insects, and so on. From these facts he concluded that the struggle for survival keeps natural populations in balance with available food supplies.

A third fact—which Darwin discovered—was that within populations of an interbreeding type (or species) plants and animals all vary slightly. No two individuals are exactly identical. The realization led to his most important conclusion. In the struggle for survival, organisms with slightly advantageous differences were those most likely to survive and reproduce. This mechanism of natural selection, over a long period of time, would cause gradual changes in living populations—organic evolution.

Darwin's work still left many questions. The cause of the slight variations in populations was not understood until twentieth-century biologists studied such matters as the chemistry and structure of the reproductive structures called chromosomes. Darwin's theory, in fact, provided a whole set of interesting new problems, which, of course, are the lifeblood of ongoing investigations in science. But today, no competent biologist or geologist doubts that life originated in the remote past and has gradually evolved during an immense span of Earth history.

---

5 Malthus's prediction was not realized in his time because of improvements in agriculture and transportation of food supplies. Today, the exploding human population, coupled with world climatic changes (as in the famine zone of Africa), could soon make Malthus's prophecy into a grim reality.

Geology and Human Beliefs

The grand schemes of geology have long been entangled with the cherished notions of people outside the profession. Who cares, except for geologists, whether streams cut their own valleys and whether volcanos build up over a long period of time?

We Live with Our Fellows As a matter of fact, during the nineteenth century especially, a great many people were deeply concerned and interested. The poet Alfred, Lord Tennyson (1809–1892), wrote: "There rolls the deep where stood the tree" and "The hills are shadows and they flow/From form to form and nothing stands." The lines reflect the impact of Charles Lyell's thinking and differ markedly from the poetic images of "eternal mountains" and "everlasting hills" of Genesis 49:26. In the nineteenth century, collecting minerals, rocks, and fossils was a popular hobby—and the specimens involved geologic interpretations that were highly controversial.

In his innocent occupation of studying rocks in the late 1700s, Dr. Hutton had unintentionally stirred up a major theological controversy because his geological theory (previously discussed) implied that the Bible was in error. A similar intellectual clash had occurred 200 years earlier when Galileo Galilei (1564–1642) was condemned for heresy during the birth of the scientific revolution. Then, science had collided with Catholic theology when Galileo supported the astronomy of Nicolaus Copernicus (1473–1543)—the theory that the Earth was not the fixed and stable center of the universe. Now Hutton challenged the reality of Noah's Flood (an extraordinary event) and envisioned the Earth as immensely older than the biblical account of its creation in about 4000 B.C.

Hutton was dismayed by the bitterness of the Neptunist-Vulcanist controversy. He was attacked as an atheist and subverter of the established order by the likes of Richard Kirwan, an articulate chemist and Neptunist, president of the Royal Irish Academy, who heaped scorn on Hutton's poor literary style and devoutly defended the literal six days of the Earth's creation recorded in the Book of Genesis.

The Neptunist collapse and the vehemence of the encounter made geologists wary of grand schemes based on insufficient evidence. Henceforth they tended to emphasize observation and the careful collection of field evidence. Thus the subsequent nineteenth-century disagreement between Catastrophists and Uniformitarianists—although underlain by the same theological issue—was a relatively calm and good-natured affair (at least in geologic circles). Lyell, the uniformitarian leader, had been a student of Buckland, a leading Catastrophist. Moreover, the argument was largely between members of the same "clubs," the British geological societies. The elders could smile at the indiscretions of youth, and the younger members could recall their own former old-fashioned views. Catastrophist geologists did not require a literal interpretation of the Book of Genesis; on the other hand, Lyell presented his case calmly and carefully avoided offense to theological sensitivities.

Since the geological contestants were all proper Victorians, the nineteenth-century contentions were never "a battle between science and religion." Buckland and his fellows clearly believed their geological findings gave proof of the Divine Providence. Lyell saw in his work "clear proof of a Creative Intelligence" and held it unthinkable that "the evidence for the beginning or end of so vast a scheme lies within the reach of our philosophical enquiries or even of our speculations." Then too, the general social climate of the times favored a more tolerant outlook. Hutton's "radical" views were being argued when the excesses of the French Revolution were on people's minds; but as the nineteenth century progressed, the British middle class came to favor the social idea of mankind's progressive improvement—an evolutionary concept. But if the physical aspects of Earth history could be treated calmly, the development of life and man's ancestry was quite another matter.

Man's Ancestry In the 1840s, *Vestiges of the Natural History of Creation* was first published. This enormously popular book was anonymous, and became a favorite conversation piece at dinner parties and in intellectual and scientific circles. Speculations as to the author ranged from Prince Albert to William Makepeace Thackeray to Charles Lyell. The book was actually written by Robert Chambers (1802–1871), a highly literate Scottish editor and self-taught naturalist, who stated that life had successively progressed from simpler to more complex animal types, culminating in man. Chambers presented the scheme of evolution as a natural theology requiring only one special creation that was followed by the operation of natural law as decreed by the Deity. Chambers believed this scheme was more befitting to the intelligence of the Divine Creator than the constant interventions and new special creations required by catastrophism.

Theologians, of course, were outraged, and geologists—who usually ignore pseudoscientific works—were aroused.[6] Adam Sedgewick "exploded" that the book was false philosophy, violated social manners, made the Bible a fable, and meant that we were not made in the image of God but were the children of apes. Even Lyell, Thomas Huxley, and Charles Darwin, who could have been sympathetic, gave the book unfavorable reviews because it contained no original research and was largely a compilation of material from a wide variety of scientific sources that were frequently misunderstood. Although improved in later editions, the book contains much science fiction; but Chambers's intuitive forecast of organic evolution touched a raw nerve of Victorian science that soon felt the

impact of Charles Darwin. And no scientist ever said that Darwin wrote pseudoscientific nonsense.

In *The Origin of Species*, a difficult and highly technical book, Darwin mentioned briefly—only once—that his work might involve mankind's ancestry. He purposely avoided the emotion-charged issue, but the mere hint aroused righteous wrath. The outrage that "Darwinism" generated erupted far and wide through Western society. Darwin himself avoided active embroilment because he was a shy person and never in good health after the *Beagle* trip. But among the scientists who sprang to battle on both sides was the brilliant biologist Thomas H. Huxley (1825–1895). As Darwin's "Bulldog" he enthusiastically argued the case for evolution. Samuel Wilberforce (1805–1873), bishop of Oxford and an eminent scientist in his own right, vehemently led the opposition.

The climax of the controversy, in intellectual circles, came in June of 1860 at a meeting of the British Association for the Advancement of Science. During formal debate, Wilberforce asked "whether Huxley's ancestral ape was on his grandmother's or grandfather's side of the family." Huxley retorted that "he accepted a monkey ancestor, but would be ashamed to be descended from a man who used his great intellectual gifts to obscure the truth." In the shocked audience, a lady fainted. But the bishop had overdone it and offended the British sense of fair play. Thereafter, Darwin's theory gained an increasingly sympathetic audience. In 1871, Darwin published *The Descent of Man, and Selection in Relation to Sex*, which further shocked Victorian society. This book presents the genealogy from fish to man and attributes human evolution to the feminine preference for the most intelligent and least hairy mates (a debatable hypothesis first suggested by Darwin's grandfather Erasmus). The furor continued, but a vast amount of evidence, much of it from fossils in the geologic record, made the occurrence of evolution a virtual certainty. Scientific interest now focuses on evolutionary mechanisms.

6 A somewhat comparable present-day phenomenon is the flurry precipitated by Immanuel Velikovsky. His stimulating *Worlds in Collision* and other books, however, represent an "advance backwards" in presenting catastrophic astronomical mechanisms for physical happenings reported in the Old Testament.

In the Grand Scheme of Things   Most of the older catastrophist-creationist geologists of the nineteenth century never accepted either uniformitarianism or organic evolution. But their cause was lost; badly weakened by the geologic work of Lyell, it collapsed from the biological impact of Darwin's theory. As the catastrophists died, none of the younger generation of geologists carried on their schemes. And except for a rearguard action by people whose textbook of geology was the Old Testament Genesis, Western society came to have a new cosmological image. Copernicus had removed our dwelling place from the center of the universe. Hutton had opened vistas of an immense Earth history far preceding man's late arrival on the scene. Darwin made us an end product of continuing natural mechanisms. Mankind seems relegated to insignificance in the vastness of space and the enormity of time. It is a disquieting image. Yet it is mankind—or a few of the most gifted members—that developed the awareness and intellectual methods needed to grasp the cosmic scheme.

SUGGESTED READINGS

Adams, F. D., *The Birth and Development of the Geological Sciences*, New York, Dover Publications, 1954.

Fenton, C. L., and Fenton, M. A., *Giants of Geology*, Garden City, N.Y., Dolphin Books, Doubleday & Co., 1952 (paperback).

Gillispie, C. C., *Genesis and Geology*, New York, Harper & Brothers, 1959 (paperback, Harper Torchbooks).

Leveson, D., *A Sense of the Earth*, Garden City, N.Y., Doubleday & Co., 1972 (paperback, Anchor Books).

Lyell, C., *Principles of Geology*, New York, D. Appleton & Co., 1872.

Mears, B., Jr., *The Nature of Geology*, New York, D. Van Nostrand Co., 1970 (paperback).

Playfair, J., *Illustrations of the Huttonian Theory of the Earth*, New York, Dover Publications, 1964 republication of 1802 book (paperback).

Walker, M., *The Nature of Scientific Thought*, Englewood Cliffs, N.J., Prentice-Hall, 1963 (paperback).

Wendt, H., *In Search of Adam*, New York, The Crowell-Collier Publishing Co., 1963 (paperback, Collier Books).

Apollo 17 photograph of Earth, December 1972. Courtesy of NASA.

2

The Planet Earth

PLANETARY FEATURES

The Earth in Space

The modern image of the Earth's place in the universe (described in the opening paragraph of Chapter One) is now so widely accepted that most people take it for granted. Yet it is far from self-evident. In fact, it goes against all common sense and personal experience. Simple observations clearly suggest that the Earth is the fixed and stable center of the universe, that the Sun, Moon, and visible planets are smaller bodies moving around us, and that the stars are points of light in a revolving sphere above us. Just such an Earth-centered astronomy was fully developed by Ptolemy of Alexandria in the second century A.D. and, for many centuries, it was the accepted system (Fig. 2-1).

Nicolaus Copernicus started a scientific revolution when he challenged the Ptolemaic Earth-centered theory for the universe by presenting his Sun-centered scheme, which has developed into the one we accept today (Fig. 2-2). With hindsight, it is easy to condemn the thinking of the Inquisitors who persecuted Galileo because he believed in the Copernican system. Aside from theological implications, however, the affair should be viewed in the context of scientific knowledge in the early 1600s. For centuries the Ptolemaic system had worked well in predicting the apparent motions of the planets (certain of its basic assumptions are still used today as working approximations in marine navigation). In the 1600s, the Copernican heliocentric concept was merely another possible hypothesis.

Ultimately the old Ptolemaic system had to be replaced. As more and more astronomical observations were made, increasingly complicated geometric manipulations were needed to fit them to the theory. Many inconsistencies developed that were better handled in the Copernican scheme. The heliocentric system was a more simple and mathematically elegant explanation (a point in favor of any scientific concept) for the observed motions of the planets. But many years

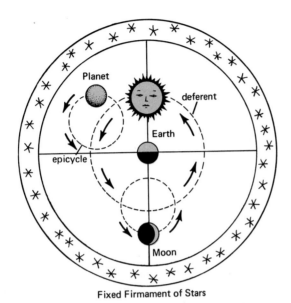

Fig. 2-1 Ptolemaic system of planetary and solar motion based on an Earth-centered universe. Adapted from a fifteenth-century German text.

passed before the theory was proven, and confirmed by critical observations and experiments.

Although educated people today can recite the "facts" of our present concept of a solar system, few have any idea as to why we think it holds true. In dealing only with the Earth, evidences of shape and especially of motions are actually based on subtle observations and inspired reasoning. The now-established view of the solar system, as shown in Fig. 2-2, was an early triumph of scientific thought over old common sense, and was a harbinger of many triumphs to come.

Rotation The Earth rotates (spins) on an axis through its north and south poles (Fig. 2-3a). The time required for one rotation is about four minutes less than 24 hours. Thus a point on the Earth's surface at the equator, where the circumference is 40,000 kilometers (about 25,000 miles), is moving on a curve at a rate exceeding 1600 kilometers (1000 miles) per hour. Exactly at the north and

south poles the rate of speed is zero, and between the poles and the equator the rate varies with the circumference of the parallels. For example, the 60° parallel is a circumference half as great as the distance around the Earth at the equator; hence a point at latitude 60° is rotating at about 800 kilometers (500 miles) per hour.

How do we know that the Earth rotates? The apparent movement of the Sun through the sky from east to west, which causes night and day, results from the Earth's rotation. But it is not proof of rotation, because the same motion would occur if the Earth were motionless and the Sun moved around it.

In 1851 the French scientist Foucault performed his famous experiment, often cited as proof of the Earth's rotation (Fig. 2-4). Inside a high building (the Pantheon in Paris) he suspended a heavy iron ball from a wire 60 meters (196.8 feet) long and carefully set it swinging. As the hours of the day passed, the path of this pendulum appeared to shift, to the right in Paris—and in later experiments in the Southern Hemisphere, to the left. There were two possible explanations. Either the pendulum's path was shifting and the Earth was fixed, or the path was fixed in space and the

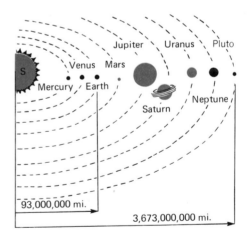

Fig. 2-2 Diagrammatic representation of the solar system by modern accounts (not to scale).

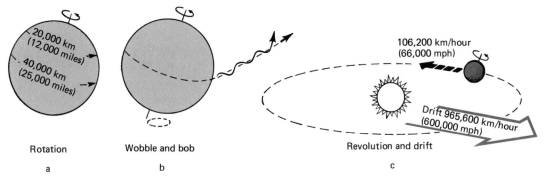

Fig. 2-3 Movements of the Earth.

Earth actually shifted beneath it. Thus, at first glance, it might appear that we are no more sure which is which than we were when we watched the Sun apparently move across the sky. But a pendulum can be manipulated and its motion studied, and it can be shown that the plane of a swinging pendulum is fixed in space. The Foucault pendulum swings, therefore, in a fixed plane, and its apparent shift is a result of the Earth's moving beneath it.

The Germans during World War I encountered a more modern proof of the earth's rotation. Long-

Fig. 2-4 M. Foucault's famous experiment in Paris (1851) wherein he used a pendulum to demonstrate the Earth's rotation.

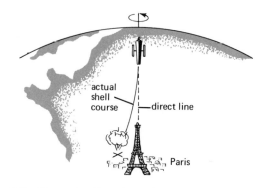

Fig. 2-5 Since the Earth's surface was rotating more rapidly at Paris than where "Big Bertha" was fired, the cannon shell hit consistently to the right. Paris was a "moving" target and should have been sighted with a lead.

range artillery shells, shot some 120 kilometers (about 75 miles) south at a target in France, were unexpectedly missing their target by about 450 meters (almost 1500 feet) to the right. As the shells did not deviate in their course from the gun muzzle, the deflection was judged to have resulted from the slightly different speed of rotation of the Earth's surface in the target area (Fig. 2-5). Although negligible for most shooting, at very long ranges the Earth's rotation affects accuracy of aim.

Revolution The Earth revolves in an orbit around the Sun. The average distance of the Earth from the Sun is about 149,000,000 kilometers (approximately 93,000,000 miles), although this varies, since the orbit is slightly elliptical rather than a perfect circle. A trip around the orbit takes 365¼ days (one year). From this (assuming, for simplicity's sake, that the Earth's orbit is circular) we can calculate the Earth's velocity in orbit as 106,000 kilometers (about 66,000 miles) per hour.

Parallax of stars is one of several proofs now available for the Earth's revolution. To demonstrate parallax, hold up a finger at arm's length, and closing your right eye, sight across the pointing finger to some distant object. Now open your right eye and close your left. Notice how the finger now points to the left of the distant object, as though it or the background had done some shifting. This effect is a parallactic shift. In Fig. 2-6 (much ex-

aggerated) an observer on a revolving Earth who sights a near star overhead at sunset, say in June, should see it shift against the background of more remote stars by December, and then shift back again by June—a kind of weaving back and forth as the Earth reaches and leaves extreme positions in its orbit. We can put this expectation in the scientific form of the Copernican hypothesis to be tested, thus: *If the Earth revolves around the Sun,* then we predict that an observer will note a parallactic shifting back and forth of some fixed celestial objects against a background of other and more distant celestial objects. If no such effect is noted, then the Earth is not revolving around the Sun (or else all fixed celestial objects are at the same distance!).

The ancient Greeks were well aware of the simple logic of this hypothesis, and the fact that they could detect no parallactic shift among the stars over the seasons convinced them that the Earth was motionless with respect to anything fixed in space. They did not realize that astronomical distances are so enormous that, with their limited means of observation, it was as if all fixed stars were at the same infinite distance. Only through refined modern instruments has it been possible to register the very slight displacement of nearer stars against farther stars over the course of the year and so confirm the hypothesis of Copernicus.

There are still other motions of the Earth through space. The Earth's axis of rotation is slowly moving around, making a circuit in about 26,000 years. This motion, called precession, is like the wobbling on its axis of a spinning top (Fig. 2-3b). Besides this, the Sun is moving at about 19 kilometers (12 miles) per second towards a point in the galaxy near where the star Vega is now. And as the Sun moves, its planets, including the Earth, drift with it (Fig. 2-3c). Thus anything at rest on the surface of the Earth (e.g., you reading in your chair) is moving rapidly through space in a motion that can be described with at least four components, to say nothing of other less prominent ones.

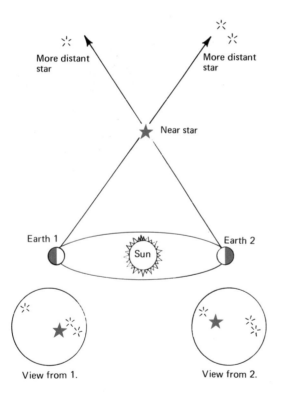

Fig. 2-6 Astronomers looked for parallax of stars a good many years before they found it because even the nearest stars—unlike the diagram—are tremendous distances away and any movement is very slight.

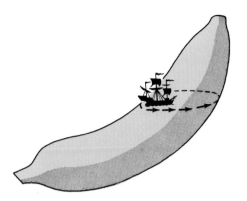

Fig. 2-7 Circumnavigation did not prove the Earth was a sphere.

The Sphere

Shape Magellan's sailors, who finished circling the globe in 1522, are generally held to have proved that the Earth is a sphere. Their voyage did indeed show that the Earth is a solid of some sort and certainly not flat and that people would not fall off into space from an underside. But Magellan's crew could have sailed around a planet shaped like a banana (Fig. 2-7); so their voyage was not after all proof that the Earth is a sphere.

A spherical Earth had been proposed as early as the sixth century B.C. by the Greek Pythagoras.

In the fourth century B.C., Aristotle, the Greek philosopher and a great investigator of scientific questions, gave simple proofs based on observation. He noted that ships passing over the horizon at sea disappear hulls first, then decks, then masts (Fig. 2-8). If the manner, rate, and distance of disappearance are always the same in any direction and place on Earth, the Earth must be a sphere. He also noted that the shadow of the Earth on the Moon during a lunar eclipse is always an arc of a circle. This shadow is cast by different sides of the Earth at different times, and the only possible shape that could do this is a sphere.

Size Once it is demonstrated that the Earth is a sphere, the next question is how big it is. Today the circumference of the Earth is known to be just over 40,000 kilometers (about 25,000 miles). A good approximation of this was made in the third century B.C. by Eratosthenes.

His method of reasoning is as good today as it was then. Eratosthenes, living in Egypt, knew that Alexandria was at a distance of about 5000 Greek stades north of Syene (near modern Aswan), and that at noon of the longest day of the year, the June solstice, the Sun was at its zenith (directly overhead) at Syene because its rays reached the bottom of a deep well there. At the same time at Alexandria, however, the Sun was 7½° south of the

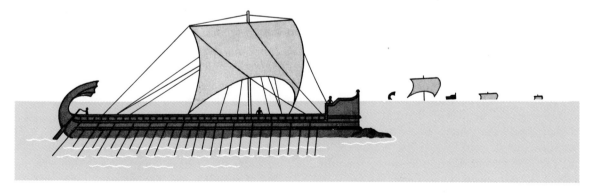

Fig. 2-8 Ancient Greek bireme seemingly dropping below the horizon.

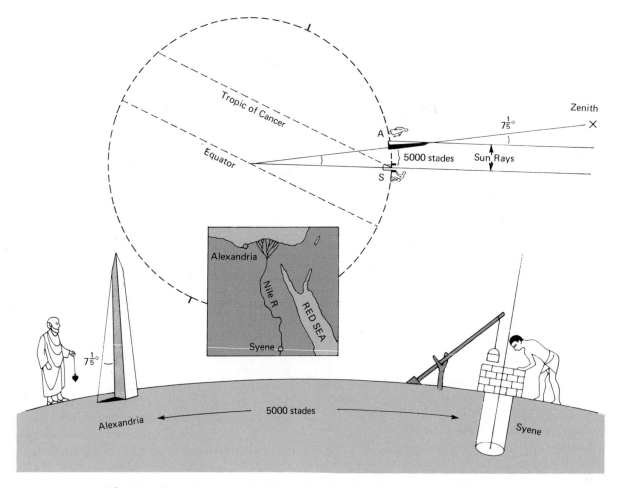

Fig. 2-9   Eratosthenes's method for determining the circumference of the Earth.

zenith as determined by observing the shadow cast by a vertical rod set in the ground (Fig. 2-9). The Greek astronomer took for granted that the Earth is a sphere, that the Sun's rays at Alexandria are parallel to the rays at Syene, and that a plumb bob points towards the center of the Earth. He reasoned that when the plumb line extended to the Sun at Syene, as it would when the bottom of the well was illuminated, it did not do so at Alexandria because of the curvature of the Earth. As the angle between the plumb line and the Sun's rays at Alex-

andria is equal to the angle between two straight lines drawn from Alexandria and Syene, respectively, to the center of the Earth, it can be determined what part of a circle the ground distance between the two cities represents. Knowing this, Eratosthenes could calculate the circumference.

Eratosthenes's method was a good one and his result, a circumference of 250,000 stades, is quite accurate if the stade he used was about 160 meters ($\frac{1}{10}$ of a mile). Of this we are not sure, for three different stades were used in ancient times.

Oblateness The Earth is not a perfect sphere. It is an oblate spheroid bulging slightly at the equator and slightly flattened at the poles. As a result the Earth's polar diameter is about 43 kilometers (27 miles) shorter than its equatorial diameter—not much compared to the diameter of the Earth of over 12,000 kilometers (7900 miles), but enough to be a factor in precise mapping and surveys.

Deductive proof of the Earth's oblateness was first given by Isaac Newton (1643–1727) from theoretical considerations of the law of gravity and the Earth's rotation. In a general way it can be explained that because of gravity all points of the Earth are attracted toward its center. This would make the Earth a perfect sphere but for one thing: the Earth is not at rest with respect to itself. It rotates. Every particle on the surface of a rotating body tends to leave the curve in a straight line— a tangent—just as mud tends to fly off a rapidly spinning wheel. The inward force of gravity (centripetal force) greatly exceeds the momentum of any particle that tends to leave the Earth at a tangent, but this momentum has an effect. Remember that the parts of the Earth's surface move at different speeds. A particle on the equator is moving in a curve at more than 1600 kilometers (1000 miles) an hour, while a particle near the North Pole may be moving in a curve at only 50 kilometers (31 miles) an hour. The momentum of the particles at the equator, tending to make them leave the curved surface on a tangent, is very much greater than the momentum of particles near the poles. Accordingly, since the Earth is not completely rigid, the region of the equator bulges outward slightly and the polar regions are a little flatter than they would be if the Earth were not rotating. As Newton predicted, the Earth is found to be an oblate spheroid rather than a sphere (Fig. 2-10).

Observations indicating that the Earth is oblate were completed some sixty years after Newton's work. In the first half of the eighteenth century, French scientists measured a number of arcs of

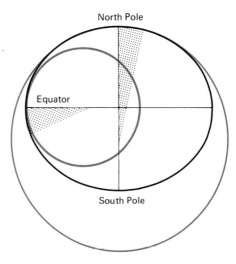

Fig. 2-10 Because of the Earth's oblateness, a circle fitting the Earth's circumference at the poles is larger than one fitting the circumference at the equator.

the Earth's surface, using the general method of Eratosthenes. They found that a degree of latitude nearer the equator represents a slightly shorter distance along the Earth's surface than a degree nearer the poles. Today a degree of latitude is taken to be 111.69 kilometers (69.407 miles) around the poles and 110.56 kilometers (68.704 miles) near the equator; that is, the curvature of the Earth's surface is more pronounced at the equator than at the poles. If the Earth were a perfect sphere there should be no difference; so the Earth must be an oblate spheroid.

There is now an ultramodern addition to our knowledge of the Earth's general form. The orbits of artificial satellites are affected by variations in the Earth's gravitational attraction, which in turn are related to the Earth's general form. Observations in 1958 of the U.S. satellite Vanguard I suggest that the Earth departs very slightly (about 15 meters, or 50 feet) from the oblate spheroid figure. It is more broadly curved around the South Pole and less so around the North Pole, resulting in a somewhat pear-shaped form. Overall then, the Earth is

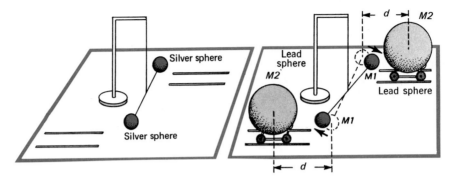

Fig. 2-11 Diagrammatic representation of how the gravitational constant was determined by the Cavendish method.

almost a perfect sphere, slightly oblate, and very, very slightly pear-shaped.

Mass Ordinarily, weighing is no great trick; we just put an object on scales which record the Earth's gravitational "pull," or force, on the object's mass. Obviously we cannot weigh the Earth itself; but mass, the amount of matter it contains, can be determined by Newton's law of gravity:

Every particle in the universe attracts every other particle with a force that is directly proportional to the product of their masses and inversely proportional to the square of the distance between them.

In mathematical form, the law is

$$F = G \frac{M_1 \times M_2}{d^2}$$

where F is the force of gravity, G is the gravitational constant (the gravitational force exerted on each other by given units of mass a given distance apart), M_1 is one mass, M_2 is another mass, and d is the distance between the centers of the two masses.

Before the Earth's mass could be calculated, the value of G had to be determined. Any method requires very careful measurement, because the force of gravity is minute for objects small enough to be convenient in this determination.

In the Cavendish method for computing G (Fig. 2-11), two small silver spheres of known weight are put on opposite ends of a small rod which is then suspended at its center from a fine wire; two large lead spheres of known mass are then brought close to the silver ones. The gravitational attraction between the silver and lead spheres will cause a slight twisting (torsion) of the suspending wire. From the amount of torsion, the force of attraction can be calculated. Thus, in the original equation, it is possible to substitute, as known values, the masses of the spheres for M_1 and M_2, the measured distances between the centers of the silver and lead spheres for d, and the force of attraction as determined from the amount of torsion for F. The equation can then be solved for G, the only remaining unknown.

Once the value of G was known for small masses, it became possible to compute the mass of the Earth. By Newton's equation, F for one of the small balls can be calculated; G is now known; M_1, the mass of the small ball, is known; and d is the radius of the Earth. Hence the equation can be solved for the one remaining unknown, M_2, which is the mass of the Earth. The mass obtained is 6×10^{24} kilograms[1] (about 6×10^{21} tons), which,

1 The kilogram is a metric unit of weight. A kilogram is 1000 grams and equals 2.2 pounds. One thousand kilograms equal one metric tonne.

TABLE 2.1 Some of the Earth's Vital Statistics

Size	Approximate	More Exact
Diameter	12,700 km (8,000 miles)	12,756 km, equatorial (7,927 miles) 12,714 km, polar (7,900 miles)
Circumference	40,000 km (25,000 miles)	40,075 km, equatorial (24,902 miles) 40,007 km, polar (24,860 miles)
Volume	1,100,000 million cu km (260,000 million cu miles)	1,083,230 million cu km (259,975 million cu miles)
"Weight"		
Mass	6×10^{21} metric tonnes (6.6×10^{21} tons)	5.976×10^{21} metric tonnes (6.587×10^{21} tonnes)
Average density	5.5 grams per cu cm	5.517 grams/cu cm
Surface		
Land area	29% 149,000,000 sq km (57,000,000 sq miles)	29.22% 149×10^{6} sq km (57.5×10^{6} sq miles)
Ocean area	71% 360,000,000 sq km (140,000,000 sq miles)	70.78% 361×10^{6} sq km (139.4×10^{6} sq miles)

if you care to write it out, is 6 followed by 24 zeros.

The total mass of the Earth is such a large number that it is difficult to grasp, so it is easier to deal with the Earth's specific gravity. This is the ratio of the weight of a given volume of Earth material, of average density, to the weight of the same volume of water. The specific gravity of the Earth has been calculated as 5.5. In other words, a cubic meter of material, of average Earth density, is about 5½ times as heavy as a cubic meter of water.

Surface Features

Recent and subtle observations of the Earth's shape bring forward very slight irregularities (e.g., the "pear shape"); but what is most impressive about this shape, rather, is its regularity and uniformity, especially as it might be seen from a distance in space. Only at close range does the detail of the Earth's surface show up as far from uniform.

Land, Sea, and Air Even from some distance the contrasts of land, sea, and air would strike an observer. The *lithosphere*, i.e., the outer solid part of the Earth, is visible on the surface as continents and islands. The *hydrosphere*, the liquid zone of water, shows up in the vast surfaces of the oceans and also in patches scattered over the land. The *atmosphere*, the gaseous envelope outside both lithosphere and hydrosphere, shows up chiefly in the bright moving patchworks of clouds—droplets of water suspended in the gaseous atmosphere and moving with currents of air (Fig. 2-12).

The hydrosphere makes up the greatest part

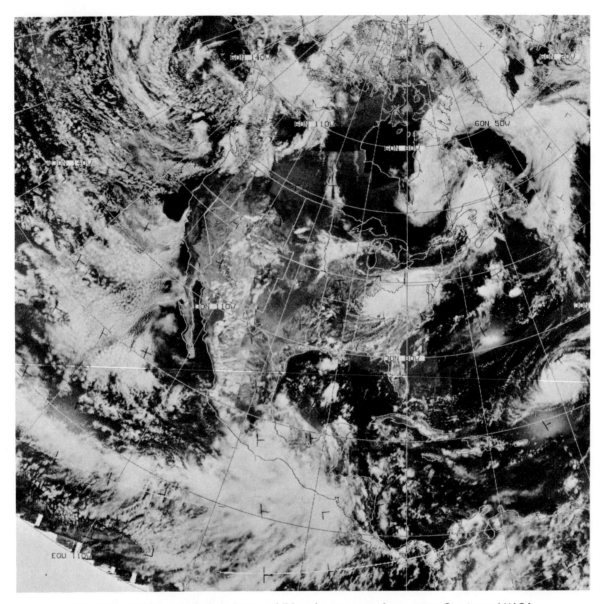

Fig. 2-12 Atmosphere, hydrosphere, and lithosphere as seen from space. Courtesy of NASA.

of the Earth's visible surface, some 71 percent. Although the oceans have an average depth of about 3 kilometers (almost 2 miles) and a maximum of about 11 kilometers (7 miles), the hydrosphere is a mere surface film compared to the 6400-kilometer (4000-mile) radius of the Earth. The atmosphere, which thins rapidly outward from the Earth's surface, has about 75 percent of its mass concentrated in the 16- or 19-kilometer (10- or 12-mile) zone directly above land and sea, and has no sharply defined outer limit. But if its extent is estimated at 600 kilometers (400 miles), then its total thickness is a mere 1/10 of the Earth's radius.

At the surface of the Earth, the hydrosphere and atmosphere are extensive and important. Both have been in constant fluid motion, attacking and reworking materials of the more stable lithosphere through the enormous range of geologic time. Atmospheric motion, in the form of mobile air masses and associated winds and currents, is generated by the Sun's uneven heating of the Earth and is complicated by the Earth's rotation. The hydrosphere, in turn, is kept in motion mainly by the winds, which cause waves and currents, and by moving air masses, which carry moisture from the oceans to be dropped upon the lands by gravity and returned to the oceans as eroding streams. Over immense lengths of time, the hydrosphere has flooded and receded, leaving its mark on broad areas of the continental masses. The center of interest for many generations of geologists has been the lithosphere, its surface and depths. But in studying it they have given close attention to interactions between the land, the waters around it, and the winds above it. The sciences of oceanography and meteorology now represent man's concern with hydrosphere and atmosphere and continue to grow in scope and detail.

Relief The surface of the Earth has obvious relief (is uneven). Local relief, the difference in elevation between valley bottoms and mountain tops, is frequently impressive to the human observer, but greater than this is the total relief of the Earth. This is the difference between the deepest known point on the ocean bottom (20 kilometers, about 36,200 feet, off the island of Guam in the Challenger Deep) and the highest mountain peak (Mount Everest, 8488 meters, or 29,000 feet, above sea level). This difference is almost 20,000 meters (more than 12 miles), or 20 kilometers (Fig. 2-13). However, it is small compared with the diameter of the Earth. A billiard ball has, for its size, as

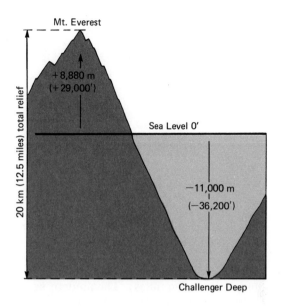

Fig. 2-13 Diagram representing the Earth's total relief.

rough a surface as the Earth. Yet the 20 kilometers (12 miles) of relief are critical. They provide the land we live on and give clues to forces working from within the Earth.

Relief features are broadly described as first-order (continents and ocean basins), second-order (the extensive mountain ranges, plateaus, and plains), and third-order (individual mountain peaks and other smaller forms) (Fig. 2-14). Third-order features reflect the destructive action of water and wind working on the surface of the lithosphere. The first- and second-order features are created by dynamic systems operating inside the Earth.

THE EARTH'S INTERIOR, DYNAMICS, AND GLOBAL ARCHITECTURE

The internal architecture of the Earth remained a mystery until the twentieth century. In the 1700s, Dr. Hutton's speculation that beneath a solid crust the Earth was completely molten was logical—in

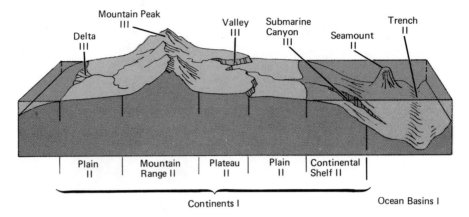

Fig. 2-14 Examples of the three orders of relief, or topographic features of the Earth's surface.

view of the only evidence available to him (from volcanic rocks and granites). The modern image of the interior became possible only after the seismograph[2] was invented in the late nineteenth century.

The Major Internal Zones

During the first half of the twentieth century, various workers suggested models of differing detail for the Earth's interior; however, most envisioned the following three major theoretical zones (Fig. 2-15).

The Core A (central) *core* is about 3500 kilometers (2200 miles) in radius. Its density increases with depth, but averages about 10.7. Most investigators think the core consists of nickel-iron, with a solid inner part and a liquid, or molten, outer part. The Earth's rotation probably causes the liquid part to circulate, thereby generating the Earth's magnetic field (like a giant dynamo).

The Mantle Since the next major Earth zone covers the core, it was named the *mantle*. This zone is

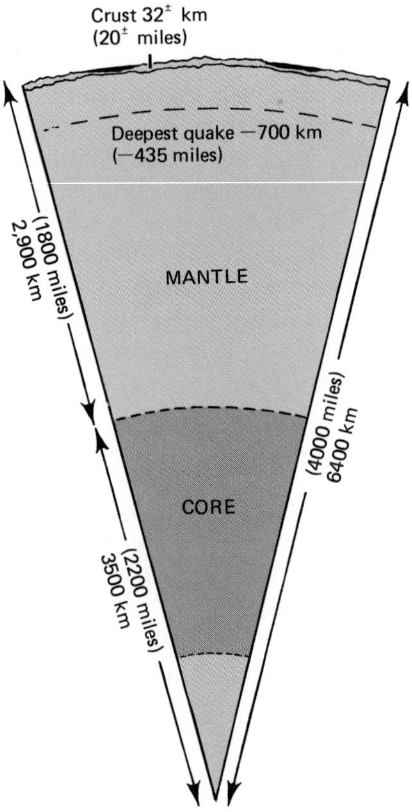

2 For the instrument and interpretation of its records, see Chapter 6.

Crust 32± km
(20± miles)

Deepest quake —700 km
(—435 miles)

(1800 miles)
2,900 km

MANTLE

(4000 miles)
6400 km

(2200 miles)
3500 km

CORE

Fig. 2-15 Internal zones of the Earth. Depth of continental crust is shown greatly exaggerated.

some 2900 kilometers thick (about 1800 miles). The density of the mantle increases with depth from the Earth's surface, but has an overall average of about 4. The mantle is generally believed to consist of dense iron-rich rock materials. Except for the outermost parts (discussed later), the mantle is thought to be in a plastic condition.

The plastic state has both solid and liquid properties. Shoemaker's wax and some candy bars illustrate the nature of plastic materials. They break like a solid if snapped suddenly, but flow in a liquid manner if slowly squeezed. Window glass provides a more relevant geological example. Careful measurement of a very old window pane may show that the bottom has become slightly thicker than the top. A plastic condition of the mantle, for which there is considerable geophysical evidence, is a basic assumption in many theories of the Earth's internal operations.

The Crust   The Earth's outermost layer, called the *crust*, ranges in thickness from as little as 10 kilometers under the ocean basins to as much as 40 kilometers under the mountainous parts of the continents (from roughly 7 miles to about 25 miles in thickness). Although it is a thin skin, when compared to the dimensions of the core and mantle, the crust contains the only rocks that can be directly observed by geologists.

The rocks in the Earth's crust have three different origins. Of these, the *igneous rocks* are solidified masses of formerly molten Earth materials. They can be compared to slag from a blast furnace or ordinary glass, which result from the cooling and solidification of melted mixtures. Igneous rocks are classed as volcanic when they originate from volcanic eruptions at the Earth's surface. Intrusive igneous rocks comprise the group that slowly cooled and solidified within the crust; they are exposed to view after erosion has carved away the overlying materials. The bulk of the Earth's crust consists of basalt and granite, the two igneous rock types that figured so prominently in the Neptunist-Vulcanist controversy. The crust beneath the world's ocean floors is mainly basalt and closely related rocks. The continental masses consist largely of granite and its close relatives.

Another major group in terms of origin, the *sedimentary rocks,* are formed on the Earth's surface from particles and fragments of still older rocks. They can be compared to concrete, a manmade mixture containing sand and rock fragments bound together by cement. A few of the natural materials forming sedimentary rocks are beach sands and river gravels, which are clearly composed of fragments of preexisting rocks. When loose deposits of such materials are bound together by cementation and other natural processes, they become consolidated layers of sedimentary rock. Although forming a relatively small part of the crust's total volume, the widespread veneer of sedimentary rocks at or near the Earth's surface provides the best record of our planet's changing geographies during past geologic times.

The third major group, the *metamorphic rocks,* mainly originate when volcanic and sedimentary rocks formed at the Earth's surface are altered by high heat and/or pressure. A homely analogy to the metamorphic process is the baking of clay to make brick or chinaware. One group of metamorphic rocks results when heat and juices from molten igneous masses cause changes in adjacent solid rock. However, the greatest volume of metamorphic rocks has been created by mountain-building processes. Their tremendous forces drive surface rocks into deep zones of extreme pressure as well as high temperature, where they literally become the roots of mountains. When exposed at the Earth's surface after millions of years of erosion, these metamorphic rocks display a pattern of swirling folds, intricate wrinkles, and other clear evidence of plastic flowage.

Overall, in terms of total volume, the crust is mainly composed of two rock types: (1) basalt and closely related rocks, in the crust beneath the ocean basins; and (2) granite and its close rela-

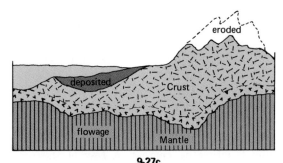

9-27c.

Fig. 2-16 Eroded mountains slowly rise and fill a basin which sinks in isostatic adjustment.

tives, which form the bulk of the continental masses. The granite masses are often called *sial* because they contain the chemical elements silicon (Si) and aluminum (Al) in abundance. The basaltic masses are sometimes called *sima,* being rich in silicon (Si), magnesium (Mg), and also iron. The other rock types in the crust are exceedingly important in geological studies, but in our preliminary examination of the Earth's major internal zones, they can be considered as thin coverings, or minor components, of the main continental and oceanic blocks.

The base of the crust is at the *Moho* (Chapter 7), a surface separating the lighter granitic sial (density 2.7) and basaltic sima (density 2.9) from the different and denser (3.2 or greater) rocks in the underlying mantle. Since the Moho lies deeper under the continental blocks than beneath the ocean basins, and is deepest under the "buoyant roots" of high mountains (Fig. 2-16), the crustal blocks are interpreted as "floating" on the mantle (the principle of isostasy, Ch. 8).

Lithosphere and Asthenosphere   Instrumental records of seismic waves from atom bomb tests recently provided an addition to the Earth model. They were used to confirm the existence of an important layer inside the upper mantle which is called the *asthenosphere.* The top of this 180-kilometer (110-mile) thick asthenopshere lies some 65 kilometers (40 miles) below the Earth's surface.

Interpreted as being weak, plastic, or partially molten, and probably of worldwide extent, the asthenosphere is thought to be the source of volcanic materials erupting through the Earth's crust and the "slippery" medium upon which slabs of lithosphere move about.

The lithosphere is now considered to be the solid encircling global layer on top of the viscous asthenosphere. Thus the lithosphere contains solid (brittle) mantle rocks as well as sial and sima of the crust (Fig. 2-17).

Global Mechanics—The Previous Schemes

Until the late 1960s, most American geologists— there were exceptions—had accepted as dogma an

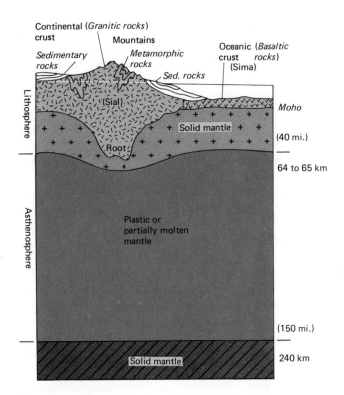

Fig. 2-17 Schematic representation of the lithosphere and asthenosphere incorporating the ideas of Barrell, Gutenberg, Suess, and others.

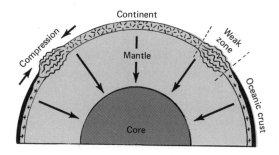

Fig. 2-18 Contraction theory of mountain making.

Earth model in which the first-order features were permanent—fixed in location for several billions of years. Now—after a revolution in geologic thinking—the accepted view is of continents and ocean basins in continuous motion, ever-changing features during the history of the Earth. But, to put the new view in proper perspective, the reasoning behind the old scheme should be appreciated.

A Shrinking Earth with Fixed Continents and Ocean Basins   The skin of a rotting apple wrinkles around its shrinking interior, and the Earth's crust might do the same (Fig. 2-18). Giordano Bruno first compared the Earth to an apple in the sixteenth century; and much later, in 1833, Adam Sedgewick suggested that mountains originated as the Earth cooled and shrank in volume. Early versions of the *contraction theory* were based on the assumption that the Earth solidified from an originally molten mass and has been cooling by conduction and radiation ever since. Once the outer part of the Earth solidified, it formed a solid crust that would no longer contract. Inside, however, the deeper materials being still molten or plastic would continue to cool and lose volume. Thus mountains originated in compressional wrinkles as the solid crust fitted itself to the shrinking interior. When the glacial concept was gaining acceptance in the late nineteenth century, the Ice Age (Pleistocene glaciation, Chapter 11), being a late episode in geologic history, seemed excellent evidence of progressive global cooling.

The *permanence of continents and ocean basins* neatly fitted into the grand mechanism of global contraction. As early as 1846, James Dwight Dana (1813–1895) stated the permanence theory: continents and ocean basins had never changed places; and, since their origin early in Earth history, had maintained the same relative positions and sizes while the Earth contracted. Shallow seas had indeed advanced and retreated across the continental blocks, leaving the cover of sedimentary rocks (as Lyell had demonstrated), and mountain-building had modified the details of continents—but no major landmass had ever become deep ocean floor, nor had any oceanic basement risen to create a continent.

The *geosynclinal theory*, introduced by James Hall (1811–1898), became part of the grander scheme. In 1857 Hall, then state geologist for New York, reported that the now-contorted sedimentary

Fig. 2-19 James Dwight Dana (1813–1895). A leading American geologist, Dana strongly influenced thinking on many subjects including the origin of mountains, continents, and ocean basins.

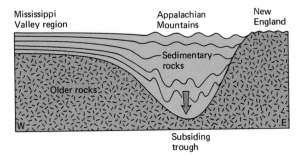

Fig. 2-20 Hall's concept of geosynclines and mountain origin. He believed folding resulted as the trough's curvature became sharper during subsidence. Later, he thought that both undeformed and deformed rocks were uplifted together by a broad continental movement. (Dana called it "a theory for the origin of mountains with the origin of mountains left out," and theorized that compression between continental and oceanic blocks buckled the geosynclinal rocks upward into high mountains.)

rocks in the Appalachian Mountains had once had a total thickness of at least 12,000 meters (about 40,000 feet), whereas the same rock strata in the Mississippi Valley of the continental interior were undeformed and less than 1200 meters thick (about 4000 feet) (Fig. 2-20). He suggested that folded mountain structures originate in belts of thick sediments, slowly accumulating in great subsiding downwarps of the Earth's crust. Dana, in 1873, named such sediment-filled troughs *geosynclines*. They came to be considered "mobile belts" along continental margins which first subsided to collect sediments. Later, the belts were crumpled into mountains by horizontal compression (like squeezing in a gigantic vise) that developed between rigid continental and oceanic blocks as the Earth contracted radially towards its center.

Thus during the 1800s and into the 1900s, geology came to have a well-accepted and successful framework of theories for explaining the Earth's global features. But as the twentieth century progressed, new discoveries complicated the original concept of a once-molten Earth gradually cooling and contracting to the present day. Evidence showed that the Pleistocene Ice Age, late in Earth history, was only one of several glacial episodes—one of which took place some 700 million years ago; so, the Earth probably had not cooled appreciably over that long span of time.

A serious challenge to the contraction mechanism developed in the 1930s, based on the discovery of radioactivity in crustal rocks. Traces of uranium, thorium, and other radioactive elements are widespread in the crust. Their decay produces enough heat, by reasonable estimates, to account for most of the Earth's present-day heat loss. Furthermore, the Earth's solid outer shell is a good insulator, slowing the escape of heat from the depths. Calculations based on these two factors suggest that the Earth may have lost no original heat from depths below 650 kilometers (over 400 miles) in the last two billion years. In fact, during the 1960s, some theorists suggested that the Earth is actually heating up and expanding.

Convection Currents in the Mantle Although the thermal contraction mechanism lost its popularity, the theory of the permanence of continents and ocean basins remained inviolate in American circles. In fact, the related schemes of the permanence of continents and ocean basins and geosynclinal mountains were neatly adapted to the radioactive findings and given greater refinement by a twentieth-century theory of convection cells in the Earth's mantle. The *convection theory* stems from unique investigations by Vening Meinesz, a Dutch geophysicist, in the 1920s and early 1930s.

In any convection mechanism, heat is transferred through liquids and gases by moving material. For example, convection develops in a beaker of water heated by a small burner. Over the flame, heated water expands; hence it becomes lighter and rises as a current to the surface. Concurrently, cooler water sinking in a descending current moves in to replace the rising mass. Cells of circulation result as water rises over the heated spot, spreads out and cools at the surface, and then descends on the cooler sides.

Using a pendulum apparatus he developed for determining the attraction of gravity from a submarine at sea, Meinesz discovered that in the troughs (which may be geosynclines) adjacent to the Indonesian island arcs, the gravity values are far lower than anticipated. The observation has since been duplicated in island arcs and troughs of the Caribbean. Meinesz reasoned from these data that light crustal rocks extended deep into the denser mantle in elongated pockets, about 50 to 160 kilometers (about 30 to 100 miles) wide, and 50 to 80 kilometers (30 to 40 miles) deep. These he christened *tectogenes*.

The light rocks are maladjusted; they should float much higher on the denser mantle beneath. The depth to which the light rocks extended suggested that, somehow, they were being pulled down. Meinesz proposed that plastic mantle rock descending in slow convection currents was the cause.

The plastic mantle will flow like a liquid under slowly applied and long-continued stresses. Thus convection currents heated at the Earth's core could rise beneath crustal blocks, move horizontally, dragging along the continental bottoms, and then descend at the continental margins pulling light sialic material downward, to form the tectogenes which trap geosynclinal rocks (Fig. 2-21). Accelerating circulation pulls the pod of geosynclinal rock deeper and compresses it, producing the oro-

genic folding and thrusting characterizing mountain structure. Later, the cell dies out when warmer material prevails at the top of the mantle and cooler at the bottom, so that a stable arrangement is achieved. Then the contorted light rocks, no longer being dragged down, bob up, and the mountains rise to their greatest height.

Drifting Continents Dana's concept of the permanence of continents and ocean basins remained the doctrine for much of the geologic profession well into the twentieth century; however, it had been challenged by a few geologists who had a radically different view of the development and history of continents and ocean basins. The jigsaw fit of continental outlines, especially South America into Africa, is impressive as seen on any schoolroom globe. Madagascar fits neatly back into the east side of the great African "skull"; Antarctica and Australia can be fitted into India, and so on. Could the continents once have been all together, then drifted apart?

In 1620 Sir Francis Bacon had noted the parallelism of the shores bounding the Atlantic Ocean. Antonio Snyder, an American expatriate in France, drew a map in 1858 showing the Americas, Eurasia, and Africa joined in a single supercontinent to explain the similarity of plant fossils preserved in coal beds throughout the Northern Hemisphere.[3] In 1908 the American Frank B. Taylor, in order to explain the distribution of mountain belts, presented the idea that the continents had once been together and then drifted apart. The *theory of continental drift* attracted little serious attention, however, until 1910—when Alfred Wegener (1880–1930), a German meteorologist, resurrected the concept and began to promote it aggressively.

Wegener and his followers believed that the Earth had indeed had supercontinents and a near-universal ocean basin. Some "driftists" envisioned

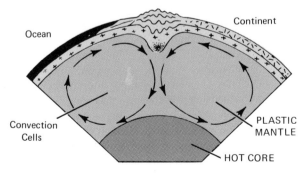

Fig. 2-21 Schematic diagram of convection mechanism for mountain making.

3 Snyder thought that the supercontinent preceded Noah's Flood and the separate continents appeared after the Deluge.

Fig. 2-22 Alfred Wegener (1880–1930). He proposed that the Earth once had a single giant ocean and a single giant landmass whose fragments drifted apart to form the existing continents and ocean basins. Some of the opposition to his theory was vehement and sarcastic. Yet if his theoretical explanation for the cause of drifting continents, based on a consideration of the then-available evidence, was clearly wrong, Alfred Wegener's concept of the changing face of the Earth was most probably right.

one original supercontinent, named Pangea by Wegener, that long ago split into a northern half, called Laurasia, and a southern half, called Gondwana. Other driftists theorized that the northern and southern supercontinents were original features. In any case, all agreed that about 200 million years ago Laurasia and Gondwana broke up and their fragments drifted apart to form the present continents. The Atlantic Ocean basin developed as an ever-widening gap between the Americas and Africa-Eurasia; the Indian Ocean gradually opened between drifting fragments represented by Antarctica, Australia, Africa, and Asia (Fig. 2-23). Mountain chains, such as the Cordilleran ranges along the west sides of North and South America, were readily explained as compressional crumplings of geosyn-

clinal troughs along the leading edges of the drifting continental blocks—rather like bow waves of a moving ship. From studies of the continents, Wegener and his advocates gathered a large amount of supporting circumstantial evidence involving the world distribution of fossil plants and animals, as well as patterns of rock types and mountain belts. On the other hand, believers in the permanence of continents could interpret much of the same evidence in a wholly different light. Essentially, the theory of drifting continents failed to convince the majority of American geologists[4] because the mechanism and driving forces Wegener proposed did not seem mechanically adequate.

Wegener started his explanation with the well-accepted geologic view that lighter continental masses are "floating" on denser underlying materials (the principle of isostasy; see Chapter 8). He then proposed that the sialic continental masses drifted across simatic ocean bottom rocks—like icebergs floating in the sea. The driving force, dramatically christened *Polflucht* ("flight from the poles"), was considered to be the Earth's rotation, which caused continents to drift towards the equator. Tidal attraction of the Sun and Moon were invoked to drag the Americas westward. Such forces do exist; but unfortunately for Wegener's mechanical interpretation, physical calculations showed them to be far too weak (by several million times) to drive sialic blocks across basalt of the sima. In the late 1920s, however, he recognized the possibility of convection currents as a driving mechanism. Alfred Wegener died in 1930 during a scientific expedition on the Greenland ice cap—the vindication of his views was posthumous.

The New World View

The answer to the riddle of the continents' locations and movements lay beneath the oceans—

---

4 Geologists living in the southern hemisphere, however, were near unanimous in supporting the concept of drift, and geologists in Europe were divided in their opinions.

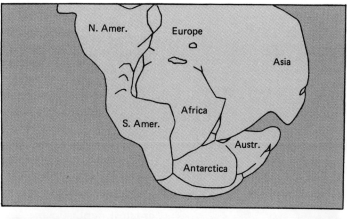

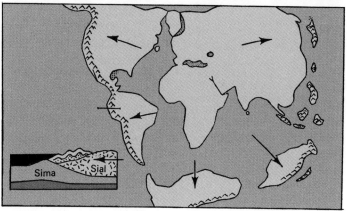

Fig. 2-23 Wegener's idea of continental drift.

in the two-thirds of the Earth's rock surface that had been virtually unknown to geologists.

Ocean Basins Among the many Earth scientists who made important contributions leading to the revolution in geologic thinking following World War II, Harry Hess of Princeton University and Maurice Ewing of Columbia University were pioneers who inspired many other investigators.

Hess, while a navy captain in the Pacific during the war, discovered (using an echo sounder) great submarine volcanos whose truncated tops lay thousands of meters below sea level. His imaginative

attempts to explain these features, which he named Guyots,[5] were involved in the reasoning by which, in 1960, he provided the theoretical breakthrough leading to the new world view. Ewing, the geophysicist who organized Columbia's Lamont Geological Observatory, was a leader in the development of techniques for measurement and in the relentless exploration of the world's submarine geology that marked the new age of discovery in the 1950s and 1960s. This surge of instrumental studies produced our first real understanding of

5 After a Swiss-American geologist, for whom Princeton's venerable geology building is also named.

Fig. 2-24 Maurice Ewing (1907–1974). Ewing lead a team of scientists whose geophysical exploration of the geology of the ocean floors resulted in many discoveries fundamental to the development of the new world view of sea-floor spreading and plate tectonics.

the nature of the rocks and the geography of the ocean floors.

The ocean floors, long considered nearly featureless because of lack of information, actually have a varied topography (Fig. 2-25). The drowned margins of the sialic continental blocks are shallow submerged shelves, cut in places by great submarine canyons, and bounded by steeper continental slopes. In places, as off the east coast of the United States, the continental slopes grade downward into less steep continental rises that lead to the abyssal depths. Elsewhere, continental slopes descend into elongated trenches, containing the greatest depths in the oceans, as off the west coast of South America. Oceanic trenches also lie close to arcs of islands, such as the Mariana Islands group. The ocean floors also have abyssal plains—the world's greatest near-featureless surfaces. The plains grade into topographies of submarine hills, Guyots, and plateaulike platforms.

Most impressive is the 1600-kilometer (1000-mile) wide, 64,000-kilometer (40,000-mile) long system of mid-oceanic ridges—by all odds the world's most continuous geologic feature. Although

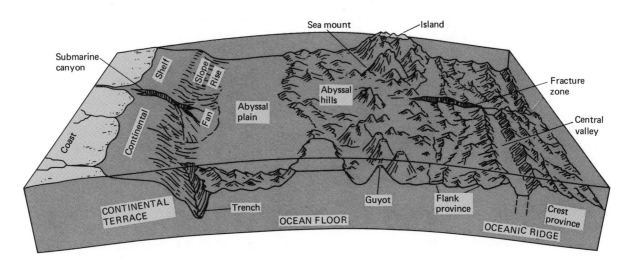

Fig. 2-25 Topographic features of the ocean floor and continental margins.

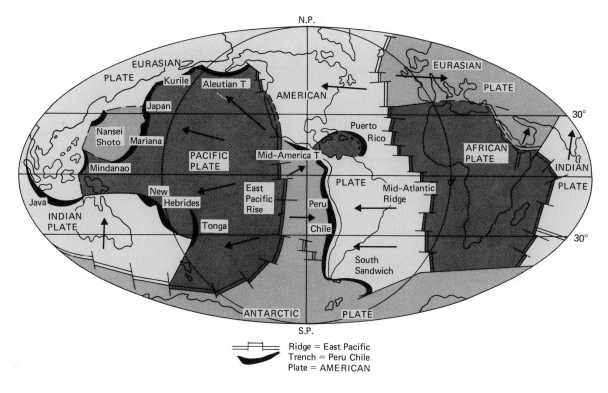

Fig. 2-26 Mollweide's map projection preserves areas but distorts shapes on margins. Mid-oceanic ridges, deep trenches, and fault lines bordering major crustal blocks are indicated in the legend. After Heirtzler et al., *J. Geophys. Res.* 73, 2119-2136 (1968).

emerging in a few places, such as Iceland and the great Rift Valleys of East Africa, the ridges are dominantly submarine volcanic mountains (quite unlike the folded mountain chains on the continents), and are often split along their axes by steep-walled central valleys.

Continuous records from ship-borne echo sounders provided the bottom profiles from which the abyssal topography was charted. The revolutionary new world view, however, is based upon several different lines of critical evidence which were obtained from a battery of other sophisticated instruments. Many of the oceanographic instruments had to be designed by the investigators to overcome difficult technical problems of location at sea, tremendous water pressures in the depths, and

the like, or to provide new sources of information as the work progressed. Dredges and piston coring devices brought up rock and sediment samples for geologic analyses and dating; other devices measured magnetism, seismicity, gravity, and heat-flow of rocks deep in the oceans. At about the same time, independent investigations with land-based seismographs led to the recognition of the asthenosphere. From the flood of new findings, some of the more theoretically minded and imaginative geologists and geophysicists sifted the clues they needed to hammer out a satisfactory new model for the Earth's global mechanics.

The Plate Tectonic Theory Today, all but the most conservative geologists accept Wegener's view of drifting continents—but they have a rather different image of the Earth's outer global architecture and associated workings (i.e., its tectonics). Harry Hess is generally considered to have fathered the new concept in a 1960 presentation. The ideas were formally published in 1962 in a wide-ranging paper, "The History of Ocean Basins," which, in view of the prevailing belief in fixed continents and ocean basins, he introduced as "an essay in geopoetry." In it, he proposed that new crust forms from upwellings of mantle material into the mid-ocean ridges, that sea-floor crust spreads laterally from the ocean ridges, and that crust is destroyed in ocean trenches. Very few geologists immediately jumped on the bandwagon, but in a few years converging lines of separate evidence convincingly supported the idea of migrating sea floors and led to the grander scheme of plate tectonics.

In the new world view of plate tectonics, the drifting continents are envisioned as surface features of immense rock plates passing through a continuous cycle of destruction and regeneration (Fig. 2-27). Some ten or more plates of lithosphere, containing sialic continental blocks and slabs of ocean-floor crust locked into outermost solid mantle, have been defined, and they are visualized as creeping across the weak surface of the underlying asthenosphere.[6] New crust is created where plates split and drift apart; molten asthenosphere erupts into the vertical fractures and hardens. The resulting seams of new-formed crust at these "divergent plate boundaries" form the mid-oceanic ridges. Gradually the new-formed crust is displaced out-

6 In a plate tectonics analogy, continents are likened to logs frozen into floating slabs of ice (representing solid uppermost mantle).

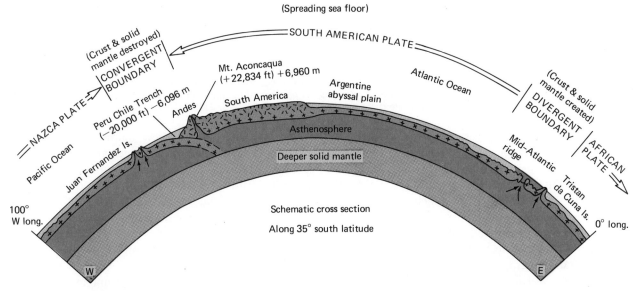

Fig. 2-27 Diagram constructed from maps, and using the plate-tectonic concept developed by many scientists. Aconcagua is South America's highest peak; north of this cross-section through Buenos Aires, Argentina, and Santiago, Chile, the Peru trench reaches a depth of at least 7,920 meters (26,400 feet) below sea level. Divergent boundaries are sites where new crust and mantle are formed by upwellings of magma from the asthenosphere. The plates are moving outward from the divergent boundaries in the process of sea-floor spreading. Convergent boundaries mark plate collisions where crust and solid mantle material of the descending plate is destroyed and reincorporated in the asthenosphere.

ward, in solid plates riding on the asthenosphere, while new material erupts into the ridges; this process was named *sea-floor spreading* by Robert Dietz of the U.S. Coast and Geodetic Survey. The mid-Atlantic submarine ridge is the present axis of sea-floor spreading for an ever-widening Atlantic Ocean basin.

Where global plates, inching out[7] from separate axes of spreading, collide at "convergent plate boundaries," the leading edge of one plate is jammed downward into the asthenosphere. Here the plate is gradually destroyed by melting; part of the molten material is reincorporated into the asthenosphere; but part may rise upward into the overriding lithosphere to produce intrusive bodies of granitic rocks, or to erupt on the Earth's surface through volcanic vents. The surface of the descending plate forms an oceanic trench.

The overriding plate may buckle upward, rising above sea level to form arcs of volcanic islands— as in the Marianas group, where two simatic ocean plates are colliding. High mountains may result where lighter sialic continental margins are deflected upward by a descending oceanic plate, as in the Andes Mountains of South America. The world's highest mountains, the Himalayas, represent a special situation where two light sialic masses are colliding. There the depressed Indian subcontinental block may provide an especially buoyant root that uplifts the overriding edge of the Asiatic continent.

Modern geosynclines may be represented by the thick deposits subsiding in the Indo-Gangetic plain along the south edge of the Himalayas, and by the deposits forming continental rises and laid down in the oceanic trenches. Overall, the new world image, based on plate tectonics, represents a drastic change in geologic thinking—and yet, it

may be just the logical expansion of Dr. Hutton's vision of an ever-changing Earth to include the first-order features, the continents and ocean basins.

The Changing Nature of Geologic Schemes

The evolution of geologists' images of the Earth gives a nice example of the dynamic nature of scientific work. Since science is human knowledge—not divine revelation—change inevitably results from the give-and-take of ideas in this ongoing endeavor. James B. Conant, chemist and educator, puts it this way: "Science is that portion of accumulative knowledge in which new concepts are continuously developing from experiment and observation and lead to further experiment and observation." But don't jump to the conclusion that scientists are determinedly avant-garde, eagerly embracing all the latest hypotheses.[8] Clearly, the theories of Copernicus, Hutton, and Darwin (and Wegener) stirred bitter resistance—because they challenged deeply ingrained patterns of thinking.

Normal Science The long, hard training required of scientists makes them a skeptical (and argumentative) lot—who put the burden of proof on the inventors of new schemes. In fact, Professor T. S. Kuhn, historian of science, suggests that natural scientists rarely set out to upset the established theories and methods of operation. Rather, they accept the collective images of nature that their scientific community considers major scientific achievements. These patterns of images—summarized in contemporary textbooks and instilled during professional education—Kuhn calls the *paradigm*. It provides a set of "conceptual boxes" into which scientists force their observations of the world of nature. Thus he compares research in *normal science* to imaginative puzzle-solving in which both

7 Calculated rates of spreading range from 1 to 5 centimeters (½ to about 2 inches) per year—trivial in human experience, but if you live to the biblical fourscore and ten years, major Earth features could have moved some 4½ meters (about 15 feet).

8 Taken out of context, Conant's definition doesn't do justice to his excellent book *Science and Common Sense*.

worthwhile problems and their satisfactory solutions depend upon accepted rules and patterns of thinking. Until the recent scientific revolution, generations of geologists had fitted their observations of rocks, fossils, and mountains into the conceptual boxes of a contracting Earth, permanent continents and ocean basins, and geosynclines. For many years this paradigm was a worthwhile and successful scheme that led to many still-valid detailed findings—because it provided problems and gave direction to geologic work.

Scientific Revolutions   Since the new plate tectonic paradigm changed a whole way of viewing the Earth, it caused a scientific revolution of the sort associated with the theories of Hutton and Darwin. But since this revolution was a highly technical affair (critical observations required the use of modern technology, dependent on physics, chemistry, and mathematics), and since it did not create a new crisis in mankind's theological and social beliefs, the excitement was mainly inside the geological community. Even there, the debates never reached the bitter proportions of earlier geologic revolutions, because the new findings came in relatively rapidly, during the 1950s and 1960s, and they were so convincing.

Surprisingly perhaps, scientific revolutions are a natural result of normal research operations. As Kuhn suggests, scientists with no intention of challenging the accepted frame of reference do discover exceptions to their paradigm. Usually the "bothersome" exceptions lead to adjustments and improvements in the accepted schemes. The discovery of radioactivity in the Earth's crust, however, posed a serious difficulty for the classical contraction mechanism—a challenge but not a cause for revolution. Thus the contraction theory lost favor during the twentieth century, but the paradigm of permanent continents and ocean basins was retained and improved with the adoption of the Meinesz convection mechanism for geosynclines and mountainbuilding.

By no means did the many contributors to the plate tectonic revolution start out as "driftists"—most of them thought Wegener's whole scheme ridiculous. For example, Ewing and his cohorts at the Lamont Geological Observatory—who discovered the real nature of the key oceanic ridges—originally demanded fixed continents and even, at times, argued convincingly against early proposals of sea-floor spreading. Yet they were some of the most productive workers in eventually gaining near-unanimous acceptance for the drifting continents and spreading sea floors of the now-dominant plate tectonic paradigm.

Except for a very few "die-hards," the mass of information from several separate lines of evidence had made plate tectonics the new "gospel" by 1970. Now we are back to an era of normal science. Geologists are enthusiastically refining and improving the new scheme, seeking yet more "proof" for it, and busily using it to reinterpret much previous work as well as extending it to new areas.

In Retrospect   The new world view has a certain beauty in neatly tying together many geologic phenomena that were poorly related by the old paradigm—in short it is an excellent unifying concept. Moreover, the scheme is open-ended, suggesting many new interesting problems. We can, for example, learnedly state that the plate tectonic machine is powered by the Earth's internal heat—both radioactive and original. But just how does it operate? Are the global plates dragged along by moving convection currents in the mantle, or are they pushed outward by the upwelling of molten material and the sliding away from the oceanic ridges, or are the plates pulled along by the weight of the parts sinking into the less dense asthenosphere? Such questions arising from new discoveries and new schemes are the essence of any vital science.

SUGGESTED READINGS

Bunbury, E. H., *A History of Ancient Geography,* New York, Dover Publications, 1959.

Conant, J. B., *Science and Common Sense,* New Haven, Conn., Yale University Press, 1961 (paperback).

Kuhn, T. S., *The Structure of Scientific Revolutions,* Chicago, University of Chicago Press, 1962 (paperback, Phoenix Books).

Marvin, U. B., *Continental Drift: The Evolution of a Concept,* Washington, D.C., Smithsonian Institution Press, 1973.

Strahler, A. N., *Physical Geography,* 2nd ed., New York, John Wiley & Sons, 1959.

Sullivan, W., *Continents Adrift: The New Earth Debate,* New York, McGraw-Hill Book Co., 1974.

Wegener, A., *The Origin of Continents and Oceans,* New York, Dover Publications, 1966 (paperback, translation of 1929 book).

Wertenbaker, W., *The Floor of the Sea, and the Search to Understand the Earth,* Boston, Little Brown & Co., 1974.

Materials of the Earth

Following our introduction to the nature and development
of geologic ideas, along with some thinking on our planet's
global aspects, we now begin an investigation of the minerals
and rocks of the Earth's crust. What they tell us of past
geologic environments and of time are the detailed concerns
of most geologists.

Gypsum crystal with sand grains from Oklahoma. Courtesy of Southwest Scientific Supply House, Scottsdale, Arizona.

3

Matter and Minerals

Suppose you were to set out to make a list of all the materials on Earth—all in the room where you are, in the scene outside, in the plants and animals there, in yourself, in the sea and the air. You could hardly bring the list to an end. The Babylonians and the Egyptians attempted to tackle the bewildering variety of matter by placing everything into one of three categories, which they concluded were the basic constituents of the world: water, air, and earth. Later, about 600 B.C., a Greek, Anaximander of Miletus, added fire to the list of elements. A burning log shows the neat logic of the scheme: sap oozing out is water, smoke marks rising air, ashes are earth, and the flame is fire. Unfortunately, the answer to the problem is a bit more complicated, but an underlying order in the hodge-podge of materials in and around us has been demonstrated by chemistry and physics.

THE NATURE OF EARTH MATERIALS

Substances and Mixtures

Matter is anything that has mass and takes up space. Chemically, matter can be sorted into two main groups: substances (including both elements and compounds) and mixtures.

Substances Chemists have discovered over 100 basic substances, of which 92 occur naturally in the Earth. These are the *elements*, which cannot be separated into simpler substances by ordinary chemical reactions because they consist entirely of the same kind of atoms. (Atoms are the infinitesimal building blocks of all substances.) Hydrogen, oxygen, copper, and gold are typical elements. They cannot be broken down into simpler things by mixing and cooking or passing an electric current through them.

The discovery of the chemical elements simplified the study of matter because it turned up a limited number of substances from which all others are derived. The great variety of substances comes about even though the elements are few, because they combine in a great variety of proportions to form multitudes of chemical *compounds*. Any compound contains two or more different kinds of atoms and can be separated into simpler substances by ordinary chemical reactions. For example, water is a compound that can be separated into hydrogen and oxygen by passing an electric current through it.

A given compound always contains the same proportion of elements. Ordinary water, H_2O, consists of two parts of hydrogen to one part of oxygen; never one to one, three to two, or any other combination. This is because the smallest particle of water that has the properties of water, the molecule, consists of two whole atoms of hydrogen bonded together with one whole atom of oxygen. Although there are vastly more molecules in a gallon of water than in a pint, the ratio of hydrogen to oxygen is the same in both because no matter how many molecules there are, each one has the same construction.

Compounds usually bear little resemblance to the elements that compose them. Table salt is a compound of two elements, chlorine and sodium. When uncombined, chlorine is a poisonous gas and sodium is a metal that catches fire on touching water. Separately, neither is recommended for seasoning food, but combined as salt they are harmless. The difference in appearance and properties between compounds and the elements that compose them gives a clue to our original problem, the confusing variety of materials around us. The possibilities of variety become clearer when we consider the number of combinations possible. Just as only 26 letters in different arrangements yield all the words in English and scores of other languages, then surely the 92 elements combined in various ways could form innumerable chemical compounds.

Mixtures Most things we see around us are not made up of a single element or even a single compound. They are *mixtures*, not pure substances. Mixtures are conglomerations of elements and compounds which retain their separate identity when blended. The molecules in a mixture do not react with each other to form new and different molecules. For example, salt and pepper thoroughly stirred together will form a mechanical mixture. The individual particles, each a mass of molecules, would be visible. Sugar dissolved in water forms a solution, another kind of mixture. Although the sugar disappears, it has merely divided into individual molecules. These are too small to see, but can still be tasted. The sugar is still sugar, the water is still water; they do not form a new compound. Also unlike a compound, a mixture may have components that vary in their proportions. A sugar-water solution can contain a very small or a very large amount of sugar.

Fig. 3-1 Pyrite crystal from the island of Elba, Italy. Courtesy of Southwest Scientific Company, Scottsdale, Arizona.

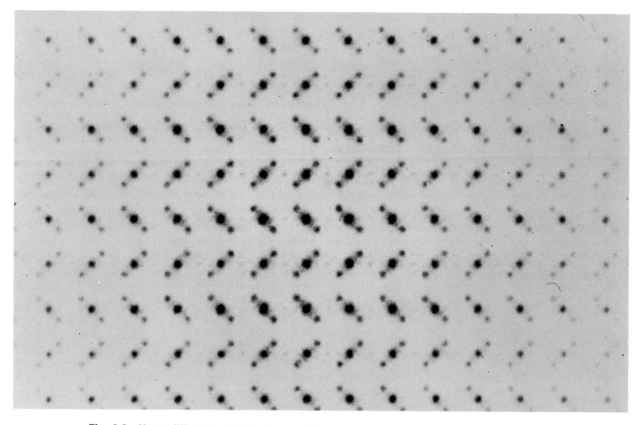

Fig. 3-2   X-ray diffraction image of the crystal structure of pyrite. Large dark spots correspond to the positions of iron atoms, smaller ones correspond to sulfur atoms. (This should not be considered a picture of atoms, but rather a shadowgraph, as when greatly enlarged silhouettes are cast onto a large sheet or screen. See Fig. 3-13 for the principle.) Magnification about 44 million diameters. Courtesy of M. J. Buerger, Massachusetts Institute of Technology.

The Structure of Matter

The atomic structure of matter—including the matter of the Earth—was reasoned out long ago by Greek philosophers of the fifth century B.C. They pointed out that if a substance could be divided into smaller and smaller pieces, eventually a minute indivisible particle would be reached. Democritus named this particle the *atom* ("that which cannot be cut"). He also worked out a theory that the differences between substances corresponded to dif-

ferences in combinations of only a few basic atomic forms. For centuries the idea was dormant. Then, in the early nineteenth century, John Dalton, an English schoolmaster, conducted experiments that clarified the concept of the atom and demonstrated its value in explaining the nature and behavior of substances. The realization that atoms are not ultimate, invisible, solid particles—as Dalton and the Greeks believed—occurred late in the nineteenth century. Ingenious laboratory investigations by Ernest Rutherford, J. J. Thomson, and other physicists demonstrated that atoms are mainly empty space containing still more fundamental particles.

Atoms   Most of us have seen models of the atom that consist of a nucleus of one or many positive charges and one or more orbiting electrons of negative charge. Such models are helpful for explanation and prediction, but the "particles in orbit" model is very far from adequate as a picture of the true relations within the atom. It is an unfortunate characteristic of submicroscopic entities such as electrons or protons to have properties that we cannot relate to anything in our experience at the macroscopic level. For instance, a rough picture of a "particle" such as an electron would be a group of waves which are concentrated to the extent that they have particle properties (i.e., mass) as well as wave properties (the electron behaves like light under certain conditions). The measurable properties of such an entity are described by a mathematical representation known as its *wave function*. The properties are affected by the environment around the "particle," and so in general the wave function will change with the environment.

This is mentioned because in an atom each electron has its own unique wave function, called an *orbital*, for a fundamental rule at the atomic level is that no two electrons in an atom may have exactly the same properties.[1] The word *orbital* is derived from orbit, but it would be truer to represent an electron in an atom as a smear around the nucleus than as an orbiting body such as a planet. Since each electron has a different orbital, it has different properties, and each is smeared around the nucleus in a different way (Fig. 3-3).

The electron has a negative charge; the proton (another "particle") has a positive charge which balances that of the electron and a mass 1837 times that of the electron. The neutron is another "particle," with no observable charge and a mass approximately that of the proton. The atom itself may be

[1] This is known as the *Pauli exclusion principle*.

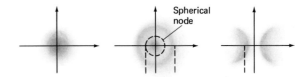

Fig. 3-3 Electron smears for three different orbitals. The intersection of the lines represents the nucleus, which is not shown because of its relatively small size.

pictured as a collection of protons and neutrons in a compact, relatively massive nucleus around which electrons are smeared at great relative distances, each electron with a different orbital.

One property that changes from electron to electron in an atom is energy. Energy is assigned to an electron in an atom in a rather peculiar way. Since unlike charges attract, and the closer together different charges are the greater the force of attraction, electrons that are smeared closest to the nucleus are bound the most strongly to the atom. Yet the most tightly bound electron is given the lowest energy state (called the ground state), and energy is assigned each electron that is smeared further away from the nucleus on the basis of the energy that it would take to change it from an electron with a ground-state orbital to an electron with the particular orbital that it has. The end result is that the most loosely bound electrons are in the highest energy state. These are the electrons that enter into chemical reactions.

It should be mentioned that electrons tend to form groups of orbitals, called *shells*, that differ from each other only a little in terms of energy. The energy difference between electrons in the same shell is small compared to the energy difference between electrons in different shells. The innermost shell carries, at most, two electrons, as in the atom of helium. The next shell beyond is filled when it contains eight electrons, and succeeding shells may carry more electrons. Hydrogen, the simplest atom, has one proton in the nucleus and one orbital electron. Helium, the next more complex atom, has two protons and two electrons; it also has two

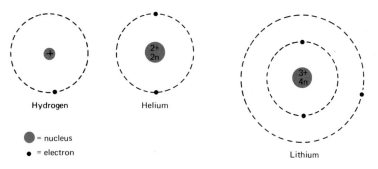

Fig.3-4 Atomic structures of hydrogen, helium, and lithium.

neutrons, but these affect neither chemical nor electrical properties. The lithium atom has three protons (four neutrons) and three electrons (Fig. 3-4). In the same way, the atoms of the other elements can be visualized as successively adding one more proton and one more electron to their structure. Uranium, the most complex natural atom, has 92 protons, 92 electrons, and 146 neutrons. (More complex atoms than uranium have been produced in the laboratory but are not known to occur naturally.)

Each ordinary atom is electrically neutral, since the number of protons in the nucleus equals the number of surrounding electrons. The chemical properties of atoms, their tendency to combine and method of combining with other atoms, are controlled by the number of electrons. But, since atoms may gain or lose electrons relatively easily (changing their chemical properties), the number of protons in their nuclei is considered the most fundamental property of atoms. Neutrons in the nucleus add to the weight of atoms, but are not reflected in the number of electrons.

Isotopes are different forms of the same element. They have the typical number of protons and electrons, hence the same chemical properties, but slightly different weights because of differing numbers of neutrons in the nucleus. For example, all forms of hydrogen have one proton and one electron. But hydrogen, which has no neutrons, has an isotope (deuterium) with one neutron in

the nucleus, and another (tritium) having two neutrons.

Ions are electrically charged atoms. They result under certain conditions when an ordinary (uncharged) atom gains or loses electrons in its outer shell. Atoms tend to seek a stable configuration wherein they have full outer shells. Sodium, for example, may give up its single outermost electron, cutting back to a stable outer ring of eight electrons. With the original 11 protons remaining in the nucleus and only 10 peripheral electrons, the ion has a positive charge. Chlorine achieves stability by acquiring an electron to bring its outer shell to eight. The added electron gives the chlorine ion a negative electrical charge (Fig. 3-5).

Bonding The tendency of atoms to achieve a stable electron configuration is responsible for joining them into molecules and crystals and for forming chemical compounds. Most geological materials, minerals and rocks, contain elements joined by *ionic bonding*. Ions of sodium (Na^+) and chlorine (Cl^-), for example, have opposite electrical charges. Since unlike charges attract, sodium and chlorine ions will join and be bonded together, forming table salt ($NaCl$).

Atoms can also achieve a stable outer configuration through *covalent bonding*. In water, for example, oxygen (O), with an outer shell of six electrons, obtains the stable number of eight by sharing each of the single electrons of two hydro-

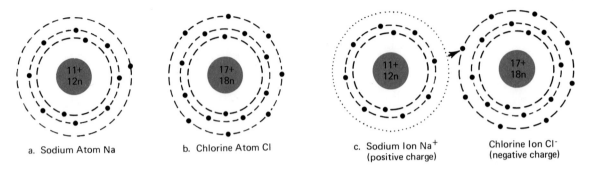

a. Sodium Atom Na   b. Chlorine Atom Cl   c. Sodium Ion Na$^+$ (positive charge)   Chlorine Ion Cl$^-$ (negative charge)

Fig. 3-5 Bonding of sodium and chlorine ions.

gen (H) atoms. In the resultant molecule (H_2O), no electrons abandon an atom to join another—they are mutually shared (Fig. 3-6). A carbon atom, which has four outer electrons, may share each of these electrons with four other carbon atoms. All these atoms achieve outer stability without becoming charged. The resultant intimately bonded crystalline structure is diamond—the hardest natural substance.

A few elements such as krypton, argon, and helium have neutral atoms with full outer shells. Being geometrically and electrically satisfied, these "noble" gases do not ordinarily enter into chemical reactions. If you have been exposed to chemistry, all this discussion has probably been too elementary, and if you have not it has probably been hard to follow. But, with a smattering of atomic

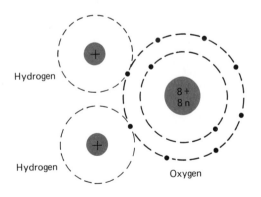

Fig. 3-6 Water molecule.

structure and substances in mind, we are set to tackle the minerals and rocks of the lithosphere.

THE EARTH'S CRUST

The Earth is composed of naturally occurring chemicals, called rocks and minerals. Rocks are the larger masses which are most evident in the land around us. On closer examination, it can be seen that rocks are composed of minerals. Since rocks and minerals are mixtures and substances, they have been studied by both chemical and physical means. Chemical analyses of several thousand rock samples show that two elements, silicon and oxygen, compose almost 75 percent of the materials in the Earth's crust. Adding the six next most abundant elements, to give a total of eight, accounts for almost 99 percent of the material in the crust. These elements in order of abundance by weight are as follows:

oxygen	46.6%
silicon	27.7
aluminum	8.1
iron	5.0
calcium	3.6
sodium	2.8
potassium	2.6
magnesium	2.1
TOTAL	98.5%

The other 84 naturally occurring elements comprise a bit more than 1 percent of the Earth's crust. Their overall rarity does not mean that these other elements are unimportant. Economically and scientifically, some of these elements are of great value and importance—for example, gold, silver, uranium, copper, and lead. But, in a study of the materials which form the bulk of the Earth's crust, they may be ignored. This greatly simplifies the problem of understanding Earth materials because the great variety of rocks can be considered as combinations, in different forms, of only eight abundant elements.

Properties of Minerals

Composition Minerals are natural, inorganic elements and compounds having crystalline structure. Being pure substances, relatively speaking, they have characteristic chemical compositions and physical properties. By definition, only naturally occurring substances are minerals. This rules out synthetic things such as artificial rubies or the compounds in an ordinary brick. Even some natural substances are not, strictly speaking, minerals because they are organic instead of inorganic. That is, they are compounds of carbon with hydrogen and oxygen that are usually formed by living organisms. Coal, for instance, is formed of substances originating in plants, so is not classed as a mineral by the geologist.

The elements in various minerals can be determined by chemical analyses which range from some that are rather easy to perform to others which are very difficult. Rather simple qualitative tests, which merely indicate the elements present, have been developed to aid in identifying minerals. These are useful mainly to prospectors and amateur mineral collectors. Quantitative analyses, which determine precisely how much and in what proportions elements are present in minerals, are difficult and require elaborate laboratory equipment and highly trained personnel. However, a great many such quantitative analyses, performed in geochemical laboratories, have provided most of the knowledge of the precise composition of minerals as well as the abundance of elements in the Earth's crust.

Minerals may consist of a single element. Gold, silver, copper, and sulfur are elements that occasionally occur in native form, pure and uncombined. However, most minerals are compounds because atoms or ions of the abundant elements in the Earth have a strong tendency to combine. As examples, silicon combines with oxygen to form silica (SiO_2), the mineral quartz; both together may combine with the abundant metallic elements to form a number of complex minerals called silicates.

Crystal Structure Minerals are solids with a definite internal structure, that is crystalline. Thus, of the three states of H_2O, ice is generally considered a mineral, whereas water and steam, being liquid and gas, respectively, are not. Some solid substances, such as opal, lack a regular and repeated internal arrangement of atoms (more commonly ions), so are properly called *mineraloids*, rather than minerals.

The crystalline structure largely determines a mineral's physical properties, which are extremely useful in mineral identification. In order to understand the nature of Earth minerals, the physical properties should be related to the structure of minerals, that is, how atoms are assembled to form different minerals. Although our knowledge of crystal structure is based on advanced physics, chemistry, and mathematics, the fundamental ideas are not difficult to understand. They give a rather neat and orderly starting point for the study of how the Earth's materials are put together in increasingly complex forms.

A crystal is a solid, bounded by smooth faces whose orderly arrangement reflects the internal structure of the crystal (Fig. 3-7). This merely means they are such forms as cubes, pyramids, and

Fig. 3-7 Some crystal forms.

flat-sided columns, as single forms or in various combinations. Furthermore, these shapes develop because a crystal's building blocks, the atoms, are arranged in a minute three-dimensional pattern that is precisely repeated throughout the whole crystal. A somewhat similar relation is shown in a standard setup of bowling pins. The setup is triangular overall because the individual pins are arranged in the same orderly way throughout. However, the rules of bowling restrict the pins to exactly 10, so a total setup is always the same size. No such rule applies in crystallography, so crystals of the same mineral may vary tremendously in size (Fig. 3-8).

You may well inquire how the internal structure of crystals is known. Atoms 1/50,000,000 of a centimeter in diameter cannot be directly observed, even with the most powerful optical microscopes. Proof of the matter required twen-

tieth-century X-ray technology. Yet long before its invention, human reasoning based on the careful examination of crystal forms and faces had shaped the essentially modern concept.

Fig. 3-9 Nicolaus Steno lived many years in Florence, Italy where he latinized his original name of Niels Stensen, a common practice in his time. Besides his important work on crystals, he also contributed the laws of original horizontality and of superposition (see Chapter 4), which are essential to the interpretation of sedimentary rocks and deformed structures.

Bowling Pins   Ions in Salt Crystal

Fig. 3-8 Shapes determined by internal geometric arrangements.

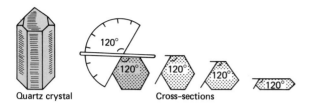

Fig. 3-10 Quartz crystal cross-sections based on Steno's work.

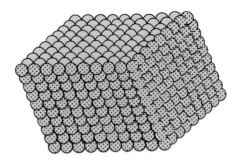

Fig. 3-11 Huygens's model of calcite structure.

The first clue was discovered by Nicolaus Steno (1638–1687), a Danish bishop, physician, and naturalist. He reported in 1669, after patiently measuring a great many crystals of the mineral quartz, that the angles between corresponding faces were always the same. The quartz crystals did vary in shape because the lengths of their smooth faces (and sometimes the number) were not always the same (Fig. 3-10), but the internal angles were constant. Other investigators found that the relation held true in different minerals (although other minerals have their own characteristic angles). These critical observations led to the "law of constancy of interfacial angles." However, Steno never proposed a hypothesis for the relation.

Shortly thereafter, in the 1600s, Robert Hooke in England and Christian Huygens in Holland—two of that century's most wide-ranging and outstanding "natural philosophers"—suggested explanations for the structure of crystalline solids. Both speculated that the symmetry of outer crystal faces reflects a regular internal packing of very minute internal particles—which they visualized as spheroidal forms—in the same repeated internal arrangements. Hooke illustrated his idea with stacks of lead shot. Huygens was impressed by minerals that break to form smooth flat surfaces that are parallel to each other—the phenomenon called mineral *cleavage*. Calcite is a mineral having three different cleavage directions that allow it to be broken into pieces resembling tilted cubes, called rhombs. Its large rhombs can be broken into

smaller rhombs which can be broken into still smaller rhombs, and so on as far as the eye can see. Huygens suggested that the rhomb forms resulted from breaking along planes in an ultimate, minute, internal structure composed of spheroidal particles, neatly stacked in sheets lying upon each other (Fig. 3-11).

Early in the nineteenth century, the French abbé René Just Haüy accidentally dropped a specimen of dogtooth spar, a variety of calcite having long sharp-pointed crystals. According to legend, the specimen broke along a cleavage plane that was not parallel to the crystal faces—which set him to thinking that minerals had ultramicroscopic

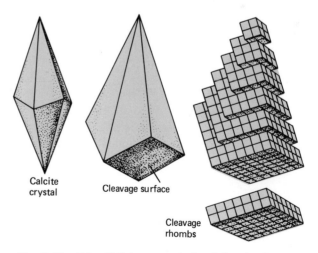

Fig. 3-12 Abbe Haüy's concept of calcite structure.

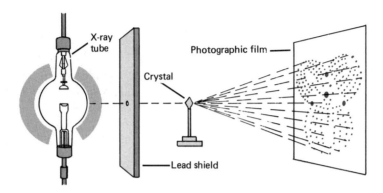

Fig. 3-13   Principle of X-ray diffraction. After Gilluley, Waters, and Woodford.

structures of neatly stacked, rhomb-shaped "in-tegrant molecules." He went further and calculated the possible surfaces that could develop between neatly stacked rhombs. Some of the calculated surfaces corresponded to faces of calcite crystals—which further supported the theory of internal orderly structure. Since these natural philosophers worked long before the twentieth-century concept of atoms, their ideas of the nature of the basic particles may seem quaint. Nonetheless, they had reasoned out the essential features of crystalline solids, based solely on common sense and on observations with the unaided eye. One of the most satisfying twentieth-century scientific achievements has been the confirmation of their early ideas.

The concept of crystalline structure received final scientific verification in 1912 when Professor Laue at Munich and his two students, Friedrich and Knipping, introduced the use of X rays for the study of crystals. They reasoned that because X rays have very short wave lengths, they might be diffracted by particles as minute as atoms. Using equipment which projected a pinpoint beam of X rays into crystalline substances, they found that regular patterns of dots were recorded on photographic paper placed behind the crystals. The dots resulted because the orderly arrangement of atoms broke up and deflected parts of the X-ray beam (Fig. 3-13). The development of this technique

permitted rapid advances in the precise knowledge of crystals, minerals, and solid materials in general. It has also provided a most reliable method for mineral identification.

The three-dimensional crystalline structure of the mineral halite, ordinary table salt, was the first to be carefully worked out by the Englishman W. H. Bragg. Halite, although not an abundant rock-forming mineral in general, is ideal for illustrating the general features of crystals (Fig. 3-14). It has the simplest crystal shape, a cube, and a simple internal structure, compared to most other minerals. The smallest particle which would have the properties of crystalline halite is an orderly cluster of sodium and chlorine atoms[2] about 1/12,-000,000 of a centimeter across. Its cubic form is an obvious result of the way its atoms are packed together. A visible crystal of salt would simply be a much larger cube made up of a tremendous number of sodium and chlorine atoms, packed, in this same manner, throughout the crystal. The variety of crystal forms that some minerals display results from the different packing arrangements, called space lattices, of the atoms. Moreover, several possible crystal shapes may develop from the same space lattice, for there are 14 lattices, but many more external forms. In all of them, the outer

2 Strictly speaking, the sodium and chlorine particles which make up the structure of halite should be called ions.

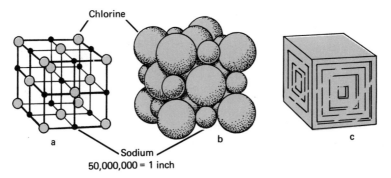

Fig. 3-14 Halite structure showing models of internal arrangement: (a) space lattice; (b) a packing diagram. (c) The external form of a visible crystal.

crystal form results from an orderly internal packing of atoms, as in the case of halite.

The halite structure is interesting as the first deciphered by X ray, but geologically another arrangement is far more important—the silicon tetrahedron. It is the basic building block of the Earth's crust. Imagine this tetrahedron, chemically

SiO_4, as a pyramid of four cannon balls, representing larger oxygen ions, with a golf ball, a silicon ion, tucked inside. Remember, however, that the whole thing is about 2.8 Ångstrom units across (Fig. 3-15).[3]

3 An Ångstrom unit is one hundred millionth of a centimeter or one four billionth of an inch.

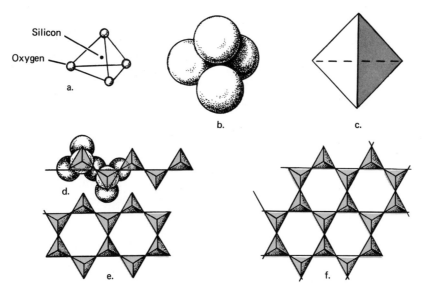

Fig. 3-15 Silicon tetrahedron: (a) Lattice diagram. (b) Packing diagram (silicon ion hidden in center). (c) Representation of tetrahedron; points are at the nuclei of ions. (d) Chain structure, as in pyroxene, showing relation of (b) and (c). (e) Double chain, as in amphibole. (f) Sheet structure as in mica.

Silicate minerals, far and away the most common in the Earth's crust, are built of silica tetrahedrons with some other ions included. In the mineral olivine, described more fully later, separate tetrahedrons are "glued" together by iron and magnesium ions so that its formula involves SiO_4. In ferromagnesian minerals, another important group, some of the oxygen ions are shared by adjacent tetrahedra, thus building chains of tetrahedra; single chains in pyroxenes, and double chains in amphiboles. In micas, the tetrahedrons form sheets by sharing oxygens. In quartz, SiO_2, all oxygens are shared by adjacent tetrahedra. Put another way, each oxygen ion serves as the corner for two tetrahedra. Thus in quartz the internal structure can be visualized as a tight framework of overlapping tetrahedra, in marked contrast to the separate ones in olivine. Feldspars have a structure resembling quartz, in which aluminum ions substitute for some of the silicons, and with various additional ions involved.

Cleavage   The manner of breaking is a physical property of minerals intimately related to crystal structure. Some minerals, such as quartz, fracture or shatter in an irregular manner. However, many minerals have cleavage. That is, they break along planes which have constant angular relations to each other. In such minerals, the number of cleavage planes may vary from one to as many as six. The planes are always parallel to actual crystal faces, or to possible crystal faces, even though a particular mineral may not actually have developed such faces. Cleavage planes result from the packing arrangement of atoms. In a few cases, such as graphite, cleavage planes result where ions are weakly bonded together. However, most cleavage planes are parallel to alternate zones of high and low concentrations of ions. If ions are closely packed in certain directions, they tend to hold together, making a cleavage surface when the mineral is broken. Because cleavage fragments tend to have characteristic shapes for specific minerals,

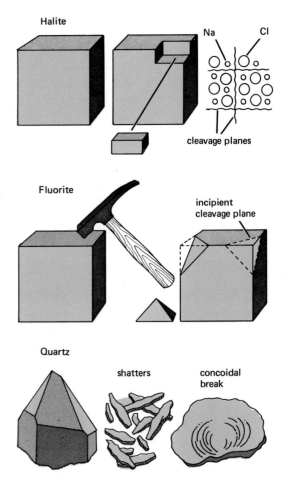

Fig. 3-16 Schematic representation of cleavage and the lack of it in certain minerals.

this property is useful in mineral identification and as a clue to the internal structure of minerals (Fig. 3-16).

Hardness   Hardness of minerals is a physical property frequently used for purposes of quick mineral identification. Commonly, hardness is measured by the ease with which a mineral may be scratched in comparison to a group of 10 standard minerals, arrayed in a hardness scale. This scale, developed

TABLE 3.1   Mohs' Hardness Scale

1 Talc
2 Gypsum
  2.5 Fingernail
3 Calcite
  3 Copper penny
4 Fluorite
5 Apatite
  5.5 Knife blade, window glass
6 Feldspar (Orthoclase)
  6.5 Steel file
7 Quartz
8 Topaz
9 Corundum
10 Diamond

by the German Friedrich Mohs, assigns numbers from 1 to 10 to each of the standard minerals, according to increasing hardness. The softest mineral, number 1, is talc, which is the base for talcum powder; the hardest, number 10, is diamond, which is the hardest natural substance known. The hardness of an unknown mineral may be determined by which of the standard minerals it will or will not scratch. Actually a standard set of minerals is rarely used because adequate determination can be made using certain easily available things of known hardness, such as one's fingernail, a copper penny, a knife blade or glass, and a steel file.

The factors controlling mineral hardness are rather complex. Involved are the properties of atoms themselves, crystal structure, bond types, and cleavage. However, a clear-cut example of the possible importance of crystal structure is illustrated by two minerals, each composed of the same pure element, carbon. Graphite, the substance in ordinary pencil lead, is exceedingly soft because its crystals consist of weakly bonded, minute sheets of carbon atoms which easily cleave and slide by each other. For this reason, graphite is frequently used as a lubricant. Diamond, also composed of carbon atoms, is extremely hard because of the interlocking arrangement of the atoms in its crystal structure.

Specific Gravity The weights of minerals also illustrate the relation of physical properties to composition and structure. In most work, "weight" is best expressed as specific gravity. This is the ratio of a mineral's weight to the weight of an equal volume of water.[4] For example, the common mineral quartz has a specific gravity of 2.65, which means that a cubic centimeter of pure quartz weighs that many times more than a cubic centimeter of water. Remember that in measuring specific gravity, it makes no difference what volume is used so long as it is the same for both substances. With experience, specific gravity can be estimated by "hefting" mineral samples. However, because many minerals are close to each other in specific gravity, simple laboratory techniques are often used for more accurate determinations.

Specific gravity of a mineral depends on two factors: the weights of the types of atom which compose it, and the closeness with which the atoms are packed in the crystal's structure. Atoms of the element iron weigh over twice as much as atoms of aluminum; therefore, minerals rich in iron have higher specific gravities than those rich in aluminum. The effect of packing is illustrated by comparing two different minerals composed of identical atoms but having different structures. For instance, diamond has a specific gravity of 3.5 whereas graphite, which has less closely packed atoms, has a specific gravity of only 2.1. In other words, fewer atoms in a unit volume results in less weight, and vice versa.

Color and Luster Among the other physical properties useful in mineral identification are color and luster. Some individual minerals have one distinctive color; however, others have a range of colors, and sometimes different minerals have the same color. Especially in normally light-colored minerals, minute amounts of foreign substances may strongly modify the mineral's color. Thus, color, although

4 Fresh water at 4°C—the maximum density of water.

useful and often very obvious, is seldom a reliable means of mineral identification. Luster is the appearance of a mineral in reflected light. It is usually described in nontechnical descriptive terms, such as metallic, vitreous or glassy, earthy, etc. Color and luster, which are dependent on a mineral's effect on light, are used mainly for quick identification of minerals. However, there are other optical properties which, when studied with a microscope, are very important in the precise identification of minerals.

Common Minerals

On the basis of physical and chemical properties, some 2000 different minerals have been recognized and described by mineralogists. However, this great number need not discourage us from attempting to understand the Earth's materials. Most of the 2000 minerals are rare and of interest mainly to specialists. Actually, fewer than two dozen are abundant, and a knowledge of only ten mineral types is an adequate basis for a generalized understanding of the bulk of rocks that are most frequently encountered (Fig. 3-17).

Feldspar The most abundant mineral type, feldspar, composes over 60 percent of the rock materials in the Earth's crust. Strictly speaking, the term *feldspar* refers to a group of closely related minerals having generally similar composition and characteristics. They are alumino-silicates of sodium, potassium, and calcium, which explains why these elements, along with oxygen, silicon, and aluminum, are so abundant in the Earth's crust. Potassium feldspar includes the minerals orthoclase and microcline. Plagioclase includes several sodium- and calcium-bearing feldspars.

Quartz Quartz is the second most abundant mineral. It is a specific mineral, the only common one of the silica group. Chemically, silica is SiO_2; thus quartz is a compound composed entirely of the two most abundant chemical elements. In large pure crystals, quartz resembles colorless glass. However, slight impurities may give it a variety of colors, and some minutely crystalline varieties, such as flint, may be opaque and of a waxy luster.

Mica Mica is the name of a group of minerals that are readily split into thin flexible sheets. This distinctive property results from a single perfect cleavage plane, which is repeated throughout the mineral. The minerals are very complex potassium, alumino-silicates with added oxygen-hydrogen combinations. Two common minerals in this group are dark mica (biotite), which also contains iron and magnesium, and colorless or white mica (muscovite), in which these two elements are absent. The atoms of all these elements are arranged in a complicated manner to produce a loosely bonded sheetlike structure within the crystal. This is responsible for the characteristic cleavage.

Amphibole and Pyroxene A number of silicate minerals and mineral groups are referred to as *ferromagnesian minerals*. The name indicates that they contain appreciable amounts of iron (from the Latin *ferum*) and magnesium. Common dark mica is, therefore, in this general class, although the other micas are not. Amphibole and pyroxene are two common groups of related, ferromagnesian minerals. Like most iron-bearing minerals, they are mostly dark colored and of comparatively high specific gravity. Both groups are rather similar in most properties, but can be distinguished in many cases by their cleavage, which results from their somewhat different internal crystal structure.

Fig. 3-17 Summary chart of general properties of some of the commonest rock-forming minerals (facing page).

Hardness Specific Gravity MINERAL Chemical Composition	COMMON CRYSTALS Color	CLEAVAGE and fragments	USES
H = 7 Sp.G = 2.67 QUARTZ SiO_2	Pure-clean, colorless Impure-pink, purple black etc.	None-shatters	Crystals in electrical and electronic work Radio quartz Optical lenses & prisms Window glass
H = 6 Sp.G = 2.5-2.9 FELDSPAR GROUP Sodium: Potassium- Calcium Aluminum silicates	Light colored: white pink, grey, cream	Two directions	Glazes and Flux enamels for ceramics
H = 2-2.25 Sp.G = 2.76-3 MICA Complex: Potassium Aluminum, Iron, Magnesium silicates	Clear, whitish brown, black, purple	Parallel, excellent One direction	In electrical and electronic equipment Heat-proof windows
H = 5-6 Sp.G = 29-3.4 HORNBLENDE (AMPHIBOLE GROUP) Complex: Potassium Magnesium, Aluminum Sodium silicates, Iron	Dark: green to black	Crystal faces Cleavage Fragments Two dir.	Some fibrous members of amphibole group used as asbestos
H = 5-6 Sp.G = 3.2-3.6 AUGITE (PYROXENE GROUP) Generally like hornblende	Dark: green,black,brown	Crystal faces Cleavage Two	
H = 3 Sp.G = 2.72 CALCITE $CaCO_3$	Usually colorless or white,Other tints common	Three	Optical Prisms (transparent crystals) Range finders Microscopes
H = 1-2.5 Sp.G = 2-2.6 CLAY GROUP Complex: Aluminum silicates combined with water	Pure: white Impure: wide range of colors		Bricks Pottery, china Drilling mud Paper slicks

Olivine   In hardness and luster, olivine resembles quartz, but the color and common occurrence in small grains are helpful for identification. Olivine is a ferromagnesian silicate characterized by a green (olive) color. Internally, olivine consists of separate silicon tetrahedra linked by iron and magnesium atoms. It is common in certain dark-colored igneous rocks and may be an important component of the rocks below the Earth's crust.

Garnet   The garnets are a group of complex silicate minerals having various colors and a hardness equal to, and in some cases greater than, quartz. Garnets have high specific gravity ranging from a bit over 3 to more than 4, and often form well-developed crystals. Because of its hardness, garnet is used as an abrasive substance; some varieties are considered semiprecious gem stones. Garnet is an important component of certain rocks formed under high pressures deep in the Earth's crust.

Serpentine   Commonly green or shades of green, serpentine is a relatively soft mineral (hardness 2.4) having a waxy or silky appearance in reflected light. Although it is usually compact or platy, a fibrous variety, called chrysotile, is the commonest mineral (of several) used in making the fireproof cloth asbestos. Some serpentine forms masses of verde antique, a handsome rock that is cut and polished to make green "marble" used as an interior building stone. Serpentine originates from the chemical alteration of ferromagnesian minerals, such as olivine and pyroxenes, by the incorporation of water into their crystal lattices. Rocks largely composed of serpentine figure prominently in some discussions of plate tectonic theory.

Calcite   Calcite is the only common rock-forming mineral that lacks silicon. All others are silicon and oxygen compounds in the form of silica or silicates. Calcite, however, is a carbonate, one of a class of compounds based on carbon-oxygen combinations. The "calc" in the word calcite comes from the Latin word *calcis* (lime), and indicates the presence of the common element calcium. Calcite has the simple formula $CaCO_3$.

At first glance, calcite might be confused with quartz, for both are clear, colorless, and "glassy." However, it is easily distinguished because, unlike quartz, calcite is quite soft, has excellent cleavage, and fizzes when drops of hydrochloric acid touch it. Calcite is a very widespread mineral in certain common rocks and in the shells of many organisms.

Clay   Clay minerals are important and abundant products of the slow chemical breakdown of rocks. Especially susceptible to this "rotting," or weathering, are feldspars and ferromagnesian minerals in rocks exposed at the Earth's surface. Pure clays, with some exceptions, are white or gray, but they usually contain impurities that stain them to blue, red, or other colors. Chemically, clays are complex aluminum or aluminum-magnesium silicates with added oxygen-hydrogen combinations. Clays, in bulk, have the property of being plastic and moldable when wet. This is because they consist of minute particles, less than 1/4000 of a centimeter across, which resemble mica in form and cleavage. Structurally, these particles are made up of sheets of atoms that become plastic when water particles slip between them.

GENERAL CONCLUSIONS

An exposure to elementary mineralogy should point up several fundamental aspects of science in general—foremost, that science is a search for explanations about nature based on careful observations. The concept of the internal structure of crystals is such an explanation. The existence of crystals was known in antiquity. But interest in them did not become scientific until certain people, who were willing to accept a natural origin for such things, asked themselves, "What causes the beautiful and regular arrangement of crystal faces?"

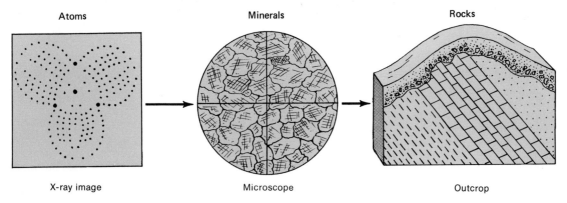

Fig. 3-18 From the ultramicroscopic to the megascopic—atoms to minerals to rocks.

Following up the question with detailed observations, logical thinking, and a bit of inspiration led to the explanation relating external crystal shape to an invisible internal structure.

An introduction to mineralogy should also suggest the importance of the development of instruments for advances in science. The idea that crystal form depended on internal structure had been developed by the early nineteenth century. The next major advance in knowledge of crystal structure did not occur until the early twentieth century, because precise information about the arrangement of atoms within crystals and the explanations of related properties was impossible to obtain until research in physics led to the development of X-ray equipment and techniques.

Mineralogy is geologic because its general objective is an understanding of the Earth; but, because its special interest is minerals, it requires the application of principles and techniques from chemistry, physics, and mathematics. Thus, mineralogy gives an excellent example of the overlapping nature of the traditional fields of science.

The most pertinent reason for a knowledge of elementary mineralogy is to give the necessary foundation for the study of geology in general. One must understand what minerals are, how composition and internal structure determine their properties, and the use of properties in mineral identification. Minerals are the substances that make up rocks, and rocks provide the evidence for most of the ideas encountered throughout geology.

SUGGESTED READINGS

Compton, Charles, *An Introduction to Chemistry,* Princeton, N.J., D. Van Nostrand Co., 1958.

Holden, A., and Singer, P., *Crystals and Crystal Growing,* Garden City, N.Y., Doubleday & Co., 1960 (paperback, Anchor Books).

Keller, W. D., *Chemistry in Introductory Geology,* Columbia, Mo., Lucas Brothers, Publishers, 1957 (paperback).

Pearl, R. M., *Rocks and Minerals,* New York, Barnes and Noble, 1956 (paperback).

Zim, H. S., and Shaffer, P. R., *Rocks and Minerals,* New York, Golden Press, 1957 (paperback).

Microphotograph of a tuff. Courtesy of Robert S. Houston, University of Wyoming.

Rocks, the Major Groups

Rocks are the documents of geology. The facts, principles, and explanations of geological studies are based on careful observation of the rocks, for they alone can preserve the evidence of forces and processes working in and on the lithosphere.

Rocks are natural aggregates of mineral matter. Most are mixtures. For example, a close look at granite, a very common rock, shows an interlocking mass of quartz, feldspar, and ferromagnesian particles. The individual particles are usually small and their crystal outlines often unrecognizable, but they can be identified because their physical properties are the same as in larger specimens. Some rocks are mainly composed of a single mineral. Limestones, for example, are sometimes almost entirely calcite. A few rocks, such as coal, are not composed of true minerals. So, we hedge our definition and say that rocks are mineral matter, not just minerals. Thus, rocks include all natural masses forming appreciable parts of the lithosphere.

In the laboratory, the petrographic microscope has been extremely useful to geologists. Many rocks are so fine grained that their minerals are difficult or impossible to recognize without a microscope, and even in coarser rocks significant details of mineral grains are often miniscule. Besides giving strong magnification, the petrographic microscope also enables the study of the optical properties of mineral crystals in polarized light. As the light passes through thin sections of rock ground to transparency, it gives distinctive patterns or colors, depending on the microscope attachment used, which are most reliable in mineral identification. Actually, much can be learned by studying hand specimens with the unaided eye or a hand lens, but the petrographic microscope greatly extends our normal powers of observation.

The geologic literature is loaded with several hundred different names for rocks. With minerals, we avoided the problem of bringing order to great variety by ignoring all but the very common ones. Scientific classification makes the study of great diversity far easier. A classification should show an orderly relation between things, so that they can be sorted into groups and better understood.

Igneous, sedimentary, and metamorphic—geologically all rocks are sorted into these three great groups. This classification is genetic, that is, based on origins. Igneous rocks result from solidification during cooling of molten rock-forming material. Their origin resembles that of ordinary glass which is made by cooling of a melted mixture. Sedimentary rocks are accumulations derived from older materials deposited as fragments or from solution. They differ from igneous and metamorphic rocks by originating at the Earth's surface, where temperatures and pressures are low. They are like concrete, a manmade mixture of rock fragments or sand, bound together by cement. Metamorphic rocks originate from the transformation of solid rock to new types by heat and pressure. The baking of clay to make chinaware is a similar process.

IGNEOUS ROCKS

Igneous rocks compose about 95 percent of the volume of the Earth's crust, and are the ancestral material from which all other rocks are ultimately derived. The origin of igneous rock is dramatically illustrated in an erupting volcano where flaming lava, flowing over the ground, cools and solidifies into solid rock. Igneous rocks, in general, originate from the solidification of molten rock-forming material. The molten material is called *magma* when below the Earth's surface. It is a hot liquid containing chemical elements which solidify into rock, gases which later escape, and solid particles of rock or mineral (Fig. 4-1).

Description

Texture and mineral composition are two important features observable in igneous rocks. Texture, in general, refers to the arrangement and size of particles in an object. In cloths, burlap has a rough or coarse texture because its threads are large; silk has a smooth or fine texture because its threads are very thin. In the same way, rocks have coarse textures if made up of large crystals and fine textures if composed of very small crystals.

Fig. 4-1 Columnar joints in basalt, a volcanic rock. Photo was taken in Columbia lava plateau in Idaho. Courtesy of U.S. Geological Survey, by H. E. Malde.

Texture Textures of igneous rocks are described in special terms. *Granular* refers to coarser textures in which crystals are all about the same size and easily visible to the naked eye. *Aphanitic* describes fine textures in which crystals exist but are not readily visible without magnification. In other words, a magnifying glass or microscope is needed to study the crystals. *Glassy* is an apt term for rock that has no crystals. *Porphyritic* describes uneven textures composed of two markedly different grain sizes, where larger crystals are surrounded by a groundmass of much finer crystals or glass. Like all technical terminology, these words substitute one word for many.

Composition Determining composition of rocks involves recognition of the minerals present and determination of the relative amount of each. This can usually be done in granular-textured rocks without special equipment, although a hand lens helps. Aphanitic rocks require laboratory study with a petrographic microscope. In porphyritic textures the large crystals are a clue to the minerals present.

Classification

Volcanic and Plutonic Rocks Igneous rocks are genetically classified in two main groups, depending on their place of origin. Magma, rising through

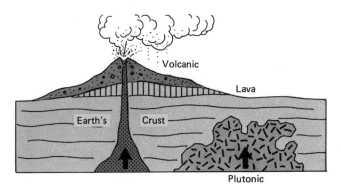

Fig. 4-2 Volcanic and plutonic rocks.

existing rocks in the lithosphere, may erupt upon the surface of the Earth; the resulting rocks are classed as volcanic. Some magma may stop rising before it reaches the surface and solidify within the Earth's crust; these rocks formed at depth are called plutonic (Fig. 4-2). The terms are well chosen. Vulcan was the Roman god of fire, and Pluto was the Roman god of the underworld. That volcanic rocks originate from magma is clear; the process can be seen. Whether all plutonic rocks originate from magma is a lively issue in geology today because they are observable only after erosion has removed their overlying rock, long after they were created.

A general classification of igneous rocks using mineral composition and texture can be approached in a more empirical way. The idea behind this classification is as simple as arranging cards by suit and value in a bridge hand and serves the same general purpose, i.e., it makes their study easier. It uses only names for the most common family groups of rocks. Be not dismayed if you hear specialists using such tongue twisters as troctolite, shonkinite, and jacupirangite; they are only trying to be more precise in referring to relatives within our broad family groups. Actually, two names will describe most igneous rocks you are apt to encounter in the field. Among the plutonic rocks, *granitic* types are more common (95 percent) than all the others combined. Of volcanic rocks, *basaltic* types are by all odds the most abundant (97 percent).

The three most common granular rock types, granite, diorite, and gabbro, all contain feldspar and ferromagnesian minerals. Of these, only granite contains quartz and potassium feldspar. Diorite is mostly plagioclase but may contain abundant ferromagnesian minerals. Gabbro is mainly ferromags but may have considerable plagioclase, usually dark colored (Fig. 4-3). Generally, from granite to gabbro, there is a loss of quartz and an increase of ferromagnesian minerals. This results in a corresponding change from light- to dark-colored rocks.

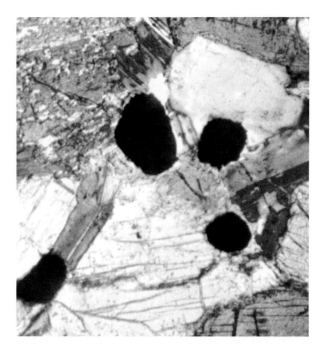

Fig. 4-3 Gabbro (norite) as seen through a microscope. Largest crystals are pyroxene; smaller banded crystals are feldspar (plagioclase); circular dark masses are crystallized droplets of sulfide minerals (pyrrhotite). Photo courtesy of Robert S. Houston, University of Wyoming.

Peridotite is a general term for igneous rocks with notable amounts of olivine, and with or without other ferromagnesian minerals. It has little if any feldspar. Being high in olivine, this rock is relatively heavy. Although not a particularly common rock at the Earth's surface, peridotite figures prominently in speculations on the subcrustal zone.

Aphanitic rocks have the same general mineral compositions as their granular cousins. Rhyolite is like granite in containing quartz, feldspar, and some ferromagnesian minerals. Andesite has the same general composition as diorite. Basalt resembles gabbro mineralogically. The aphanitic rocks usually reflect compositional changes by their colors, which range from light color in rhyolite, to intermediate color in andesite, and to dark color in basalt.

Glassy-textured rocks are not truly composed of minerals. They are actually noncrystalline solids in which atoms are randomly dispersed, instead of being organized into mineral crystals. However, chemical analyses show the same amounts and variations of chemical elements, such as silicon, iron, magnesium, and aluminum, as in related aphanitic and granular rocks. Color is not a clue to composition in obsidians; all are dark, largely because of finely disseminated iron minerals. Pumice and scoria do not have a glassy look because they are so porous. Since gas-filled magma in eruption behaves like warm, well-shaken soda pop when the bottle is uncapped, pumice is literally a frozen bubbly froth of glass.

The light-colored igneous rocks, which contain much feldspar and usually quartz, are referred to as

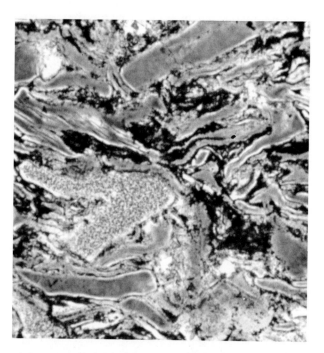

Fig. 4-4 Tuff, much enlarged, viewed through a microscope. Fragments of volcanic ash are flattened and fused together by intense heat of eruption to form a welded tuff. Photo courtesy of Robert S. Houston, University of Wyoming.

TABLE 4.1 Generalized Classification of Igneous Rocks

Texture	Composition			
	Felsic*		Mafic†	
	Quartz, feldspar hornblende dark mica (generally light colored)	Feldspar hornblende dark mica (intermediate)	Dark feldspar augite (generally dark colored)	Much olivine
granular	Granite	Diorite	Gabbro	Peridotite‡
aphanitic	Rhyolite	Andesite	Basalt	
glassy		Pumice Scoria		
		Obsidian		
pyroclastic		Breccia		
		Tuff		

*Felsic means abundant feldspar and silica.
†Mafic means abundant ferromagnesian minerals.
‡Peridotite has no aphanitic equivalent.

felsic. Rocks dominated by dark ferromagnesian minerals are described as *mafic*. If rocks are composed of minerals containing a very large proportion of iron and magnesium, such as certain pyroxenes and olivine, they are called *ultramafic*. These terms are often used in geologic literature discussing rock assemblages of the mantle and crust.

Pyroclastic Rocks Magmas heavily charged with gas may explode, causing volcanic eruptions, showering the area with rock fragments. These accumulate as pyroclastic rock.[1] Volcanic ash is a loose deposit of glass fragments, crystals, or pumice particles that sometimes are no larger than fine dust. Consolidated to lightweight rock, it is called tuff (Fig. 4-4). Volcanic breccias consist of fragments, ranging from pebble-sized to some large blocks ripped from the volcano or surrounding rock. Pyro-

[1] In Greek, *pyro* means "fire" (igneous), *clast* means "fragment."

clastic fragments are unlike the intergrown crystals or glass of similar composition, which form as magma freezes. Pyroclastics are texturally distinctive igneous rocks because they are fragments, heaped together. In fact, some geologists prefer to classify them as sedimentary rocks.

Explanations

Crystal Growth We can observe the general rule that slow cooling leads to large crystals; fast cooling leads to small crystals or glasses. This can be experimentally demonstrated by controlled cooling of silicate melts in ovens. Also, accidental slow cooling of mixtures used in making ordinary window glass has produced crystals. A great many observations show that plutonic rocks have granular textures, and volcanic rocks have either aphanitic or glassy textures.

In the formation of plutonic rocks, magmas cool slowly. At depth, the surrounding rocks act as

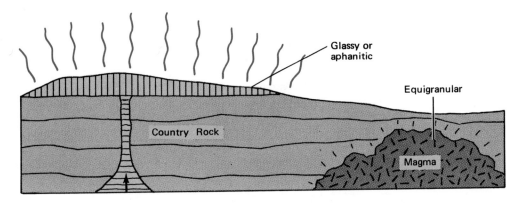

Fig. 4-5 Extrusive magmas cool rapidly in air. Intrusive, being insulated, lose heat slowly.

insulation, preventing the rapid escape of heat, since there is no radiation and conduction is very slow. Thus, larger crystals and coarser textures result (Fig. 4-5). In eruption, magmas are chilled more rapidly because their heat is rapidly lost into the air. Volcanic rocks, therefore, have fine (aphanitic) textures or are glassy if cooling is very rapid. But answering one question in cause and effect leads to another: How does slow cooling cause large crystals?

To explore this basic question, visualize magma as a liquid. This means that its ions are initially evenly dispersed and in contact, but free to move around, like a dish full of ball bearings. During cooling and solidification, ions of the different chemical elements move to randomly distributed centers throughout the magma. There they "freeze" into solid atomic crystal structures which slowly grow, as the liquid disappears, until the magma has become rock (Fig. 4-6). The growth of crystals can be observed by dangling string into concentrated sugar syrup for several days.

Growth of crystals in a magma depends on two main factors: rate of cooling and viscosity. A slow-cooling magma, remaining liquid for a relatively long time, allows many ions to migrate to relatively few centers, so that large crystals can build. We can imagine the centers, or nuclei, as having a

magnetic attraction for related ions. In a faster-cooling magma, less time is available during which ions are free to move, so fewer ions gather at more centers and smaller crystals result. If magma chills very rapidly, the mass solidifies before ions can migrate into crystal structures. This produces a glass.

Viscosity of magmas also affects the buildup of ions into crystal structures. Viscosity refers to the "thickness" of liquids. A highly viscous one is stiff and pasty; one with low viscosity is thin and fluid. Differing viscosity of magmas is shown by the flow

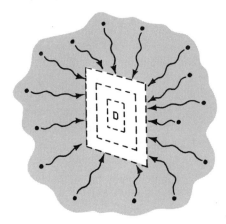

Fig. 4-6 Ions migrating through a magma to build crystals.

of lava during volcanic eruptions. Fluid lavas may move quite rapidly; on a comparable slope, highly viscous ones move slowly. High viscosity hinders ion movement through a magma, so crystals grow slowly. Low viscosity allows more rapid ion migration into crystal structures. As a rule, therefore, if other conditions are equal, viscous magmas develop rocks with finer textures, and less viscous magmas develop rocks with coarser textures. This explains even textures, but what about the marked difference of crystal sizes in porphyritic texture?

Change of environment is one possible explanation for porphyritic textures. A magma, cooling slowly at depth, would start to develop some large crystals. During cooling, the magma would become a slush, that is, a mixture of solid crystals and still-liquid magma. Such a magma could start to rise again, for various reasons, and might even erupt at the surface of the Earth. In its new surroundings, the magma would chill more rapidly, so that the remaining liquid would solidify as a fine-textured mass around the larger crystals previously formed at depth (Fig. 4-7).

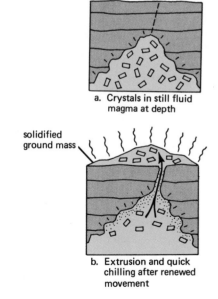

a. Crystals in still fluid magma at depth

solidified ground mass

b. Extrusion and quick chilling after renewed movement

Fig. 4-7 A possible origin of porphyritic texture.

These explanations about igneous rocks are the barest outline of complex matters. They suggest the sort of reasoning used in attempting to explain what went on millions of years ago, in magmas then miles below the Earth's surface, as well as in observable volcanic eruptions of today.

SEDIMENTARY ROCKS

Sedimentary rocks are the most widespread on the continental masses, forming some 75 percent of their surfaces. However, the volume of sedimentary rocks in the outer 16 kilometers (10 miles) of the Earth's continental crust is small, a mere 5 percent. Yet, this thin rock veneer has given the main clues for the central theme of geology—the history of the Earth. Sedimentary rocks are the record of the ceaseless action of air and water on solid rock, through the enormity of geologic time (Fig. 4-8).

Origin

In large measure, sedimentary rocks are the debris of older rocks. They derive from preexisting rocks through the workings of weathering, erosion, transport, and deposition—the never-ending processes reworking the Earth's crust (Fig. 4-9). *Weathering* is the crumbling of the crust, in place, through exposure to atmospheric "elements." It paves the way for *erosion*—the carving away of rock by streams and glaciers, wind and waves. *Transportation* is the carrying of fragments or dissolved particles by moving air or water. *Deposition* is the settling out, or precipitation of materials—the defining feature of a sediment. Loose gravel, sand, and silt along a river, the cobbles and sand of beaches, and muds of sea bottoms are sedimentary rocks in the making.

Many geologists would consider unconsolidated deposits as rock, but to most people a rock is something hard and solid. In any case, still another

Fig. 4-8 Panorama of characteristically bedded sedimentary rocks as exposed in the north rim of Grand Canyon. U.S. Forest Service photo by L. J. Prater.

process, *lithification*, following deposition, can convert loose fragments into solid sedimentary rocks. Sand and larger fragments are bound together by cementation. Water, seeping through sediments, deposits bonding material from solution around individual particles, just as "hard water" leaves a lime coating inside pipes. Calcite, silica, and iron oxides can be seen with a microscope as common cementing materials in the spaces between the original sedimentary particles. Compaction by itself lithifies some sediments. In fine-grained materials, such as clays, the added weight of more and more deposited materials compresses them, squeezing out water until they cohere in a solid mass.

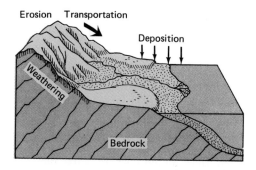

Fig. 4-9 Processes active on the Earth's surface that produce sediments and sedimentary rocks.

Classification and Description

Sedimentary rocks can be divided into two general groups: clastic, and nonclastic or chemical-organic.

Clastic Rocks Clastic rocks are masses of cemented fragments—like man-made concrete. Their fragments are piled together with much pore space in between. Under a miscroscope, sandstone, a solid clastic rock, looks like a heap of loose beach sand except that its pores contain cementing materials

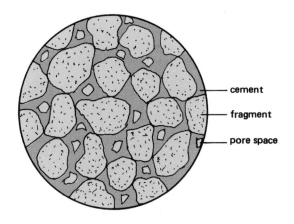

Fig. 4-10 Magnified view of grains and cement as found in quartz sandstone.

Fig. 4-11 Conglomerate. Photo by Rhodes Fairbridge.

(Fig. 4-10). Clastic rocks are subdivided and named according to fragment sizes.

Conglomerate, the coarsest-textured clastic rock, has easily visible rock fragments ranging from boulders to small pebbles (Fig. 4-11). Fragment shapes vary from rough and jagged to smooth and rounded. Increasing smoothness, along with decrease in size, is a measure of the distance and violence of transport before deposition. In moving water, edges and corners are knocked off as particles strike and grind against each other. This has been demonstrated in rotating drums and other devices that churn particles about.

Sandstone is just what its name implies—lithified sand. Although it can be composed of any sand-sized mineral particle, there are three typical sandstone types. Many sandstones are composed largely of quartz. This mineral is exceedingly resistant to chemical change, and quite resistant to mechanical wear in water, after it is reduced to

TABLE 4.2 Sedimentary Rock Types

Clastic	Chemical-Organic
Conglomerate	Limestone
Sandstone	Chert
quartz sandstone	Rock salt
arkose	Coal
graywacke	
"Shale"	
siltstone	
claystone	

sand size. Experimental work on sand grains shows that angular ones are quite rapidly rounded when blown along the ground in wind, although not during transportation in water; however, once rounded their reduction is very slow, as they act like well-lubricated ball bearings. Thus, rounded quartz grains are almost indestructible, whereas other common minerals such as feldspar are broken down, dissolved, and winnowed away by extended transport in moving air or water.

Arkose, a second type of sandstone, contains mainly feldspar grains, although it can also contain some quartz and various other mineral grains. Its composition indicates erosion of nearby bare granite whose feldspars have not been completely weathered to form clays. (Extended reworking by waves, streams, or wind would break down and remove its feldspars and less resistant minerals, and ultimately produce a quartz sandstone.)

Graywacke, a third type, is a hard, dark, "dirty" sandstone (Fig. 4-12). It is a mixture of sand and mud, often containing tuff and particles of rock debris. The appearance of graywacke indicates that it was rapidly deposited in deep submarine basins, so that winnowing of silts and clays by wave action has been minimal.

The name shale is loosely applied to the finest-textured clastic rocks originating as muds. Strictly speaking, shale is a rock that breaks into platy chips. If lacking shaly structure, the fine-grained clastic rocks are best called siltstone and claystone, depending on their composition. Silt, greatly magnified, shows angular particles apparently unmodified by transport in either wind or water. Some may be chips knocked off sand grains, but of this we cannot always be sure. Clay is produced by the weathering of several minerals, including the abundant feldspars. Its particles are far too small and light to be shaped by impact during transport. These fine-grained rocks often indicate deposition in quiet or protected waters where current and wave action is not too pronounced.

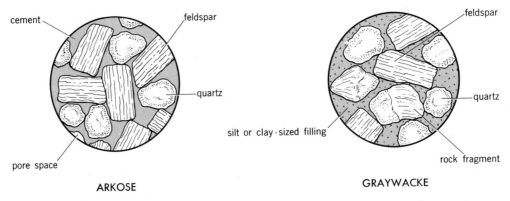

Fig. 4-12 Diagrammatic magnified views contrasting two sandstones: an arkose and a graywacke.

Chemically and Organically Deposited Sedimentary Rocks Besides the clastic group, some sedimentary rocks are deposits of material originally dissolved from older rocks. The particles are carried in solution as ions, which, being too small to be seen with the best optical microscope, are invisible during transport, in contrast to clastic fragments. The lime coating in water pipes and tea kettles is chemically precipitated; the material in a clam shell is an organic deposit.

Textures of chemical-sedimentary rocks are sometimes akin to those formed in cooling magma in that both are crystal masses formed in liquids. In chemical sediments, crystals form when water becomes saturated with their chemical elements. Fine crystals settle to the bottom of the water forming a loose layer, typical of lime muds in some tropical seas. Later compaction and continued growth of crystals in the layer can produce a dense interlocking mass of crystals.

Chemical-organic rocks generally lack clastic texture, so in a simple classification are named from their composition. There is a considerable variety of such rocks, but a few common ones illustrate their various origins. Rock salt, composed of the mineral halite, is precipitated from solution as water evaporates. It is a common deposit in and around salt lakes, such as the Dead Sea and the Great Salt Lake of Utah, and in restricted or isolated bodies of ocean water. Gypsum (calcium sulfate) is another common salt deposited in the same manner.

Chert is an abundant sedimentary rock, composed of aphanitic silica which is, perhaps, deposited as a gel or, perhaps, as an ion-by-ion replacement of other materials. On the deep ocean floors, generally far from land areas, chert originates from organic oozes. Such deposits represent countless siliceous skeletons of microscopic one-celled organisms called diatoms and radiolarians which are very slowly settling from the open ocean above down into the depths.

Coal is characterized by combustible carbon. It is a noncrystalline rock which often preserves the woody structure of trees or other plants from which it is derived.

Limestone, which makes up about 20 percent of the world's sedimentary rock, is mainly composed of the mineral calcite. It may be chemical in origin, a direct precipitate from solution in water. Much is also organic, precipitated by animals which take in dissolved ions and convert them to solid shell materials. These progressively accumulate as sedimentary rock as the animals die. The attack of waves along a sea coast may break chemical or organic limestone into fragments which are later cemented into clastic rock—a calcareous sandstone.

Some widespread deposits of limestone originate on the floors of the open oceans as calcareous oozes. These deposits accumulate from the slow settling of foraminifera, microscopic one-celled animals, from the open ocean above. Foraminiferal oozes are widespread and abundant to depths of about 4000 meters (a little over 13,000 feet) but at greater depths they are extremely rare. The minute calcareous skeletons are apparently dissolved in the high pressures of cold, deep waters that are relatively rich in dissolved carbon dioxide.

Sedimentary Structures

Layering Stratification (layering or bedding) is the most characteristic structural feature in sedimentary rocks (Fig. 4-13). The rock tends to be separated along bedding planes because of its origin as broad sheets of debris laid down on the surface of the Earth by moving fluids such as in streams, waves, and wind. The first sediments laid down cover irregularities in the surface beneath them. Then, sediments are deposited in nearly horizontal layers. Individual layers vary from paper-thin to beds hundreds of meters thick. The pattern is always the same—layer upon layer, like papers piled in order on a table.

The first laid down, the oldest, is at the bot-

Fig. 4-13 Relatively thin sedimentary stratification is well displayed in Green River lake beds of western Wyoming at Teapot Rock. Note figure under "spout." Photo taken in 1869 by W. H. Jackson, famous pioneer photographer of the Old West. U.S. Geological Survey photo, National Archives.

tom, and the rest fall in order with the last deposited, or youngest, at the top. These relations, so obvious in stacking a pile of papers, were revolutionary when first clearly stated as early as 1669 by Nicolaus Steno (Niels Stensen) of Denmark (Fig. 3-19). The truth is that they are readily seen only if the origin of sedimentary rocks is appreciated. Steno's two laws are stated below.

LAW OF ORIGINAL HORIZONTALITY

Water-laid sediments are deposited in strata that are not far from horizontal and parallel or nearly parallel to the surfaces on which they are accumulating.

LAW OF SUPERPOSITION

In any pile of sedimentary strata that has not

been disturbed by folding or overturning since accumulation, the youngest stratum is at the top and the oldest at the base.

They are the key to unraveling sedimentary layers which have later been folded or broken.

Bedding in sedimentary rocks usually originates from fluctuations and changes during deposition. In some cases, the type of material may change. In a mountain lake, silt and clay, normally carried in by streams, may alternate with sand pebbles and leaves washed in during times of flood, to give a layered sequence. On a larger scale, limestone depositing on an ocean floor may be periodically covered by clays and silts swept far out to sea from the mouths of rivers made especially muddy during major floods. Over great spans of geologic time, shallow seas have advanced and retreated on the continents, leaving blankets of deposits contrasting with the layers laid down on land by streams, wind, and glaciers (Fig. 4-14).

Deposition itself is periodic. A desert lake, a playa, receives water only during rare storms. Most of the time it is dry, allowing the materials washed in to harden. Later, more debris washes in to form additional layers on the older ones. Pauses in deposition range from extremely short, as in the pulses of a continuously flowing stream, to extremely long. For extended periods, areas may emerge from water so that sediments are no longer deposited, and erosion may set in. If deposition is resumed later, the interruption is marked by a surface that forms a bedding plane.

While layering suggests sedimentary origins, it is not proof positive. Successive eruptions give layering to volcanic rocks, and layerlike structures occur in plutonic igneous and metamorphic rocks. The recognition that rocks are sedimentary, and the reconstruction of their origin and environment, depends on the study of minor structures inherited from unconsolidated sediments.

Other Depositional Features   *Mud cracks* are common in dried-out puddles, lake bottoms, and marshes where clays and silts shrink as water evaporates (Fig. 4-15). As a muddy surface dries, it can be imagined as shrinking in circular areas towards evenly dispersed centers. Cracks form where contraction circles overlap (Fig. 4-16). Although the causes for the contraction differ, this shrinkage theory of cracking in mud applies to far more spectacular forms, such as giant columns of basalt formed by contraction of freezing lava, and large areas of cracks forming great networks on arctic tundras, which result from contraction during intense freezing of water-saturated ground.

Assuming that ancient mud cracks formed like those of the present day, those preserved in solid rock allow us to visualize a geographical setting that disappeared in some long-gone time. Most mud-cracked rocks probably originated in mud flats occasionally covered by shallow water and periodically dry—just as they do today.

Ripple marks on slabs of sedimentary rock are duplicated in modern sediments (Fig. 4-17). They

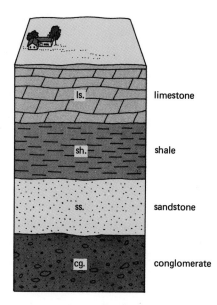

limestone

shale

sandstone

conglomerate

Fig. 4-14 Bedding may result from deposition of a succession of rock types.

Fig. 4-15 Mud cracks with worm trails. Colorado River, Glen Canyon, Utah. Photo by Tad Nichols.

are common on sandy stream bottoms, tidal flats, or beaches, on some parts of the deep-sea floor, and on dunes of windblown sand. Asymmetric ripples with the steeper side to the lee (downwind), called current ripples, develop under the action of water or wind moving in a constant direction. Water sloshing back and forth, as it does on a lake bottom or near-shore sea floor where there is little current, develops symmetrical, sharp-crested oscillation ripples (Fig. 4-18). These features are evidence for the specific movement of water or wind millions of years ago.

Cross-laminations are sloping layers within larger sedimentary beds (Fig. 4-19). They reflect the buildup of sediment coming from a specific direction. Water and wind spill material down advancing slopes in the same way that bulldozing dumps dirt into a ditch, eventually giving a flat-

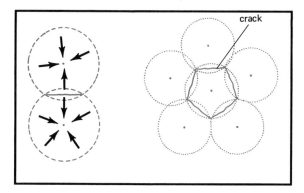

Fig. 4-16 A theory for the mechanics of shrinkage cracks.

topped fill (Fig. 4-20). Although cross-laminations are often beautifully exposed on rock in cliff faces, their origin in sediments is difficult to observe in progress because they form by burial. However, by running water loaded with sand into glass-sided tanks, their formation can be seen.

Graded bedding is characterized by coarse particles at the bottom of a layer and progressively finer ones toward the top. Pebbles or coarse sand at the base grades upward into fine sands, silts, and clays (Fig. 4-21). Experiments in laboratory flumes indicate that graded bedding results from *turbidity currents,* muddy clouds of debris-charged water spilling down slopes in standing bodies of water such as lakes. As the currents slow down,

Fig. 4-17 Ripple marks in sedimentary rocks. Photo by Rhodes Fairbridge.

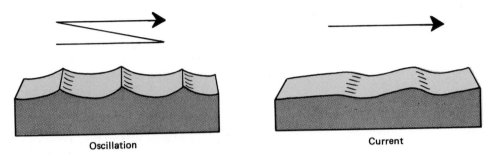

Oscillation

Current

Fig. 4-18 Two types of ripple marks.

Fig. 4-19 Perhaps these are "fossil" dunes or perhaps they were deposited in water. In any case, erosion is now exposing impressive cross-beds in the Sand Creek area, Wyoming. Photo by the author.

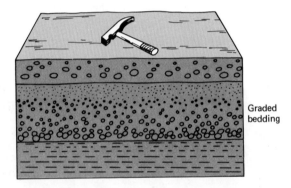

Graded bedding

Fig. 4-21 Graded bedding is common in eugeosynclinal deposits of continental slopes and of deeply subsiding basins in tectonically active volcanic island regions.

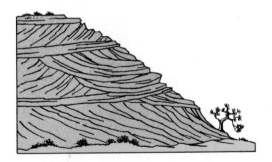

Fig. 4-20 Sketch of several cross-bedding types: inclined planes above and below with curved scour-and-fill in the middle layer.

the coarsest particles drop out first and successively finer material settles on top of them. Graded bedding is a common feature of unconsolidated sediments recovered in cores from ocean bottoms below the continental slopes. In sedimentary rocks it is characteristic of thinner layers in graywacke sandstones.

Some features of sedimentary rock form within a sediment after it has been deposited. *Concretions* are lumps, or nodular masses, of materials seemingly different from the beds in which they are found. They have a variety of shapes, such as spheres, grapelike clusters, disks, and irregular forms (Figs. 4-22, 4-23).

Most concretions are hard and well-cemented

Fig. 4-22 Concretions photographed in Antarctic by Larry Lackey.

lumps resulting from deposition by water percolating through porous rock layers. Centers of organic materials in sediments may create chemical conditions around themselves which encourage precipitation. Concretions themselves are inorganic in origin but are sometimes mistaken for fossils.

Geodes are concretionlike forms, but hollow, and often lined with beautiful layers of calcite or quartz crystals. The crystals are deposits from water filtering into cavities commonly weathered out of limestone. They may resemble concretions until cracked open to reveal their crystal linings.

Past Worlds Recorded in the Sedimentary Rocks

The study of sedimentary rocks introduces a fundamental rule of geologic detective work: "The present is the key to the past." The great contribution of Dr. Hutton's principle of uniformitarianism (discussed in Chapter One) is the concept that physical processes have operated in the same manner throughout geologic time. Applied to ripple marks, for example, it means that ancient ones preserved on sedimentary rocks were formed under

Fig. 4-23 A rather weird collection of concretions exposed in the old bed of the Salton Sea. Most, I understand, have now been collected and grace rock gardens in the older parts of Los Angeles. U.S. Forest Service photo by Hutchinson.

similar conditions to those of modern sediments. Ripple marks may seem trivial; yet, much of geology is based on the careful study of similar minutiae viewed in the light of uniformitarianism.

Paleogeography In large measure, *historical geology* is geography,[2] because it deals with the face of the Earth and with once-living communities. But it is geography of a special sort, a geography in the fourth dimension that involves the patterns of par-

ticular bygone ages, and their progressive change through an enormity of time. As you might suspect, present-day geography is a model for the past, for *paleogeographic reconstructions*. Any geography is a composite of many environments, local scenes, or surroundings, each influencing communities of life.

As a preliminary to historical reconstruction, the Earth's many surface environments that are observable today can be viewed as three broad groups. *Continental*, or terrestrial, environments make up the world's landscapes, which we shall discuss in some detail (Chapters 9–11). They occur in mountains, plains, and plateaus, and the lesser features made by the destructional external forces working on them, all of which have certain variations ac-

2 Geographers will complain, mumbling something about the effect of physical environment on man, which is a somewhat restricted field of human geography; and geologists, who resent being classed as geographers, will consider the statement a gross oversimplification.

cording to their occurrence in particular climatic zones. Geologists are much concerned with *marine* environments because the sedimentary rocks blanketing three-quarters of the continental surface, and which have been geology's most important historical documents, were largely deposited in shallow seas. For ours is a time of rather notable emergence, and, in the last half-billion years for which there is an adequate record, only the interval about 200 million years ago (late Paleozoic–early Mesozoic), was at all like it. During most of geologic time the continental platforms have been more widely flooded. Marine environments vary with depth from shallow seas to the deep ocean floors and in water temperature from arctic to tropic seas. *Transitional* environments are represented by beaches, deltas, and other features developed by shore processes and the interplay of land and sea.

In reconstructing paleogeography, geologists must of necessity start by looking at individual outcrops. Later this local information can be compiled and related to broader patterns; but first, what can be read from rocks at a single place?

Primary Structures Depositional features of sedimentary rocks are often useful indicators of the conditions under which they formed. Mud cracks preserved in shales suggest a former mud flat that was alternately covered by water and then dry. Ripple marks indicate the nature of waves or currents creating them. Cross-bedding may indicate whether beds were formed as windblown dunes, stream deposits, or on the sea floor. The largest cross-laminations are generally thought to be formed in windblown sand. Those of streams, if cut across the channel axes, may show a pattern of concave curves cutting across each other. Considerably more needs to be learned about the origin of cross-bedding, but used in conjunction with other evidence it may indicate environment, and it is useful as evidence of the former direction, strength, and variability of moving air or water.

Volcanic rocks can also indicate past environ-

ments. Those that cooled in air are frequently jointed into vertical polygonal columns (Fig. 4-1) while those cooling in submarine surroundings often have pillow structure. This develops when lavas, usually basaltic, erupt under water. Pillow lavas look like a mass of tightly packed pillows (Fig. 4-24). Their cross-sections may be round, elliptical, or convex on top and concave on the bottom. They result when erupting magma is quickly chilled by water to create a plastic skin which fills with still-molten material before the whole mass solidifies. Pillow lavas are very common in sea-floor eruptions. These and similar physical features are records of ancient environments where we know enough of the modern processes to read them.

Color In some cases the color of rocks tells something of environmental conditions. *Black shales* may owe their color to abundantly preserved organic material, which suggests a stagnant ocean bottom of "stinking muds" or rapid deposition, for if long exposed to well-oxygenated waters the dark organic material would be decomposed. These shales contain mainly fossils of swimming and

Fig. 4-24 Pillow lavas near the crest of the Mid-Atlantic Ridge. After Ewing, Ewing, and Talwani, 1964; courtesy of Lamont Doherty Geological Observatory.

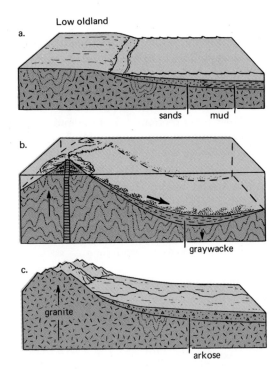

Fig. 4-25 Environments inferred from different types of sandstones: (a) quartz sandstone; (b) graywacke; (c) arkose.

floating organisms that sank to the bottom on death; bottom dwellers[3] are rare, which suggests a rather unhealthy sea floor in a sinking zone of a geosyncline or shelf sea. *Gray shales*, on the other hand, suggest a more stable floor where the water circulation is better.

Red beds include conglomerates, sands, and shales whose striking color comes from iron oxide (the mineral *hematite*), which may coat quartz or other clastic grains, or sometimes from abundant particles of red minerals such as red feldspar. In some places red beds are not happy hunting grounds for fossils; however, in others footprints, amphibians, fish, and plant remains have been

3 Swimmers are called *nekton*, floaters *plankton*, bottom dwellers *benthos*.

found, indicating a dominantly continental environment. The red beds are flood plain, alluvial fan, and deltaic deposits, which are sometimes carried into the open sea, in a strongly oxidizing environment. Red beds have posed a classic geologic problem—whether they do or do not represent a particular climate.

Some workers maintain that red beds indicate a desert climate because of their frequent association with layers of salt and gypsum. These precipitated rocks, called evaporites, are thought to form in partially closed basins, where restricted arms of shallow seas are periodically cut off and evaporated to near dryness. Other geologists suppose a semiarid climate with marked wet and dry seasons, like the savanna grasslands in present-day Africa. Still others envision the red beds as the red soils,

Fig. 4-26 Leonardo da Vinci (1452–1519), a near-universal genius. Da Vinci's notebooks show that he clearly appreciated the nature of fossils, of erosion, transport, and deposition, and of shifting seas and land. Free of dogmatic misconceptions, he was a naturalist far ahead of his time. Unfortunately, his geologic ideas were lost for several centuries.

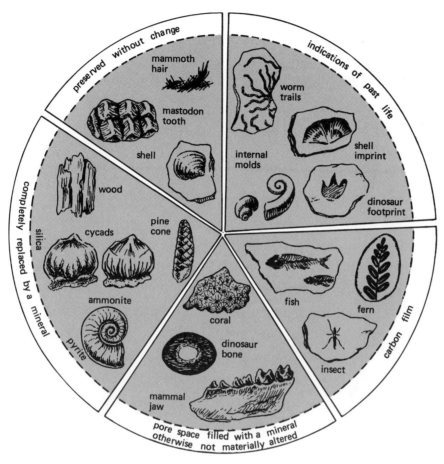

Fig. 4-27 Fossils and some types of preservation. Based on an exhibit in the University of Wyoming Geology Museum.

called laterites, of wet tropical areas, the jungles. It may well be that red beds have several climatic origins.

Rock Types Rocks are products of their heredity as well as the environment in which they formed. This is evident in those metamorphic rocks which have not been so drastically altered by deep-seated environments that relic features of original sedimentary and volcanic surface rocks have been obliterated. Slates bear evidence of once being shales, marbles of being limestones, and so on. Less

directly, perhaps, the composition of sedimentary rocks reflects the regions from which preexisting materials were derived. For although the only positive environmental record is in rocks at the sites of deposition, we can, from a knowledge of existing erosional landscapes, make informed guesses as to whether the crust was stable, or being actively uplifted in nearby plains and mountains which erosion later destroyed.

Well-sorted quartz sandstones, shales, and limestones suggest relatively stable areas of deposition and source regions of low to moderate

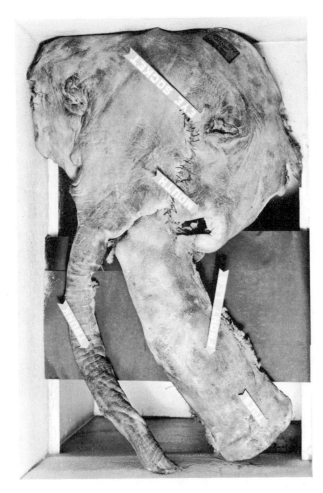

Fig. 4-28 Actual remains of a baby wooly mammoth preserved in frozen ground of Alaska until discovered. Courtesy of the American Museum of Natural History.

relief whose rocks were extensively weathered. The insoluble end products of chemical weathering are quartz and clays. Wave action, and in some cases wind, winnow these products, leaving "clean" sandstones in and near shore surroundings, and carry muds farther offshore where they settle beneath the zone of wave stirring. Pure limestone, derived from the soluble products, suggests precipitation in warm, clear, and shallow seas.

Conglomerate is taken as evidence of relatively rapid uplift and rugged topography in a source area of mountains drained by swift streams, or of wave-attacked headlands and promontories. Some, however, may represent submarine landslides. A poorly sorted conglomerate containing snubbed and scratched boulders could well be a lithified glacial deposit.

Graywacke also suggests crustal instability. It has been described as having a "poured-in" look, meaning it accumulated rapidly, with little reworking by waves and currents, which would have removed the finer materials. Thus graywackes indicate rapidly subsiding geosynclinal areas and orogeny, marked by rising mountains or islands and deeply subsiding basins into which sediments were spilled rapidly with almost no winnowing or sorting.

Arkose is literally a "granite wash," suggesting rapid uplift and rugged relief in a source area of granite, little affected by chemical weathering (Fig. 4-25).

Fig. 4-29 Limestone composed of a mass of shell fragments. Photo by Tad Nichols.

Fig. 4-30 Partially excavated dinosaur bones from Bone Cabin Quarry, Wyoming. Courtesy of the American Museum of Natural History.

Fossils Since fossils are, by definition, "evidences of past life preserved in rock"—they are important clues in unravelling the ancient geologic environments in "past worlds." Some fossils occur in igneous rocks where volcanic eruptions once overwhelmed plants and animals; and some fossils are found in metamorphic rocks where the heat and pressure transforming the parent rock materials was not too extreme. By and large, however, most fossils are features of sedimentary rocks that formed in depositional environments where burial of organic remains was relatively rapid and complete decay was inhibited.

Fossils come in many different varieties (Fig. 4-27). They may be the actual remains of plants and animals: a mammoth carcass complete with flesh and hair frozen in an arctic tundra (Fig. 4-28); an oyster shell buried in an ancient shore (Fig. 4-29). They may be stone replicas from slow molecule-by-molecule replacement of original organisms by minerals: an agate log, a silicified bone (Fig. 4-30). They may be impressions, molds left in sediments after organic material has disappeared; or later fillings of such molds forming the reproduction of a leaf or shell in sandstone or shale (Fig. 4-31). They may be mere indications that an ani-

Fig. 4-31 Fossil fern from Petrified Forest, Arizona. Photo by Tad Nichols.

mal was once there: dinosaur tracks, worm borings in now solid rock (Fig. 4-32).

We assume that the ancient plants and animals, represented by fossils, were adapted to particular environments just as are now-living types. Oysters, a sedentary lot, never left the sea so their shells indicate a marine environment. A lizard's tracks on a lamina of a cross-bedded sandstone[4] suggest a dune sand, rather than one laid in water. Coal, the fossil record of a forest, in the icy wastes of the

[4] An uncommon situation, but it illustrates the point.

Antarctic indicates a far warmer climate when this rock formed. Massive coral reefs, so far as is known, form only in tropical ocean waters. Fossils can also be used to reconstruct more detailed elements within the major environments, as for instance the shore, shallow water, and deeper ocean bottom.

A single feature of a rock is rarely conclusive evidence of the rock's particular origin; it takes a number of features to make a strong case. One observation, like a statment taken out of context, can lead to erroneous conclusions. For instance, a fossil tree entombed in a rock might well mean a terrestrial environment. Yet the rock could be marine if the original tree drifted to sea before it became waterlogged and sank. In this case, closer examination might disclose clam shells, fish remains, and other evidence of an ocean bottom.

Stratigraphic Columns Although the contacts between rock units in a vertical column may be quite sharp, they need not—as early Catastrophists assumed—represent any sudden flood or drastic change of environment. The slow migration of a sea across a particular area is an adequate cause, in line with uniformitarian thought. *Transgressive seas,* advancing over an area, would leave a vertical rock sequence such as the following: terrestrial red beds at the bottom, overlain by a sandstone laid down at the margin of an encroaching sea, above which is a marine shale of deeper waters (Fig. 4-33). The shifting environments as a *regressive sea* withdraws over an area might be marked by a sequence including from bottom to top: a limestone marine facies, a shale of the deltaic facies, and a sandstone of a terrestrial sand dune facies.[5]

Restorations Changing patterns of shifting seas, rising mountains, and other episodes in the geologic drama through time are largely reconstructed

[5] *Facies* means "aspect," or "look." There are also igneous and metamorphic facies, but let's not get involved in these.

Fig. 4-32 Mr. Utterback digging out a Daemonelix (Devil's Corkscrew) which is a cast of rodent burrow, Sioux County, Nebraska. J. B. Hatcher photo, courtesy of the Carnegie Museum, Pittsburgh, Pennsylvania.

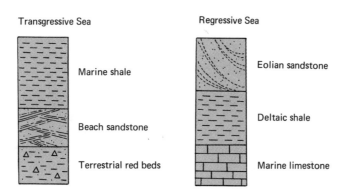

Transgressive Sea

Marine shale

Beach sandstone

Terrestrial red beds

Regressive Sea

Eolian sandstone

Deltaic shale

Marine limestone

Fig. 4-33 Local rock columns indicating advancing and retreating seas.

ENVIRONMENTS

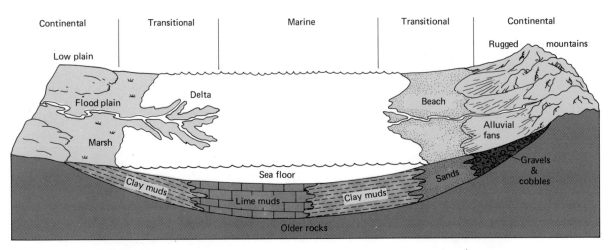

Fig. 4-34 The concept of sedimentary facies illustrated in present-day environments.

from restored *cross-sections*. Any cross-section is a geologist's interpretation of what a vertical slice into the Earth would show. A restored cross-section is an attempt, after eliminating the effects of later erosion and later structural contortion, to show rock relations as they must have been at the end of some particular interval of time. Because of the spotty nature of outcrops, restored sections must normally be compiled from rock columns measured and described at many separate places across a wide region. They are based on both lithologic correlation, to demonstrate the continuity of particular rock types, and also time correlation, from place to place, to establish time lines marking the slow migration and fluctuation of continental, transitional, and marine environments. For, if traced far enough, rocks of the transitional shore zone, for instance, do tend to cross time lines.

Facies Because of the variety of environments, different sediments are being laid down side by side at that particular time we call the present day. The fact is self-evident; they can be walked (skin-dived, bathyscaphed, or more simply bottom-

sampled) from one to another. Boulders and gravels in alluvial fans at the foot of mountains may grade into sands of a beach, the sands may grade into muds of the offshore bottom, and the muds into limey reefs in clear waters remote from the shore.

Lateral changes in the overall "look" of sediments or sedimentary rock units that reflect different depositional environments are known as sedimentary facies (Fig. 4-34). Thus piecing together facies relations in rocks—which often takes considerable ingenuity—is important in the construction of paleogeographic maps for a particular time in the geologic past. Such maps, however, are like single frames on a reel of motion-picture film; they are still shots of an ever-changing scene.

There is a noble ring to the story of historical geology, wherein great mountains rise and then are worn low, and the sea comes in and goes out across continental platforms. Such grand episodes are largely read from changing patterns of facies through time, for the geography of the present is not that of the past nor, for that matter, will it be unchanging in the future.

METAMORPHIC ROCKS

Origin

Metamorphic rocks, the third great genetic group, form where great heat and pressure have altered solid preexisting rocks. Lesser metamorphic belts are found along the margins of igneous intrusions. Greater metamorphic masses, ranging from relatively small exposures to broad tracts, such as those extending over half of Canada, represent the crumpled roots of mountain ranges now exposed after deep erosion.

In *contact metamorphism,* the heat of an intruding magma is conducted outwards into the surrounding wall rock, which is "cooked" by the heat, and possibly contaminated by hot solutions or vapors from the magma (Fig. 4-35). The heat, in any case, speeds up chemical reactions. *Regional metamorphism,* which involves broad areas, results from crustal deformation wherein surface rocks may sink to great depths in the Earth's crust. Here, under tremendous squeezing pressures and high temperatures, minerals of the original rock are altered to types better adjusted to high temperatures and pressures (Fig. 4-36). Flat or platy minerals tend to develop that are oriented

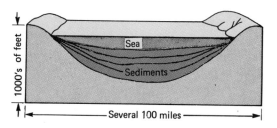

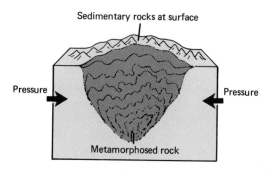

Fig. 4-36 Regional metamorphism resulting when strata (above) are subjected to mountain-making forces (below) that alter deeper sediments in the mountain roots.

perpendicular to the directions of pressure (Fig. 4-37). Many of the rocks develop contorted layers as if the rock had been kneaded like bread dough. At the temperatures and pressures deep in the Earth, rocks become plastic and flow. Closer to the surface, the same rocks would be rigid and would break under pressure.

Classification and Description

Metamorphic rocks range from those with many sedimentary or igneous characteristics, through those with mere vestiges of their parent rock, to those bearing no hint of ancestry. Therefore, although it would be nice to classify them

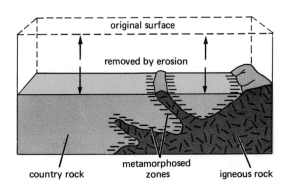

Fig. 4-35 Metamorphic zones resulting originally from alteration of rocks intruded by a magma and exposed much later by erosion.

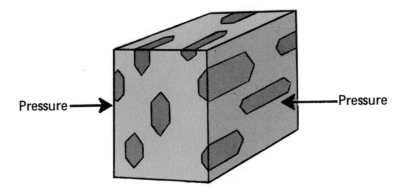

Fig. 4-37 Block shows how minerals tend to be oriented to pressure directions in metamorphic rock.

genetically according to their parents, as altered igneous or altered sedimentary, this is not always possible. Moreover, one metamorphic rock type may result from the metamorphism of several different igneous or sedimentary types. Commonly, the metamorphic rocks are classified empirically into two broad textural groups: unfoliated or weakly foliated, and strongly foliated.[6] Foliation is a rather general term now used to describe a tendency of metamorphic rocks to part along a plane or surface.

6 Since most rocks subject to regional metamorphism have some degree of foliation, we cannot just divide them into foliated and unfoliated groups—a case of scientific hedging.

Unfoliated and Weakly Foliated Rocks   Hornfels is a very tough rock produced by contact metamorphism of shale, tuff, lavas, and other fine-grained rocks. It is unfoliated, for although the rock may have relics of bedding or other oriented structures, it will not break along these structures. Hornfels has a very fine texture, similar to aphanitic lava. Under a microscope, it is a mosaic of recrystallized equidimensional grains. Hornfels may contain a variety of minerals (Fig. 4-38).

Quartzites and marbles may have foliation, but since this feature is not as clearly diagnostic of them as of the strongly foliated group, they are classed with hornfels. Quartzite is a dense sugary-

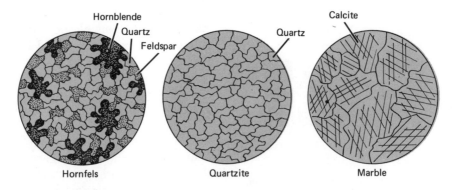

Fig. 4-38 Diagrammatic views of non- or weakly foliated metamorphic rocks as they would appear under the microscope.

TABLE 4.3 Metamorphic Rock Types

Non- or Weakly Foliated	Markedly Foliated
Hornfels	Slate
Quartzite	Schist
Marble	Gneiss
Serpentinite	Eclogite

textured rock that is extremely hard because it is almost pure quartz. Quartz sandstone completely cemented by silica is hard to distinguish from metamorphic quartzite. The metamorphic rock, however, has interlocking crystals rather than grains with silica cement between, a distinction that requires a microscope to be seen.

Marble has clearly visible, interlocking crystals of calcite derived from shell fragments or, originally, from fine grains of limestone. Its glistening crystal faces make it a handsome ornamental building stone. Pure marble is white. The wide range of colored marble results from slight impurities in the rock. Marble results from either contact or regional metamorphism of limestone.

Serpentinite (serpentine rock) is produced by the alteration of peridotite and some other rocks composed mainly of mafic and ultramafic minerals. Such minerals as olivine and the pyroxenes are relatively unstable and readily convert to serpen-

tine, the mineral, in the presence of reactive water solutions expelled by crystallized magmas. The resulting rock, serpentinite, is a metamorphic rock in the sense that it did not directly crystallize from a molten magma; however, in some classifications serpentinite is considered an igneous rock. It is closely related to magmatic processes, and does form some large intrusive masses into the Earth's crust. In any case, it is a plutonic rock that is prominent in the contorted cores of some eroded mountain belts and plays a part in some thinking on the composition of tectonic sea-floor plates.

Strongly Foliated Rocks Well-foliated rocks are all produced by regional metamorphism. Slates are characterized by excellent, smooth, and parallel foliation, called slaty cleavage,[7] which is largely the result of pressure (Fig. 4-39). Its mineral grains are too fine to be seen with the unaided eye. Good slate has economic value because it can be split along its foliation planes into smooth slabs used for shingles, blackboards, and the tops of billiard tables.

Schist has distinctly visible grains, mostly of platy and bladelike crystals, such as mica and hornblende (Fig. 4-40). The flat crystals have

[7] Not to be confused with mineral cleavage, which is splitting within crystals, slaty cleavage is determined by parallel arrangement of many crystal faces.

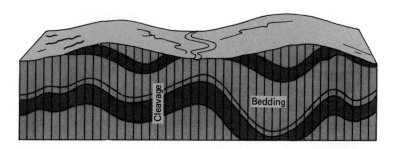

Fig. 4-39 The characteristic cleavage slabs of slate are sometimes confused with bedding. Cleavage planes are oriented with respect to directions of pressure applied to the rock and may cross true bedding at any angle.

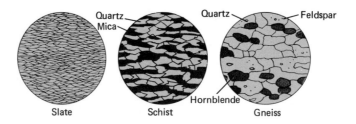

Fig. 4-40 Diagrammatic, microscopic rerpresentations of some foliated rocks.

nearly parallel arrangements, giving a finely banded, wavy foliation to the rock. Schists are named according to their predominant mineral—for example, biotite schist and hornblende schist. They originate mainly from further metamorphism of slate, under high temperature as well as pressure, or from alteration of pyroclastic rocks, such as tuff.

Gneiss, a very abundant metamorphic rock, resembles granite in composition and crystal size. It differs in having coarse and irregular streaks and layers of oriented minerals, such as biotite and hornblende, separated by masses of equidimensional minerals, mainly quartz and feldspar (Fig. 4-41). Gneisses, like schists, are differentiated by their composition—for example, granite gneiss or hornblende gneiss. Some gneisses, when traced by mapping into regions of low metamorphism, are transitional into such rocks as shales and sandstones. By and large, it seems that the parent materials of gneiss are sedimentary. On the other hand, gneisses have been observed grading into granite. So, could granite, that once "classic intrusive igneous rock," be instead the end product of the regional metamorphism of sediments?

Eclogite, although not a particularly common rock at the Earth's surface, figures prominently in some theories about the nature of the mantle. Eclogite's main mineral components are a pale pink garnet and a variety of pyroxene. The rock is gen-

erally classed as foliated, metamorphic; however, some geologists think it is of igneous origin. In either case, this rock is formed under the high-pressure environment existing deep in the Earth's crust or in the mantle.

Fig. 4-41 Folds in gneiss; specimen about one foot across. Photo by R. B. Parker.

Explanations

In volcanic eruptions, muddy streams, and wave-worked beaches, we see igneous and sedimentary rocks in the making. Here, the present is the key to the past. But the origin of metamorphic rocks, like intrusive igneous, is hidden in the depths. It must be deciphered from purely circumstantial evidence.

Alteration How is it established that metamorphic rocks are altered sedimentary and igneous ones? Assume you are having your first field experience with rocks. You examine a buff-colored, fine-grained rock. With a hand lens, small shell fragments can be seen. A drop of hydrochloric acid fizzes on the rock. You conclude it is limestone, composed of calcite. Walking along the bed, the appearance of the rock changes. It becomes whiter and has larger, shiny crystals; a lens does not show any shell fragments. The crystals can be scratched with a knife; a drop of acid on them fizzes—the mineral is still calcite. Nearby, you notice another rock which you identify as igneous. Recalling the origin of igneous rocks, you can visualize it as originally intruding into the limestone as molten magma.

From these observations you infer that the magma baked the limestone, recrystallizing the fine calcite grains into a coarser marble texture. Logical reasoning, but does heat recrystallize calcite? You do some reading on the matter and find that as early as 1805, Sir James Hall of England worked on the problem. He experimented by stuffing powdered chalk, which is a soft limestone, into a porcelain tube. He then slid it into a gun barrel, which he tightly sealed and heated. On cooling, he found a rind of granular calcite crystals around the chalk, indicating that heat can recrystallize limestone. From all of your research, it seems logical to conclude that your marble is thermally metamorphosed limestone.

In another situation, hard, dark, aphanitic rock surrounding an igneous intrusion grades into typical shale. We infer that the shale is baked to a natural bricklike rock, a hornfels. Through a microscope the hornfels exhibits small crystals, which are not present in unaltered shale. Analysis shows similar chemical elements in shale and hornfels. Baking shale in an oven reproduces effects of thermal metamorphism. Hornfels, therefore, may be baked shale.

Oriented Crystals Shale may be altered to slate. Slaty cleavage resembles bedding, but close observation shows that cleavage planes cut across fine layers of silt and clay, the original bedding features. Over broad regions, this bedding is usually bent and folded as if the rocks were contorted by squeezing. The cleavage, however, remains generally parallel and constant in direction. Thus, slaty cleavage is not bedding, and is developed in folded rocks.

Although slate appears to be as fine-grained as shale, a microscope shows that it contains minute flaky minerals, mainly micas. Furthermore, the flakes are all parallel to the cleavage planes. The smooth slabs into which slate breaks result from the oriented minerals.

What causes alignment of the mineral flakes? Around 1853, H. C. Sorby of England mixed flakes of iron ore in soft clay. When he squeezed the mixture, the iron flakes became oriented just as the mica flakes in slate (Fig. 4-42). The clay mix-

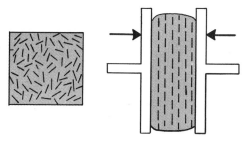

Fig. 4-42 The effect of pressure on clay and mica flakes in Sorby's experiment.

ture could also be split into slabs, suggesting slaty cleavage. But the mica flakes associated with slaty cleavage are not present in the parent rock, shale. They develop later during metamorphism, so Sorby's demonstration does not reveal the whole story.

In 1906, F. E. Wright of the United States cut several glass cubes having the chemical composition of different minerals. These he compressed at temperatures that allowed the blocks to remain solid, but were high enough to let recrystallization begin. In all cases, the resulting crystal fibers were at right angles to the directions of opposing pressure. This suggests that these crystals aligned themselves normal to the pressure directions by growth, rather than by rotation after they had formed. The new orientation favors the development of cleavage planes.

Because schist has easily visible minerals, wavy cleavage surfaces, and often different minerals than slate, these two rocks might seem unrelated. In the field, however, some slates, when followed across country, merge into schist. Chemical analyses of slates and schists show the same chemical elements, even though the minerals differ. Schists result from continued regional metamorphism of slate or other fine-grained rocks. Their larger crystals are the result of a more intense pressure and temperature—a higher grade metamorphism—than that producing slate.

Gneiss may result from metamorphism of granite, a reasonable conclusion where quartz and feldspar grains are streaked and elongated from crushing and deformation. But, many gneisses contain bands and ghostlike remnants of schist. These rocks appear to be formed from schists penetrated by magmatic juices. The fluids bring new elements that react with the minerals of schist to produce zones of feldspar and quartz crystals characteristic of gneissic texture.

Crystal Growth   How can crystals grow, or new minerals form, in solid rock? That metamorphic rocks remain solid is shown by relic structures: bedding planes, fossils, porphyritic textures, and other features inherited from their parent rocks. Complete large-scale melting would obliterate these features; when it does occur, magma forms and igneous rock develops. In igneous rock, mineral growth can be conceived as ion migration to crystal centers in fluid magma. This mental image does not apply to metamorphic rocks.

The regrouping of ions in solid rock to form larger or different crystals requires a solvent. Commonly, rocks contain small amounts of water in pores and minute fissures. The water comes from several sources. It may be imprisoned in the rock on burial, be driven from mineral structures such as clay on heating, or have escaped from nearby magmas.

Small amounts of water allow ions to be dissolved from the surface of solid particles. The ions are then redeposited in crystal structures of new minerals, better adapted to prevailing conditions of temperature and pressure. The process is slow and the water constantly reused, so that the amount required is small. If too much is present, large-scale solution occurs and the metamorphic structure is destroyed.

In the metamorphism of limestone to marble, the ions in calcite regroup into larger crystals. Ions escape more readily from smaller particles and are redeposited on larger crystals. Thus the rich get richer, and the poor get poorer, until a once fine-grained rock is composed entirely of larger crystals.

ROCK ASSEMBLAGES RELATED TO THE NEW WORLD VIEW

Certain groups of rocks—including igneous, sedimentary, and metamorphic types—nicely fit the grand plate tectonic scheme.

In the Ocean Basins

At divergent plate boundaries, where new rocks are being created, the rocks result from eruptions from the asthenosphere. Thus mid-oceanic ridges largely contain surficial submarine basalts, mainly in pillow lavas, and deeper igneous intrusions of gabbro and peridotite (they are mafic and ultramafic rock assemblages made up of iron and magnesium-rich minerals).

Under the open oceans and far from landmasses, the new igneous rocks formed at the divergent boundaries of the great spreading plates are thinly blanketed by very fine-grained sediments called *pelagic deposits*. They include red and brown clays consisting of dust-sized particles of quartz, feldspar, and clay minerals that have been carried far from land—by ocean currents or wind—and are slowly settling into the oceanic depths. Pelagic deposits contain appreciable amounts of fine ash from volcanic eruptions and also meteoric dust from outer space. Supplementing the clastic muds are organic oozes that accumulate from the slow settling of myriads of microscopic skeletons of one-celled organisms: calcareous foraminifers and siliceous diatoms and radiolarians. Deposits on the deep ocean floors also include some larger particles—such things as shark's teeth, whale earbones, and rocks dropped from floating ice. Concretionary nodules of manganese are widespread on the deep sea floor. If pelagic deposits become lithified, the resulting rocks would largely be shales, limestone, and chert.

Near continents and islands, sediments on the deep ocean floor are mainly poorly sorted accumulations of sand, silt, and clay—called *turbidites*. They are derived from sediments in shallow waters on the continental shelves—where materials are stirred and passed along by waves and currents, or suddenly jolted by earthquakes to spill into the deep waters. The descending sediments move down submarine canyons or continental slopes as turbid suspensions, forming density currents, or in submarine landslides. Turbidites may create deposits which are thousands of meters thick in oceanic trenches and continental rises, and may spread far out on the deep ocean floor to create the smooth surface of the abyssal plains. The rock assemblages represented by such deposits include graywackes, marked by graded bedding, and relatively unfossiliferous shales, as well as some interbedded pelagic limestones and chert beds and some interspersed volcanics.

Thick accumulations of turbidites and related rocks in continental slopes and oceanic trenches have recently been interpreted as *eugeosynclines*. These outer geosynclines, of the continental margins and ocean floor, are important features in reconstructions of continental evolution (Fig. 4-43).

The deep ocean-floor rock succession (peridotite, gabbro, and basalt, capped by pelagic and turbidite deposits) is called an *ophiolite suite*. Where such suites are exposed on land, many geologists consider them remnants of deep ocean floor raised above sea level during the collision of global tectonic plates. The deep-water origin of the suites now exposed on land is, in part, indicated by the presence of radiolarian cherts which are interpreted as lithified, pelagic, ocean-bottom deposits (Fig. 4-44).

On the Continents

Sedimentary rocks deposited on the sialic continental blocks are varied, but two assemblages are very important in historical reconstructions.

When continental shelves are dominated by quietly subsiding geosynclines—of the sort recognized by James Hall—the deposits are largely well-washed quartz sandstones, shales, and limestones. Although such deposits may become several thousands of meters thick as the crust slowly subsides beneath them, they originate in shallow waters and do not differ in kind from the thinner deposits left

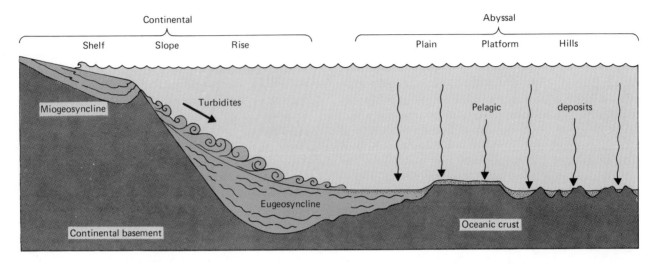

Fig. 4-43 Some deposits and sites of deposition in the new world view. Eugeosynclines generally contain clastic sediments such as graywackes, indicating rapid spurts of deposition. As shown here, the eugeosyncline is forming on a continental slope. In other cases, eugeosynclines develop in tectonically active regions of deeply subsiding basins and rising volanic islands. In such environments, the turbidites would be intermixed with basalts and other volcanic rocks.

by the flooding and ebbing of seas across the stable continental interior. The remains of shellfish and other marine animals are fairly abundant in the beach sandstones of the transitional zone and in the silts, clays, and lime-muds of the shallow offshore sea bed—in contrast to the scarcity of life remains in sediments of the deep ocean floors. Such geosynclines with their characteristic rock assemblages are called *miogeosynclines*. The coastal plain and associated continental shelf along the Gulf and East coasts of the United States probably represent a modern site of this type of geosyncline (Fig. 4-45).

When the continental margins become actively involved in orogeny (mountain-building), the rising highlands shed erosional debris towards the continental interior and bury the miogeosynclinal rocks beneath an assemblage of deposits—collectively called a *clastic wedge* (Fig. 4-46). It represents giant coalescing fans of stream, swamp, and lake deposits which, as mountains rise, expand and

progressively replace the deposits of shallow receding seas. From their maximum thickness, which may reach several thousand meters near active mountains, the wedge deposits gradually thin towards the continental interior. The rocks representing ancient wedges are typically fossiliferous and cross-bedded conglomerates, sandstones, and shales. Many have abundant red beds, as in the rock sequence in New York State called the Catskill "Delta." Vast plains extending eastward from the foot of South America's impressive Andes Mountains are the site of a modern wedge.

Where Continents and Ocean Basins Collide

In the new world view, mountains result from the clash of global plates. Although generations of geologists have studied mountain structures, their incredibly complicated patterns of deformed rock

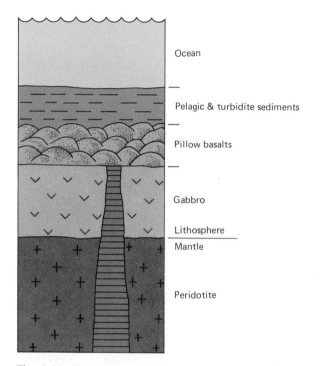

Fig. 4-44 The ophiolite suite, consisting of mafic and ultramafic igneous rocks often veneered by deep-water sediments, is characteristic of ocean-floor rock assemblages. Where similar rocks are found in contorted mountain belts, the rocks are interpreted as relics of ancient oceanic plates.

are still not fully understood.[8] Yet from a welter of detailed investigations, some general rock assemblages can be sorted out and fitted to the grand tectonic scheme.

Extensive tracts of broken, sheared, and generally jumbled rocks, which are found in mountain-building belts around the world, have been christened *mélanges* (French for "mixture"). They are large mappable masses of moderately metamorphosed, fine-grained, shaley material which generally contain angular blocks or slabs of foreign (exotic) rock. The exotic rocks are often masses of

[8] It is a geologic platitude that "more work needs to be done."

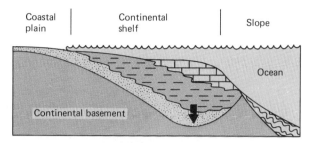

Fig. 4-45 Miogeosyncline developing in the region of the continental shelf.

ophiolite and occasionally of such rocks as limestones and granites, sliding down from continental assemblages. These jumbled rocks of mountain belts often contain extensive intrusions of serpentinites. Whether these altered, igneous peridotites originated under and then rose to intrude eugeosynclinal deposits, or whether they were original components of colliding sea-floor plates, remains an intriguing geological problem. The chewed belts of mélange usually contain graywackes, with graded bedding, and they are generally devoid of fossils (except for some microscopic forms in cherts). Mélanges include: the California belt of rocks called the Franciscan group; rocks in the Alps called the wildflysch; a group of rocks in Italy called argille scagliose; mountain belts in Turkey;

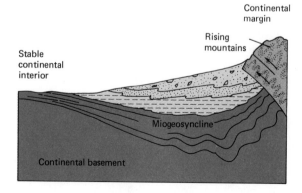

Fig. 4-46 Rising mountains along a continental margin shed a clastic wedge towards the continental interior.

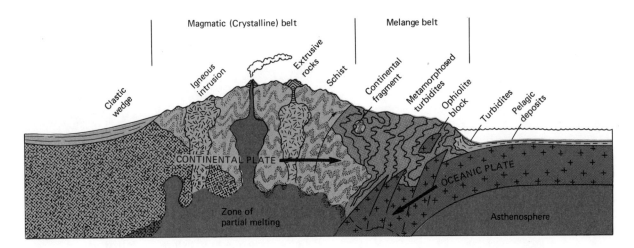

Fig. 4-47 Most of the world's major mountain ranges (such as the Alps and western Cordillera of North America) have remarkably complex structures. Essentially they consist of the wreckage of geosynclines which, in the new world view, were crushed by colliding tectonic plates.

the Zagros Mountains of Iran; Indonesia; and other far-flung localities.

In the light of plate tectonics, mélanges are believed to be wrecked eugeosynclines, where deposits in oceanic trenches and continental rises have been crushed between global plates (Fig. 4-47).

The masses of ultrabasic rock incorporated in the chaotic turbidite deposits are interpreted as being broken off from descending oceanic plates. Enclosed blocks of limestone, granite, and similar rocks are attributed to landsliding off higher continental masses. Mélanges seem to originate in zones of moderate metamorphism, involving relatively low temperatures but high confining pressures developed between colliding tectonic plates where rapid burial is common and subsequent elevation (as yet poorly understood) follows.

Parallel to mélanges lie adjacent *magmatic belts* containing complicated assemblages of plutonic rocks—both igneous and metamorphic. They include several grades of metamorphic rock, but most notably schists—often called greenschists or blueschists from the colors given by certain minerals in the amphibole and pyroxene families. Such crystalline belts—in contrast to mélanges—have been extensively invaded by magmas producing large bodies of intrusive granitic and related rocks, as well as extrusive andesitic and rhyolitic rocks erupted from volcanos. Recent theorizing suggests that the magmatic belts develop above the deeper parts of descending tectonic plates. High temperatures in the asthenosphere partially melt the plates to create magmas that rise as buoyant "blobs" into the overlying crust. The ascending hot molten masses metamorphose the surrounding crustal rocks as they penetrate towards the Earth's surface. Thus the magmatic belts of crystalline rock assemblages exposed at the Earth's surface are interpreted as resulting from a high-temperature but low-pressure metamorphic environment.

SOME CONCLUSIONS FROM ROCKS

It would be wrong to leave the impression that the whole story of rocks is told. They offer many problems, great and small, which challenge imagination, skill, and training. Is granite entirely igneous, formed from magma? Or is it metamorphic, the final alteration of solid rock soaked in juices coming from the depths? Being a scientific problem, the question is naturally argued with all the open-minded calmness of a hot political debate. There are even party labels: "pontificators" for the conservative magma disciples; "soaks" for the metamorphic radicals who believe more in granitization. In part the problem is that what goes on in the depths of the Earth can only be inferred from indirect evidence. The making of plutonic rocks is never seen; they come to light only after thousands of feet of rock have been removed—increasing the difficulty of geologic detective work.

Another problem—nature is seldom neat and tidy. Rocks are no exception; they defy easy pigeonholing. Still, most rocks can be easily fitted to our simple classification, which is quite adequate for ordinary purposes. But you would be misled to think that all rock types are separate and distinct. Classifications are man-made distinctions to bring simplicity and order; rocks in nature merge.

Igneous rocks grade in mineral composition from granite into diorite into gabbro, with no sharp breaks. Aphanitic rocks, such as rhyolite, merge imperceptibly into coarse-grained equivalents, in this case granite. In sedimentary rocks, there are clean sandstones and obvious shales. But some are shaly sands or sandy shales. Shale grades into slate, a metamorphic rock, slate into schist, and schist into gneiss.

The problem arises from the nature of rocks.

It is in part because most are mixtures—varying assortments of minerals, in all possible proportions, and grading through wide ranges in grain size. Also, rock types are not static and everlasting.

If geology has any message worth remembering after the detailed facts have fled, it is the constant change of seemingly substantial things. Nowhere is this better shown than in rocks, symbols of the enduring. For rocks can be visualized as passing through a cycle of physical processes. In the rock cycle, igneous rocks change to sedimentary by weathering, erosion, transport, and deposition; sedimentary rocks change to metamorphic by mountain-building and intrusion; if melting is complete, metamorphic rocks go back into igneous. Shortcuts occur, in the large cycle, if metamorphic rocks are exposed to weathering, or igneous rocks are involved in mountain-building (Fig. 4-48).

Change occurs because rocks adapt to new environments. Different conditions of temperature and pressure, in the presence of waters and gases, cause chemical reactions in the minerals of a rock. Feldspar, crystallized from magma, weathers to clay under the low temperature and pressure of surface conditions. Compressed and heated in the depths, clay may alter to mica, which eventually may be melted back into a magma. Throughout the cycle, the atoms, remaining unchanged, are shuffled and rearranged to form new mineral structures, as the chemical elements are circulated from the depths to the surface, and back to the depths. The rock cycle neatly ties together a raft of seemingly unrelated processes.

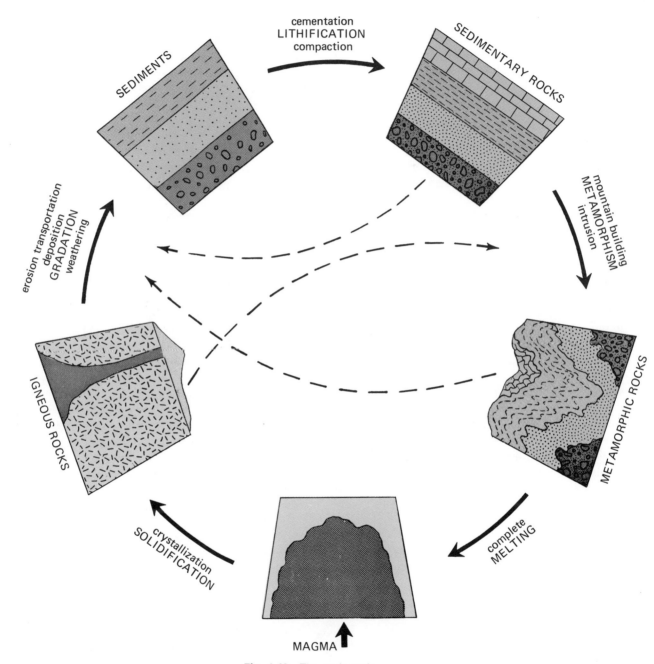

Fig. 4-48 The rock cycle.

SUGGESTED READINGS

Dietz, R. S., "Geosynclines, Mountains, and Continent Building," in *Scientific American,* March, 1972.

Dunbar, C. O., and Rogers, J., *Principles of Stratigraphy,* New York, John Wiley & Sons, 1957.

Fenton, C. L., and Fenton, M. A., *Rocks and Their Stories,* New York, Doubleday & Co., 1951.

Mather, K. F., and Mason, S. L., *A Source Book in Geology,* New York, McGraw-Hill Book Co., 1939. Contains excerpts of original statements by James D. Dana, James Hall, James Hutton, Nicolaus Steno, and many other early geologists.

Spock, L. E., *Guide to the Study of Rocks,* New York, Harper & Row, 1953.

Cross-bedding at Sand Creek, Wyoming. Photo by the author.

Rocks and Fossils as Records of Geologic Time

The central theme of geology, which sets it apart from other sciences, is Earth history—the succession of past worlds recorded in the rocks. Historians rely on written documents and human monuments for their accounts of people and nations during the last few thousand years. But geologists have faced special problems in deciphering the several billions of years in Earth history. The reckoning of geologic time has required the development of proper methods.

TELLING TIME

In general, time determinations are of two sorts. *Relative dating* merely gives the order of events: which happened first, which came next, which followed that, and so on. It is represented in ordinary affairs by words like yesterday, Tuesday, April, and the like. One of the major accomplishments of nineteenth-century geologists was the establishment of a still-valid relative geologic succession for the history of the Earth. *Absolute time determination,* or finite dating, is the measure of time in units related to a reference point—as, for example, June 5, 1948, A.D., which is a measure in years, months, and days since the birth of Christ. Absolute geological time determinations of how many years ago something happened or how many years an event lasted(which always involve some experimental error) are mainly a twentieth-century development based on instrumental measurements of radioactive isotopes.

Relative Time

In unraveling the order of events in a particular place, geologists are sometimes faced with the end results of a complicated physical history. They may have to sort out the relative times of deposition and erosion, volcanism and igneous intrusion, folding and breaking (faulting) of strata—often from scattered exposures of bedrock that is largely covered by soil. To establish a relative dating of geologic events in such cases requires field experience, a certain flair for geometric rock puzzles, and a thorough grounding in physical geology. Once the local structure—whether complicated or simple—is worked out, however, something more is needed. The local history should be fitted to broader regional patterns—based on the master plan of relative geologic time (see end papers).

Timepieces Fossils are worldwide timepieces. They can be used for determining order, but more importantly they date the particular interval of relative time during the last 600 million years of Earth history when the rocks containing them were laid down.

Fossils are evidence of past life, which is the reason they can be used for telling time. The discovery that fossils date rocks, which paved the way for a coordinated geologic history, was inde-

pendently made around 1800 by Georges Cuvier (1769–1832) and Alexandre Brongniart (1770–1847) working as a team in the Paris Basin, and William Smith (1769–1839), a canal engineer in England. These pioneers showed that layers of comparable age in different places have similar fossils, and that fossils in layers above and below are different. Fossils can be used for dating because the groups of plants and animals living together, the floras and faunas, have gradually changed with time.

Fossils show evolution in the same way that cars, houses, furniture, and clothing date old photographs because of changing styles. If the remains of automobiles were buried in successive layers, as they would be in a city dump, students of Ford cars, for example, could date the layers by the different models. Scraps and pieces of Model Ts would be found in older layers, Model As in somewhat younger ones, simple V-8s above them, and near the top would be bizarre forms with ornate fins and monsters with four headlights. Because this make of car is common and widespread, it would be possible to determine which layers had been put down at the same time in widely separated places.

In the case of fossils, dating is based on two fundamental laws: the *Law of Superposition*, that undisturbed strata are in the order of their deposition, and the *Law of Faunal and Floral Succession* (Fig. 5-2). The essence of the latter law is that plants and animals have progressively changed through time, and that each period of time is characterized by distinctive fossil groups. Intensive study has shown that faunas of a given time have never occurred in any other geologic time, and that comparable faunas of a given time have existed on all the continents. That species have not recurred has been a major advantage in using the paleontologic "clock."

Some fossils are better timepieces than others. Conservative beasts that persist through ages with little change are of limited value. One stock of marine organisms, whose members are a kind of shellfish called Lingula, has persisted for hundreds

Fig. 5-1 William Smith (1769–1839). "Strata" Smith's firsthand observations while surveying canal routes in England showed him that different rock layers have distinct fossil groups. He used this idea of faunal succession in correlating rocks to produce his remarkable colored geologic map of England. His introduction of careful mapping was a major factor in the development of geology.

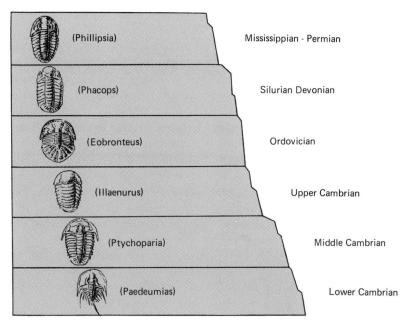

Fig. 5-2 Faunal succession. Selected members of the long-extinct group, trilobites—indicating changing styles with time.

of millions of years with scarcely any change that can be recognized from their shells (Fig. 5-3). The best fossils for dating, called *index fossils,* are the remains of rapidly changing stocks of plants and animals whose members existed during a short span of geologic history, but at the same time were geographically widespread. They should also be abundant and easy to recognize. Most paleontologists (students of fossils) prefer to work with assemblages of fossils because even if good index fossils are missing, the chance that last, or first, appearances of certain types will be found is greatly increased. The overlapping of several "tops and bottoms" of life ranges for conservative fossil groups allows a finer time determination than from a single form.

Techniques for dating by fossils have been used for over 250 years to compile an orderly history for the Earth's crust. However, they give relative times only; reliable methods of finite dating are a twentieth-century development.

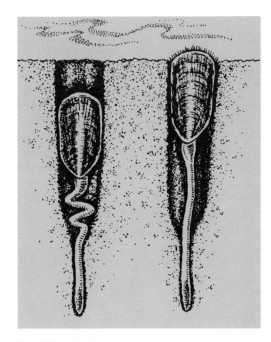

Fig. 5-3 Lingula, a most conservative animal.

Finite Time

Being so far removed from ordinary experience, the length of geologic time is difficult to comprehend. Saying the Earth is at least four and a half billion years old is one thing; sensing the enormity of its history is another. If a dime, ⅛ of a centimeter (1/20 of an inch) thick, were set atop the 381-meter (1250-foot) Empire State Building in New York City, the thickness of the dime would more than equal the span of recorded human history in proportion to the building, which would represent geologic time at the same scale. As a scientific revelation, geologic time ranks with astronomical concepts of the distances to the stars in space, and the microcosm of atomic structure.

Older Techniques Attempts have long been made to establish absolute geologic time relations. In 1654 Archbishop Ussher of Ireland proclaimed that the Earth was created in 4004 B.C.[1] Such *theological determinations* were based on the Book of Genesis wherein a careful chronology gives the years of the various periods, from the time of Adam to the building of the Temple in Jerusalem, an event recorded on existing calendars. Scientists also did considerable ineffectual groping.

The *saltiness of the oceans* was the basis of a late-nineteenth-century calculation that the Earth was 100 million years old. It was assumed, reasonably enough, that the oceans were originally fresh water and have become progressively saltier, through time, because of the workings of the hydrologic cycle. The salt is derived from weathering of rocks on land, and is carried by streams to the sea where it is left behind, while fresh water is distilled to fall in rain or snow, during the phase of circulation from ocean to land. By calculating the total salt in the oceans and the total salt brought in by streams each year, and dividing the first

1 Dr. John Lightfoot, vice-chancellor of Cambridge University, arrived at a more precise determination of 9:00 A.M., October 23, 4004 B.C.

figure by the second, an age of about 100 million years results. Unfortunately, this answer is far too low; somehow the assumptions of this simple, neat, and plausible scheme are astray.

In a somewhat similar way, using simple arithmetic, the *rate of accumulation* of sedimentary rocks has been tried for the later history of the Earth. If an overall maximum thickness of sedimentary beds is totalled in meters, or any suitable units, and the average number of years required to deposit 1 meter, 10 meters, or any particular interval can be determined, then the total thickness multiplied by the number of years it takes to deposit 1 meter gives the time in years for the whole deposit.

The trouble with this scheme, as applied to any considerable length of geologic time, is the difficulty in determining a total thickness of continuous deposition. Intervals of time when deposition ceases are common and often unrecognized. The rate of deposition varies widely from place to place, and there is good reason to believe it has been far from constant through time, being rapid during episodes of mountain-building and very slow in the long intervals when lands were low and largely flooded by the seas. So, even if the rate of deposition is well established for hundreds or even a few thousands of years in any place, there is no guarantee it represents any sort of average for most of geologic time.

Deposition can be reliably used, however, in special situations where thin layers, or laminae, mark a given length of time. For instance, *varves* laid down in glacial lakes are paired laminae, forming annual deposits. During the summer when the lake is open, sands and silts washed in by streams settle to the bottom; in winter when the lake freezes over creating a lid, deposition of this coarser sediment ceases, and a thin layer of clay that was in suspension settles out. Thus a combination of one thicker silty layer and one thinner clayey layer indicates one year; so, by counting varves, the finite duration in years of the former glacial lake can be determined. Some other deposits of non-glacial origin are also varved. The Green River lake beds in western Wyoming and Colorado contain a varved shale (Fig. 4-13), which has been used to obtain an estimate of six and a half million years for the duration of this ancient lake, a figure verified by radioactive methods. Varved deposits are not common, however, and even where they do exist it is not always possible to be sure what time interval each lamina represents.

Just prior to 1900, Lord Kelvin, the British physicist, calculated the age of the Earth as, most likely, 20 to 40 million years. He assumed that the Earth was originally molten and had progressively cooled to its present state with no addition of heat except from the Sun. Using principles of physics he calculated the expected *rate of cooling*, and, from this, the time required for the Earth to reach its present temperature. Many geologists were unhappy, believing that the Earth must be far older to account for the great changes and complicated sequence of events in its history, but Kelvin's view prevailed for a while because his scientific eminence and authority were so great. His age determination came to naught, despite the seemingly flawless quantitative reasoning, when his assumptions were invalidated. Heat has been added to the Earth by radioactive disintegration of minerals in the mantle and crust.

Radiometric Methods   The radioactive minerals provide the most precise geologic clocks and are the basis for the modern belief in the tremendous length of geologic time and the great antiquity of the crust. In 1896, Antoine Becquerel, a French physicist, exposed photographic plates by leaving them in a drawer with uranium salts—an accident that revolutionized our understanding of the physical world. Two years later the Curies, in France, isolated radium as a prelude to the discovery of the other radioactive elements. Geology is especially indebted to Bertram Boltwood, an American chemist working with Lord Rutherford's group of phy-

sicists in England. When he found, in 1907, that the *ratio of lead to uranium* was constant in rocks of the same age, the measuring of geologic time in years could begin.

Although the laboratory determinations require highly sophisticated techniques and exceedingly precise equipment, the general principle of radiometric dating is easily understood. Radioactive elements in minerals are timepieces working somewhat like an hourglass. For example, once uranium atoms are locked in a crystal structure, they start breaking down at a known and steady rate by expelling particles, rays, and helium gas to form new and different radioactive atoms that also decay, until finally, the transmutation ends with the creation of the stable end product, lead.

All of the almost countless radioactive atoms[2] in a newly formed uranium mineral do not start to break down at once; a few disintegrate almost immediately, some follow shortly, others are slow in starting, and in some the breakdown is almost indefinitely delayed. It is impossible to predict when any particular atom will start to decay; however, over a long period of time, the number of transmuted atoms in the total population increases at a steady and predictable rate, so that the amount of uranium in a sample becomes progressively less, while the lead increases proportionally. Thus by determining the ratio of lead to uranium in the minerals of a rock, its absolute age in years can be obtained.[3] The principle is like that of the life expectancy tables used by insurance companies to determine their policy rates. Their statisticians do not know when any particular person will die, but out of 100,000 people born in a certain year, they

2 Actually we should be speaking of isotopes, slightly different kinds of atoms having the same chemical properties but very slightly different weights. There are two isotopes of uranium and four of lead.
3 Perhaps relatively absolute is a better expression. Because of technical problems, dates are usually given with a plus-or-minus factor of so many years; however, the margin of error is usually only a few percent.

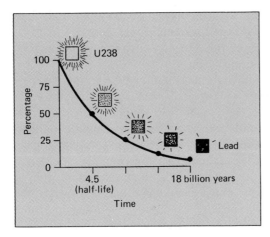

Fig. 5-4 Radioactive disintegration of uranium to lead.

can predict what percent will survive and what percent will be dead in each successive year.

Unlike an hourglass, where sand flows at a constant rate, or curves of life expectancy, where the mortality rate increases rapidly after a certain point, the change in the overall proportion of uranium to lead becomes increasingly slow with time (Fig. 5-4). The change is greatest during the time when the first half of the uranium atoms are changing to lead. Then it takes the same length of time for half of the remaining atoms to break down, increasing the total amount of lead by only a quarter; thereafter, the same length of time for half of the still-remaining atoms to break down, increasing the amount of lead by one-eighth, and so on, half of a half of a half, *ad infinitum.* It is virtually impossible to determine when the element is totally "dead." For this reason, the duration of an element's activity is customarily given as its half-life, the time required for half the material to decay. The half-life of uranium is 4.5×10^9 years (4,500,000,000 years), which makes it a very long-lived element.

The uranium-lead method was the first to be used extensively; now, however, other methods have largely replaced it because they may be simpler, more accurate, better for certain ranges of geologic time, or may be used on elements in more abundant minerals. All, however, are based on obtaining ratios between parent radioactive elements and their offspring. Helium, the gas, is a radioactive byproduct of uranium decay, and its ratio to lead has been used in dating. The element thorium, which also produces lead, has been used, and so has the element rubidium, whose stable daughter product is strontium. The proportion of radioactive potassium to its decay product, the gas argon, is now widely used in dating. This method can be applied to the abundant mica minerals and also sylvite, which is a salt occurring in sedimentary deposits. Radioactive potassium is by far the most abundant radioactive element in the Earth's crust, and the breakdown is a simpler action involving less chance of loss or addition of critical elements.

Uranium, thorium, rubidium, and radioactive potassium are all elements with long half-lives that are thought to have originated at the same time as the Earth itself. They do give rise to radioactive daughter elements such as radium, which are being formed today, but the question arises as to how the primary radioactive elements can indicate anything but the time of their origin, or the age of the Earth, because they must have started to break down when they formed. As a matter of fact, the age of the Earth has been estimated from the breakdown products. Although all lead has the same chemical properties, there are several kinds of lead atoms, or isotopes, which are slightly different in weight and can be distinguished by certain physical techniques. Nonradioactive lead (Pb^{204}), which is not a decay product, is lighter than the types representing the residues (Pb^{206}, Pb^{207}, Pb^{208}) of uranium and thorium. So, assuming that only nonradioactive lead was present in the Earth at the beginning, and its amount has remained the same while the radioactive leads have steadily increased at a known rate through time, a maximum geologic age for the

Earth has been calculated as 5.6×10^9, or 5,600,-000,000 years.

The reason radioactive elements can be used to date rocks formed at various geologic times relates to the way minerals form by crystal growth during solidification of a magma, recrystallization associated with metamorphism, or precipitation of salts. In magmas and solutions, all sorts of atoms (ions, isotopes) may be present, including those of radioactive elements and decay products, and the atoms are randomly dispersed and free to move about. When crystallization occurs, however, they aggregate into crystal structures of different minerals, each having the same chemical composition. The structures of certain minerals would tend to take on atoms of radioactive elements and exclude lead and other daughter products,[4] which would combine in separate minerals. Thus, the sorting action during crystallization sets the starting time of the radiometric clocks. The oldest rocks dated so far have an age of 3.1×10^9

4 Some lead does get into uranium minerals, but there are ways to handle the problem, and a few minerals, like zircon, are believed to exclude all lead when they are formed, which makes them the most reliable indicators.

years, which gives a minimum estimate for the age of the Earth.

Radioactive carbon, C^{14}, has a uniquely important spot in finite dating. Having a relatively short half-life of about 5500 years, it is invaluable for dating within the last 40,000 years. The elements with long half-lives cannot be used to date rocks younger than one million years because the relative change in proportions of parent elements and daughter products during that length of time is too slight to be reliably detected. Unlike the Earth's primordial elements and the radioactive elements produced by their decay, C^{14} is continuously created in the air, even today.

Cosmic rays bombarding the upper atmosphere convert nitrogen there to C^{14}, which combines with oxygen to form carbon dioxide. This gas, bearing radioactive carbon, becomes mixed in the atmosphere with carbon dioxide containing ordinary carbon atoms, C^{12}. The mixture is taken in and incorporated into living plants and animals (Fig. 5-5). So long as the organism lives, its C^{14} and C^{12} content remains in balance, the radioactive carbon being replaced as fast as it decays to ordinary carbon. After the organism dies, however, no new

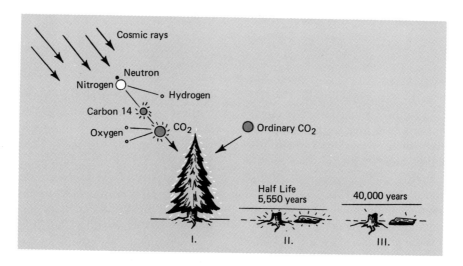

Fig. 5-5 Production and decay of carbon 14.

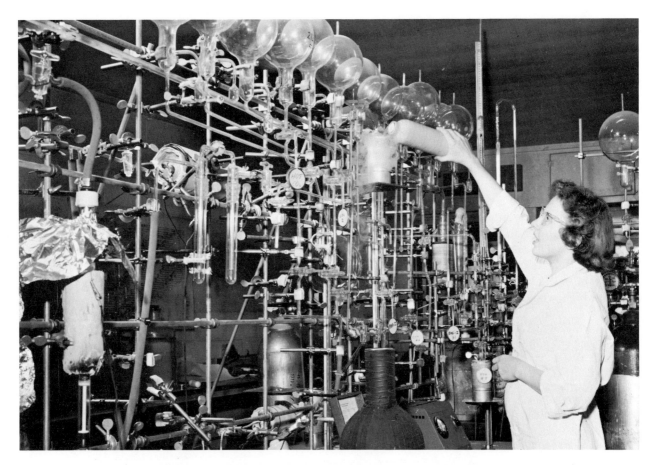

Fig. 5-6 Laboratory equipment used in carbon 14 dating. Sample purification system. Courtesy of Isotopes, Inc., Westwood, New Jersey.

radioactive carbon is assimilated, and the C^{14} slowly breaks down to the ordinary variety. Hence the less the amount of C^{14} in such organic materials as charcoal, wood, shell, bones, and teeth, the older the material is. The C^{14} method has been a great boon to geologists, and especially to anthropologists and archeologists studying the "prehistoric" history of man.

Radiometric dating has given such striking results that it may seem a panacea for all problems of geologic time; yet it is not perfect. First of all there is a gap, from about 40,000 to almost 1,000,-000 years ago, between the oldest possible C^{14} dates and the youngest reliable age determinations using long-lived elements—a blind spot inaccessible to present radiometric methods. Clastic sediments present a special problem because their minerals are largely derived from preexisting rocks. So, although radioactive potassium and rubidium do occur in the grains of many sedimentary rocks, absolute age determinations would reflect the time the parent rock originated and not the time the fragments were deposited as a sediment. Often where an intrusive mass can be absolutely dated, the age of surrounding rocks must still be determined by cross-cutting relations.

Samples must come from fresh-looking rock, showing no signs of chemical decay, and be carefully selected. From a practical point of view the methods are expensive (Fig. 5-6). Although the rate of radioactive decay is unchanged by heat, pressure, and chemical reactions, weathering and leaching remove elements, upsetting critical ratios, and metamorphic or igneous action may introduce new materials contaminating a sample. Gases af-

fecting helium ratios may escape if minerals are minutely fractured. Carbon dates may be in error if present-day plant roots and soil organisms are not thoroughly removed from a sample.

Since World War II, however, great progress has been made in techniques and instruments (Fig. 5-7). Most of the earlier measurements are now considered unreliable. Students in the early 1940s were taught that the age of the Earth as revealed by uranium-lead determinations was incredibly greater than anyone had thought—two billion

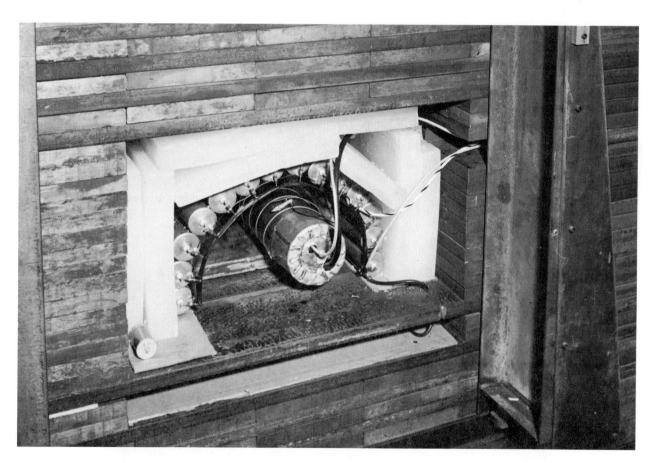

Fig. 5-7 Radioactive determination equipment. Sample in central cylinder is surrounded by radioactive counters and the entire apparatus is shielded by paraffin blocks inside steel shields to eliminate outside influence of cosmic rays. Courtesy of Isotopes, Inc., Westwood, New Jersey.

years. Now the accepted figure is at least twice that—from four and a half to five and a half billion years—on the basis of improved methods and cross-checking of samples using several different elements.

CORRELATION

So far we have largely stressed the nature and the reading of geologic clocks to tell the passing of time in a sequence of rocks. Yet if you stop to think of it, time determinations are of two sorts; besides time order, there is the matter of time equivalence in different places, which is of paramount importance in reconstructing the history of any appreciable part of the Earth. Correlation, in general, means to show the connection between related things; however, to geologists it usually means establishing that rocks in different areas originated at the same time, whether locally as between northern and southern New Jersey, or between continents—North America and Australia, for instance. In a broader sense, since rocks record

geologic events, was the episode of mountain-building responsible for the Urals contemporaneous with that of the Appalachians? Was the eastern United States a scene of coal swamps at the same time as England? Were equatorial Africa and South America both glaciated in the same ancient Ice Age?

The only reliable means for far-flung correlations are fossils and radioactive minerals, which are essentially synchronized watches; more locally, it may also be possible to use similar physical features of rocks. However, no single method is applicable to all rocks. For instance, fossils, the key to correlating sedimentary rocks, are almost totally absent from igneous and highly metamorphosed rocks. Circumstances determine the method or methods used—we do the best we can with what we have.

Fossils

The major breakthrough in correlation was made by William Smith when he recognized and applied faunal succession. It soon became evident

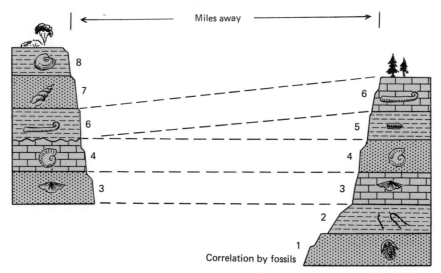

Fig. 5-8 Correlation by fossils. Certain index fossils are keys to matching of sedimentary strata in widely separated outcrops.

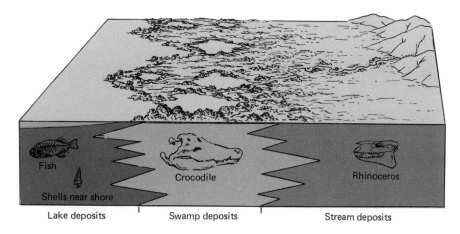

Fig. 5-9 The record of the rocks in a prehistoric Wyoming lake indicates how quite different fossils are of the same age but from different environments. Based on an exhibit in the University of Wyoming Geology Museum by P. O. McGrew.

that not only could separate outcrops be correlated in the relatively limited area of Great Britain, but they could also be correlated across the English Channel with rocks in France (Fig. 5-8). It was eventually realized that the overall succession of life had been similar on all the continents; hence fossils are universal time labels for sedimentary outcrops. However, correlation by fossils is not without some inherent difficulties.

It was fortunate for Smith, Cuvier, and Brongniart—and geology—that these pioneers worked in parts of Europe where the strata were rather neatly laid out and good index fossils were relatively abundant. Had they lived in New England, Switzerland, northern India, and many other localities, geochronology would have been founded by someone else. Even today, it is often difficult to establish the time and order of deposition where sedimentary strata are metamorphosed or highly jumbled from deformation. Some thick sequences of sedimentary rocks are notably devoid of fossils, and if fossils are present there may be complicating factors.

Living groups of plants and animals, which are

a basis for uniformitarian thinking, indicate a problem in correlation by fossils. Modern sea shells on a Florida beach are quite unlike many in New England, and, more locally, those found south of Cape Cod, Massachusetts, are noticeably different from the shells along beaches to the north. Changes are even more pronounced at right angles to a shore, going from land to sea; as a result, the correlation of terrestrial with marine deposits in the geologic record has often been difficult. The animals entombed in a river flood plain may be quite different from those of the sea (Fig. 5-9). Considering only a single type of marine shellfish, the clam family, those that live in the tidal zone are apt to be mainly burrowing sorts, while in shallow waters further offshore many may be attached forms on the ocean bottom, like oysters, which have a rather different look. Fortunately, there are some widespread forms, such as the swimming and floating animals in the sea, that, after death, may be washed onto the beach or settle to the bottom in any one of the several offshore life zones. Fossils reflect different habitats, which has complicated correlation; however, by studying widespread forms, it has been possible to discover which of the local assemblages indicate the same geologic time. Marine and continental beds are interlayered in places, and where beds containing marine fossils are sandwiched between beds with distinctive terrestrial fossil types, a tie between the age of land and sea faunas has been established.

The first rocks in which fossils are abundant and varied enough for correlation have been absolutely dated at about 570 million years, whereas the oldest known rocks have an age of about three and a half billion years. Thus some 85 percent of geologic time, recorded in the Precambrian basement complex (Table 5.1), is inaccessible to fossil correlation. Local sequences of events reconstructed from cross-cutting relations, buried erosion surfaces, and other physical relations in Precambrian rocks had been compiled into relatively simple continental and even world-wide histories based on similarity of events and rock types. However, the recent application of improved techniques of absolute dating has shown some great errors in correlation. For example, two widely separated Canadian granites, which had been considered age equivalents because they looked alike, were found to be two billion years apart in time of origin, a magnificent miscorrelation. Radiometric dating, the only reliable method for correlating basement rocks, indicates that the first three billion years of recorded Earth history were far more eventful than previously thought.

Physical Criteria

Lacking useful geologic clocks, physical features have been used to establish the time equivalence of rocks. *Continuity*, the simplest and most direct method of physical correlation, is possible where beds can be directly traced from one place to another. It may be a matter of "eyeballing," a sweeping glance along a cliff face, or walking beds out on the ground, or tracing them on an aerial photograph. This method, which is something of a special case, requires continuous well-exposed outcrops—the cliffs and slopes of the Grand Canyon, for example. *Lithologic similarity* is used in correlating isolated samples, outcrops, or drill cores, where rock units have rather unique characteristics, such as distinctive minerals, sand grains or pebbles, concretions, cross-beds, or similar features, and in some instances overall color. *Similarity of sequence* is often a more obvious clue to rock correlation than the features of a single layer or bed. A gray limestone sandwiched between a red shale above and a brown sandstone below might make a dis-

TABLE 5.1   Geologic Time Chart

Era	Period	Epoch	Millions of Years Ago*
Cenozoic	Quaternary	Recent (Holocene) Pleistocene	
			2
	Tertiary	Pliocene Miocene Oligocene Eocene Paleocene	
			65
Mesozoic	Cretaceous Jurassic Triassic		136 190
			225
Paleozoic	Permian Pennsylvanian Mississippian Devonian Silurian Ordovician Cambrian		280 320 345 395 430 500
			570
Precambrian	The 4.7 billion years now estimated for the age of the Earth makes the length of the Precambrian time far exceed that of later eras.		

*Absolute times after U.S. Geological Survey.

tinctive sequence readily identified as the same in scattered outcrops.

Thus, similar physical features can be used like fossils as empirical "trademarks" of certain rocks; however, differences in fossils reflect organic evolution, a one-way process of continuous change. Lithologic features and rock types result from physical processes that, harking back to uniformitarianism, have operated in the same way and repeatedly formed the same products through time. So the features of sandstones, schists, granites, and other rocks are distinctive of no one geologic time. Actually most physical methods show the equivalence of rock, but not necessarily of time.

As to whether a particular rock unit is everywhere the same age, the answer is that some are and a good many are not. Volcanic eruptions may blanket wide areas with ash beds that, being the same age throughout, are excellent for time correlation as far as they extend. On the other hand, a beach sand, continuously deposited along the shore of a sea that has slowly spread from a marginal geosyncline over the interior of a continental platform, may well be millions of years younger in the heart of the continent than where its deposition started. The pitfalls of correlating time by similar lithology of widely separated rocks have already been mentioned in the case of the Canadian granites. All this does not mean that lithologic similarity is useless. It may be the only possible method in certain cases, and used with proper caution in limited areas it may well give a time correlation; but for far-flung correlations, the only reliable indicators of time equivalence are fossils and radioactive minerals.

THE UNIVERSAL CALENDAR

Once fossils were recognized as telling time, the next step was to make a universal calendar, a systematic arrangement of time divisions for worldwide dating of events. The geologic time scale is just such a gargantuan calendar whose eons, eras, periods, and epochs are grand and less-grand units for dividing the geologic versions of ancient, medieval, and modern time. The years involved in these various divisions are a late addition from radiometric techniques; the time scale itself was established by purely relative means.

Constructing the Time Scale

The modern time scale was a major accomplishment of the nineteenth century based, as things geologic must be, on the record of the rocks. Nowhere is the record complete; so to compile it, north European geologists in England and on the continent had to piece together a master geologic column combining local rock successions from different places. Had the whole column been laid out in some gigantic cliff of fossiliferous and uncontorted strata, unbroken by intervals of nondeposition (unconformities), there would have been less of a problem.[5] But the normal order of separate fossiliferous sequences had to be established, originally by walking out barren intervening beds to determine whether datable sequences lay above or below each other. It was a major task because the rocks in places were structurally complicated. There were and are other problems, here glossed over, such as the nature of the boundaries between rock sequences, and the fact that more complete sections have since been found outside the classic areas; nonetheless, with certain modifications, the composite rock column established in Europe provides the world-wide standard.

Thus the time scale is based on a master rock column that is in turn based on local rock units. *Rock units* are the sandstones, shales, and other types forming the observable sequence in any

---

5 Boundaries might have been difficult to select, however, without unconformities in between a wealth of fossiliferous strata because evolutionary changes are characteristically transitional.

given place. Of the rock units, the most basic is the *formation,* which is essentially a distinct mappable unit and the sort most commonly dealt with by geologists in the field.[6] A formation is a rock mass, most commonly sedimentary, although it can be metamorphic or igneous, that is usually formed by a single process, making it readily distinguishable from adjacent rocks.

The lithologic features distinguishing a formation could be such things as color, rock type, the presence of coal, or some peculiar concretions—in short almost anything that sets the formation apart. A formation customarily has two names: a geographic one from the place the unit was first described in the geologic literature and another from its dominant rock type; although if several types are present the word formation is used. The Potsdam Sandstone is a formation named from a city in New York State; the Goose Egg Formation has a type locality near an old Wyoming ranch. In Europe where their usage originated before people worried about consistent naming, famous formations are called such things as the Old Red Sandstone and the Coal Measures, for obvious reasons; to avoid confusion, these names have largely been retained. Although its name comes from a single locality, a formation may be present over vast

6 Geological "formation" is an almost universal misnomer among the uninitiated, who use the term to describe fossils, concretions, stalactites, and other features which do not qualify.

areas. For instance, the Niobrara Formation, named for exposures in northeastern Nebraska, is also found in the Dakotas, Montana, Wyoming, Colorado, Kansas, and New Mexico.

A single formation may not have the same age in all places it is found. Formations are correlatable by lithologic similarity (p. 116), and may represent the same environment of deposition. But the environment may migrate with time. For example (Fig. 5-10), one formation might represent a sandy beach, another the muddy near-shore bottom, and a third clear waters of a transgressing sea. Over millions of years, the sea might gradually advance from a geosyncline onto a continental platform. In this case, each formation would become progressively younger towards the continental interior—a fact that might be determined from fossils.

Time-rock units, which make up the composite geologic columns of either particular regions or the master world column, are somewhat generalized units representing all rocks laid down during a certain time, as read from fossil evidence. For this reason they are *not* defined by any lithology since many different rocks—sandstones, shales, schists, basalts, and others—have originated at the same times. Time-rock units are represented by one, several, or portions of physically distinguishable rock units such as formations. Lastly, *time units,* the most abstract of all, are the subdivisions of the geologic calendar, or time scale.

The fundamental time-rock units are called

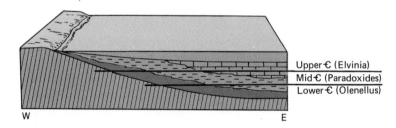

Fig. 5-10 Rock types deposited in the advancing Cambrian sea cross time lines, indicated by fossils.

systems, and they correspond to the purely time units of the geologic calendar called *periods*. Both are named the same; hence if we speak of the Devonian System we mean rocks, and if of the Devonian Period we mean time alone. All this may seem like boring terminology, yet it is essential to an understanding of how the universal geologic time scale was constructed, and it is the time scale which holds together the grand patterns of Earth history.

The Geologic Time Scale

The sudden appearance of abundant and varied fossils, an explosive event in geologic terms, makes a worldwide time line for dividing the geologic time scale into two main parts. These divisions are most disproportionate—in length of years, and in our detailed understanding of each. The first, and by far the longest, is commonly referred to simply as *Precambrian* time. As the Precambrian rocks, most of which are deformed, metamorphosed, and intruded, are relatively barren of fossils, and those fossils that have been found are valueless as time-pieces—this part of the geologic calendar has not been satisfactorily subdivided. Radiometric dating may help in the future, but because even a small plus-or-minus factor of long-lived radioactive minerals amounts to a good many million years, it is improbable that Precambrian time can ever be subdivided in any great detail.

Historical geology was founded on the second major division, whose abundant fossils and sedimentary rocks allowed the application of dating and correlation to produce a finely divided time scale. There are no generally accepted names for the two greatest divisions, although it has been suggested that, with appropriate grandeur, they be called eons: Cryptozoic, meaning "hidden life," for the older; and Phanerozoic, meaning "evident life," for the younger. Having mentioned these terms, let us ignore them and concentrate on the commonly used divisions after Precambrian time.

The major subdivisions are called *eras*, of which there are three, set apart on the basis of their fossil records of dominant life. The first is the *Paleozoic* (in Greek, *paleo* means "ancient" and *zoic*, from which the word *zoo* is derived, means "life"). In this era, lasting from about 570 million to 225 million years ago, the highest forms of animals were first invertebrates and later vertebrates, including fish, amphibians, and primitive reptiles. The second great era, the *Mesozoic* (meaning "middle," or "medieval life"), extending from 225 million to 65 million years ago, is the age of dinosaurs. The third era, the *Cenozoic* ("recent life"), is marked by the ascendancy of the mammals. Each of the eras contains less-grand divisions called *periods* based, as you recall, on the time-rock units called systems.[7] They are fundamental units because the master column and time scale were largely set up from the study of systems. Their given names, which are the same for equivalent systems and periods, originated in several ways.

Some represent a locality where the rock sequence was first worked out. For instance, Cambria was the Roman name for Wales. A system first studied there by Adam Sedgewick is called *Cambrian*, and rocks throughout the world of the same age belong to the Cambrian System, which records events of the Cambrian Period of time. Other names describe the rocks, such as *Cretaceous*, which refers to the dominantly chalky nature of a system of rocks where they were first studied in Europe. *Triassic* is derived from a sequence of rocks that in Germany includes three main divisions. *Tertiary* and *Quaternary* are holdovers from an earlier attempt to establish a master rock column. Its "Primary" and "Secondary," which referred to igneous-metamorphic complexes and well-consoli-

---

7 The time-rock units equivalent to an era have no special name and are simply called Cambrian rocks, Mesozoic rocks, or Cenozoic rocks.

dated sedimentary rocks, respectively, are now defunct terms.[8]

As the column became known in more detail, lesser divisions were established. Periods are subdivided into *epochs* corresponding to time-rock units called *series*. Except for the Cenozoic era, the epochs and series cannot easily be applied from one continent to another, so local names are used. Geographical names are given to these divisions before the Cenozoic; for example, the *Cincinnatian* refers to an epoch and a series represented by upper Ordovician rocks in the United States. The only epochs we will be concerned with are those of the Cenozoic because the historical record of these rocks is more detailed and complete than those of any earlier era. Cenozoic epochs and series were originally named by Charles Lyell, according to how recently their fossil populations appeared. The names he coined reflect the observation that remains of modern animal types become fewer and fewer in progressively older rocks. The subdivisions of the Cenozoic, which are used worldwide, all end in *cene* which means "recent," and their prefixes are such things as *Pleisto*, meaning "most," and *Oligo*, meaning "few"; hence Pleistocene means "most recent" and Oligocene means "few recent." So much for how geologists order events and tell time.

8 It has been proposed to standardize the terminology by getting rid of Tertiary and Quaternary, but geologists are creatures of habit in whom the terms are ingrained. Even if a new system of naming were introduced, one would still have to know the old terms to understand the geologic literature, and the existing patchwork naming has a certain merit in showing how our ideas evolved.

Time Perspective

A word of warning before going on. Short summaries of geologic history almost of necessity have two flaws: first, they tend to be pontifical, for the evidence and controversies are largely ignored to make the story flow; second, they may leave the impression that the history is like a variety show where scenes and actors whisk on and off. Keep in mind that geologic changes, which can be described in a few words, are incredibly slow. Saying that Mount Everest rose from the sea to its existing height of 8,848 meters (29,028 feet) in Cenozoic time implies no sudden event in ordinary terms. A rate of uplift averaging $\frac{1}{40}$ of a centimeter ($\frac{1}{100}$ of an inch) per year would accomplish it in somewhat less than 40,000,000 years, which is shorter than the time geologists believe was involved. If the Colorado River deepened its channel $\frac{1}{16}$ of a centimeter (about $\frac{1}{40}$ of an inch) per year in cutting the Grand Canyon, in 1600 years it would down-cut one meter (3.28 feet); continuing at this rate the 1.6-kilometer (one-mile) deep canyon would result in something over two million years. A sea, advancing over a subdued continental platform at a rate of about 15 centimeters (almost 6 inches) per year, would take two and a half million years to reach 300 kilometers (189 miles) inland. So when you read that the seas came in and the seas went out, that mountains rose, that great canyons were cut, such changes can be viewed as rapid only in the perspective of geologic time.

SUGGESTED READINGS

Dunbar, C. O., and Rogers, John, *Principles of Stratigraphy,* New York, John Wiley & Sons, 1957.

Hurley, P. M., *How Old Is the Earth?* Garden City, N.Y., Doubleday & Co., 1959 (paperback, Anchor Books).

Kulp, J. L., "Geologic Time Scale," in *Science,* Vol. 133, No. 3459, pp. 1105–1114, 1961.

Earth Structures and Evidence for the Global Architecture

Having considered minerals and rocks as geologic materials, let us now examine how they are assembled to make various geologic structures. First, we shall consider earthquakes and what they tell of the Earth's internal architecture; next, volcanos and related features, the igneous structures and phenomena; then, deformational structures, folds, and fractures and their relation to the framework of continents and ocean basins; and last, the nature of the evidence—and some arguments—for the past and present unifying geologic concepts.

The ROCK
POWERFUL BIBLE DRAMA
Here Sunday Evening MAR.12 - 7:30

California earthquake. U.S. Information Agency photo, National Archives.

The Vibrant Earth and Its Interior

The Earth quivers. Its constant vibration, detectable only by sensitive instruments called seismographs, consists of microseisms that reflect storms and atmospheric pressure changes, as well as random "noise" from nearby blasts, falling objects, moving vehicles such as trucks and trains, and even stamping feet. Rising above this muted background are frequent earthquakes, distinctive ground tremors noticeable to anyone near the source.

EARTHQUAKES

Earthquakes reflect the deep-seated forces that break and crumble the Earth's crust into deformational structures; moreover, their instrumental records have helped provide the only satisfactory information about the Earth's internal architecture. Hundreds of thousands of earthquakes occur each year. Most, being small or from distant sources, are indiscernible to us except by seismographs; about a hundred are capable of damaging any nearby buildings; and perhaps one or two of these are potentially devastating.

Statistics can only suggest the terror, destruction, and personal tragedies in populous areas struck by a major earthquake. The 1775 earthquake in Lisbon, Portugal, caused 60,000 fatalities and countless injuries. The earthquake of 1906 in the San Francisco Bay area of California was the most destructive in the history of the United States (Fig. 6-1). Some 700 people were killed. The financial loss from the fire that followed was $400,000,000. In 1908 the cities of Reggio and Messina in southern Italy and Sicily were devastated, with a death toll of some 100,000. The list is long, including Kansu in western China, where 100,000 were killed in 1920 and an equal number in 1927 (mostly in the collapse of cave dwellings cut into soft deposits of windblown silt called loess). In Japan the 1923 earthquake in the Tokyo-Yokohama region left 142,809 dead or missing and 103,733 injured, and destroyed 576,262 buildings. Fortunately, most earthquakes are less intense, and many occur in sparsely populated regions or on the ocean floor. Some of the devastation caused by the "Good Friday" earthquake in Alaska (March 27, 1964) is shown in Figures 6-2 through 6-5.

Fig. 6-1 "Good Friday" earthquake in Alaska. Partial collapse of new J. C. Penney store in Anchorage. Note adjacent glass building still intact. U.S. Coast and Geodetic Survey photo.

Surface Effects

A strong earthquake is, to say the least, unnerving. The "psychology of the observer" should always be considered in evaluating the accounts of survivors, which often blend fantasy with fact. What phenomena can be expected?

Buildings Well-built modern buildings withstand most earthquakes with little damage. Structures on

Fig. 6-2 Good Friday quake. Settling of this Anchorage Theater left the marquee at street level. U.S. Coast and Geodetic Survey photo (facing page).

Fig. 6-3 Slumping caused by earthquake destroyed many homes in the Turnagin area of Anchorage. U.S. Coast and Geodetic Survey photo.

solid bedrock receive less shaking than those on soft or swamp ground because the different materials behave like a bowl of jelly. Tapped on the outside with a spoon, the solid bowl transmits the shock but it is not thrown into waves, such as pass across the soft jelly surface. Earth materials react to strong vibrations in much the same way (Fig. 6-5).

Chances of survival during an earthquake in cities are better indoors, under stairs, doorways, or stout tables, for the death toll soars when people rush into the streets and are crushed by falling masonry of old or poorly bonded stone buildings. Fires following an earthquake are often more destructive than the shaking. Tremendous fires accompanied the initial shocks in San Francisco (Fig. 6-6) and Tokyo.

Fig. 6-4 Great ocean waves, called tsunamis, generated by the earthquake severely damaged the railroad yards at Seward. U.S. Coast and Geodetic Survey photo.

Shocks The duration of an individual earthquake shock is surprisingly short considering the damage it can do. The actual period of shaking ranges only from a few seconds to about a minute, or in the case of the Tokyo disaster a minute and a half. However, a major shock is usually followed by sporadic aftershocks that may continue for months, with decreasing intensity and frequency until several hundred or perhaps a thousand have occurred; the majority of the later ones are usually detectable only with seismographs.

Scarps and Cracks Frequently the ground is suddenly offset by the appearance of a scarp, where one side of a fault has moved up relative to the other (Fig. 6-7). A single displacement in Assam,

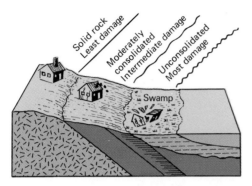

Fig. 6-5 Relation of earthquake damage to foundation material.

India, in 1897 created a scarp 10.7 meters (35 feet) high; others about 6 meters (20 feet) high are fairly common. Rumors persist of great cracks opening and swallowing people or even whole villages. Cracks do form; however, they are relatively shallow features caused by the shaking of unconsolidated soil or dirt (Fig. 6-8). Solid rock does not gape open in abyssal chasms; certainly the faults, which constitute major bedrock fractures, must remain tightly jammed together to produce the quaking. Fear of burial is justified, however, at the base of mountain slopes, when rocks have been dislodged or landslides triggered.

Ground Roll and Sounds Waves a foot or so high have been reported moving across the ground during earthquakes. These are a possibility on loose or swampy ground. One scientist, however, observed such waves moving through a concrete basement floor that did not show cracks after the shaking ended. Thus some ground roll is considered an optical illusion resulting from vibration of the fluid in the eye. Some earthquakes produce sounds because their seismic waves emerge from the ground in an audible frequency range. Loud snappings are reported near the center of an earthquake; farther away the sounds are like the rumbling of heavy traffic or distant explosions.

Distribution

Strength The strength of an earthquake is measured by various scales. Some are based on instrumental readings, but because the number of seismograph installations is limited, qualitative scales based on readily observed phenomena are commonly used to give better coverage in mapping earthquake *intensity* (see Table 6.1). These scales, printed on postcards, are sent after a shock to reliable citizens, often the postmaster, in the affected area. On their return, intensity zones are plotted on a map using the reported surface effects. Ideally the zones should form a bull's-eye pattern of decreasing intensity outwards from a center. However, the effects of seismic waves are strongly influenced by local differences in the consolidation of bedrock and other complicating factors, so the pattern is irregular.

The *magnitude* of an earthquake, often reported in newspapers, refers to a quantitative scale

TABLE 6.1 Modified Mercalli Scale of Earthquake Intensities (after Trefethen)

I. Instrumental: detected only by instruments.
II. Very feeble: noticed only by people at rest.
III. Slight: felt by people at rest. Like passing of a truck.
IV. Moderate: generally perceptible by people in motion. Loose objects disturbed.
V. Rather strong: dishes broken, bells rung, pendulum clocks stopped. People awakened.
VI. Strong: felt by all, some people frightened. Damage slight, some plaster cracked.
VII. Very strong: noticed by people in autos. Damage to poor construction. Alarm general.
VIII. Destructive: chimneys fall, much damage in substantial buildings, heavy furniture overturned.
IX. Ruinous: great damage to substantial structures. Ground cracked, pipes broken.
X. Disastrous: many buildings destroyed.
XI. Very disastrous: few structures left standing.
XII. Catastrophic: total destruction.

Fig. 6-6 San Francisco, following the great earthquake of 1906. Note damage to buildings on the right from the shaking. The great fire resulting can be seen down Sacramento Street. Library of Congress Collection.

originally devised by C. F. Richter of the California Institute of Technology. It is quite different from the qualitative, human, and damage-oriented scales such as the Mercalli. The numbers of the *Richter scale* are a measure of the energy of an earthquake at its source as calculated from instrumental records of seismographs (p. 135). It is a logarithmic scale in which each successively higher number represents a bit over a 30-fold increase in energy. Thus an earthquake of 8 on the Richter scale has over 30

Fig. 6-7 Fourteen-foot fault scarp formed during the 1959 earthquake around Madison Canyon and West Yellowstone, Montana. The trees remained upright although their roots were cut. U.S. Geological Survey photo by I. J. Witkind.

times the energy of an earthquake of 7 and some 900 times the energy of a 6. A magnitude of 2.5 represents a noticeable earthquake; 4.5 causes some damage; 7 is a major earthquake; and 8.9 is the greatest magnitude so far recorded.

Two centers of maximum strength are determined for an earthquake. The *focus*, the place of greatest magnitude, is at varying depths in the Earth. The *epicenter* is the point directly above it on the Earth's surface (Fig. 6-9).

Fig. 6-8 Fence compressed by shifting ground along small fracture marked by shovel. U.S. Geological Survey photo by J. R. Stacy.

Regions No region is immune to earthquakes. Even Boston, Massachusetts, a city noted for stability, has experienced a strong seismic disturbance which, by a peculiar historical coincidence, centered near Cambridge in 1775. However, when epicenters are plotted on a map for any length of time, well-defined earthquake belts become apparent.

The continental margins and associated arcs of islands completely rimming the Pacific Ocean are especially active. The circular belt coincides with the Andean mountain chain along the west coast of South America, then passes through the Isthmus of Panama, with a somewhat isolated branch through the West Indies, into the coastal ranges of North America. It continues through Alaska and the Aleutians to the islands of Japan, the Philippines, New Zealand, and eventually to the Antarctic. Another active belt, branching from the circum-Pacific zone at the Celebes, extends through Indonesia and the great Himalayan-Alpine mountain ranges into the Mediterranean region.

One system of extensive seismic activity is largely submarine. It starts at the north in Iceland and follows the continuous system of submarine ridges (which apparently mark global fractures) through the Mid-Atlantic Ridge, around Cape Horn into the Indian Ocean. Here one branch passes into East Africa and the Middle East along a great system of fault valleys; the other trends off into the Pacific Ocean. These major seismic belts clearly correspond to major zones of mountain-building and deformation, where the Earth's crust is being fractured and folded.

Depth Depending on the depth of the focal point, earthquakes have been classified as: *shallow*, less than 70 kilometers (about 43 miles) below the Earth's surface; *intermediate*, 70 to 300 kilometers

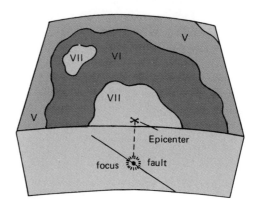

Fig. 6-9 Earthquake centers and intensity zones.

down (about 43-186 miles); and *deep,* greater than 300 kilometers (186 miles). The deepest earthquake foci yet reported are about 700 kilometers (about 435 miles) below the ground's surface. The number of recorded earthquakes decreases noticeably with depth. Using broadly rounded figures, almost 4600 earthquakes are recorded in the world each year. Of these, not quite 3500 have shallow foci, about 850 are classed as intermediate, and some 250 are deep. Thus there are about four times as many shallow earthquakes as intermediate-depth ones, and about three and a half times as many intermediate-focus earthquakes as deep ones (the figures do not change much if you manipulate them to use the top and bottom of the asthenosphere for "intermediate-depth" earthquakes). Generally, the magnitude of earthquakes also decreases with depth in the Earth.

In relation to global earthquake belts, the mid-oceanic ridge system has many shallow-focus earthquakes, but no intermediate or deep ones. The observation neatly fits into the plate tectonic scheme, because the ridges are considered to be regions where the solid crust is thin and is being created by rising masses of magma from the asthenosphere.

The seismic belt rimming the Pacific Ocean—marked by volcanos, active mountain-building, and adjacent subsiding ocean-floor trenches—is a region of shallow-, intermediate-, and deep-focus earthquakes. Deep foci were reported in the 1920s by H. H. Turner of England—but few took him seriously until the suggestion was confirmed in the 1930s by Beno Gutenberg, the German-American geophysicist. Thereafter Gutenberg and Richter discovered a pattern in the Pacific epicenters that corresponded with foci at depth. Shallow-focus earthquakes are mainly along the landward margins of oceanic trenches, and under continents (or island arcs) the foci become progressively deeper inland (Fig. 6-10). In the 1950s, Hugo Benioff of the California Institute of Technology proposed that the deepening earthquake foci clustered along great

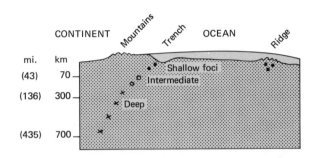

Fig. 6-10 Depth of foci. That earthquake foci are shallow beneath oceanic trenches and become progressively deeper under tectonically active continental margins (such as South America) had been known for decades before the plate-tectonic revolution.

Earth fracture planes (called thrust faults, p. 194) where landward crust jammed over that of ocean basins. In light of plate tectonic theory—or hindsight—these seismic zones, now called *Benioff zones,* are interpreted as marking the surface of overridden global plates descending into, and being consumed in, the asthenosphere (Fig. 6-11). The regions of descending plates of lithosphere are now referred to as *subduction zones.*

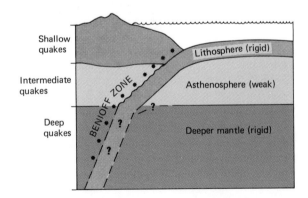

Fig. 6-11 Benioff zones. In the plate-tectonic era, the progressively deepening earthquake foci are called Benioff zones and are related to subducting (descending) sea-floor plates.

HOW WE KNOW THE EARTH'S INTERNAL ARCHITECTURE

Instrumental Observations

Seismographs Seismology, the science of earthquakes, blossomed in the late nineteenth century after the Englishman, John Milne, while in Japan, developed the first useful seismograph for recording Earth vibrations. Subsequent improvement of the instruments has resulted mainly from the efforts of Wiechert in Germany, Galitzin in Russia, and Benioff in the United States.

A seismograph is essentially a pendulum, and no matter how complicated by mechanical and electrical gadgets, it contains three basic elements: a fixed base, an inert mass, and a recorder (Fig. 6-12). In larger installations, the fixed base is a concrete pier set in bedrock to form a single solid unit. The inert mass is merely a heavy weight, the pendulum, suspended from the concrete pier on springs or a pivoting boom (Fig. 6-13). The recorder is a slowly rotating drum wrapped in a paper strip that, in the simplest seismograph, is marked by a pen connected to the inert mass.

Fig. 6-13 Three units containing inert masses (the single and split cylinders) mounted on a single concrete pier in a modern seismograph installation. University of Wyoming photo by Herb Pownall.

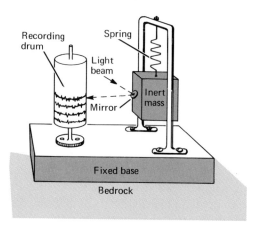

Fig. 6-12 Elements of a simple seismograph. In this case, recording is done by a pinpoint light beam on photographic paper, a system used on many modern instruments.

When earthquake waves are received, the fixed elements vibrate with the bedrock. The suspended mass remains motionless in space because of its inertia. When the base and inert mass are moving in relation to each other, the recorder plots a wiggly line as the marking device moves up and down on the rotating paper strip.[1]

Seismic Waves Modern ideas of the internal architecture of the Earth became possible in 1897, when

1 If the motion were large enough, it would seem to you that the suspended mass was moving and the concrete pier was stable. This may sound confusing, but remember that you would be moving with the ground.

R. D. Oldham of Great Britain distinguished three fundamental types of seismic waves. In the case of earthquakes, infinitesimal vibrating particles in rock pass the energy along in waves. Although the motion of the particles is at most only a small fraction of an inch, their quick trip-hammer action can cause the far larger and potentially destructive swaying of the ground. The different waves (Fig. 6-14) result from the kinds of motion affecting the particles.

The types originating at the earthquake focus are called *body waves* and they radiate in all directions through the Earth. *Primary, P,* waves are also called "push-pull" waves because they are compressional. Their motion is like the "bump" that passes through a string of railroad cars when another unit is added to the train. In rock a particle is driven into its neighbor and bounces back, the neighbor strikes the particle beyond and rebounds, the next particle is disturbed in the same way, and so on *ad infinitum* as the wave speeds away from the focus. The primary are the fastest waves, traveling at about 4.8 kilometers (some 3 miles) a second in the Earth's outer rocks. Their speed increases as they penetrate into materials at progressively greater depths within the Earth, so that 2900 kilometers (about 1800 miles) below the surface

they travel at almost 15 kilometers (more than 8½ miles) per second.

The *secondary, S,* called "shake" waves, have transverse motion. The rock particles move up and down. This motion can be demonstrated by tying a rope to a door handle, extending the rope, then snapping the free end so that waves roll down the rope. Secondary waves also move at rates of kilometers per second, but only about two-thirds as fast as *P* waves. The paths of both *P* and *S* waves, passing through the Earth's body, tend to curve upwards towards the surface, in response to changing properties of materials at depth. Significantly, the waves may be bent (refracted), or bounced (reflected), when they encounter a boundary between layers in the interior.

The *surface, L,* waves are complex and have at least two different simultaneous motions, producing a "shimmy." They originate where body waves (the *P* and *S*) from the focus reach the Earth's surface. Because they move around the outside of the Earth, following the crust at rather shallow depths; they are classed as surface waves. They are the slowest and broadest of the three wave types and travel at a nearly uniform speed. Each wave type creates distinctive jogs in the lines being continuously recorded on the seismograph paper to form a seismogram.

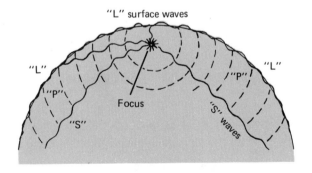

Fig. 6-14 Types of earthquake waves.

Locating Earthquake Centers Finding the epicenter of a distant earthquake requires cooperative effort among at least three seismograph stations. At each, the distance from the earthquake is first determined by using the difference in the time of arrival of the first *P* and the first *S* waves. Because both waves start from the focus at the same time, and because the *P* travels faster than the *S*, the time interval between their arrivals becomes progressively greater the farther they travel. Thus the interval recorded at a station can be referred to a travel-time table that gives the distance to the earthquake.

Next, the direction to the epicenter is determined. A circle is marked on a globe, using the position of the local station as the circle's center and the known distance as a radius. The epicenter must be somewhere on the circle's circumference. Then, using data obtained from a second station, another circle is constructed which should cross the first one at two points that are both possible epicenter locations. An arc based on a third station should cross one of these two points, giving the location of the epicenter (Fig. 6-15).

A major accomplishment of seismologists was the establishment of reliable travel-time tables. About 1940 Jeffreys and Bullen in England and Gutenberg and Richter in the United States independently derived tables in close agreement. Such tables, to which the arrival times of seismic waves are referred, are necessary for the accurate location of foci, as well as epicenters, of earthquakes. They are essential for studying the variations of velocity in P and S waves at depth, in relation to the Earth's deep internal zones.

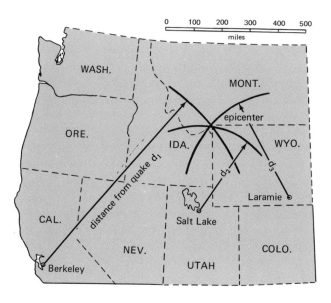

Fig. 6-15 Locating the epicenter of an earthquake.

Earthquake Causes and Warnings

Earthquakes are the end product of a complicated chain of cause and effect. The most fundamental cause is the little-understood operation of the Earth's internal heat on mantle materials. Whatever its nature, the effect of this deep-seated operation is to power the plate tectonic "machine" whose splitting and colliding slabs of lithosphere generate the worldwide earthquake belts. The direct cause of seismic waves, the vibrations we call earthquakes, is the deformation and slippage of masses of rock caused by the jostling of global plates.

The Elastic Rebound Theory The long-accepted mechanism for the generation of major earthquakes was "elastic rebound" of strained rock masses along active, deep earth-fractures called faults. This mechanical theory was proposed by H. F. Reed of Johns Hopkins University from studies following the 1906 San Francisco earthquake which was caused by movement on the great San Andreas fault of California (Fig. 6-16). He noticed that a series of carefully surveyed points, and such reference lines as fences and roads crossing the fault at right angles, became warped during periods of years free of earthquakes. He assumed that the ground was being strained (deformed) by stresses building up within the Earth—but that no slippage occurred because of the tremendous friction between opposing rock masses jammed tightly together along the fault. Ultimately, he reasoned, the stresses increased until they overcame the friction and caused a sudden slippage, in which fences and other reference lines were sheared and offset as the ground snapped to a position of no strain (Fig. 6-17). Rock masses grinding by each other during the sudden displacement generated the vibrations that move out as earthquake waves.[2] Thereafter the stresses, which were temporarily relieved by the

2 Something like the vibration in your hands when you draw two coarse files across each other.

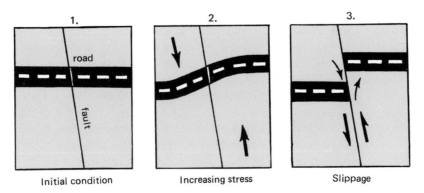

Fig. 6-16 Elastic rebound mechanism.

slippage, slowly accumulated over the years, and the process is repeated causing sporadic earthquakes. Although the basic concept still has merit, later investigations indicate that it is an oversimplification.

A few faults have had gradual movements that were not accompanied by earthquakes, as indicated by the buckling and breaking of oil pipes and well casings. Thus quakes seem to require that faults resist slippages until forces build up, finally producing a sudden release. Although the classic concept of elastic rebound may apply to shallow-focus earthquakes where rocks on both sides of a fault are solid and rigid, deeper-focus quakes occur where some of the rock materials are deformed plastically. A possible explanation for the deep-focus quakes involves the idea that the solid descending plates in subduction zones encounter enough resistance in the plastic—but viscous—asthenosphere to cause periodic fracturing and slippage—just as shoemaker's wax will flow slowly under gradual pressure, but snaps when forces are applied more quickly.

The Dilatancy Mechanism and Earthquake Prediction Until quite recently, specific predictions of earthquakes—their exact time and potential devastation—were the province of California religious mystics and science-fiction writers. Lately, however,

studies initiated by Russian seismologists—since confirmed and followed up by American and Chinese investigators—provide an encouraging basis for future scientific prediction of earthquakes. Careful examination of seismic records from dynamite blasts, nuclear bomb tests, and minor or distant earthquakes provide significant information for interpretation by seismic laboratories in earthquake-prone areas.

It now seems evident that before a notable seismic event the time interval between the arrivals of P and S waves becomes shorter and then returns to normal just before an earthquake occurs. Since the S-wave velocities remain constant, the phenomenon reflects changes in the travel times of the P waves. Moreover, the length of the interval of abnormally closer arrival times seems proportionate to the strength of the ensuing earthquake. That is, the decrease in P-wave velocities signaling an incipient quake may last for a decade before a devastating event of 8 on the Richter scale, for about a year before a quake of magnitude 7, and for some months before a shock in the range of 6. Other indications of an impending earthquake include an increase in electrical resistance of rocks in a susceptible region, as determined by geophysical instruments. Also, an increase in the radioactive gas radon in well water preceded the Siberian earthquake of 1949, which was studied in detail by

Fig. 6-17 Fence offset 8½ feet horizontally by the main fault during the 1906 San Francisco earthquake. U.S. Geological Survey photo by G. K. Gilbert.

Russian seismologists. An interesting observation by Chinese scientists, which is not to be dismissed lightly, is that animals change their behavior, apparently sensing an earthquake, just before a major seismic shock.

An hypothesis that may explain the physical observations stems from investigations of rock properties by scientists led by William Brace of the Massachusetts Institute of Technology in the 1960s —although the work attracted little attention at the time. As rock is subjected to great pressure and approaches its breaking point, the strain (change in shape and volume) causes microscopic fracturing, rotation of grains, and other changes that give a more open structure to the material. Such increases in volume are called dilatancy. Experimental evidence indicates that the change initially *increases* rock resistance to earthquake-generating slippages. It is now theorized that *P* waves, which travel most rapidly through dense and rigid materials, decrease their velocities on encountering rocks which have increased in volume by developing countless minute fractures. The return to normal velocities of P waves is thought to reflect the filling of the microscopic voids by underground waters. The water, in turn, may so weaken and lubricate the affected rocks that they suddenly give way to the accumulated forces built up inside the Earth—and generate earthquakes.

The introduction of water into rocks can produce earthquakes. In 1962 the Department of Defense drilled a 3600-meter (12,000-foot) deep well at the Rocky Mountain Arsenal in Denver, Colorado, into which were injected toxic wastes from the manufacture of poisonous nerve gas. Shortly thereafter, the city had its first earthquake in some 80 years, and in the following three and a half years over 700 nondevastating (only three exceeded a magnitude of 5) quakes rocked the Denver area. The deep disposal technique was stopped and the number of shocks decreased markedly— after David Evans, now a geological engineer at the Colorado School of Mines, demonstrated a pronounced statistical correlation between the rates of waste injection at the arsenal and the number of Denver's quakes. In the late 1960s, U.S. Geological Survey scientists began monitoring small earthquakes in the Rangely oil field of western Colorado where water was being forced down played-out oil wells in order to increase oil recovery in other wells. They found that many small earthquakes were clearly related to the rates of water injection. Based on such studies, it has been suggested that earthquakes might be controlled by carefully planned injection of water into wells purposely drilled along potentially dangerous faults in order to dissipate the buildup of deep-seated geologic forces in multiple small earthquake shocks—rather than allowing the stresses to rise to potentially devastating levels. The technique should, of course, follow some study and trials in uninhabited regions before use in heavily populated areas. An excellent candidate for such control is the San Andreas fault near San Francisco, which some seismologists believe is due for a major event in the future, which, in view of the growth of the city, could cause far more death and destruction than did the 1906 disaster.

An Earth Model

The major internal zones of the Earth have been briefly described (Chapter Two). Yet any such model of the Earth's hidden interior must come from human interpretation. So how have the bumps and squiggles of lines on a seismograph record been supplemented by various logical deductions and calculations to produce our present ideas about the core, mantle, crust, and asthenosphere and lithosphere?

The Core   A major discovery in the exploration of the Earth's interior was made in 1906, when Oldham demonstrated that the Earth has a large central core. Its size was determined in 1914 by Gutenberg, who calculated that the outer boundary of the core lay 2900 kilometers (1800 miles) below the Earth's surface, which gives the core a radius of about 3500 kilometers (2200 miles).

The core was discovered from the seemingly anomalous behavior of body waves. Up to a distance of 11,300 kilometers (7000 miles) from an epicenter, both P and S waves have well-ordered travel times. Beyond this distance, the S waves disappear and are not recorded by seismographs. Transverse waves, S waves, are damped out and not transmitted through a liquid. Thus, the outer part of the core is assumed to be in a liquid state, probably molten. The P waves, which also disappear at a distance of 11,300 kilometers (7000 miles), reappear very strongly at seismographs more than 16,000 kilometers (10,000 miles) from the epicenter, but their arrival is as much as three minutes late. The intervening Shadow Zone can be explained by an inward bending of compressional waves, caused by entrance into a medium where their velocity is less. Thus, it is assumed that the core bends and concentrates the P waves, in the same way, and for

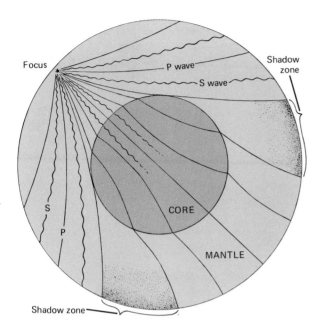

Fig. 6-18   Effect of the core on seismic waves.

core, but they are the best existing circumstantial evidence.

The Moho   The sharp boundary between the mantle and crust is called the Mohorovičić discontinuity (the "Moho" for short) in honor of its discoverer, a Croatian (Yugoslavian) seismologist. He noted, in 1909, that two distinct sets of both *P* and *S* waves were recorded from earthquakes originating less than 800 kilometers (500 miles) away. From records of earthquakes at various distances within this range, it was evident that one set moved more slowly but directly through the outer part of the Earth. Within 160 kilometers (100 miles) or less, this set arrived first; but beyond that distance fell behind the other, arrived progressively later, at increasing distances, and died out about 800 kilometers from the source (Fig. 6–19).

The explanation of this behavior is best introduced by an analogy with auto travel, on direct local roads or superhighways a moderate distance away. For short trips, the necessarily slower rate of travel on the more direct local roads saves time. But on longer ones, the time lost getting to and from the superhighways is more than gained back by the faster rate of travel there, so that despite a longer trip in miles you reach your destination sooner.

Mohorovičić applied similar reasoning to the travel of seismic waves. Once they reach a certain depth their speed increases, so that the deeper-moving waves, although they lose time getting to and from their "superhighway," soon pass their slower-moving surface counterparts. Thus, he deduced a boundary, called a discontinuity, above and below which the rates of travel differ considerably. Because the velocity of seismic waves is influenced by the elasticity of the materials through which they move, a crust and mantle of different physical properties must exist.

The Mantle   The rock in the uppermost mantle could be either peridotite or eclogite. Both these

the same reason, that a glass lens focuses light (Fig. 6-18).

Two lines of evidence suggest that the core is nickel-iron. First, the overall average density of the Earth is 5.5 compared to the average density for surface rocks of 2.7. Thus, the interior must have some materials of higher density than the overall average to compensate for the much lighter surface rocks. If the volume of the core, which can be calculated from its radius, is assumed to contain nickel-iron, then the density relations are nicely explained. A second totally different line of evidence bears on the case. Meteorites, which some believe are fragments of a shattered planet, are mainly of two types: stoney ones that are high in iron silicates, not too different from the assumed compositions of the mantle and simatic part of the crust, and nickel-iron ones that indirectly suggest a possibly similar composition for the core of the Earth. Neither line of evidence proves a nickel-iron

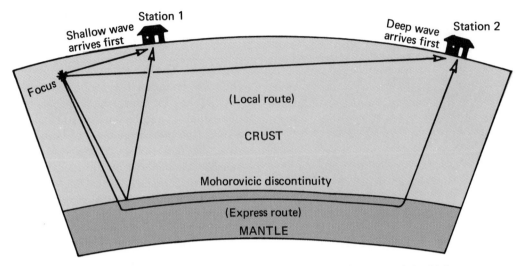

Fig. 6-19 Evidence and interpretation leading to the discovery of the Moho.

rocks have the correct densities, between 3.2 and 3.4, and laboratory experiments on their physical properties suggest a close similarity to materials in the outermost mantle whose properties are indicated by their transmission of seismic waves. Peridotite, as you may recall (p. 68), contains considerable amounts of the mineral olivine and also pyroxene minerals (some varieties of peridotite are almost entirely olivine). Eclogite consists of garnet and various pyroxenes. If the mantle is composed of peridotite, then the Moho (or M-discontinuity) marks a boundary between rocks of different chemical composition. That is, minerals in rock underneath the Moho contain more iron and other heavier chemical elements than occur in the minerals making up the basaltic rocks of the overlying crust. Peridotite is a likely candidate for upper-mantle rock. Partial melting of peridotite results in a magma that, upon eruption, would cool to produce basalt. Fragments of peridotite have been found in basalt lavas, and it occurs in some intrusive masses which came from great depths. If on the other hand, the upper mantle is eclogite, then the Moho represents a phase change.

Phase changes can be imagined as a compaction and reshuffling of ions in which one group of internal crystal structures is converted into a more dense group of crystal structures. In other words, one group of minerals is changed into another group of minerals that are better adjusted to the great pressures in the depths of the Earth. During such changes, the materials involved remain in the solid state.

The phase-change concept implies that rocks above and below a discontinuity have the same chemical composition (the same elements in approximately the same amounts), but that the minerals are different (ions are arranged in different crystal structures). Whether or not the Moho represents a phase change, the concept is important in considering other discontinuities deeper within the mantle. In any case, basalt and eclogite contain different minerals, but chemical analyses of whole-rock samples are almost identical in the various elements present and their amounts. Thus at high pressures and temperatures, such as exist at the depth of the Moho, the ions in basalt (or gabbro) could be rearranged into the more compact crystal lattices characterizing the different minerals found in eclogite. This rock, having a density of 3.3, is better adjusted to the high pressures at great depths than are basalt and its intrusive equivalent gabbro, whose densities are about 2.95. Because rocks remain in a solid state during such conversions, phase change can be considered a metamorphic process (complete melting to form a magma is not involved).

The phase-change theory is supported by laboratory work. The General Electric Company pioneered the manufacture of commercial diamonds using equipment that creates pressures of 100,000 atmospheres and temperatures of 15,000°K.[3] Under these conditions, the loose packing of carbon atoms in the very soft mineral graphite is converted into the dense compact arrangement of diamond. Recently, in a geochemical laboratory, George C. Kennedy of U.C.L.A. used high-pressure equipment to convert basalt into crystalline gabbro, and then with still greater pressure some of the minerals typical of eclogite were created.

It was once planned to drill a hole through a thin part of the Earth's crust into the mantle. This plan was called the "Mohole" project. Had it been completed, samples of the mantle could have been obtained and used to settle the problem of whether the mantle consists of peridotite (representing a change of composition) or eclogite (representing a phase change). Unfortunately, the project was abandoned.

Although the mantle was once considered a relatively homogeneous major Earth zone, seismic evidence now indicates a more complicated structure. Intensive seismic investigations indicate that P waves travel at distinctly faster velocities in outer mantle under the eastern United States than they do beneath the mountainous regions of the West. Thus mantle materials underlying the Moho may well differ laterally from one region to another. Moreover, the mantle seems to have major layers

3 One atmosphere is a pressure of about 1 kilogram per square centimeter (14.7 pounds per square inch), standard atmospheric pressure at sea level; °K means degrees Kelvin, a temperature scale like the Celsius except that zero is equal to −273° Celsius.

at depth. Discontinuities, marked by changes in seismic-wave velocities, are reported at about 400 and about 650 kilometers (250 and about 400 miles) below the Earth's surface. These boundaries may represent either changes in the chemical compositions of the rocks or phase changes resembling the basalt-to-eclogite transition (but involving other rocks with different minerals that require higher pressure to change their crystal structures).

Based on the seismic records, the mantle is now thought to have three main zones: the upper mantle, extending from the Moho down to the 400 kilometer (250 miles) deep discontinuity; a transition zone, extending beneath this to the 650 kilometer (400 mile) deep discontinuity; and under that the lower mantle, whose lower boundary is at the Earth's core (Fig. 6-20). The transition zone and lower mantle are, as yet, little known; most recent interest has been concentrated on the low-velocity seismic zone in the upper mantle which is related to the plate tectonic theory.

The Asthenosphere and Lithosphere A key concept in the modern plate tectonic theory is that of the existence of a weak asthenosphere beneath a rigid outermost lithosphere. These global shells, however, did not appear in most presentations of Earth models until the 1960s, after seismic records finally confirmed the presence of a low-velocity zone (the asthenosphere) lying from about 70 to 250 kilometers (some 45 to 155 miles) below the Earth's surface. Yet the idea is not a new one.

As early as 1914, Joseph Barrell gave the name "asthenosphere" to a "thick, hot, basic [mafic], rigid yet weak shell" which he reasoned lay about 100 kilometers (some 60 miles) below the Earth's surface. He also thought that beneath the asthenosphere and extending down to the Earth's core was a deeper rigid shell (which he called the "centrosphere"). Above the weak asthenosphere, he envisioned a strong outer shell, the lithosphere, which contained granitic and basaltic rocks. Thus Barrell was probably the first person to suggest the Earth

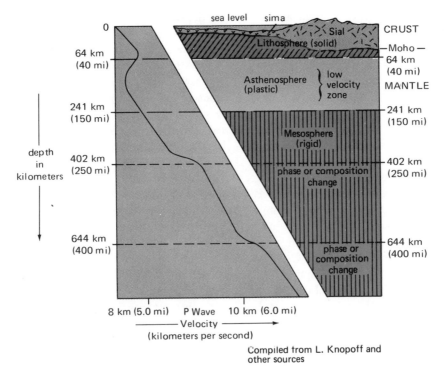

Fig. 6-20 Structure of the crust and mantle as interpreted from the velocity of *P* waves.

zones now used in the plate tectonic model. He did not, however, believe in continental drift or laterally moving plates.

From the 1920s into the 1940s, Reginald A. Daly—an imaginative and wide-ranging theoretical geologist—also put great emphasis on the concept of a weak internal shell in his theories of continental movements and mountain-building. Although he did not use the term asthenosphere[4] in his two major books, *Our Mobile Earth* (1926) and *Architecture of the Earth* (1938), Daly theorized that under the force of gravity, plates containing granitic continental rocks and underlying solid basalts slipped laterally across a deeper, weak layer of glassy basalts (which he later considered to be peridotite). He thought mountains originated where

4 Perhaps because Daly **taught at** Harvard and Barrell was a Yale professor.

one plate plunged under another down into the less dense, weak global shell and crumpled the geosynclinal sediments into orogenic structures. Thus—unlike most of contemporary American geologists—Daly looked favorably on the ideas of continental drift proposed by Alfred Wegener. Daly's scheme of slipping plates may well be considered the ancestor of the modern plate tectonic theory.

The first instrumental evidence for a weak shell within the upper mantle was obtained by the German-American seismologist Beno Gutenberg in the 1920s. Most other seismologists, however, remained skeptical about his data from the seismograph records until some 40 years later. The zone was confirmed by seismic records from nuclear bomb tests whose strength, location, and time were accurately known and gave more clear-cut results.

Gutenberg had originally noticed that *P* waves

were markedly weakened when received at stations from about 100 kilometers (roughly 60 miles) to about 1000 kilometers (about 600 miles) from an earthquake center. The waves had much stronger records at distances up to 100 and beyond 1000 kilometers. The behavior was interpreted as indicating a partial Shadow Zone, like the more pronounced one associated with the discovery of the Earth's core. Bending of the waves into a low-velocity zone was thought to produce the shadow, although in this case weak waves did emerge in the shadow because of inhomogeneities of the rocks and irregularities of the wave paths in the zone. The low velocities were attributed to a change of mantle rock from a solid to plastic state, with possibly some degree of melting (Fig. 6–21).

The plastic state has both liquid and solid properties. Shoemaker's wax, mentioned earlier, and some candy bars are plastic, for they break if suddenly snapped but flow if slowly squeezed. As the mantle transmits both primary and secondary seismic waves, it responds like a solid to sudden shocks, but under long-continued stresses it apparently flows like a highly viscous liquid. That rocks do flow under temperatures and pressures at depth in the Earth is amply demonstrated by the contorted structures in metamorphic rocks. The velocity of seismic waves is dependent in part on the elasticity of rocks, or the readiness with which they return to their former shape after deformation. Thus the crustal rocks, which are rigid and more elastic, transmit waves faster than a plastic material, which tends to flow and not return to its original shape when deformed. Increase of temperature tends to lower strength as a material approaches its melting point; on the other hand, increasing pressure increases strength by raising the melting point.

At about 64 to 80 kilometers (40 to 50 miles) depth, temperature apparently dominates so that the rock material becomes plastic—and hence the velocity of seismic waves decreases and they are deflected. Deeper than 240 kilometers (150 miles), pressure increase is thought to more greatly influence the rock materials so that they become more elastic, and the waves beneath the low-velocity zone speed up again with marked changes at the

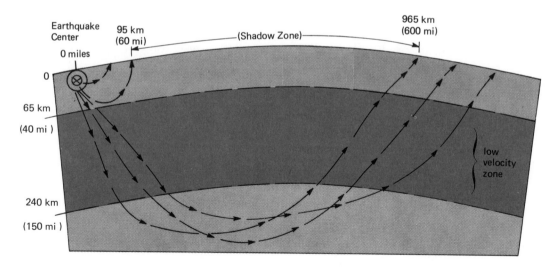

Fig. 6-21 Paths of points on fronts of P waves, illustrating the origin of the upper-mantle shadow zone. Based on D. L. Anderson.

400- and 650-kilometer (250- and 400-mile) depths of the deeper mantle.

The Crust The crust is now considered the thin global skin that, combined with underlying rigid rock of the mantle, forms the Earth's outer solid shell—the lithosphere. In thickness the crust ranges from some 65 kilometers in continental mountains to as little as 10 kilometers in ocean basins (from about 40 to 6 miles); the crustal cross-section has a mirror image. Evidence that the continent-forming sial is largely granite-gneiss (perhaps 75 percent) and granite is based on the velocities of transmitted seismic waves as recorded by seismographs. Sial is directly observable only where its thin but widespread sedimentary veneer has been stripped by erosion—as in the shield regions where the ancient rocks of continental blocks are exposed, and in deeply breached mountains and uplifts. Under ocean floors, sial is absent and only basaltic rocks beneath a thin carpet of sediments are indicated by the travel times of seismic waves. In older concepts—such as Wegener's theory of drifting continents—oceanic basalts, the sima, were envisioned as extending under the continents as a basal layer of crust. Recent geophysical findings, however, indicate that deep continental crust is a complex jumble of sialic and basaltic rocks—not a homogeneous and continuous layer of sima.

Now that, from the viewpoint of plate tectonics, the Moho is no longer considered the most significant structural boundary in the outer part of the Earth, the crust may seem less important. Despite this, the crustal rocks still remain our only directly accessible source of information for the operation of the Earth's internal and external, dynamic, mechanical systems through the vast spans of geologic time.

SUGGESTED READINGS

Anderson, D. L., "The Plastic Layer of the Earth's Mantle," in *Scientific American,* Vol. 207, No. 1, pp. 52–59, 1962.

Bullen, K. E., "The Interior of the Earth," in *Scientific American,* Vol. 193, No. 3, pp. 56–61, 1955.

Hamblin, W. K., Chapter 2 in *The Earth's Dynamic Systems,* Minneapolis, Burgess Publishing Co., 1975.

Hodgson, J. H., *Earthquakes and Earth Structure,* Englewood Cliffs, N.J., Prentice-Hall, Inc., 1964 (paperback).

Knopoff, L., "The Upper Mantle of the Earth," in *Science,* Vol. 163, No. 3873, pp. 1277–1287, 1969.

Press, F., and Siever, R., Chapters 19 and 20 in *Earth,* San Francisco, W. H. Freeman and Co., 1974.

Paricutin and church tower of San Juan village, standing in 6 meters (20 feet) of lava. Photo by Tad Nichols.

7
Volcanism

Spectacular, often violent, and sometimes catastrophic eruptions give volcanos a special interest. Besides being newsworthy, they clearly show geologic forces in action, in contrast to the more subtle and slower processes such as rock decay and the work of streams and glaciers. Moreover, they are windows to the Earth's interior, for magma, erupting at the surface, gives tangible evidence of the rocks and processes at depths far below any probed by the deepest drilled holes or brought to light by long erosion into the roots of former mountain ranges.

OBSERVATIONS

Volcanos and Related Structures

Volcanos and lava plateaus are the principal structures composed of extrusive igneous rocks: the rhyolites, andesites, and basalts. Volcanos are hills or mountains built around pipelike openings, or conduits, through which magma erupts from the depths. Ordinarily the conduit mouth forms a funnel-shaped crater at the top of the volcano. If the central depression is especially large, and several times wider than it is deep, it is considered a *caldera* (Fig. 7-1). Calderas occupying a large fraction of the structure result from former violent explosions and internal collapse when supporting magma is blown out or drained away at depth. They mark decapitated volcanos.

Volcanic Ejecta Liquid lava and solid particles of various sizes create the volcanic structures; however, ejecta during eruptions also include abundant gases. Nitrogen, hydrogen, rotten-smelling hydrogen sulfide, acrid sulfur dioxide, and a variety of other gases, many highly poisonous, are expelled. Volcanic exhalations also include large amounts of steam. Some may represent surface water that has seeped down to magma bodies; however, there is good evidence that much is an original component of the magma and has come from the Earth's interior.

Lava, the liquid ejecta, flows from an eruption until it cools and hardens (Fig. 7-3). Dark basaltic lavas are typically fluid and may flow considerable dis-

Fig. 7-1   Crater Lake, Oregon. Water-filled caldera with small more recent volcano (cinder cone), Wizard Island. Spence Air Photo.

tances, even on gentle slopes, before congealing, whereas rhyolitic and andesitic lavas are usually quite viscous and harden before moving very far. The rock of frozen lava flows often gives evidence of its molten past. Many lavas are shot through with vesicles, the frozen holes left by gas bubbles, to such an extent, in the case of pumice, that this rock is light enough to float on water. Many basalts are broken into polygonal columns, in some case several meters across at the top and a hundred or more meters in length. These structures are bounded by large, deep, contraction cracks, produced by shrinking of the basalt mass as it cools and hardens. Other basalts, looking like a jumbled mass of pillows, are evidence of eruption under water (Fig. 7-4). Here, rapid chilling quenches lava outpour-

ings, producing rounded, frozen skins on successive lava tongues as they continue to pile over each other.

Solid ejecta, the pyroclastics, are thrown into the air by explosions. The fragments may be smashed or pulverized rock from the volcano or clots of lava, which become at least partially solidified in the air before falling to the ground. Larger fragments are called blocks and bombs (Fig. 7-5); the progressively smaller particle sizes include cinders, ash, and dust. Although some of the terms might imply that volcanic rocks have burned like coal (and, in fact, volcanic eruptions were attributed in earlier days to burning subterranean coal), such is clearly not the case.

In total volume, basalt is by far the world's

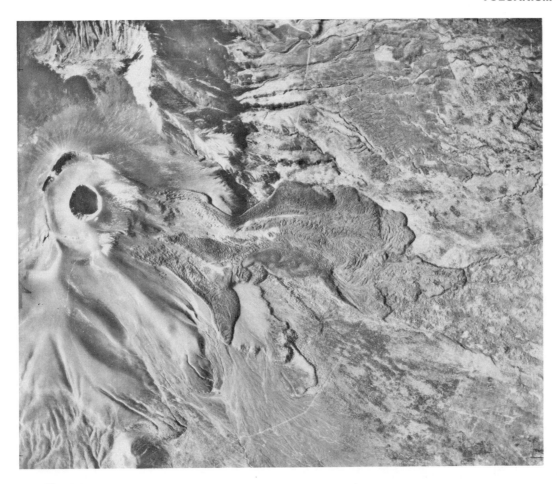

Fig. 7-2 Active Asama Volcano, Honshu Island, Japan. Ash covers most of the surface near the crater, which is partially encircled by the rim of a caldera. A lava flow is conspicuous. Courtesy of U.S. Geological Survey.

most abundant lava. It erupts into oceanic ridges and islands to form the rock of the sea-floor plates; and, in places, on continents to form vast lava plateaus. The fluid nature of its magma and the resulting ease with which it flows accounts for the abundance of basalt as lava. In marked contrast, the largest volumes of pyroclastic rocks are associated with andesitic and rhyolitic volcanism whose silica-rich magmas are gas-charged and

viscous, hence erupt explosively. They blast out some larger fragments and extensive clouds of glowing gas mixed with volcanic ash that rush down the flanks of volcanos in turbulent flows, called "nuées ardentes," which fuse into masses of *welded tuff* (Fig. 7-6). Vast quantities of welded tuff, supplemented by great quantities of higher airborne ash and dust, create rocks called *ignimbrites*. They may be 100 meters (330 feet) thick and cover tens

Fig. 7-3 Smooth, ropey-surfaced basaltic lava (pahoehoe). Kilauea crater, Hawaii. Photo by Tad Nichols.

of thousands of square kilometers, as in Utah and Nevada; and they form much of the rock in Wyoming's Yellowstone Park.

Types of Volcanos   Continuing explosive eruptions of pyroclastics produce *cinder cones*, steep-sided conical piles of solid ejecta around the central vent or crater (Fig. 7-7). Although the cones may grow on top of lava—and lava may break out from their bases—the cones themselves contain only pyroclastics. They are the smallest type of volcano, rarely exceeding a few hundred meters in height, and may occur in swarms, producing volcano fields. Mount Suribachi on Iwo Jima, Sunset Crater in Arizona, and the Mexican volcano Paricutin, in its early development at least, are representative cinder cones (Fig. 7-8). Most are basaltic in rock type although some consist of rhyolite.

Most impressively beautiful in their size and shape, and also the most violently explosive, are *strato-volcanos*, sometimes called composite cones (Fig. 7-9). Deep erosion into them exposes layers of pyroclastics irregularly interspersed with lava flows. The pyroclastics erupt largely from the main crater, or in parasitic cones on the main volcano's flanks; however, lava may erupt on the sides from deeply penetrating cracks. Fujiyama, the sacred mountain of Japan, Mayon in the Philippines, and the great volcanos of the Cascade Mountains in the western United States are examples of these impressive structures that often reach many thousands of meters in height. They are composed mainly of andesite and rhyolite (Fig. 7-10).

The *shield volcanos*, whose profile suggests a Greek warrior's shield, are largely great outpourings of mainly basaltic lava which, because it flows and spreads, produces gentle slopes (Fig. 7-11). Though their flanks are not precipitous, shield volcanos may rise to great heights. Mauna Loa, on the island of Hawaii, rises over 9000 meters (30,000 feet) above the surrounding ocean bottom, and nearly 4300 meters (14,000 feet) above sea level. Thus the island of Hawaii is the greatest volcanic

Fig. 7-4 Pillow lavas west of Othello, Washington. Photo by Vernon Anderson.

pile erupted upon the Earth's crust. The alignment of the Hawaiians and other island groups in the central Pacific Ocean suggests that they are localized upwellings of lava along extensive rifts or cracks in the ocean bottom.

The structure of volcanos reflects the nature of their eruption and, because some change their habits, *compound volcanos* are produced that are combinations of the three basic types. Mt. Etna, on the island of Sicily, represents a large shield volcano that is capped by a later, and exceptionally large,

cinder cone, which is at least 3000 meters (1000 feet) high (Fig. 7-12).

Although not strictly volcanos, because they lack central conduits, *lava plateaus* are eruptive structures. They are volcanic table lands, often of tremendous expanse, that develop where fluid magma erupts along extensive, great fractures in the crust (Fig. 7-13). Fluid basaltic magma flows out in a succession of nearly horizontal sheets. Individual flows may be only a few tens of meters thick, but where many flows are periodically erup-

Fig. 7-5 Early-morning photo of Paricutin Volcano, Mexico, during eruption (1944). Light streaks are made by incandescent volcanic bombs. Photo by Tad Nichols.

ted upon each other, over a long period of time, lava plateaus a thousand or more meters thick, and covering hundreds of thousands of square kilometers, have developed. The Columbia Plateau of the northwestern United States and the Deccan Plateau of India are among the most extensive volcanic features on Earth.

Distribution of Volcanos Today some 450 volcanos are active. These, along with many more so little eroded they cannot be long extinct, lie in definite belts corresponding generally to the major earthquake zones (Fig. 7-14). Great andesitic strato-volcanos of the "Circle of Fire" rim the Pacific. An offshoot from this belt passes through

Fig. 7-6 Eruption of nuées ardentes on Mt. Mayon in the Philippine Islands. Photo from U.S. Information Agency, U.S. National Archives.

Indonesia, and thence, with many sizeable gaps, along the Alpine-Himalayan mountain belt into the Mediterranean region. Volcanic activity is also prominent along the global system of largely submarine ridges extending from Iceland along the Atlantic Ocean floor and into the Pacific. Other largely submarine belts in the Pacific are marked by islands, notably the Hawaiian group. In contrast to the circum-Pacific belt and its offshoot to the Mediterranean, the submarine systems of volcanos seem dominantly basaltic.

Intrusive or Deeper Magmatic Structures Igneous structures formed at depth are exposed only after prolonged erosion. Long after a volcanic cone has disappeared, its former existence is recorded in a

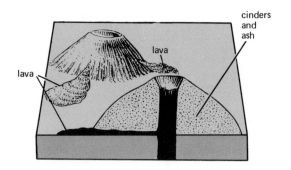

Fig. 7-7 Cinder cone.

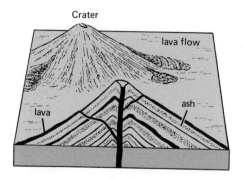

Fig. 7-9 Strato-volcanos.

Fig. 7-8 New lava emerging from the base of Paricutin in an early stage of its development. Photo by Tad Nichols.

Fig. 7-10 Chilean volcano Osorno from Lake Llanquihue. Photo by E. A. Carter.

volcanic neck composed of solidified magma in the deeper part of a volcanic conduit (Fig. 7-15). *Dikes* and *sills* are tabular, or sheetlike, igneous intrusions into older rock. Dikes are discordant, meaning that they cut across the structures in the older, or country, rock (Fig. 7-16). Before solidification, they may have fed lava flows or other igneous structures including sills. Sills resemble dikes except that they are concordant, following the structure of the country rock (Fig. 7-17). They may be horizontal, or nearly so, as in the Palisades sill along the Hudson River across from New York City, but they

can have any attitude, so long as they conform to the surrounding structure. Both range in size from a few inches to many feet in thickness. *Stocks* are vertical pluglike masses of igneous rock extending from depth. They are formed by magma masses that bow up the rocks into which they are intruded. *Laccoliths* are concordant igneous masses of lenticular shape, like a giant blister, that have bowed up the rocks into which they were intruded (Fig. 7-18). They may be fed by a dike or a sill, and form a simple flat-bottomed lens, or, in some cases, a more complicated structure whose cross-section looks like

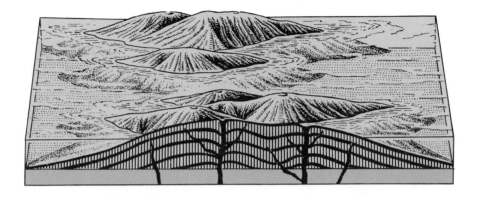

Fig. 7-11 Shield volcanos.

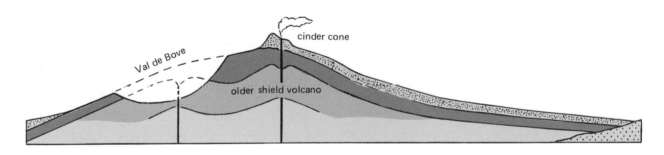

Fig. 7-12 Compound volcano. Cross-section of Mt. Etna (from Sir Charles Lyell).

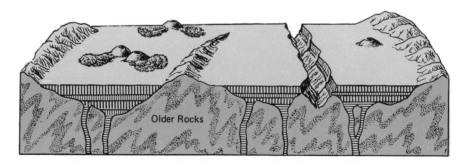

Fig. 7-13 Lava Plateau.

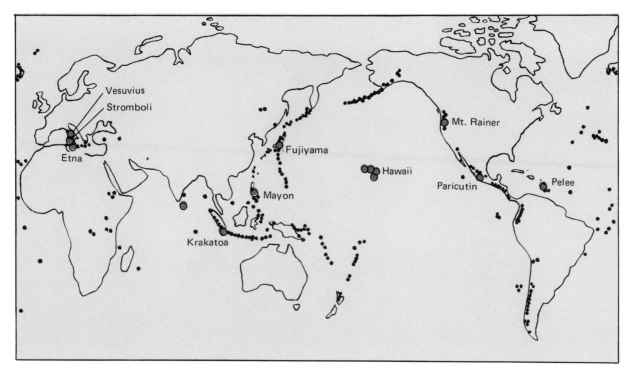

Fig. 7-14 Location of some of the Earth's major volcanos. Labeled volcanos are open circles. Note the "Circle of Fire" around the Pacific.

a Christmas tree. Laccoliths may be offshoots of large stocks, as in the Henry Mountains of Utah, which are several kilometers in diameter and more than 300 meters (1000 feet) high.

Eruptions of Some Famous and Infamous Volcanos

Vesuvius To the Romans, Monte Somma was just a large mountain close by Herculaneum and Pompeii, resort cities on the coast of the Mediterranean. There was no record in classic times of any volcanic activity, that is, until 79 A.D. During the preceding 16 years the region had received a series of strong earthquakes, and then, in August, Monte Somma awoke with devastating eruptions. Pompeii was

overwhelmed by clouds of white-hot ash, and Herculaneum was buried under streams of mud. As Vesuvius, the mountain has been erupting with periodic violence ever since, and has built a modern cone inside the relic rim of Monte Somma's caldera.

During the 79 A.D. eruption, the Roman naturalist and historian Pliny the Elder was admiral of the Roman fleet stationed nearby in the present Bay of Naples. To satisfy his scientific curiosity and to rescue panicky friends, he went ashore, where he lost his life. The story of his death and an excellent description of the eruption are preserved in letters to the historian Tacitus from the admiral's nephew, Pliny the Younger, who prudently declined his uncle's invitation to go ashore because he had some studies to complete.

Pompeii and Herculaneum, buried by ash and

Fig. 7-15 Aglathla, a volcanic neck near Kayenta, Arizona. Photo by Tad Nichols.

mud, were forgotten for more than 1000 years. Rediscovered, they were excavated to expose Roman cities in excellent states of preservation, along with the story of their violent destructions (Fig. 7-20). Household goods, personal belongings, and even food have been found in Pompeii, and molds of volcanic ash formed around the bodies of people and animals have been recovered and filled with plaster to form rather macabre statues (Fig. 17-5).

Pelée Mount Pelée, named for the goddess of fire, on the island of Martinique in the West Indies, came to life in 1902 after lying dormant for some 50 years. Eruptions began with explosions in the crater, which eventually ceased and were followed by periodic eruptions from the flanks that sent dense swirling clouds of incandescent ash, lubricated by intensely heated gases, avalanching down the mountain's sides. Most Pelêan clouds were directed down valleys leading away from St. Pierre, the capital of Martinique, but on May 8, a flaming outburst slashed through the city, smashing walls, uprooting trees, capsizing ships in the harbor, and incinerating over 30,000 people, almost in an instant. The sole survivor was a prisoner deep underground in the city jail.

In October, a lava plug was slowly extruded from Pelée's throat to form a spine rising almost 300 meters (1000 feet) above the crater's rim. Within a year, however, the soft rock of the spine,

Fig. 7-16 Vertical dikes cutting sedimentary beds in New Mexico. U.S. Geological Survey photo by N. H. Darton.

jarred by explosions, was destroyed. Around 1930, the volcano was active again, producing more flaming clouds and lava spines. This time the Pelêan clouds all followed a valley north of St. Pierre, sparing the reoccupied city.

Krakatoa The greatest explosion in recorded history was the eruption of Krakatoa in the Sunda Straits between Java and Sumatra in 1883. Dormant for two centuries, it commenced erupting in May, culminating in four tremendous explosions, one of which was heard in Australia, 4800 kilometers (3000 miles) away. The ash-filled air brought total darkness at mid-day in the adjacent islands during the height of the eruptions. Fine dust, blown high into the atmosphere, encircled the globe bringing exceptionally brilliant sunsets everywhere for some two years after the eruption. Where island masses had stood several thousand feet above sea level, a caldera appeared whose bottom lay 300 meters (984 feet) below the sea. The death toll on Krakatoa was zero—it was uninhabited—but over 36,000 lives were lost on Java and Sumatra, where giant sea waves, called tsunami, generated by the blast swept the low coastal flats. The waves were recorded by tide gauges around the world.

Stromboli Not all volcanos are given to violent eruptions blowing away and collapsing their tops. Stromboli, forming an island north of Sicily, has had almost continuous mild eruptions at intervals from a few minutes to an hour or so apart throughout recorded history. Mild explosions throw clumps of lava into the air where they solidify as volcanic bombs or scoriaceous chunks. The glow of the frequent eruptions in the crater, reflected by steam clouds overhead, make this volcano the "Lighthouse of the Mediterranean."

Hawaiian Islands In the shield volcanos, such as those now active on the "Big Island" of Hawaii, eruptions are comparatively quiet. They are mainly lava flows breaking out of the sides of the moun-

Fig. 7-17 Salisbury Crags in Queens Park at Edinburgh, Scotland, is a large sill studied by Dr. Hutton (photo by Brainerd Mears, Jr.).

tains, or jets and fountains shot up from lava lakes in their craters, which are steep-sided fire pits several kilometers across (Figs. 7-21, 7-22). Occasional explosions are caused by steam, when downward-seeping surface waters penetrate to molten lavas.

Santorini (Thera) The legend of Atlantis, whose roots seemed lost in antiquity, may have a dramatic geologic explanation. The first recorded account of the "land that disappeared beneath the sea" was written in Greece by Plato in 355 B.C. He used the legend merely as a background theme for a discourse on his philosophical ideas. Plato described a great empire, whose capital city of Atlantis was built on a circular plan some 24 kilometers (15

miles) in diameter and was on an island that sank beneath the sea during a war with the Athenians. He located the Atlantean archipelago west of the "Gates of Hercules"—i.e., in the Atlantic Ocean beyond the modern Straits of Gibraltar. Historians of classical times paid little attention to Plato's story, but in later centuries down to the present the "lost continent" theme has spawned a vast—and mostly fantastic—literature, in which "evidence" of Atlantis and its culture has been found in such far-flung places as Africa, Asia, Sweden, the arctic islands of Spitzbergen, and Mexico. Although not a lost continent, the "Atlantean Empire" could have been in the islands of the Mediterranean Sea.

The ancient Minoan civilization, source of such Greek myths as the human-eating Minotaur and

Fig. 7-18 Bear Butte, South Dakota, a laccolith surrounded by the eroded edges of upturned sedimentary beds. U.S. Geological Survey photo by N. H. Darton.

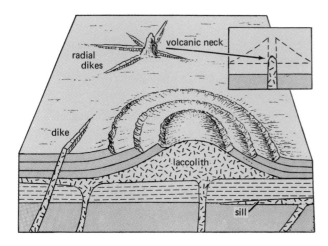

Fig. 7-19 Intrusive igneous structures.

the labyrinth, was known and admired by historians and writers of ancient times including Homer, who described the island of Crete as being densely populated and having ninety towns. Since 1900 A.D. archeological excavations on Crete have revealed the high development of this first European civilization. The Minoan culture, a contemporary of ancient Egypt, flourished from about 2600 B.C. until around 1500 B.C. During this time, more than a thousand years before Plato wrote, the Greeks were still barbarians. The Minoans, a seafaring people who traded throughout the Mediterranean region, were apparently a peaceful lot, judging from their unfortified, rambling palaces (whose sanitary plumbing systems were not equaled until Roman times). They had a system of writing, engaged in sports and games, and produced

Fig. 7-20 Excavated streets of Pompeii with Vesuvius in the background. Photo by E. A. Carter.

impressive works of art including large mural paintings, sculpture, pottery, jewelry, and other sophisticated items. Living in a geologically active region, the Minoans on Crete were apparently plagued by natural calamities, as indicated by several periods of repair and even rebuilding of their palaces. A serious episode of damage around 1500 B.C. preceded the rapid decline of their civilization, which by 1100 B.C. had been absorbed by the early Greeks.

About 105 kilometers (65 miles) north of

Fig. 7-21 Night view of glowing lava fountains caused by bursts of gas in the caldera of Halemaumau, Hawaii. Bursts reach 90 meters (300 feet) in height and average about 23 meters (75 feet). National Park Service Photo.

Crete is Santorini, a large crescent-shaped island composed mainly of outward-sloping layers of volcanic rock (Fig. 7-23). From its curved outer shore, Santorini rises gradually to the brink of an inner arcuate cliff reaching 270 meters (some 900 feet) above the sea. Along with two smaller islands to the west, Santorini encircles a lagoon, as much as 355 meters (1000 feet) in depth, whose roughly circular form encloses two other islands. The central small islands have been active volcanos from at least 186 B.C. down to the present day (Pliny the Elder reported the 186 B.C. eruption; the latest eruption was in 1950). Charles Lyell discussed the Santorini island cluster in his *Principles of Geology*

and concluded that it represented a former, large, single volcano which had a "paroxysmal explosion" creating the central gulf, where subsequent eruptions slowly built the smaller islands. Modern geologists agree with his interpretation and consider Santorini a large caldera.

In 1967 archeological excavations on Santorini disclosed extensive Minoan ruins containing many artifacts and the largest Mediterranean wall fresco yet found—beneath thick layers of volcanic ash that have been dated by modern techniques at about 1500 B.C. It has been suggested that the ruins are part of a great city extending across a large volcanic island (then known as Kallisti) which existed before

Fig. 7-22   Vents of shield volcanos, called calderas, are much broader than they are deep. The caldera of Halemaumau (Hawaii) when photographed in 1952 had a floor of some 61 hectares (150 acres). National Park Service Photo.

the major eruption.[1] The dated ash corresponds to the time of major destruction of palaces on Crete.

Thus we can envision the downfall of a whole civilization caused by a Krakatoa-type eruption, accompanied by caldera collapse in which the heart of a densely populated island disappeared beneath the sea. The resultant great sea waves (tsunamis) and concurrent earthquakes devastated Crete and adjacent islands. Perhaps Santorini is not the source of the Atlantis legend, but the 1500 B.C. volcanic

event probably represents—in its human aspects—the world's most devastating volcanic eruption.

INFERENCES

The Nature of Volcanic Eruptions

Much is known of the surficial effects of volcanism, but explanations of its origin and causes are tentative at best. The materials and shapes of volcanos are readily observable, and their structures are exposed after deep erosion, although never as neatly as we would like. Eruptions are also observ-

---

1 Since only two human remains have been found in Santorini excavations, it is possible that the people had fled when alerted by preliminary eruptions or earthquakes—possibly to be overwhelmed at sea.

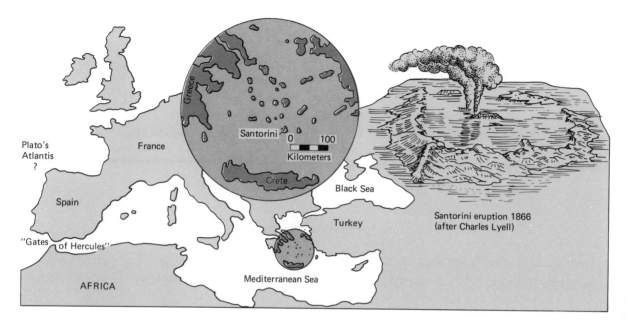

Fig. 7-23 Mediterranean region of the Greek island of Santorini, the possible location of Atlantis.

able, but their study poses problems because they are relatively infrequent, certainly uncontrollable, and cannot be summoned on demand like some reaction in a test tube. One must admire such volcanologists as Frank Peray, who stayed in an observation tower high on the western slope of Vesuvius throughout its 1906 eruptions, and also ascended Mount Pelée to study its eruptions from 1929 to 1932. Consider, too, the dedication to their science of Day and Shepherd, who walked out onto the crusted lava over the fire pit of the Hawaiian volcano Kilauea to thrust iron pipes into the cracks of flaming blisters above lava fountains, in order to collect volcanic gases uncontaminated by air. Despite all of this, our knowledge of the deeper plumbing of volcanos and the causes of volcanic action remains highly speculative.

Periodic Violence Why are strato-volcanos like Krakatoa violently explosive, others less so, and

some such as Stromboli rather well behaved? Apparently Stromboli's conduit, which never solidifies, acts as an open safety valve allowing continuous small eruptions, so that dangerous pressures do not arise. On the other hand, violent volcanos like Krakatoa and Vesuvius become plugged when lava in their conduits "freezes," causing tremendous internal pressures to build up. Recall that these volcanos all had long periods when they were seemingly extinct before their most violent explosions. Moreover, the spine extruded by Pelée is direct evidence that a stout plug had formed in the throat of this malevolent strato-volcano. In contrast to the fluid basalt associated with shield volcanos and lava plateaus, andesitic magmas are notably viscous and gassy. Thus, strato-volcanos are especially prone to plugging with a resultant development of great internal pressures that can be relieved only by breaching the volcano's sides, producing Pelèan clouds, or violently decapitating the struc-

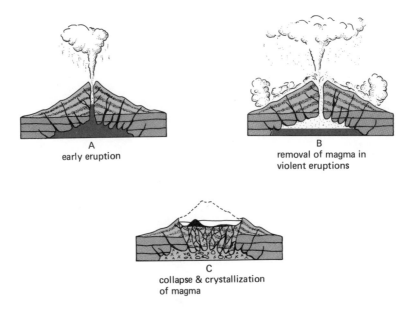

Fig. 7-24   Formation of a caldera such as Crater Lake. After H. Williams.

ture by explosion and collapse from loss of magma at depth (Fig. 7-24). They are like old steam boilers having faulty safety valves.

Gas Production   The generation of gas in volcanic magmas can be explained by a scheme involving the chemistry and physics of magmas. In hot magma, volatiles are held in solution so that little gas pressure exists. As magma cools and starts converting to igneous rock, however, solid crystals form that reject dissolved gases from their structure. Thus the remaining magma becomes progressively richer in volatiles. When the liquid reaches saturation and can hold no more, some is expelled as free gases including steam. At the high temperatures within a volcano, these confined gases have a strong tendency to expand, creating great pressures. The pressures force magma up the conduit in a more normal type of eruption. If this outlet is plugged, the pressures eventually become tremendous and cause violent explosions.

The Rise of Magma   Geological knowledge of the deeper volcanic plumbing that feeds conduits and fissures in the lithosphere is very sketchy. However, recent instrumented studies of the Hawaiian volcano Kilauea indicate a pattern of magma movements as an eruption develops. Several months before a surface eruption, the seismic observatory records tremors originating some 60 kilometers (about 40 miles) below the volcano. These tremors suggest that deep-seated magma is moving from the asthenosphere into fractures or conduits in the lithosphere. As the time of surface eruption approaches, the number of tremors increases and the depths of origin become progressively shallower. The rise of magma into the volcanic throat is further confirmed by tiltmeters—sensitive instruments recording slight changes in ground slope. They record that just before an eruption, the volcano's flanks bulge appreciably; then, while lava is actively erupting, the central area of the shield subsides. Hence, at a relatively shallow depth, the volcano seems to

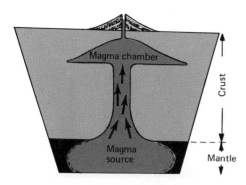

Fig. 7-25 Highly schematic diagram of deeper volcanic "plumbing." The shapes of the chambers and nature of the connections are not really known. Suggested by a diagram by G. A. MacDonald.

contain a bladderlike *magma chamber,* which inflates as it fills and deflates during surface eruptions (Fig. 7-25). Excessive emptying of such chambers after strong eruptions probably causes the collapse of some volcanic cones and the creation of large central calderas. In any case, such chambers are only storage tanks; the source of magma is at greater depths.

The Origin of Magmas

When the Earth's interior was thought to be completely molten, the source and origin of magmas was no problem. But twentieth-century geophysicists have complicated things by ruling out any continuous molten zone above the Earth's core. The core is an unlikely candidate for magma because of its depth and the fact that volcanos do not erupt a nickel-iron melt (which is thought to compose the core). Just how solid or plastic rock materials are melted to create bodies of magma is not well established. Important factors, however, include the heat, pressure, and chemical composition of the Earth's interior.

Temperatures in the Earth The increase of temperature with depth in the outer part of the Earth's crust has been directly measured in deep mines and

many drilled holes. Although differing from place to place, the increase is about 20° to 30° C per kilometer (58° to 87° F per mile). If such increases continued to the Earth's center, the mantle would be mainly molten—as most nineteenth-century geologists believed—and the temperature of the core would reach some 25,000° C (45,000° F). Since seismic evidence indicates that the deep mantle is solid, the temperature increase must become considerably less in the depths below the asthenosphere. The core, it is theorized, may reach temperatures somewhat over 4000° C (about 7200° F), based on laboratory experiments on temperature-pressure relations that would explain the outer molten part of the core just beneath solid mantle. To account for the high temperature increase recorded in the Earth's outer part, it is theorized that heat-generating, radioactive minerals are concentrated in crustal rocks—notably in granites and to a lesser extent in basalts. This explanation is strongly supported by new heat-flow and radioactivity data. At one time, it was suggested that magmas originated in localized places of especially concentrated radioactive heating in the deep crust or outer mantle. Since Geiger counters have recorded no excessive radioactivity in active Hawaiian lava eruptions or their associated gases, this hypothesis seems unlikely.

The Role of Pressure Along with increasing temperatures, the confining pressures on rock rise markedly within the Earth because of the increasing weight of overlaying materials. Pressure raises the temperature at which rock melts. Thus, under high confining pressures, rock in the depths remains solid or plastic at temperatures that would readily cause melting at the Earth's surface. A former hypothesis for the generation of magma in belts of mountain-building and crustal unrest held that upwarping or deep crustal fractures might reduce the pressure on deep "super-heated" rock to a point where melting resulted. Now, however, most investigators doubt that such a mechanism could sufficiently reduce pressure to cause melting.

Fig. 7-26 Mt. Etna in eruption, 1669. (From an old painting in the cathedral at Catania, Sicily.) Monte Rossi threw out a vast volume of lava that destroyed much of Catania, 48 kilometers (30 miles) away.

The top of the partially molten asthenosphere, some 70 kilometers (about 44 miles) below the Earth's surface, may represent the depth at which rising temperature overcomes the internal pressures tending to keep rock solid (Fig. 7-28). The lower limit of the asthenosphere, beneath which rocks re-gain some rigidity, may reflect a smaller tempera-ture increase at a depth of 250 kilometers (155 miles), where pressure again comes to dominate and keep rock materials solid. Whatever its origin, many theorists consider the asthenosphere to be a rock "slush" whose molten portion is an attractive

Fig. 7-27 Leopold Von Buch (1774–1852). Although Von Buch studied paleontology and alpine mountain structure, and produced the first extensive geologic map of Germany, he is best remembered for his work on volcanos. Originally taught by Professor Werner at Freiberg that the Earth's crustal rocks were precipitates from a universal sea, Von Buch reluctantly rejected the erroneous grand theory after his far-reaching travels convinced him of the igneous origin of balsalt.

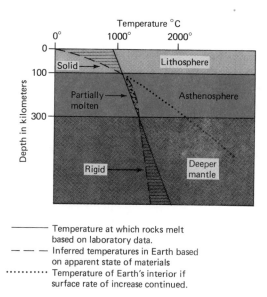

Fig. 7-28 Schematic representation of possible explanation for the athenosphere.

source for magma. If so, the question remains: How is the dispersed melt concentrated into sizeable magma bodies?

Mantle Convection A popular hypothesis in plate tectonic speculation now attributes the origin of much magma to "plumes"—rising convection currents in the mantle (Fig. 7-29). Although the scheme presently has several different versions, they broadly resemble the mantle circulation associated with the older Meinesz hypothesis (p. 35). Highly viscous upwellings of plastic rock from deep in the mantle could carry superheated materials to levels of much-reduced pressures that allow melting. Basaltic magmas erupting at spreading centers in mid-oceanic ridges and in fracture systems within

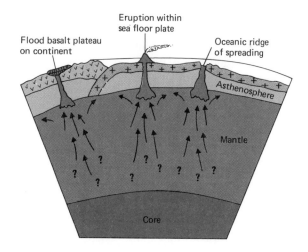

Fig. 7-29 Plume hypothesis for basaltic eruptions at tensional zones in the lithosphere.

sea-floor plates (as represented by the Hawaiian Islands) might have such an origin. Plumes are inferred from "hot spots"—zones of very high heat flow recorded at the surface of the lithosphere. Although a useful and reasonable hypothesis, the plume mechanism is based on speculation because there are as yet no instrumental records indicating the actual movement of rising materials inside the mantle.

Subduction Zones Magmas generated in zones of mountain-building at margins of colliding global plates require different explanations. Here, some melting of a rock slab might reflect its descent into the hot asthenosphere. Another source of heat could result from the tremendous friction along the surface between colliding plates of lithosphere. Moreover, laboratory experiments indicate that the presence of water lowers the melting point of rocks. Since the veneer of sediments on descending sea-floor plates is probably saturated with water, magmas could be generated at lower temperatures than in "dry" materials.

Some investigators, however, point out that although water is easily added to a laboratory rock melt, its relative importance in the "real world" of colliding global plates remains to be demonstrated. On the other hand, the mantle contains much water chemically locked in crystal structures of certain minerals. As evidence, Kimberlite, a complex variety of mica-rich peridotite, forms a pipelike intrusion into the continental crust of South Africa which is one of the world's major sources of diamonds. The diamonds (and included eclogite fragments) indicate that some of the rocks in the pipe at Kimberly came from zones of tremendous pressure—well down in the Earth's outer mantle. The groundmass in the Kimberly pipe contains much serpentine (an alteration product of peridotite in the presence of water), mica (whose crystal structure contains water), and calcite (a source of carbon and oxygen). Thus the mantle contains the ingredients of water; in fact a widely accepted

theory (p. 324) holds that the hydrosphere has gradually seeped out of the mantle in volcanic exhalations. Whether carried down in subducting plates or derived directly from the mantle, water involved in the generation of magma by the partial melting of plates seems to be no real problem.

Why Are Magmas Different?

The rocks created by volcanic eruptions range in chemical composition from basalt to andesite to rhyolite (intrusive magmatic equivalents are gabbro, diorite, and granite). Conjectures on the origin of different magmas and their resultant rocks involve geochemical considerations.

The Solidification of Magma Unlike the freezing of water—the simple compound H_2O—the solidification of magma into rock is a relatively complicated matter. In the early twentieth century, N. L. Bowen, at the Geophysical Laboratory of the Carnegie Institution in Washington, D.C., helped to unravel the history of rock melts. By quickly chilling carefully prepared melts to quench further chemical reactions, he found that different minerals crystallize at different temperatures (Fig. 7-32). In a melt of basaltic composition, the first crystals to form at high temperatures are of the ultramafic and dense mineral olivine, and also of calcium-rich plagioclase feldspars. As the "slush" (a mixture of solids and liquid) cools, the first-formed crystals react with the remaining molten liquid to form pyroxenes and also plagioclases containing progressively more sodium. Complete solidification of such a melt produces basalt—a seemingly obvious result. But from his laboratory experiments, Bowen came to a not-so-evident conclusion.

In his theory of *fractional crystallization,* Bowen explained how an original basaltic magma could produce a variety of different igneous rocks. If somehow the first-formed crystals, which lock up iron and magnesium, are removed from a gradually

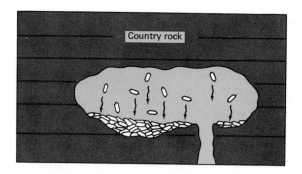

Fig. 7-30 Bowen's theory for separating crystals from a magma. Settling of first-formed, high-temperature olivine crystals to the bottom of a magma chamber.

cooling magma, the remaining liquid would change to a more felsic and silica-rich composition. Bowen suggested that a natural separation of crystals from a melt would occur in a magma chamber (Fig. 7-30). Here, the progressively forming crystals could settle to the bottom and thus be removed from

further reaction with the remaining molten liquid. It was suggested that in zones of mountain-building separation of crystals from a melt resulted from a process called "filter-pressing" (Fig. 7-31). Here, orogenic forces might compress the solid fraction of a slush into solid rock while driving out the remaining liquid to form silica-rich magma bodies—something like squeezing water out of a sponge.

If, as a basaltic magma solidifies, all crystals remain to react with the melt, the end product (as previously mentioned) is basalt. But if the minerals forming at high temperatures, such as olivine and sodium-plagioclase, are removed from reaction, they lock up much magnesium and calcium but little silica—thereby enriching the remaining liquid in sodium, potassium, iron, and silicon. These remaining elements combine into crystals of pyroxene, amphibole, and plagioclases having more sodium—minerals forming andesite or diorite. Should these andesite-forming minerals leave a melt before it completely hardens, the liquid residue would contain much silicon, potassium, and other elements

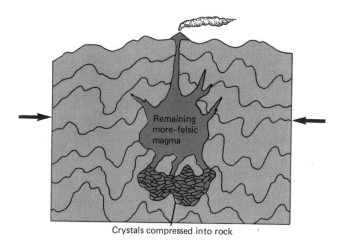

Fig. 7-31 Filter-pressing mechanism. Mountain-building forces compress a chamber of partially crystallized magma "slush," compacting the first formed ferromagnesian crystals into rock and driving out the still-molten fraction to create a magma of more felsic composition.

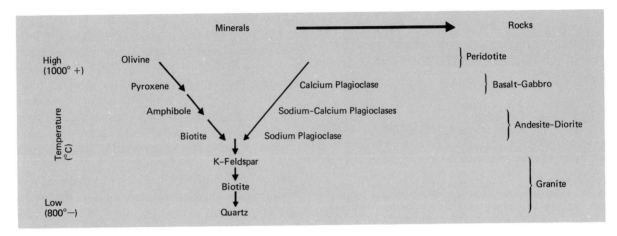

Fig. 7-32 Bowen's scheme for the order of mineral crystallization and the resulting igneous rocks.

that crystallize at relatively low temperatures into amphibole and biotite, K-feldspar, and lastly quartz—the minerals comprising rhyolite and granite (Fig. 7-32).

Field evidence for Bowen's theory is provided by some local, layered igneous rock bodies, such as the Palisades along the Hudson River near New York City, where an olivine ledge lies near the bottom of an intrusive sheet of gabbro. Other large sheetlike intrusions of highly mafic rock—many containing valuable metallic mineral deposits—give even more convincing evidence of fractional crystallization in an originally basaltic magma. Like the Palisades sill, they are layered intrusions having peridotite, consisting largely of the mineral olivine, towards their bases. Higher in some complex intrusions, such as the Stillwater complex of Montana, alternating layers of lighter and heavier igneous minerals (for example, plagioclase and pyroxene) form layers resembling those of sedimentary rocks; and some layers have minor features like the ripple marks, cross-bedding, and graded bedding of sedimentary rocks. The minor igneous features seem excellent evidence for Bowen's concept, in that convection currents of still-molten magma seem to have flowed across loose accumula-

tions of previously settled crystals—like water over sand grains.

Nonetheless, most geologists doubt that the mechanism accounts for the vast sialic rock bodies comprising the continental crust—because only the final 7 percent of a basaltic residue could solidify into granite. Today geologists rely on many of Bowen's findings, but in light of plate tectonic theory they reverse his scheme by emphasizing the partial melting of solid rock to form differing magmas.

Plate Tectonics and Igneous Rocks In terms of global patterns, magmas can be roughly sorted into three broad (and overlapping) assemblages: those that are dominantly basaltic, those that are largely basaltic-andesitic, and those that are mainly andesitic-rhyolitic (Fig. 7-33). Eruptions of large volumes of basalt seem related to zones of crustal tension (stretching), as in diverging plate boundaries at mid-oceanic ridges, in fracture zones within sea-floor plates (as represented by the Hawaiian Islands), and also within some continental masses, such as the Columbia Plateau of the northwestern United States. Basalts, which make up over 90 percent of the world's volcanic rocks, are generally considered as upwellings of molten material from the mantle—possibly from melted eclogite or from partial melting of peridotite.

The basaltic-andesitic and andesitic-rhyolitic provinces seem related to island arcs and orogenic belts. In 1912 Philip Marshall proposed an "andesite line," since extended around the Pacific, marking the outer limit of the great andesitic volcanos that encircle the dominantly basaltic central ocean basin. Although once interpreted as a global boundary between continental and oceanic crust, over 1500 kilometers (some 900 miles) of basaltic ocean bottom separates the andesites of the Mariana Island arc from the Philippine Islands, which, in turn, are separated from the continent of Asia. An even greater distance lies between the andesites of the Fiji Islands and Australia. Today the andesite line seems best explained by plate tectonic theory.

The basalt-andesite province coincides with belts of island arcs where two mafic oceanic plates collide. The basalts could well up directly from the mantle through fractures in the plates. The andesitic eruptions are now commonly attributed to *partial melting* of a subducted sea-floor plate descending into the asthenosphere (Fig. 7-34). In a sense, the concept of partial melting reverses Bowen's approach—by relating magmas to the progressive fusion of rock subjected to gradually increasing temperatures. In a descending plate having a mafic crust of gabbro-basalt and a veneer of

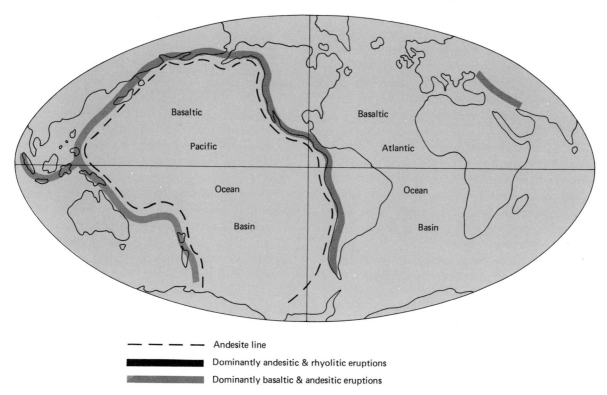

— — — — Andesite line

▬▬▬ Dominantly andesitic & rhyolitic eruptions

▬▬▬ Dominantly basaltic & andesitic eruptions

Fig. 7-33 Basaltic, basaltic-andesitic, and andesitic-rhyolitic global provinces. The ocean basins are mainly basaltic. Where two sea-floor plates collide the resulting rocks tend to be basaltic and andesitic. Where sea-floor and continental plates collide, the igneous rock produced tends to be andesitic and rhyolitic.

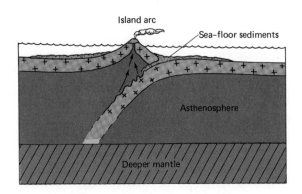

Fig. 7-34 Schematic diagram of hypothesis for generation of andesitic magma from partial fusion of a subducted plate.

wet sediments, the water could lower the temperature at which melting begins. The basaltic melt, somewhat enriched by silica and water present in the sediments, could create a magma having a chemical composition and temperature (based on Bowen's studies) that would erupt as andesite. This rock occurs in island chains developed on the overriding sea-floor plates.

The andesite-rhyolite province coincides with mountain belts where sea-floor plates are subducted beneath overriding continental margins. Here the andesites could come from partial melting of the subducted plate, the same mechanism as in island arcs. Another reasonable hypothesis involves *contamination*—a process wherein the very hot basaltic magmas, rising from the mantle, melt and absorb material from the continental crust. Dilution of a mafic magma by additions from sialic crystalline rocks and siliceous sediments could change the melt to an intermediate composition—that of andesite.

Several origins have been proposed for the siliceous magmas producing rhyolitic eruptions. Once an andesitic magma is created—whether by partial melting or contamination—its fractional crystallization could produce a magma of rhyolitic composition. Since rhyolites are mainly associated with continental plates, however, it is possible that some magmas originate from the melting of deep parts of sialic plates. Another likely origin for the rhyolitic magmas is the melting of sedimentary rocks in the depths of geosynclines. This hypothesis is supported by laboratory analyses which show that the chemical composition of shales and graywackes is remarkably similar to the chemical composition of rhyolitic volcanic rocks.

The Granite Problem The source of the rhyolitic and andesitic magmas is related to a long-standing geologic controversy—the origin of granite. In the cores of many mountain belts exposed after deep erosion, great masses of granitic rock form bodies called *batholiths*. To be classed as a batholith, at least 100 square kilometers (40 square miles) of plutonic rock must be exposed at the Earth's surface. Batholiths are composite masses made up of numerous smaller plutonic bodies, some of which may be granite but most of which are granodiorite —a rock containing plagioclase as well as potassium-feldspar, quartz, and hornblende or biotite. They extend to such depths that prolonged erosion has never exposed their floors. The rocks adjacent to batholiths are contorted gneisses, schists, and other dynamically metamorphosed rocks.

Some batholithic masses have sharp contacts with the surrounding country rock, and these granites appear to have been intruded as magmas. In such regions, dikes of granite intrude the older wall rocks, and in places blocks of the country rock (called xenoliths) are surrounded by granite, strongly suggesting that a chunk of solid rock was quarried by rising magma and then settled into the molten liquid. Such observations, dating back to the late 1700s, were used by James Hutton to demonstrate that granite was an igneous rock (Fig. 7-35).

On the other hand, the Norwegian geologist B. Keilhau discovered in the early 1800s that some granites very gradually merge into the surrounding metamorphosed country rock. In the merging zones,

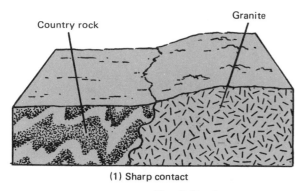

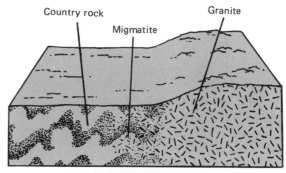

(1) Sharp contact

(2) Gradational contact

Fig. 7-35   Batholithic contacts as exposed after deep erosion.

the rocks are often *migmatites* (mixed rocks) composed of gneisses and schists containing stringers and veins of granite. Keilhau inferred that granites originated from highly altered siliceous sediments and not from some magma coming from the Earth's interior. An extreme version of this theory of *granitization*, promoted in recent times by the eminent British geologist H. H. Read and others, holds that granite is a metamorphic rock. That is, the conversion of gneiss into granite occurs mainly in the solid or plastic state in the presence of fluids—granite magma would be a secondary development. Many other geologists, who also believe granite is an alteration product of sediments, disagree and suggest that granite is an igneous rock whose magma resulted from the final melting of highly metamorphosed rocks.

In any case, the great size of some clearly intrusive batholiths indicates that large volumes of granite are igneous (consolidated from a magma). Thus the gradational batholithic boundaries could represent exposures of deep levels in crumpled geosynclinal belts where highly metamorphosed rock was melting to form magmas. Sharp contacts between granites and country rock may represent shallower levels in orogenic belts where buoyant magmas rose in domelike masses (called igneous diapirs) that plastically deformed overlying rock materials (Fig. 7-36). Moreover, since shales, gray-wackes, and other geosynclinal deposits have a

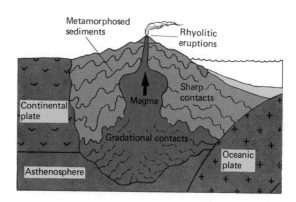

Fig. 7-36   Hypothesis for the origin of granite by metamorphism and melting of geosynclinal sediments deep in the roots of an orogenic belt. In this theory, granitic magmas result from the fusion of sedimentary rocks.

siliceous composition, their metamorphism and melting could produce the felsic magmas that create plutonic batholiths as well as the eruptions of rhyolitic volcanos.

Some geologists still stoutly maintain that the silica-rich granitic-rhyolitic magmas originate from mantle materials far beneath the Earth's crust. In fact, Bowen's theory well explains some rhyolitic eruptions after a long interval of quiescence in the generally basaltic island of Iceland, the emergent part of the Mid-Atlantic Ridge. Rhyolites and granites, however, are conspicuously missing from the

Pacific Ocean basin where, based on evidence from Hawaiian volcanos, magma chambers are available for crystal settling and magma differentiation. Somehow batholiths are continental features. Some modern geologists, who suggest a mantle source for rhyolites and granites, note that many batholiths consist of granodiorite, a rock transitional between granite and diorite, whose chemical composition is not far from andesite. Thus they suggest that the partial fusion of subducting plates could be the source of magmas forming rocks in batholiths.

Clearly much remains to be learned about volcanic and related igneous and plutonic processes. Many present-day hypotheses are little more than informed guesses—speculations showing our present inadequate understanding of the Earth's interior. They provide a stimulating topic for future data-gathering and investigation.

SUGGESTED READINGS

Bullard, F. M., *Volcanos in History, Theory, and Eruption,* Austin, Tex., University of Texas Press, 1962.

Reed, H. H., "Interventions from the Depths," in *Geology, an Introduction to Earth History,* New York, London, Oxford University Press, 1958, pp. 82–121.

Williams, Howell, "Volcanos," in *Scientific American,* Vol. 185, No. 5, pp. 45–53, 1951.

Syncline and West Henrietta fault, SW Tyrone Quadrangle, Pennsylvania. Courtesy of John S. Shelton.

The Deformed Crust

If the Earth's crust were static, the lands would have long ago disappeared beneath a universal sea. But the external destructive forces of weathering, gravity, and the agents of erosion are counteracted by the internal dynamics of the Earth. New rock builds up in volcanic eruptions, and the crust is uplifted by deformational forces. Folded, contorted, and broken sedimentary strata are clearly evidence of past deformation. If the crust were static, mud cracks, ripple marks, and many more features of clastic sedimentary and other layered rocks would not be found on steeply sloping surfaces where they often appear today.

Moreover, there is direct evidence of crustal movement in historic time. Most obvious and dramatic is the sudden appearance of scarps and horizontal displacements during earthquakes. Clear-cut examples of slow warping are also common. Abandoned irrigation systems of ancient Near Eastern civilizations now go up and down over hills that must have risen up to warp the ditches after they were dug. The rise and fall of coastal land is often marked by warped, wave-cut terraces that can be related to existing sea level, the ideal reference plane.

The classic example of coastal deformation in historic time is the Temple of Jupiter Serapis, described by Sir Charles Lyell in his *Principles of Geology*, the leading textbook of the nineteenth century. The "temple" was probably a Roman market place, but in any case stone columns, now high and dry, are pitted by rock-boring clams up to 6 meters (18 feet) above the temple floor. Thus, after the temple was built, the coast must have submerged to allow the clams to attack the columns, and then, later, reemerged to expose the columns as they are seen today. In Lyell's time the case needed proving, but now the accumulated evidence is overwhelming that solid rocks are being deformed in the Earth's mobile crust.

LOCAL STRUCTURES

Preliminary Matters

Rocks are deformed by compression, or squeezing; tension, a pulling apart; and torsion, a twisting resulting when two forces act in opposite directions, but

Fig. 8-1 Upturned strata forming the dome at Sinclair, Wyoming. U.S. Geological Survey photo by J. R. Balsley.

not on the same line. Up to a point, solid rocks are elastic, i.e., they regain their original shape and volume when an applied stress is released. Under greater stresses they are permanently deformed by folding or fracturing. Deep-seated rocks, especially in zones of dynamic metamorphism, lose rigidity, and flow into the often highly contorted structures characterizing plastic deformation.

Dip and Strike   Technical terminology, a bane to many, actually simplifies description for the initiated. Conversations among structural geologists are liberally sprinkled with "dip" and "strike," terms describing the attitude in space of a plane, such as the surface of a rock layer or fractures of various sorts (Fig. 8-2). *Dip* is the inclination of a plane or bed measured from the horizontal in a vertical

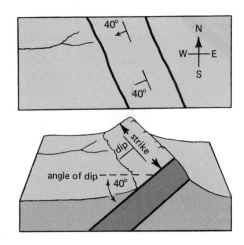

Fig. 8-2 Map and block diagram illustrating dip and strike.

plane. Put another way, it is the maximum tilt that can be measured on a sloping plane. Dips are recorded in degrees, so that a horizontal bed has a 0° reading, no dip at all; a vertical bed has a 90° dip; and sloping beds have various readings in between. *Strike* is the compass direction of a line formed by the intersection of a rock surface and a horizontal plane. It should correspond to a horizontal line drawn on the surface of a sloping bed, and it is always at 90° to the dip direction (Fig. 8-3).

Maps and Cross-Sections Folds, and other geologic structures as well, are usually illustrated by means of maps and cross-sections. Block diagrams are excellent for simple or generalized representation, but detailed and complex structural relations must of necessity be presented through two-dimensional drawings. A geologic map is essentially a scale drawing of the distribution of outcrops—a floor plan of the rocks. Most are made in the field by plotting *contacts*, the boundaries between different rock masses, and recording symbols for dips and

Fig. 8-3 The stream surface gives a reference plane for the dip (toward the left) of these strata on the South Branch of the Potomac River. U.S. Forest Service photo by T. C. Fearnow.

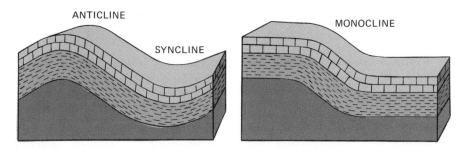

Fig. 8-4 Basic fold types.

Fig. 8-5 Sheep Mountain in north central Wyoming, an anticlinal fold cut through by the Bighorn River. Courtesy of the Jersey Production Research Company.

strikes and other pertinent data on a base sheet. The map should also include roads, property lines, streams, and other features that help in determining location. As outcrops are often patchy, being covered by vegetation, soil, and other surficial materials, many bedrock relations shown on a map are inferred, i.e., informed guesses. Cross-sections, representing vertical slices, are drawn along selected lines across the map to show the visualized relations of rock at depth. This mild digression seems essential because any worthwhile understanding of geologic structures involves an exercise in solid geometry, requiring three-dimensional thinking.

Folds

Geometry Folding warps rocks into upfolds called *anticlines* or downfolds called *synclines*. These two basic fold types are usually adjacent to each other, like wrinkles in a tablecloth. The *monocline*, a different type of flexure characterizing some regions, can be described as half a fold. It develops where near-horizontal rocks locally dip more steeply and then flatten out again. In size, folds range from minute crinkles, 3 centimeters or less across,

to structures forming mountain ranges (Figs. 8-4, 8-5).

The diversity of folds in nature is most economically described by referring to their individual geometric parts (Fig. 8-6). The *axial plane* is visualized as dividing a fold into halves that are approximate mirror images. The *fold axis* is an imaginary line formed by the intersection of the axial plane with the curved surface of the fold. The flanks of a fold on either side of the axial plane are called its *limbs*. Thus in an anticline the dips diverge from the axial plane, and in a syncline they converge towards the plane.

The tightness of folds varies greatly. In *open folds* the limbs have a gentle to moderate dip (Fig. 8-7). *Closed folds*, resulting from more intense deformation, have steeper limbs dipping more than 45° in relation to the axial plane. Tightly compressed folds with parallel limbs, common in dynamically metamorphosed rocks, are called *isoclinal*. The attitude of folds can be described by reference to their axial planes (Fig. 8-8). *Upright folds* have vertical axial planes and opposing limbs of approximately equal dips. *Asymmetric folds* have one limb whose dip is notably steeper than the other (Fig. 8-9). In *overturned folds*, the axial plane is tilted, and one limb turns under the other. The axial planes of *recumbent folds* are horizontal; the

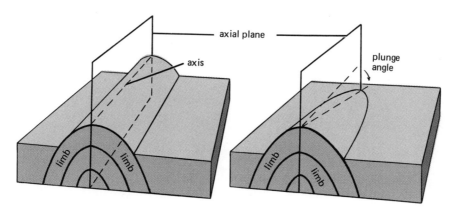

Fig. 8-6 Descriptive nomenclature of folds.

Fig. 8-7 Open and upright fold. Brown's Mountain anticline, "The Devil's Backbone," West Virginia. U.S. Forest Service photo by L. J. Prater.

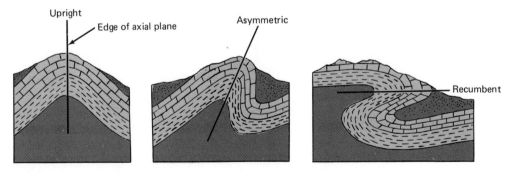

Fig. 8-8 Cross-sections showing attitudes of folds.

Fig. 8-9 Asymmetric fold, Wind River Mountains, Wyoming. Photo by D. Kisling, courtesy of R. B. Parker.

Fig. 8-10 Recumbent fold in Rheem's Quarry, Pennsylvania. Courtesy of R. B. Parker.

fold lies on its side, and one limb is above the other (Fig. 8-10).

Outcrop Patterns The surface expression of folds and the outcrop pattern shown on a map are controlled by the geometry of the deformed beds and also the ground surface eroded across them. The interpretation of geologic structures is predicated on Steno's Laws of Original Horizontality and Superposition. If strata are folded, and the topmost bed is not breached by erosion, it will extend across the area in anticlinal hills and synclinal valleys.

Commonly, however, folds are eroded so that the ground surface is beveled across the structures, exposing a series of beds in banded outcrop patterns. An anticline produces parallel outcrops having the oldest bed in the middle, flanked by progressively younger corresponding beds on either side (Fig. 8-11). An eroded syncline has a similar pattern except that the order is reversed, so that the axis in the center follows the youngest bed, which is flanked by progressively older beds on either side (Fig. 8-12).

However, folds do not extend indefinitely. Folded zones may die out where dips decrease, in regions where the deformation was less. Within

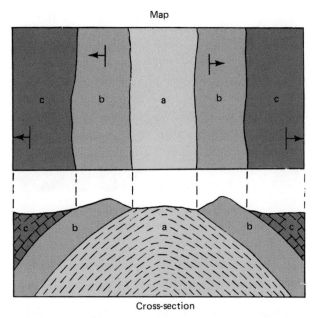

Fig. 8-11 An eroded anticline in map and cross-section. Bed (a) is oldest, bed (b) is younger, (c) is youngest.

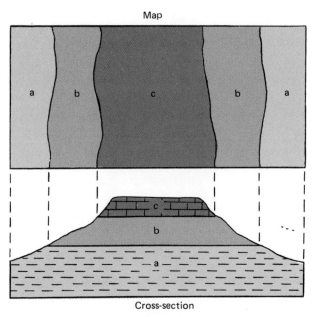

Fig. 8-13 Map pattern and cross-section of deep valley eroded through horizontal strata.

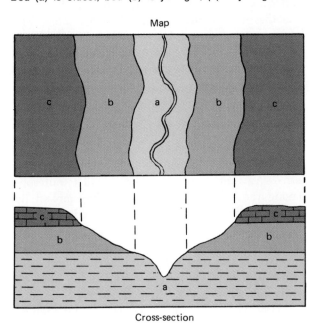

Fig. 8-12 Eroded syncline. Oldest exposed layer is (a).

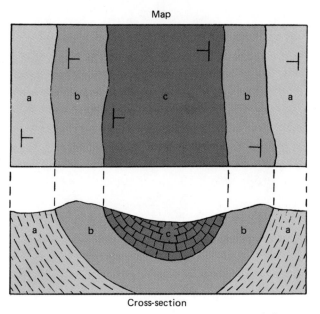

Fig. 8-14 The youngest layer (c) reflects an eroded mesa or similar ridge.

folded regions, the individual folds *plunge* where a whole rock sequence goes deeper into the ground. The outcrop pattern of a plunging fold wraps around the axis producing a *nose*. Where many adjacent folds rise and plunge along their axes, a zig-zag pattern may result, somewhat like the wrinkles in a rumpled tablecloth.

Banded outcrop patterns, roughly resembling those of anticlinal and synclinal structures in dipping beds, may develop where nearly flat-lying beds are deeply eroded. A valley cut through several horizontal beds creates a map pattern suggestive of an anticline, since progressively older beds are exposed towards the center (Fig. 8-13). A synclinelike pattern can develop on a hill whose sides expose several flat-lying beds (Fig. 8-14). In general, the basic map relations provide the bases for unraveling structural areas where the individual elements are interwoven in complex overall patterns.

Faults

Faults and folds are not mutually exclusive results of deformation, as they often occur together and may grade into each other. A fault is a fracture along which rock masses have been displaced. Along many faults, great displacements measuring thousands of meters, or even many kilometers, are certainly not the result of a single colossal slippage. Instead, such displacements are the sum total of many small slippages over a great length of time. This pattern of movement characterizes active faults. Their periodic slippages are at most 10 meters or so (about 33 feet), and usually much less, as is clearly indicated by the vertical scarps or horizontal offsets appearing during the attendant earthquakes. Moreover, most faults are now inactive. For example, the southern Appalachians, highly faulted mountains, have few earthquakes and lack any topographic or other evidence of recent movement (Fig. 8-15). Not all faults show great displacement; some grade into joints, fractures in rock having no displacement (Figs. 8-16, 8-17). And not all faults are distinct planes; in some, the movement is distributed through a fractured zone.

Nomenclature The terms *upthrown* and *downthrown* apply to the vertical movements of opposing blocks along a fault. The terms are relative, as both blocks may have actually moved downward, closer to the Earth's center, with one side getting ahead of the other; or both may have risen, but at slightly different rates. A fault surface exposed on the face of an upthrown block forms a cliff known as the *fault scarp. Key* or *marker* beds are parts of a distinctive unit whose separation indicates the amount of fault displacement. *Hanging wall* refers to the face of a fault block that would form an overhang if the blocks were somehow pulled apart. The *footwall* is the opposing block's surface that forms a complementary slope. They are old mining terms, originating because tunnels driven through mineralized zones of faults exposed these surfaces: the hanging wall sloped overhead; the footwall was underfoot. Although scarps characterize recent faults, erosion often destroys them on inactive ones so that the ground surface extends uninterrupted across both fault blocks. With these terms in mind, an economical description of fault classification is possible (Fig. 8-18).

Classification Genetic classifications are ultimately the most desirable in geology, and faults can be classed genetically according to the tensional, compressional, or other forces that created them. Unfortunately, such classification is not always possible because the forces producing a given fault are often unknown or controversial. Geometric classifications, being empirical, are the most generally useful and applicable. In these, relative movement, the attitude of the fault plane, and other observable features are used.

Fig. 8-15 Exposed fault in Johnson County, Tennessee. Crumbling of rocks on fault zone produced gulley and talus in center. Beds on the right dip steeply right parallel to the fault. Beds on the left dip left, except close to the fault where they are bent, or dragged, near vertical because the rock mass on the right moved up. Courtesy of U.S. Geological Survey.

Fig. 8-16 Normal fault of slight displacement. Oak Creek Canyon, Arizona. Photograph by the author.

Faults are arbitrarily classed as *high angle* if their fault plane dips 45° or more, and *low angle* if the dip is less. Using the direction of movement along the fault plane, three basic classes of faults are recognized. In *strike-slip faults,* like the San Andreas rift, the movement is largely horizontal, in the direction of the fault plane's strike; as a result no scarp is formed, although a valley may be eroded into the crushed fault zone. *Dip-slip faults* have dominantly vertical motion, in the direction of dip of the fault plane, so that a scarp is likely on active faults. The movement on some faults com-

Fig. 8-17 Well-developed joints in sandstone hogback at Muddy Gap, Wyoming. Photo by Herb Pownall.

bines both strike and dip-slip components about equally; hence they are called *oblique-slip* faults (Fig. 8-19).

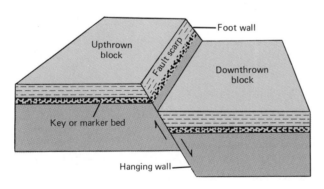

Fig. 8-18 Fault nomenclature.

Three names, commonly used in the field, are applied to varieties of dip-slip faults. In *normal faults,* a common result of tension, the footwall is upthrown relative to the hanging wall (Fig. 8-20). The name "normal" originated in early European mining districts where such faults were the most common. However, in many regions normal faults are not "normal"; the reverse is true. In *reverse faults,* usually a result of compression, the hanging wall is upthrown (Fig. 8-21). Recent reverse faulting might be expected to form an overhanging scarp; actually, such clifflets are almost never formed, because the rocks along joints collapse, producing a scarp resembling that of a normal fault. Because of the mechanics of rock fracturing, normal faults tend to be high angle. Reverse faults

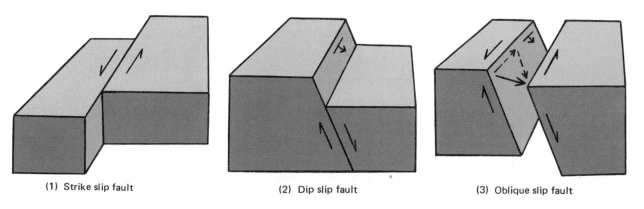

(1) Strike slip fault   (2) Dip slip fault   (3) Oblique slip fault

Fig. 8-19 Fault movements.

Fig. 8-20 Normal fault at Coyote Springs, Wyoming. Shale (upper right) is downthrown against sandstone. Slope (lower right) is shale talus. Road cut is 9 meters (30 feet) high. Photograph by the author.

Fig. 8-21 Reverse fault (normal faults shown in Figs. 8-17 and 8-19, part 2).

show no such preference, and the abundant low-angle reverse faults are commonly called *thrust faults* or simply *thrusts*. Associated thrusts and folds characterize many of the great contorted mountain ranges, which may be the result of major compression with great crustal shortening or, in some cases at least, the product of gravitational sliding.

Outcrop Patterns In eroded folds, the same beds appear at the ground surface in several different places, but they are always adjacent to beds de-

posited immediately before or after them and are of approximately normal thickness. Faults, in contrast, generally play havoc with the sequence of beds, may produce an apparent thinning or thickening of individual beds, and often abruptly truncate the outcrop of a whole sequence of beds. Although there are exceptions, relating to the dip of strata and the character of the surface topography, certain outcrop patterns are characteristic. Thrust or reverse faults commonly cause *omission of beds*, in map view, because the upthrown hang-

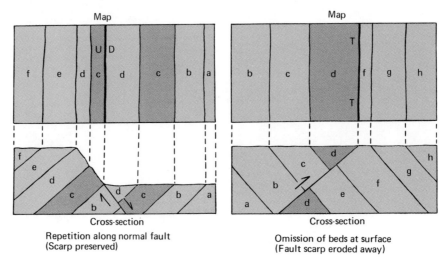

Fig. 8-22 Repetition and omission.

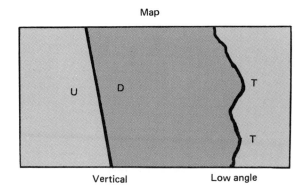

Fig. 8-23 Fault traces.

ing wall jams older rocks over younger ones that, as a result, do not appear at the surface. *Repetition of beds* characterizes many normal faults (Fig. 8-22).

The line a fault makes across the ground is called its *trace*. The sinuosity of a fault trace, whether it is straight or crooked, is controlled by the steepness of the fault plane where it emerges at the surface and the nature of the topography. A vertical fault plane has a straight trace regardless of the dissection of the land surface. A low-angle fault, across any but completely flat ground, is sinuous; the lower a fault plane dips, and the rougher the topography, the more irregular is the trace (Fig. 8-23). This elementary discussion may

suggest that the geometry of folds and faults can become a complicated matter, especially in contorted mountain ranges. Its study, however, gives clues to the nature of the forces deforming the crust, as well as to more practical things such as the search for oil.

Unconformities

James Hutton first comprehended the historical significance of those structural features called unconformities. To his creative mind, they conjured up "vestiges of lost worlds," a dramatic allusion rooted in facts. The appearance of an unconformity is deceptively simple. As seen in many roadcuts, it is soil or rock of one sort lying across different rocks—a buried erosion surface separating rocks of different ages (Fig. 8-24).

In types called *disconformities*, strata above and below the erosion surface are essentially parallel. Their history is largely one of deposition interrupted by an extended period of nondeposition or erosion, a *hiatus* that produced the buried erosion surface. Because *angular unconformities* involve deformation as well as the processes creating the disconformity, the layers above and below the erosion surface are not parallel, which makes these features easier to recognize.

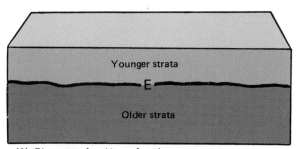

(1) Elements of an Unconformity

E = Buried surface of erosion or non-deposition

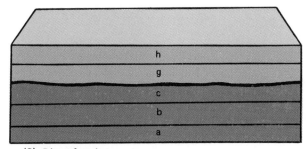

(2) Disconformity

Fig. 8-24 (1) Elements of an unconformity. (2) Disconformity.

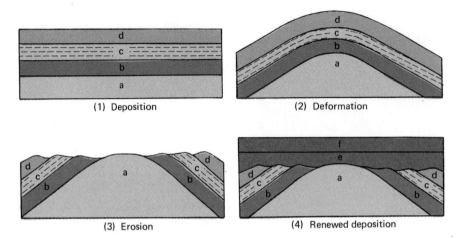

(1) Deposition   (2) Deformation

(3) Erosion   (4) Renewed deposition

Fig. 8-25 Origin of angular unconformity.

An angular unconformity indicates at least four major geologic events: (1) an initial period of deposition in which older strata are laid down near-horizontal and in order; (2) a subsequent period of folding and faulting that disturbs the then-existing beds; (3) an ensuing hiatus when the contorted beds are truncated; (4) finally, a period of renewed deposition that buries the erosion surface beneath the younger set of rocks (Fig. 8-25).

The relations in an angular unconformity have been important in dating past deformation where the beds involved can be dated by fossils or other means. The deformation must be later than the youngest layers it affects, and earlier than the first layers deposited thereafter. The precision with which the deformation can be bracketed in time depends on the closeness in age of the dated rocks. If the time interval is long, the deformation cannot be closely dated for it could have occurred at any time within the interval.

The "lost worlds" suggested by an angular unconformity may well involve shifting seas of long duration, the rise of mountains as impressive as the Himalayas, their slow destruction and burial under broad plains of their own debris, and finally a quiet reinvasion by the seas. Angular unconformities record, in outline form, the history of the Earth's great chains of deformational mountains (Fig. 8-26).

CONTINENTAL STRUCTURES

Individual folds, faults, and associated unconformities are countless, and often incredibly complex, details in grander structural patterns of a continental scale. Broadly viewed, the continental faces show two topographic elements: vast low interiors of flat or gently rolling land outside of which lie the high or mountainous regions. Within each element are differing geologic provinces whose arrangement is roughly comparable in all the continents (Fig. 8-27).

Stable Interiors

Shields At the heart of each continent is a *shield*, a broad rolling lowland across a largely crystalline basement complex. Its granites and metamorphic rocks belong to the oldest geologic eras,

Fig. 8-26 The angular unconformity of Siccar Point, Berwickshire, Scotland, that suggested "lost worlds" to Dr. Hutton. H. M. Geological Survey photograph; Crown copyright, reproduced by permission of the Controller of Her Majesty's Stationery Office, London.

Precambrian time, and are essentially the exposed sial whose generally contorted and metamorphosed nature strongly suggests the deeply eroded roots of extremely ancient mountain ranges.

Interior Plains The interiors continue as broad plains, underlain by a relatively thin veneer of nearly flat, little-disturbed, sedimentary rocks. These rocks bury extensions of the beveled shields, or, in some cases, similar, but younger, basement rocks.

Locally, the sediments may thicken considerably into broad basins where the basement has subsided, or they may thin over gently upwarped welts and domes. Such structures are so broad and gradual, however, that associated dips are virtually unnoticeable.

The higher parts of continents lying beyond the stable interiors form plateaus and mountains. Which is which, geologically at least, depends on their internal structure rather than topography.

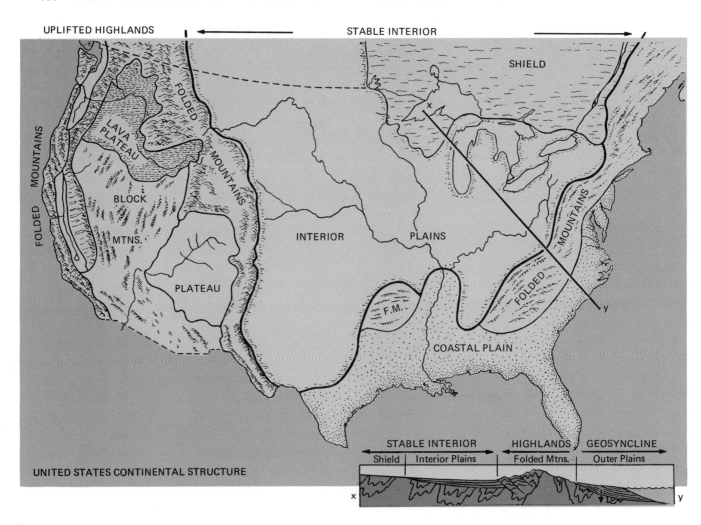

Fig. 8-27 The United States is part of a rather symmetrical continent having a stable interior flanked by highlands. The Atlantic coastal plain may represent a geosyncline.

Plateaus Structurally, plateaus are highlands characterized by the near-horizontal attitude of their sedimentary or volcanic layers (Fig. 8-28). Most do form tablelands, but when deeply eroded, plateaus have a distinctly mountainous appearance. Many plateaus of great elevation contain fossiliferous rocks deposited in the sea, most notably the vast central plateau of Tibet which lies some 4900 meters (about 16,000 feet) above sea level. Thus they indicate considerable crustal deformation of a remarkable sort; for aside from some high-angle faulting and monoclinal flexing, the uplift of plateaus results in no pronounced folding or crumpling.

Block Mountains Mountains, in general, are those elevated structures—if volcanos are excluded as

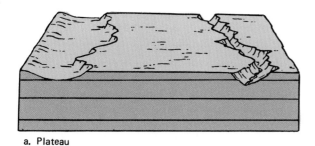

a. Plateau

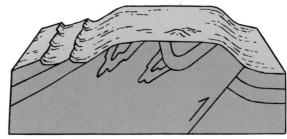

b. Mountains having a
plateau-like upland

Fig. 8-28 (a) Horizontal strata forming a plateau; (b) plateaulike upland eroded across deformed rocks of mountain structure.

being a special eruptive type—that are characterized by marked folding and faulting. Although a saw-tooth topography is typical, many mountains rise to a plateaulike upland, an erosion surface beveled across their contorted rocks. Structurally, mountains are of two distinctive types.

High-angle faulting may fracture the crust into angular mountains and valleys bounded by fault scarps (Fig. 8-29). The block mountains are called *horsts* if elongate; the valleys, especially long narrow ones, are called *grabens;* the many long uplifts in the Basin and Range country of the western United States are horsts. Death Valley, California, and the upper Rhine Valley are grabens, and so is the valley containing Palestine's Dead Sea (Fig. 8-30).

Active Margins

Folded Mountains Most attention has centered on the spectacular folded mountains, represented in the modern and imposing Alpine-Himalayan and Andean–North American Cordilleran chains and in older worn-down ranges such as the Appalachian in America and the Caledonian and Hercynian in Europe. The modern ranges follow continental margins in worldwide chains corresponding to the circum-Pacific and Alpine-Himalayan belts of earthquakes and volcanism. It seems possible that many of the older ranges, when they formed, were also marginal to the stable continental interiors.

Though called folded mountains, these structures also involve much reverse faulting, both high angle and thrust. Their folds range from open and upright types, to overturned, to recumbent; and in more deformed areas, especially those affected by metamorphism, isoclinal folding is common. *Nappes* are great slabs of complexly deformed rock, often in stacks, that are either great recumbent folds or extensive sheets of rock bounded by thrust faults of very low angle (Fig. 8-31). Nappes have been attributed to tremendous compression in which rock has behaved very plastically, and also to great, slow landslips from high and actively rising mountain masses. Whatever their origin, nappes give incredibly complicated structure to some folded ranges, notably the Swiss Alps.

Coastal Plains The Atlantic seaboard of the United States is a broad *coastal plain,* the emergent part of the continental shelf, that extends from New England to Florida, thence along the Gulf Coast into Mexico. Unlike plains in the continental interior, it is an outer plain lying, for much of its length, beyond the worn-down Appalachian folds. Moreover, seismic evidence indicates no thin veneer on shallow basement but, rather, tens of

Fig. 8-29 The Panamint Range, near Death Valley, California, is an eroded fault scarp. The mountain is a horst and the valley an alluvial-filled graben. Photo by John H. Maxson.

thousands of feet of sedimentary rock in what seems a gigantic, linear, crustal downwarp along the margin of the continental platform. Here, some geologists suggest, is a modern geosyncline filling with clastic sediments eroded from the periodically rejuvenated stumps of the Appalachians, and supplemented along the Gulf Coast by materials from the continental interior and limestones that have been deposited in warm seas.

ON DEEPER CAUSES

Overall, the continental framework exhibits two different styles of deformation. *Orogeny*, which means mountain-making, produces the contorted structure of folded mountains in the relatively narrow and elongate mobile belts. *Epeirogeny*, meaning continent-making, involves broad, gentle crustal warps that may be accompanied by block faulting

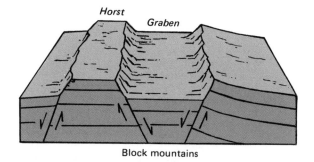

Horst
Graben

Block mountains

Fig. 8-30 Block mountains produced by high-angle faulting.

and associated volcanic eruptions. This deformation affects the more rigid continental platforms, warping shields and interior plains into welts and basins as well as cracking them along normal faults, which may create horsts and grabens. Plateaus, whose rocks are not contorted, show that the total amount of epeirogenic uplift can be great.

Epeirogeny usually affects folded mountains after the orogenic phase is complete. Commonly, folded mountains are broadly upwarped in their later history, and their folds and thrusts are offset along younger high-angle faults, frequently accompanied by volcanic eruptions. It seems as if the deformation and metamorphism resulting from orogeny consolidate the sedimentary rocks in mobile geosynclines, which thereafter react as rigid additions welded to the continental plates. Because

the contorted rocks of shields have the look of ancient mountain roots, and because new geosynclines, such as the one beneath the coastal plain, come into being along the margins of continental platforms (previously enlarged by orogeny), a grand proposal has been made—structurally, the continents have expanded into the ocean basins through geologic time.

Isostasy

Most considerations of mountains, or of continents and ocean basins, involve the concept of isostasy (from the Greek: *isos*, "equal"; *stasis*, "standing"). One might assume that the load of high mountains is supported by the strength and rigidity of the Earth's crustal rocks. However, Earth scientists do not hold this idea today. Rather, solid continental blocks and mountains within them are considered to be "floating" on the denser, plastic mantle beneath—in isostatic equilibrium. Geologic observations and geophysical measurements provide the evidence.

The discovery leading to the concept of isostasy was made in the 1850s. The British, under Sir George Everest (for whom the mountain was named), conducted a survey to locate reference points for the mapping of India. Precise latitudes were determined by triangulation, with plumb

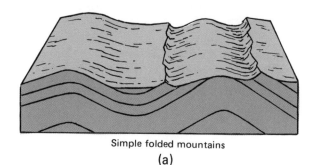

Simple folded mountains

(a)

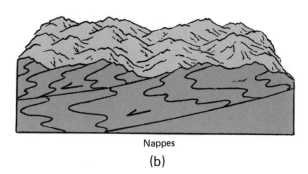

Nappes

(b)

Fig. 8-31 (a) Simple folded mountains. (b) Complex recumbent folds, of considerable size, forming nappes.

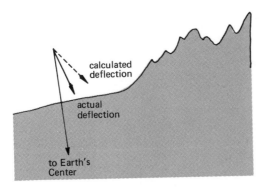

Fig. 8-32 Schematic diagram of plumb bob deflection (greatly exaggerated) caused by gravitational attraction of nearby mountain mass.

bobs used to establish the direction of the Earth's center (as in Eratosthenes's method). Some points located were in error when checked against independent astronomical observations.

J. H. Pratt, Archdeacon of Calcutta, assumed that plumb bobs caused error because they were gravitationally attracted by the mass of the nearby Himalaya Mountains. Estimating the volume of the mountains and knowing the approximate density of their rocks, Pratt calculated the expectable deflection of the plumb bob, assuming that the Earth's crustal rocks were of uniform density. The results showed that the actual deflection was only about one-third of the theoretically expectable deflection (Fig. 8-32). Pratt attributed the difference to less dense rocks in and beneath the mountains than in the adjacent plains. He further proposed that the rocks of the mountains were "floating" on denser material beneath. In Pratt's hypothesis, crustal blocks of the same weight and surface dimensions, but of different densities and thicknesses, had their bases at the same level within the Earth; however, the tops of the less dense blocks stood higher, forming mountains.

At the same time, in 1855, G. B. Airy, Astronomer Royal of England, made an alternate proposal that accounted equally well for the observed deflection of the plumb bob from the center of the Earth. In his view, all the floating crustal blocks had the same density. Assuming comparable lengths and widths, thicker blocks would be heavier, hence their bases would be at greater depths to achieve equilibrium while their tops would stand higher than thinner blocks. This "roots of mountains" hypothesis has been generally substantiated by twentieth-century seismic studies. They indicate thickest crust under high mountains, thinner crust beneath lower continental surfaces, and thinnest crust under ocean basins; however, Pratt's concept of different densities may well be involved in the contrast between lighter, higher-standing sialic rocks of continents and the denser simatic rocks of ocean basins (Fig. 8-33).

Isostasy is an important geologic concept. Vertical movements probably result where removal of weight by erosion of their tops allows lightened mountain masses to rise and come to isostatic equilibrium, and also where the weight of thickening deposits depresses parts of the crust, as in geosynclinal basins. Uplifted shorelines in the Great Lakes and Baltic regions probably represent isostatic uplift after the melting and removal of continental glaciers. In such cases, vertical movements of the crust are thought to be compensated for by slow flowage of the underlying plastic mantle.

However, isostasy is not an orogenic force capable of initiating folded mountain belts. Rather, it is a condition of equilibrium responsible for the higher elevation of continental masses in contrast to the denser simatic ocean basins. Where isostatic equilibrium is upset, as by erosion or deposition, epeirogenic movements may result. But the creation of folded mountain belts involves forces within the Earth.

Evidence for the New World View

In the era of normal scientific investigations following the geologic revolution of the late 1960s,

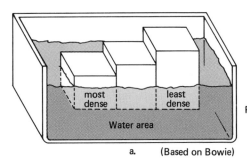

Pratt hypothesis: all blocks same surface
dimensions, same weight,
but different densities
(weight per unit volume).

a.   (Based on Bowie)

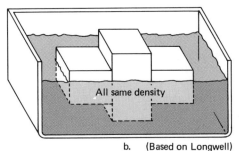

Airy hypothesis: all blocks same density,
same surface dimensions
but different depths.

b.   (Based on Longwell)

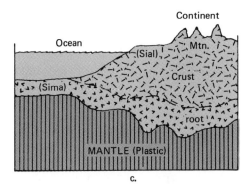

c.

Fig. 8-33 Isostasy models made by floating wooden blocks in water demonstrate the Pratt hypothesis in (a), and the Airy hypothesis in (b). A schematic diagram of crustal thicknesses is shown in (c).

theories for the origin of continents and ocean basins, mountains, and other crustal deformation are generally fitted to the unifying plate tectonic scheme. Folded mountains and associated volcanism are considered products of great compression between colliding global plates, squeezing and contorting geosynclinal sediments into deformed mountain structures. Block mountains, basaltic islands, and mid-oceanic ridges are attributed to tensional zones where the lithosphere is splitting with volcanic eruptions of material derived from the mantle. Although the plate tectonic theory has been described (Ch. 2), we now present some of the physical evidence—a different matter—that led to the near-universal geologic acceptance of the grand new view of the Earth.

Paleomagnetism   The Earth behaves like a great dipole (bar) magnet whose lines of force align a compass needle with the north and south magnetic poles. What causes the Earth's magnetic field is not surely known; perhaps its rotation affects the molten nickel-iron core to make the Earth a giant dynamo. In any case, paleomagnetic studies assume that certain minerals are magnetized by the Earth's existing magnetic field at the time the minerals are locked into rock. Both igneous and sedimentary rock can be used to determine the directions to the Earth's magnetic poles when they were formed. In basaltic lava flows, which preserve the strongest relict magnetism, magma first solidifies to rock, which continues to cool until at a certain temperature (called the Curie point) the minerals become magnetized parallel to the Earth's existing lines of force. In sedimentary rocks, the relict magnetism is far weaker (1/10,000) than in basalts; however, magnetic minerals settling through water do become aligned with the Earth's magnetic field before the rocks are consolidated. In both cases the magnetic minerals become minute compass needles showing their latitude (by the steepness of their dip) and the direction of magnetic north at the time they were incorporated in rock.

Since the Earth's magnetic field quite possibly results from the Earth's rotation, it is generally assumed that the magnetic and geographic (rotational) poles have never been far apart.

Paleomagnetic studies, notably by P. M. S. Blackett and K. Runcorn of England, suggest that the Earth's magnetic poles have wandered through geologic time—their past positions were quite different from those of the present day. Furthermore, measurements from rocks on different continents give different paths of polar wandering. That is, rocks of comparable age on a single continent give the same locations for ancient poles; but the different continents yield different pole locations for the same geologic times. The predicament of multiple north and south magnetic poles can be avoided if continental drift is assumed. By theoretically sliding the continents into different positions from those they have today, the mineral compass needles on different landmasses can be reoriented to point to single north and south poles in the past (Fig. 8-34).

Not everyone accepted this paleomagnetic evidence because of the necessary assumption that magnetic and geographic poles have always been close and because the technique is difficult and may have flaws. Physicists were unimpressed, for they argue the Earth is a gigantic gyroscope that strongly resists any change in its rotational axis. But perhaps the lithosphere slips as a unit over the stable gyroscopic interior. Thus one could accept or reject the paleomagnetic evidence; but if it is valid, the permanency of continents and ocean basins is denied.

Reversing Poles and Sea-Floor Bands   Still later paleomagnetic work showed that the north and south magnetic poles seem to have periodically reversed themselves during geologic time. The cause of the reversals, which probably relates to the origin of the Earth's magnetic fields, is not yet understood, but magnetic minerals in old lava flows are sometimes normal (like those of today), and

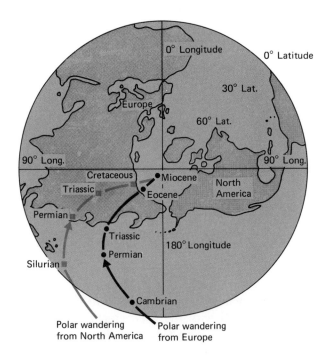

Fig. 8-34 Locations of the Earth's former magnetic poles, as determined today from different continents, do not agree for past geologic time periods. If, however, present continents are "slid back" together, for example Europe and North America, the paths of pole migration come into coincidence. After K. Runcorn.

in other samples indicate a north magnetic pole in almost the opposite direction. Allan Cox, while with the U.S. Geological Survey, and others constructed a time scale for geomagnetic reversals based on potassium-argon dating of a large number of lava flows having either normal or reversed polarity of magnetic minerals. This time scale covers only the last four million years—as older samples cannot be adequately dated because of increasing errors with time that are inherent in the potassium-argon dating method. The geomagnetic time scale (Fig. 8-35) contains longer intervals, called *epochs*, of dominantly normal or dominantly reversed geomagnetic fields (named after scientists who made important contributions to understanding Earth magnetism). The longer intervals, in

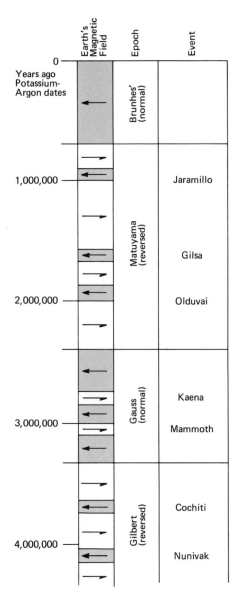

← normal field, as it is today.
→ reversed field, opposite to present-day.

Fig. 8-35 Geomagnetic time scale, based on Hoare, Condon, Cox, and Dalrymple.

turn, contain shorter-term reversals, called *events* (named from type localities). The scale had to be compiled from widely separated localities because the ideal situation—a single great exposure containing all epochs, one above the other—has not been found. Aside from providing a dated stratigraphic succession for lava flows on land, the scale provided the key for dating volcanic rocks of the ocean floors.

By the early 1960s, the global extent of the mid-oceanic ridge systems was becoming apparent; moreover, the ridges were revealed as seismically active zones with relatively high heat flow from the Earth's interior. Studies of the ocean floors in general added two more significant observations: sediments on much of the floor were either absent or much thinner than anticipated if deposition had persisted, even at a very slow rate, since Precambrian time; and—remarkably—no sediments older than the Jurassic had been found. In the early 1960s, H. H. Hess of Princeton suggested that the apparent youth of the ocean floors might mean that the floors were moving out from the major oceanic ridges where rising currents in the Earth's mantle brought up new material that was intruded in dike-like masses and then solidified to make new ocean bottom.

The idea, presented tentatively, soon received support. Vine and Matthews of Cambridge, England, suggested a test. If consolidated ridge material was being forced outwards by the intrusion of new material in the center, then the progressively older rocks on the ridge flanks should show paleomagnetic differences reflecting the normal or reversed magnetic field existing when the rocks solidified. The suggestion was soon confirmed when J. R. Heirtzler and others from Columbia University's Lamont Geological Observatory, using shipborne magnetometers, discovered parallel bands of magnetic differences along the Mid-Atlantic Ridge just south of Iceland (Fig. 8-36). Similar discoveries in other ridge areas confirmed the banded patterns, which were assumed to represent vertically

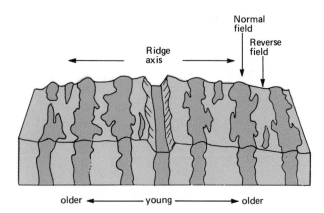

Fig. 8-36 Magnetic variations in oceanic ridges considered to be dikelike masses corresponding to the geomagnetic time scale.

intruded igneous masses; and also showed that the bands correlated with the geomagnetic time scale (determined from horizontal lava flows on land).

Sediments and Transform Faults Other new lines of evidence supported sea-floor spreading. Deep-sea cores taken various distances from the ridges show magnetic reversals in horizontal sediment layers. The closer the cores are taken to the ridge, the younger is the lowest sediment on basement rock —which confirms a spreading sea floor, as sediment sequences should be progressively younger towards the ridges.

A quite separate line of evidence involves the transform faults. Recent seismic studies show that earthquakes on the ocean bottom occur mainly along those parts of the *transform faults* between the offset ends of oceanic ridges, and that all earthquake foci are shallow. The shallow centers of origin suggest that rigid rock, whose slippage causes earthquakes, forms only a thin lithosphere in the ridge zones. The first movements of crustal blocks along the faults,·which can be determined from seismic records, are just the opposite of the movement on strike-slip faults—that is, the rocks always move away from the ridge axes. Thus along

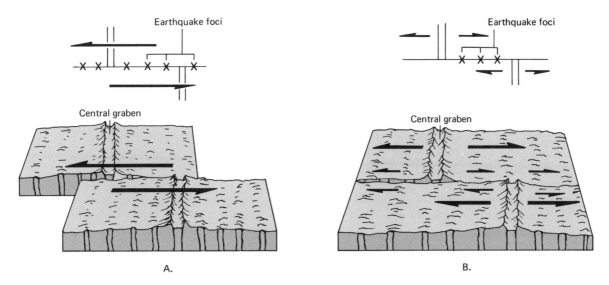

Fig. 8-37 Transform faults in mid-oceanic ridges. If the faults that were offsetting the mid-oceanic ridges were strike-slip faults, blocks would be moving in opposite directions, as shown in A. The preferred explanation, shown in B, is that the ridges are transform faults resulting from spreading from the ridge axes. Thus opposite movement (and earthquake foci) occurs only between the offset ends.

the transform faults, the rocks slip in opposite directions along the fracture between the offset ridge ends; however, beyond the offset ends, rocks on both sides of the fracture would move in the same outward direction, thereby minimizing any slippage that causes earthquakes. This situation is an expectable consequence of sea-floor spreading (Fig. 8-37).

The Geomagnetic Time Scale The rates of sea-floor spreading can be calculated for the four million years of the dated geomagnetic time scale. In different areas the rates range from about one and one-half centimeters (half an inch) to about six centimeters (two inches) a year. Though this might seem small by ordinary standards, it is a very appreciable amount in geologic time—about equal to the rate of displacement on the San Andreas fault. An extension of the time scale far beyond the radiometrically dated interval has been proposed. By assuming that the spreading rate during earlier

geologic times is the same as that of the last four million years (which is far from certain), a projected time scale has been established for over 170 field reversals that have been determined. The extended scale goes back over 70 million years. It fits rather well with Wegener's and Du Toit's earlier ideas on continental drift, wherein the breakup of Gondwanaland starting in late Paleozoic to mid-Mesozoic time produced the ever-widening south Atlantic Ocean basin and the movement of Australia away from Antarctica produced the Indian Ocean during the Cenozoic.

Subduction Zones Thus the evidence seems good for spreading away from the mid-oceanic ridges. But if new crust is continually growing in the ridges, where does the oldest crust go? Calculations show that shortening of the crust by crumpling in folded mountain ranges does not nearly balance new growth in the ridges. The suggestion that the Earth is simply expanding was considered and soon rejected as an explanation. The key to the disposal of excess lithosphere seems to be in the region of the deep oceanic trenches—for instance, off the Chile-Peru coasts and the Mariana Islands—where shallow to intermediate to deep earthquake foci slope away from ocean centers down under the trenches. Here, the deep foci long presented a problem because they are in the asthenosphere, whose plastic rocks should lack the rigidity to generate seismic waves by slippage. The answer was suggested by Oliver and Isacks, of the Lamont Geological Observatory, from work in the Pacific near the Tonga Trench, south of Samoa. Their observations suggest that a slab of lithosphere descends deep into the mantle beneath the trench. Such solid material moving downward could cause the deep earthquake centers there, and probably the same occurs near other oceanic trenches as well. Though not proven, it is now assumed that the downthrust lithosphere is reabsorbed in the plastic or partially molten asthenosphere. The belts of active volcanos, such as the "Circle of Fire" rimming the Pacific and branching into Indonesia, seem related to the deeply descending slabs of lithosphere. Sedimentary materials are possibly dragged downward, at such places as the west coast of South America, and become fused and mixed with basaltic materials to produce the andesitic magma that erupts in the world's great strato-volcanos.

Thus sea-floor spreading may involve a grand geologic cycle, wherein new lithosphere forms by injection of mantle material into the mid-oceanic ridges, is carried outward in solid slabs riding on the plastic asthenosphere, and finally descends under trenches to be reabsorbed in the plastic or partially molten mantle.

In plate tectonic theory, the American and Eurasian plates are visualized as slabs moving out from the Mid-Atlantic Ridge; the northern projection of the Antarctic plate and the Pacific plate move out from the East Pacific rise. Two plates meeting in oceanic regions create such features as the Tonga and Mariana trenches and their associated island arcs. If an ocean-floor plate thrusts under the margin of a continent, as along the west coast of South America, marginal mountains, such as the Andes, may result parallel to offshore trenches. Where the Indian subcontinent thrusts under the Asian mass, two continent-bearing plates seem to come together to create the exceptionally high Himalaya Mountains and the plateau of Tibet (Fig. 8-38).

Unanswered Questions

The evidence for the plate tectonic paradigm is convincing; however, the grand scheme represents human knowledge, not Divine Revelation. Thus future investigations will almost certainly lead to new interpretations of parts of the scheme—which still presents some unsolved problems.

The Driving Mechanism   Just how the Earth's internal heat operates the plate tectonic machine is unknown. Most hypotheses, however, stress plumes of mantle material rising into the mid-oceanic ridges. It has been suggested that injection of magma into the ridge centers pushes the global plates outwards. In another interpretation, the plumes are thought to bulge the ridges upwards, causing the plates to glide down and outwards across the unstable asthenosphere under the force

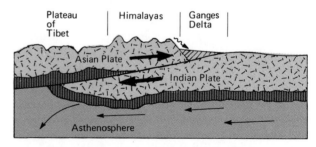

Fig. 8-38 Meeting of two continental plates to produce the exceptionally high Himalayas and the Tibet Plateau. After Argand, 1924.

of gravity. In yet another interpretation, the plates are envisioned as pulled away from the oceanic ridges by the weight of the leading plate edges, which are descending into the asthenosphere. In such a hypothesis, involving gravity, tensional splitting along spreading axes allows basaltic magma to erupt and form new crust (Fig. 8-39).

Convection currents are considered the driving mechanism in several hypotheses. In the late 1920s, decades before the plate tectonic revolution, Arthur Holmes of Great Britain defended continental drift. He suggested that convection cells, rising from the depths of the mantle, spread laterally in opposite directions along the bottom of the oceanic crust, whose "stretching" widened the ocean basins and rafted the continental blocks outward. He also suggested that crust was absorbed in the mantle where two cells met. Here, a phase change converted crustal basalt into denser eclogite, which sank into the mantle. Although generally rejected then, his concept has a distinctly modern look— he was a man ahead of his time.

Early in the plate tectonic era, during the late 1960s, the asthenosphere was proposed as the site of convection cells which provided the energy to drive the global plates. Since cells rising under oceanic ridges and descending at subduction zones would be extraordinarily flat if limited to the thickness of the asthenosphere, the idea is now seriously questioned. Today, convection cells rising from deep in the mantle are again in vogue. One such concept, resembling the older Holmes and Meinesz hypothesis (p. 35), assumes the cells spread beneath the lithosphere and drag the plates outward until plunging at subduction zones. Another version assumes that the spreading sea-floor plates themselves are the solidified tops of the cells, which descend in subduction zones where they are reincorporated in the mantle. Both schemes require that plastic materials rising from the depths somehow pass through the phase-change zones assumed to exist in the deeper mantle.

Many of the present schemes would delight Aristotle and the medieval scholars because the hypotheses are necessarily highly deductive, even though reinforced by mathematical calculations and the latest chemical and physical theory. Yet the clash of concepts gives direction to the search for explanations by pointing up the need for new and better data to test the hypotheses.

Benioff Zones, Asthenosphere, and Deeper Mantle One of the principal lines of evidence for subducting sea-floor plates is the presence of Benioff zones (p. 134) marked by progressively deeper earthquake foci. The origin of shallow earthquakes, where plates of rigid lithosphere slide past each other, can be explained by elastic rebound or dilatancy mechanisms. Quakes originating at intermediate depths, however, occur where sea-floor plates descend into the asthenosphere. The asthenosphere is considered a weak and partially molten zone where some melting of plates takes place; but earthquake generation seems to require the slippage between rigid rock masses. Perhaps the asthenosphere is viscous enough to cause slippage within a descending plate, but some other seismic mechanism may be operating here. Deep earthquake foci along Benioff zones occur at depths exceeding 650 kilometers (about 400 miles), far below the 250-kilometer (155-mile) deep bottom of the asthenosphere and well inside the rigid deeper mantle. Thus subducting plates are not completely absorbed in the asthenosphere. How then do leading edges of partially melted plates plunge into, deform, and thrust aside materials in the rigid deeper mantle?

Vertical Movements of the Crust Along the margins of continents, the world's great folded mountain belts and associated batholiths can be neatly fitted to the plate tectonic paradigm. Here, colliding tectonic plates, acting like the jaws of a giant vise, crumple geosynclines. The geosynclinal strata, compressed into narrow belts, buckle upward into surface mountains and also downward to form deep

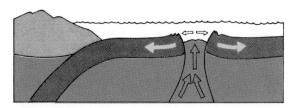

Plates pushed apart by
eruptions into ridge center

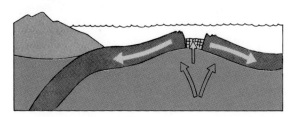

Plates slide off ridge

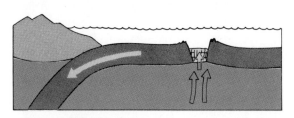

Descending edge of plate
pulls plates out

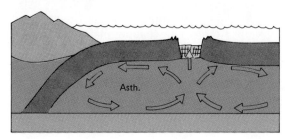

Asth.

Shallow convection
currents in asthenosphere

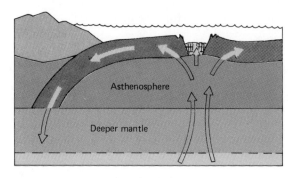

Asthenosphere

Deeper mantle

Deep convection currents
rising from mantle
traveling through plates

mountain roots. The light mountain roots, jammed into denser mantle rock, might eventually buoyantly uplift mountain belts by the isostatic mechanism.

Major vertical movements uplifting mountains to great heights could reflect the generation of magma in subduction zones. If rock deep in crumpled geosynclines melts to form magma, the resulting expansion might somewhat bow up the overlying crust. Perhaps more important, the buoyancy of vast magma bodies rising into the crust to form batholiths could cause major vertical uplifts and also plastic deformation in surrounding deep-seated metamorphic rocks. Deep erosion into these rock assemblages would expose the magmatic mountain belts.

Although opinions differ, the vertical rise of great mountainous welts might produce adjacent belts of nappes and low-angle thrust faults. Through the mechanism of gravity tectonics, near-surface sedimentary slabs containing contorted rocks may have slid down from the crests of high uplifts onto adjacent continental margins (Fig. 8-40). Most geologists admit this possibility where a jumbled mass like the Pre-Alps near Lake Geneva in Switzerland lies near a high, bare core of uplifted crystalline basement rock. However, many moderately folded ranges, where the crystalline basement is not exposed, show no loss of rock or gaping cracks, expectable if rock pulls away from the crest of a mountainous uplift. In such cases, compressional squeezing or some other primary mechanism seems necessary.

Unstable Parts of Continental Plates A major question—as yet unanswered in plate tectonic theorizing—involves the uplifts and basins that develop on continental platforms. The Rocky Mountains from southern Montana through Wyoming

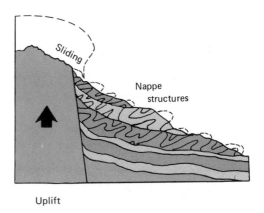

Fig. 8-40 Vertical uplift and downsliding mechanism.

and Colorado and into northern New Mexico are mainly elongate, broad uplifts having cores of ancient (Precambrian) continental basement rock. They are broad anticlines, in places bounded by high-angle reverse faults, and in other places flanked by sharply upturned sedimentary strata. The strata have been stripped by erosion from the high-standing mountain cores. Many peaks carved from uplifted crystalline basement rocks range from about 3860 to 4270 meters (roughly 13,000 to 14,000 feet) above sea level. On the other hand, broad basin floors, a kilometer more or less above sea level, which separate the crystalline-cored mountains contain thick sedimentary rock sequences resting on the same crystalline basement rocks that have locally subsided 6000 to 9100 meters (roughly 20,000 to 30,000 feet) below sea level (Fig. 8-41). The Black Hills of South Dakota form a dome-shaped outlier of the nearby higher Rockies. In the Colorado Plateau, large flat-topped uplifts bounded by monoclines and high-angle faults are structures comparable to those in the adjacent

Fig. 8-39 Various hypotheses have been proposed to explain the forces that drive the plate tectonic mechanism. Since none of them are universally accepted, they pose worthwhile problems, and give a direction for ongoing and future scientific investigations (facing page).

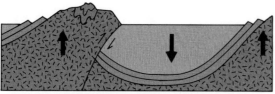

Rocky Mountain uplift and basin

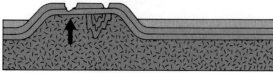

Colorado Plateau uplift

Fig. 8-41 Mountains, uplifts, and basins developed on thick granitic continental platforms seem to involve dominantly vertical movements of the crust. They have not been adequately explained by the plate tectonic mechanism.

Rocky Mountains, except that erosion has rarely stripped their sedimentary-rock covers from the underlying basement of ancient crystalline rocks.

What causes mountains in regions underlain by thick, sialic, continental platforms? The standard plate tectonic explanations of mountain-building—involving subduction zones with collapsing geosynclines and batholithic instrusions—seem unrelated to mountains originating far from boundaries between colliding global plates. Geophysical records indicate a thickening of continental crust beneath the southern Rocky Mountains and Colorado Plateau, keeping these uplifts in isostatic balance. But the origin of the roots is unexplained.

It has been suggested that material swept in by convection solidifies beneath the crust to form light mountain roots that buoy up the structures—an interesting speculation, lacking any evidence. A reasonable mechanism, proposed by George C. Kennedy of U.C.L.A. in the 1950s (and largely forgotten during the excitement of the plate tectonic revolution), involves phase changes (p. 142). The fundamental assumption is that the Moho dis-

continuity at the base of the crust marks a phase change from solid basalt to solid eclogite. Mountain-building could begin when temperature increase at the base of the crust converts denser eclogite of the underlying mantle into basalt—thereby creating a light mountain root. The deepened buoyant root would cause the overlying crust to rise isostatically, creating a pronounced uplift

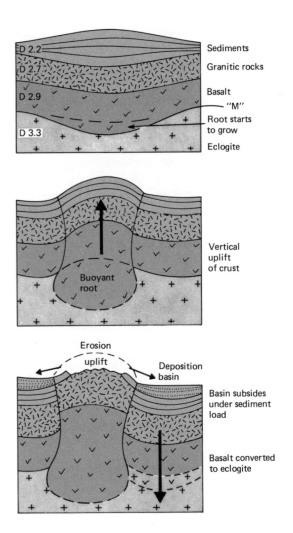

Fig. 8-42 Phase-change mechanism for mountain-building.

of the Earth's surface. Intervening basins could result where sediments eroded from the uplifts create a load that bows the crust downward. The load, in turn, increases pressure at the Moho beneath the downwarp. Here, the light crustal rock, basalt, is converted to eclogite, whose greater density and lesser volume causes further sinking (Fig. 8-42).

The plate tectonic paradigm is an excellent, new, unifying concept that is supported by several converging lines of evidence. Despite this, Dana's statement of 1855 still has elements of truth: "I am led by the conflicting views of the best authorities with regard to the conditions of the Earth's interior to hold very loosely to any theory of mountain making." Today it seems clear that future geologists will find no scarcity of interesting problems.

SUGGESTED READINGS

Dietz, R. S., "Geosynclines, Mountains, and Continent-Building," in *Continents Adrift,* readings from *Scientific American,* San Francisco, W. H. Freeman and Co., pp. 124–132, 1972.

Eardley, A. J., "The Cause of Mountain Building—an Enigma," in *American Scientist,* Vol. 45, No. 3, pp. 189–217, 1957.

Hills, E. S., *Elements of Structural Geology,* New York, John Wiley & Sons, 1963.

Holmes, A., Chapters 30 and 31 in *Principles of Physical Geology,* New York, Ronald Press Co., 1965.

Kennedy, G. C., "The Origin of Continents, Mountain Ranges, and Ocean Basins," in *American Scientist,* Vol. 47, No. 4, pp. 491–504, 1959.

Meyerhoff, A. A., and Meyerhoff, H. A., "The New Global Tectonics: Major Inconsistencies," *Bulletin of the American Association of Petroleum Geologists,* Vol. 56, pp. 269–336, 1972.

Press, F., and Siever, R., Chapter 22 in *Earth,* San Francisco, W. H. Freeman and Co., 1974.

Vine, F. J., "Sea-Floor Spreading—New Evidence," in *Journal of Geological Education,* Vol. 17, No. 1, pp. 6–16, 1969.

Wear and Tear

Antagonistic forces working on rock materials have given the Earth's crust its form and structure. The surface is constantly built up by inner forces, while, just as constantly, external forces attack it. Here and there construction or destruction dominate, but overall they are in balance—as highlands are torn down, somewhere else new mountains rise. Yet, if the seemingly substantial scenery is ever-changing, the forces acting on it are not. We now examine how the energy of gravity and the Sun working on air and water destroy the "eternal" hills.

The Bridge on Bridge Mountain, Zion National Park, Utah. National Park Service photo by E. T. Scoyen.

9
Decay and Collapse

lthough the process of rock decay is exceedingly slow in terms of a human lifetime, evidence of its progress is not wanting. An observation in an Edinburgh cemetery was made in 1880 by Sir Archibald Geike, who noted a marked crumbling of a marble tombstone erected in 1792, less than 90 years before. Cleopatra's Needle, a granite obelisk that had survived 3500 years with little change in Egypt's hot, dry climate, was brought to Central Park in New York in 1880. By 1950, some of its deep-cut hieroglyphics were obliterated by the 70 years of damp, cold winters and acid-forming city fumes. The Kamenetz fortress, built of local limestone slabs in the Ukraine in 1632, was abandoned about 1700. In 1930, V. V. Akimtzev reported that the slabs on top of a large tower had decomposed to form about 30 centimeters (a foot) of soil in the intervening 230 years.

WEATHERING

Weathering, an important factor in soil formation, is the decay of bedrock, in place, from exposure to the atmosphere. It initiates the surface phase of the rock cycle. Like metamorphism, its counterpart at depth, weathering reflects the adjustment of rocks exposed to new conditions. Minerals in rocks are in balance with the conditions where they originate; if exposed to a new environment, they slowly alter to different forms that are stable under the new conditions. In weathering, rocks and their minerals adjust to low pressures, low and fluctuating temperatures, and the waters that prevail at the Earth's surface. Thus plutonic rocks, the products of magmatic or metamorphic processes at depth, are generally most susceptible to chemical changes of weathering (Fig. 9-1).

Weathering is of two sorts that usually occur together, although one may overshadow the other in different climatic zones. *Mechanical weathering* disintegrates rocks into smaller pieces, as when quartzite breaks into smaller quartzite fragments. *Chemical weathering* is a decomposition, or rotting, of rock. New minerals, usually softer, form from the original ones; for example, feldspar alters to clay.

Fig. 9-1 A "mushroom" rock about a meter high caused by strong weathering on the lower part of a knob of coarse granite exposed in the Laramie Range, Wyoming. The pronounced undercutting develops where moisture—an aid to weathering—lingers around the base of the rock. Photo by Garrie Tufford.

Mechanical Weathering

Temperature Change (Without Water) Rock, like most solids, expands on heating and contracts on cooling. The volume changes are exceedingly small; however, theoretically at least, mechanical weathering can result from such changes in dry bedrock. The mechanism is effective under extreme heat, as in forest fires. Rock, being a poor conductor of heat, spalls off in curved slabs when the surface expands from heating, while at slight depth the rock remains cool and unchanged.

Moderate temperature changes, as between night and day or summer and winter, when repeated over many years, were once thought to be an important cause of the surface breakdown of granitic rocks into a gravelly litter. The different mineral particles in such equigranular rocks should expand and contract at different rates for several reasons, including their color. Dark minerals absorb heat and expand more than light-colored ones, which reflect heat. Thus, uneven stresses are set up, so that individual grains or crystals might be eventually freed, or "pop out," from the bedrock surface. Laboratory experiments, however, cast considerable doubt on this once-popular mechanism. Polished granite showed no change, under microscopic examination, after being heated and cooled in a dry oven from freezing to boiling temperatures enough times to equal several hundred years of natural exposure. When the experiment was repeated with the presence of moisture, however, marked disintegration occurred.

Frost Action The freeze and thaw of wet materials is a most important form of mechanical weathering in the colder climates. Broken water pipes, burst milk bottles, and fractured auto engine blocks during a cold spell are compelling evidence of disruption by frost action. A volume increase of about 9 percent accompanies the freezing of water into ice. Whether the disruptive force is simply a matter of volume increase or is caused by ice crystals whose growth exerts considerable force, provides an intriguing item for technical debate. In either case, water freezing in cracks and pockets disrupts the most solid rock. The jumble of frost-rived blocks, called *felsenmeer*, which mantles many mountains above the timberline, is a striking product of freeze and thaw.

Unloading Unloading fractures rock by pressure relief. Rocks deep in the Earth are slightly compressed by the static weight of overlying materials. When thousands of meters of overburden are eroded off, a slight expansion in the exposed rock may split off large slabs (Fig. 9-2). Evidence of this action occurs during granite quarrying. The removal of a large block is sometimes followed by a sharp shock which jostles heavy equipment and accompanies the sudden appearance of a horizontal fracture, creating a new slab under the quarry floor. Rock bursts from the walls of deeply drilled tunnels, and actual measurements of quarried blocks also demonstrate expansion from pressure relief. The resultant fractures, or sheeting joints, are roughly parallel to the ground. Unload-

ing gives a characteristic rounded appearance to the hills in many granitic and crystalline terrains. *Exfoliation domes*, like Stone Mountain in Georgia and the domes of California's Yosemite Valley region, are the most spectacular landforms from this process.

Organisms Charles Darwin suggested that incredible amounts of earth have passed through the alimentary tracts of worms—an interesting cause of rock diminution. Plants are also important agents of both mechanical and chemical weathering. Roots of the ordinary garden pea, grown experimentally between glass plates, create impressive forces equal to 15 or 20 kilograms per square centimeter (15 or 20 tons per square foot). Individual roots may seem trivially small, but the work of countless ones—including the sizeable roots of trees—probably causes much rock disintegration; moreover, some plants encourage chemical weathering by creating acid soil waters. Overall, mechanical weathering assists chemical processes by breaking rocks into progressively smaller particles, thereby greatly increasing the surface area available for chemical reaction.

Chemical Weathering

All outdoors is a chemical laboratory, albeit the reactions there are exceedingly slow. Water is essential for chemical weathering reactions. Most outcrops are periodically wet, and water, in addition to its well-known solvent action, is a very reactive chemical.[1] The chemical equations of weathering are usually exceedingly complex, but four broad groups of reactions can be observed.

Hydration Water may combine with compounds to form new compounds. The process is not a mere

---

1 We may not ordinarily think of it as a reactive chemical because our bodies, being largely water by volume, are not corroded by it.

Fig. 9-2 Curved slabs of granite, resulting from unloading and marked by weathered rills, have slid or crept into a jumble in the valley bottom in the Vedauwoo area of southeast Wyoming. Photo by the author.

soaking or absorption of water, as by a sponge, but rather a chemical change in which water enters into chemical reactions, producing new products. Hydration is important in chemical weathering, for it acts on the very abundant feldspar and ferromagnesian minerals whose decomposition results in clay.

Desilication   Various complex chemical reactions cause desilication, the removal of silica from silicate minerals. In the process, the original silicate minerals decompose, SiO_2 is leached, and the insoluble clays and iron oxides that form are left behind. Desilication produces the deeply weathered soils characterizing tropical rain forests.

Strangely, quartz, which is pure SiO_2, is little affected by desilication. The process is restricted to silicate minerals, such as ferromags and feldspars, whose silicon tetrahedrons are associated with a variety of other elements.

Carbonation   The reaction called carbonation yields soluble carbonates. It affects rocks whose minerals contain calcium, such as calcite and certain feldspars, and is most active in regions where groundwaters are slightly acid. Carbonic acid forms when carbon dioxide gas from the air and from plant activity in soils becomes dissolved in water. The acid then reacts with calcium-bearing minerals. Although limestone is already composed of calcite, which is a carbonate, this mineral is relatively insoluble in pure water. Groundwater, if weakly acid, reacts with calcite to form calcium bicarbonate, a different and far more soluble carbonate. Caves, sink holes, and many of the other solutional features that abound in humid regions underlain by limestone result from the corrosive effect of carbonation. Whereas limestone is subject to weathering and erosion in humid regions, in arid regions it is often a resistant ridge former. Where vegetation is sparse and acid waters are lacking, carbonation is at a minimum, and limestone is little affected by other types of chemical weathering.

Oxidation   The rusting of an iron pipe results from oxidation. Chemists would define it a bit more broadly, but, for our purposes, oxidation is simply described as the union of oxygen with other elements, a process which requires the presence of water. Oxidation in weathering is less important than the other methods, since iron is the only element of any abundance in rocks that is affected by this process. Oxidation produces hematite and limonite, minerals giving a dark red or brown color to many soils and rocks. Where locally forming large deposits, these two oxides have furnished much of the better grades of iron ore.

Weathering Products

Because chemical weathering creates both soluble and insoluble products, it initiates a sorting of the crustal elements. Sodium, potassium, and calcium become tied up in soluble mineral compounds, largely carbonates. Potassium, an important plant food, tends to be absorbed and held on the surfaces of clay particles; the others are constantly leached and carried off in waters flowing eventually to the sea. Here, much calcium, compounded with carbonate as calcite, is eventually deposited as chemical or organic limestones. The more soluble sodium tends to remain ionized in solution, adding to the "saltiness" of the oceans.

Waste Mantle   The insoluble weathering products form a part of the waste mantle, or regolith,[2] the unconsolidated material that blankets most bedrock. Except in artificial excavations, steep mountain slopes, rugged coasts, or glacially scoured terrains, outcrops of bedrock are often few and far between (Fig. 9-3). The waste mantle is of two kinds. *Transported mantle* is the debris dumped on bedrock by streams, glaciers, wind, and waves. Its fragments may be exotic (unlike the rock on which they lie) because they have been carried from sources some distance away.

2 The term "mantle" is frequently used, but since an internal zone of the earth is also called the mantle, the cumbersome terms "waste mantle" or "regolith" may avoid confusion.

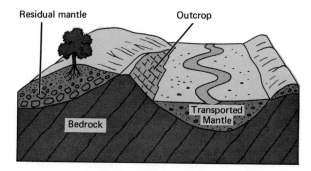

Fig. 9-3   Mantle and bedrock relations.

Weathered bedrock that remains in place forms the *residual mantle*. Its bulk is largely disintegrated bedrock fragments in deserts, arctic tundras, and above timberline where mechanical weathering dominates. In humid regions, where chemical weathering prevails, it consists mainly of insoluble weathering products. Chemically inert quartz commonly forms sand and silt-sized particles. Clay minerals, decomposition products of feldspars, and other silicates compose most of the finer fraction of waste mantle. Limonite and hematite are insoluble residues of iron-bearing bedrock's minerals.

The Soil   Whether weathering should be defined as a destructive process is a matter of opinion. In the gradation of the Earth's crust, a central geologic theme, it clearly destroys bedrock; yet from another point of view, weathering is constructional because it creates soil. Soil provides food for plants, plants are food for vegetarians, and vegetarian animals are food for meat eaters. So, fundamentally, land life, as we know it, could not have developed on bare rock or even unrefined waste mantle; it ultimately depends on chemical elements extracted from the soil.

To civil engineers concerned with excavations and the bearing strength of Earth materials, soil is any unconsolidated rock material—essentially waste mantle or regolith. But the soil as we shall consider it is not just weathered bedrock. It also contains decayed and living organic matter, moisture, and some gases that play important roles in making it a medium for plant life. Strictly speaking, soil is formed in place by further modification of waste mantle that results in a characteristic profile of three layerlike horizons, roughly parallel to the surface of the ground.

The uppermost, or *A*, horizon contains decayed plant particles forming a dark-colored humus (Fig. 9-4). This is a zone of leaching wherein seeping waters carry soluble substances and insoluble clays downward, at least in humid climates. Materials removed from above accumulate to make the

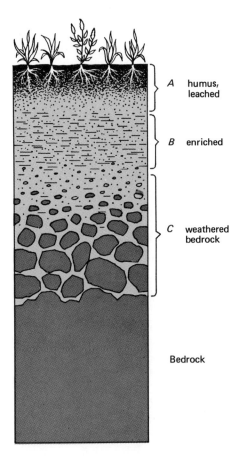

Fig. 9-4   Soil horizons as ideally developed in a cool humid region.

denser *B* horizon, usually enriched in clays. Together, the *A* and *B* horizons form the true soil. The *C* horizon lying directly on bedrock is waste mantle, or parent material, as yet little affected by soil-forming processes.

Aside from generally similar horizontal zonation, soils differ considerably in the development of their horizons and in their chemical, physical, and organic makeup. The variety in soils, as first suggested by the Russian Dokuchaiev in the late 1800s, results from the dynamic interplay of five main factors: climate, organisms, topography, parent material, and time. The *parent material,* waste mantle of the *C* horizon, was once thought to be all-important in controlling the kind of soil that develops. Bedrock is important in some conditions, and its weathered products form most of the bulk of soil; however, Russian soil scientists found that similar soils extend across different rock types in a given *climatic zone.* Thus the same soil, given enough time, tends to develop on a variety of igneous, sedimentary, or metamorphic rocks in the same climate. In contrast, on similar bedrock (granite, for example), the broadly different soils will develop that characterize deserts, tropical rain-forests, humid temperate regions and arctic tundras.

The *organic factor* includes such things as vegetation and the micro-organisms that live in the soil. Soils in the forests are usually more acid and leached than those in adjacent grass-covered areas. In tropical jungles, micro-organisms are so active that plant debris is destroyed before any significant humus, which is common in cooler temperate zones, can accumulate. *Topography* causes local variations in soils. Those on uplands tend to be thick and well drained. Soils on valley sides are generally thinner and poorly developed because less water seeps into the ground, and the greater surface runoff removes loose material. In valley bottoms, thick and poorly drained soils contrast with those on steeper slopes and rolling uplands.

The factor of *time* is most important in soil development. The time required to produce a mature soil, in equilibrium with surface conditions, ranges from a few hundred to thousands of years depending on differences of weathering rates in various climates and other factors. As to when waste mantle becomes soil, there is no general agreement. Some workers hold that fresh waste mantle is newborn soil, just beginning to develop. Others insist that material is not soil until weathering and other soil-forming processes have created noticeable horizons. In any case, there are immature soils whose horizons are in the process of de-

veloping, for until a soil is mature it is constantly changing. A mature soil is the end product of the slow adjustment of rock to the environmental conditions at the Earth's land surface.

Thus weathering is involved in many geologic phenomena. It initiates the chain of events that wears down the crust. For once bedrock is disintegrated and chemically decomposed into softer products, the active processes of erosion, transportation, and deposition are accelerated. These processes, in turn, expose fresh rock to atmospheric corrosion in an unending attack on the Earth's solid crust.

LANDSLIDES AND RELATED MOVEMENTS

Between the Canadian towns of Crow's Nest and Pincher Junction, the highway passes through a great jumble of rock blocks marked by this sign erected by the Province of Alberta:

FRANK SLIDE:
Disaster struck the town of Frank at 4:10 A.M., April 29, 1903 when a gigantic wedge of limestone 2100 feet high, 3000 feet wide and 500 feet thick crashed down from Turtle Mountain. Ninety million tons of rock swept over a mile of valley, destroying part of the town, taking 70 lives, and burying an entire mine plant and railway in approximately 100 seconds. The old town was located at the western edge of the slide where many cellars are still visible.

Gravity is the great leveler. Indirectly, it works through the erosive geologic agents, such as streams, glaciers, wind, and waves. Working directly, it creates mass movements whenever surface materials lose support beneath or become less rigid and are subject to flow. Mass movements include a multitude of phenomena, ranging from the fall of a minute grain weathered from a slowly rotting boulder to the sudden collapse of a mountain face (Fig. 9-5). The many kinds of mass movement, the landslides and related phenomena, were classified by C. F. S. Sharpe in 1938. The bases of this gen-

Fig. 9-5 Falling rocks still raise dust 10 days after the Montana quake of 1959. U.S. Geological Survey photo by J. R. Stacy.

erally accepted classification are: the type of movement, the rate of movement, the kind of material, and the water content involved.

The movement is classed as *slippage* if materials move on a definite shearing plane and the particles in the moving mass retain the same position relative to each other, as in a brick sliding down a board. In *flowage* movement, a distinct slip plane is lacking, and the particles within the moving mass shift in their relative positions, like tar spilled on a sloping roof. Slippage and flowage

require a free slope, such as a hillside or cliff. *Subsidence,* which can occur in flat land, is settling: a dominantly vertical sinking exemplified by a surface collapse over some abandoned mine workings (Fig. 9-6). The rate of mass movements grades from the rapidity of a free-falling boulder to movements so slow they are imperceptible. The moisture content, as water or ice, influences the type of mass movement, because some require saturation to a mud, and others can occur even if the material is perfectly dry. Also important is the type of material

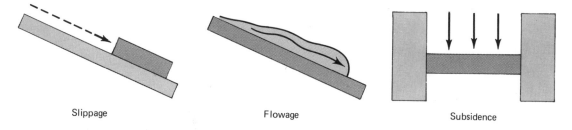

Slippage   Flowage   Subsidence

Fig. 9-6 Types of mass movement.

involved, whether bedrock, coarse rock debris, or finer waste mantle. Let us examine a few mass movements to see how the classification is set up, and something of the variety of phenomena caused by the direct action of gravity.

Flowage

Slow Phenomena Soil creep is a very slow downslope movement of rock waste. Although the actual movement is imperceptible, its effects are observable after a period of years (Fig. 9-7). Telephone poles, fence posts, and other originally vertical objects become tilted, and tombstones have toppled on hillsides undergoing creep. Trees are tilted like poles, but since trees tend to grow straight up they

develop curves in their trunks.[3] Retaining walls and house foundations may buckle and crack, and, in general, manmade structures may be damaged if the subtle evidences of creep go unrecognized and suitable provisions are not made. *Rock creep* affects bedrock, especially shales. Often where the beds are vertical, their eroded tops will slowly bend downhill under the constant pull of gravity.

Rapid Phenomena In contrast to the imperceptible action of creep, flowage movements are moderately rapid, or even precipitous. *Mudflow* is a rapid flowage of a streamlike mass of saturated waste mantle (Fig. 9-8). Mudflows are common on moderate to steep slopes having a sparse vegetational cover that ineffectively anchors the earthy materials. They are common in semiarid and volcanic regions where rain, which may be heavy at times, soaks into unconsolidated ash, silt, or clays which are rapidly converted to mud. In the Mediterranean climate of California, mudflows are frequent on canyon walls whose brushy vegetation has been destroyed by flash fires. Here, the occasional heavy rains on bared slopes churn up mud that descends to the valley bottoms where it concentrates into rapidly flowing streams of mud. These often come crashing out of canyon mouths, spreading the mud, broken trees, and even large rafted boulders over any farmland or city streets beyond.

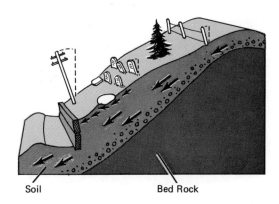

Soil   Bed Rock

Fig. 9-7 Soil creep. After C. F. S. Sharpe.

3 The same effect is sometimes caused by snow load crushing down young trees that later recover, so other things should be taken into account as evidence of creep.

Fig. 9-8 Mudflow emerging from canyon mouth onto a valley flat.

Mudflows are a frequent by-product of volcanic eruptions. The heated air and steam rising over an erupting volcano can generate thunderstorms whose heavy rain, pelting down through ash-filled air and onto ash-covered slopes, creates fluid masses of mud. In the famous 79 A.D. eruption of Vesuvius, the town of Herculaneum was buried under mudflows, while the nearby city of Pompeii was overwhelmed by hot clouds of falling ash.

Slippage

True landslides are the slippage phenomena, wherein the material may be relatively dry. Three

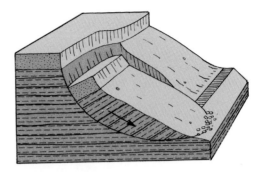

Fig. 9-9 Diagram of larger slump block showing curved slip plane and backward rotation of top.

distinct types of slippage are distinguished: slumping, sliding, and falling.

Slump If an intact block slips to a lower elevation along a curved surface, which usually causes backward rotation, the movement is classed as slump (Fig. 9-9). Slump blocks in bedrock are often very large; those in unconsolidated materials may be quite small. Where sandstone or other massive rock rests on weak rock, such as shale, the blocks may be several kilometers long and several hundred meters

Fig. 9-10 Terracettes, or "cattle walks," common on many steep hillsides.

Fig. 9-11 Point Firmin landslip. Slump along California coast. Courtesy of Spence Air Photos.

wide, as along the Echo Cliffs north of the Grand Canyon of Arizona. In the waste mantle on steep hillside slopes, small steplike terraces, referred to as "cattle walks," are multiple slump blocks (Fig. 9-10). Whether they result from the stamping of cattle is a debatable point because they are not normally located in the choicest grazing spots; it seems most likely that the steps originate first and any cow paths on them come later. Large-scale slumping in bedrock often accompanies basal excavation of a slope, either by natural agencies, such as streams and wave action along a rugged

coast, or by ill-advised engineering activities of man (Fig. 9-11).

Slides Although rivaled by some mudflows, *rock slides* are generally the most devastating mass movements. A rock slide is a rapid movement of bedrock that commences as a massive slab sliding down an inclined plane of weakness. Usually the slab breaks up into a churning jumble of large blocks. More recent than the Frank slide was the Rock Creek slide along the Madison River Gorge west of Yellowstone Park (Fig. 9-12). Here, on

Fig. 9-12 Slide in Madison River Gorge under which 19 people in the Rock Creek campground are believed to be buried. Quake Lake, foreground, resulted from the slide blocking the river. U.S. Geological Survey photo by J. R. Stacy.

August 17, 1959, 30,000,000 cubic meters (35,000,000 cubic yards) of rock, jarred loose by an earthquake, shot down the valley wall, crashed across a public campground, and climbed 120 meters (400 feet) up the steep opposite wall. Behind the resulting natural dam, a lake formed that was several hundred meters deep and extended several kilometers back into the gorge. The causes of slides are many. The Rock Creek slide occurred because of the geologic conditions when the earthquake struck. Schist and gneiss, weakened by weathering,

were supported by a strong mass of relatively unweathered dolomite.[4] When the shaking broke the dolomite mass, the mountain face slid off.

The Turtle Mountain slide at Frank, Alberta, is a textbook case of the conditions responsible for rock slides in general (Fig. 9-13). The setting was precarious. The offending mountain front was precipitous, steepened by the carving of an Ice Age

4 Similar to limestone, or marble, except that its carbonate minerals contain magnesium as well as calcium.

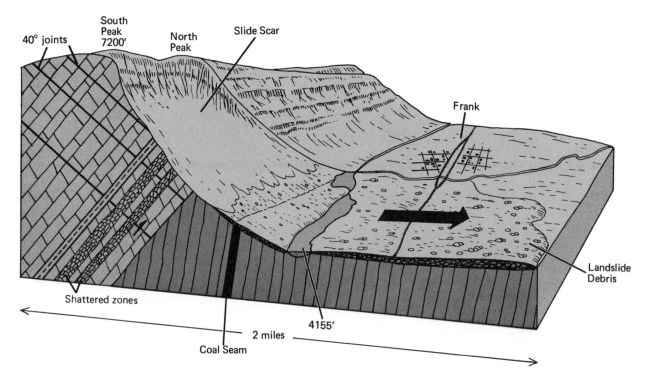

Fig. 9-13 Rockslide on Turtle Mountain that swept across Frank, Alberta, in 1903. After Canadian Geological Survey cross-section.

glacier. Structurally, the mountain was a large mass of heavy limestone that had been forced over weak shale along a fault millions of years before. Moreover, because an extensive set of cracks in the limestone mass, called joint planes, sloped more steeply toward the town than did the topography, the joints intersected the steep mountain face. Thus Turtle Mountain consisted of a series of inclined slabs, held only by friction, that were poised above the town beneath.

The actual sliding was triggered by a series of late events. An earthquake in 1901, two years earlier, could well have set it off, but instead the jarring merely loosened the whole mass. Just before the slide, frost had caused further loosening, and spring rains and melting snow had lubricated the surfaces of the limestone slabs. Finally, the col-

lapse of an abandoned mine shaft in the underlying shale administered the *coup de grace*. Thus, the Frank slide is classic because its multiple preconditioning and triggering causes neatly summarize the principal reasons, any one of which would be adequate, for most rock slides.

Fall   Perhaps it merits a separate designation (for a slippage plane is absent) but *rock fall* is classed as a type of slippage in which rock masses mix free-fall, bouncing, and sliding down cliffed faces. In the Swiss Alps in 1881, the collapse of the Plattenbergkopf, a mountain near the town of Elm, sent three great rock falls into the valley, only minutes apart, killing a score of people and destroying three houses and an inn. In Norway, fishing villages have been washed away by waves set

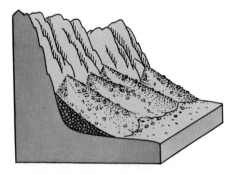

Fig. 9-14 Talus piles accumulating at cliff faces, mainly from rock fall.

up by the plunge of large rock masses into narrow steep-walled embayments that characterize its glaciated coast. Probably much rock fall is of isolated blocks loosened by weathering at sporadic intervals. Although hardly rivaling the spectacular falls and rock slides, it is quantitatively important, as measured by the aprons of fallen rock, called *talus*, at the foot of many barren cliffs. Earthquakes, however, may suddenly jar loose a multitude of blocks (Figs. 9-14, 9-15).

Subsidence

Loss of volume in material underground produces subsidence, a dominantly vertical settling movement. Although hardly rivaling the catastrophic mudflows and rock slides, the subtle nature of subsidence often delays the recognition of its serious consequences.

Caves and Mines In some cases, subsidence results from collapse into cavities where nature or man has opened up underground passages. In limestone regions, the ceilings of caves may give way and relatively rapidly produce steep-sided depressions, called *collapse sinks* (p. 268), in the land surface (Fig. 9-16). The collapse of abandoned

underground workings in coal-mining districts is a recurring problem. Many millions of dollars of damage to streets, water lines, and buildings and other structures have resulted where the workings are underneath cities, as in Wilkes-Barre and Scranton in Pennsylvania and Rock Springs in Wyoming (Fig. 9-17).

Fluid Removal Subsidence does not require the collapse of large underground cavities; an insidious form of settling results from pumping out large volumes of fluids from sediments or poorly consolidated sedimentary rocks. In this case, the removal of buoyant water from minute pores in saturated clays and silts causes the loss of underground volume. Although the resulting compaction is a slow process, the results are often impressive.

Most interesting is the opera house (Palacio de Bellas Artes) in Mexico City. Since it was built in 1904, this structure has settled some 3 meters (10 feet); on approaching the building one sees the first floor nearly at ground level. Many other buildings show interesting subsidence features in this city built largely on land reclaimed from a lake the Aztecs called Texcoco. Its clayey and sandy beds have lost great volumes of water to pumping. Houston, Texas, was built on flat land. In 15 years of intensive pumping of water for expanding industrial use a broad dish-shaped depression as much as a meter deep developed under the city, causing damage to buildings, pavements, and flood-control systems. Geologic study of the situation suggested that the removal of water from a layer beneath the city had also triggered movement on faults in the region.

Another fluid pumped out of the ground in large quantities is oil. That it can cause subsidence was recognized after a classic study, presented in 1949, of the highly productive Wilmington oil field on the coast of California. Here, the subsidence created a depression several kilometers wide and as deep as 9 meters (about 30 feet) at the center.

Fig. 9-15 Great talus piles at the foot of the 300 meter (1000-foot) cliffs of the Snowy Range above Lake Marie, Wyoming. Photo by the author.

Fig. 9-16 Giant Alabama collapse sink hole that formed when the roof of a large underground opening subsided suddenly. It is about 130 meters (425 feet) long, 106 meters (350 feet) wide, and 46 meters (150 feet) deep. Department of the Interior, U.S. Geological Survey.

An estimated $100,000,000 of damage was done to power plants, shipyards, and other facilities of the port city of Long Beach, California, and the subsidence created flood problems in the low areas along the coast.

Fig. 9-17 Subsidence resulting from collapse in a mine.

SUGGESTED READINGS

Birkeland, P. W., *Pedology, Weathering, and Geomorphological Research,* New York, Oxford University Press, 1974.

Eckel, E. B., Editor, *Landslides and Engineering Practice,* Highway Research Board, Special Report 29, National Academy of Science, National Research Council Publ. 544, 1958.

Jenny, Hans, *Factors of Soil Formation,* New York, McGraw-Hill Book Co., Inc., 1941.

Keller, W. D., *Principles of Chemical Weathering,* Columbia, Mo., Lucas Brothers, Publishers, 1957 (paperback).

Ruhe, R. V., *Quaternary Landscapes in Iowa,* Ames, Iowa State University Press, 1969.

Sharpe, C. F. S., *Landslides and Related Phenomena,* Paterson, New Jersey, Pageant Books, 1960.

Crescent-shaped sand dunes in Saudi Arabia. Courtesy of Arabian American Oil Company.

10

Seas of Air and Water

Adust devil ruffles a prairie, a rainstorm notches plowed ground, a wave-lashed cliff collapses. Some incidents seem trivial, and more impressive ones may be forgotten during long intervals of little change. Yet every day the Mississippi River dumps into the Gulf of Mexico almost 1.8 million metric tons (2 million tons) of dissolved salts, silt, and sand—the equivalent of 40,000 freight car loads[1] carved and carried from the land. Rivers alone carry off enough to lower the whole face of the United States about one meter every 27,000 years. Wind-blown dust, whose volume is hard to calculate, adds appreciably to this figure.

Through geologic time, moving air and water have attacked the Earth's face. Unlike passive weathering and direct gravitational reduction, wind, waves, glaciers, and streams are sculpturing agents that actively carve, carry, and eventually lay down materials from the crust. *Erosion*, whose Latin root means "to gnaw away," is a carving action. *Transportation*, which is difficult to separate from erosion, is the carrying of materials whose ultimate end is *deposition*, their dropping.

WIND

In one way or another, all the destructional mechanisms actively sculpting the land are related to wind. It blows away dry soil, sandblasts rock, lays down blankets of dust, and makes shifting hills of sand called dunes (p. 243). However, its greatest importance is indirect. Wind sets the oceans and other standing water bodies in motion, creating erosive waves. Perhaps most important, it carries water vapor inland that condenses and falls as snow or rain and then drains back to the oceans in highly erosive glaciers and streams. Let us briefly review the origin of winds.

---

1 Assuming open railroad cars each carrying 45.5 metric tonnes (50 short tons of 2000 pounds each); the metric tonne of 1000 kilograms is 1.1 times as heavy as the short ton commonly used in the United States and Canada.

Circulation of the Atmosphere

Solar energy, gravity, and the Earth's rotation cause the global swirlings in the sea of air. From its vague outer limit, 65,000 to 95,000 kilometers (roughly 40,000 to 60,000 miles) high, the atmosphere becomes denser towards the Earth's surface because the rapidly moving gas molecules are increasingly compressed by gravity and the weight of overlying air.

The Lower Layers of the Atmosphere   More than three-quarters of the atmosphere's mass is concentrated in its bottom layer, the *troposphere*, which ranges in thickness from some 16 kilometers (about 10 miles) above the Earth's surface in the tropics to as little as 8 kilometers (about 5 miles) over polar regions. The troposphere contains most of the winds, clouds, storms, and other phenomena of weather and climate that are important in the geological evolution of the Earth's surface features.

The turbulent troposphere is separated from the overlying *stratosphere* by an atmospheric surface called the *tropopause*. This sharp boundary at the top of the usually hazy and cloudy troposphere is often plainly visible from jet airliners flying in the clear dry stratosphere. The tropopause results from an inversion where temperatures, which become colder with increasing distance above the Earth's surface in the troposphere, start to become warmer at the base of the stratosphere. The warming is caused by the presence of ozone layers which absorb the sun's ultraviolet radiation.[2] Ozone (oxygen molecules of O_3) is created in the stratosphere by the bombardment of ordinary oxygen molecules (O_2) by the ultraviolet solar radiation. The tropopause indicates a stable situation in which the warmer air of the strato-

2 The ozone also shields life on Earth from damaging amounts of radiation. It is currently a matter of concern that flurocarbons, gases used as propellents in aerosol spray cans, may be rising and disrupting the ozone layers, and thus destroying the protection they provide.

spheric layer lies on the denser cold air in the uppermost troposphere.

The Greenhouse Effect   Although the sun provides the energy that eventually heats the troposphere, the direct source of heat is the Earth's surface. The lowermost atmosphere is heated like a greenhouse, or a car with the windows closed whose glass readily lets solar energy through to warm the interior but will not let the resulting heat escape (Fig. 10-1). Solar radiation (called

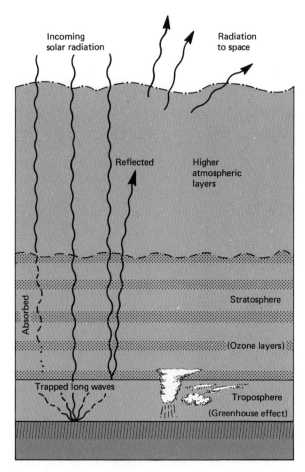

Fig. 10-1 The atmospheric layers and heating (with greenhouse effect).

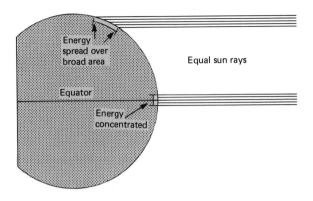

Fig. 10-2 Uneven heating of the Earth's surface. The equator, where the sun's rays strike the Earth at near right angles, is intensely heated. Polar regions are less intensely heated because an equal parcel of solar energy strikes the Earth's surface at an oblique angle and is distributed over a much wider area.

insolation) has relatively short wave lengths that readily penetrate the transparent atmosphere. The incoming waves strike and heat the opaque ground; the heated ground (or water) then reradiates less-penetrating, longer heat waves. The longer waves are trapped and absorbed by carbon dioxide and water vapor in the troposphere, thereby heating its air.

Because the incoming solar rays are parallel, and the Earth is a sphere, its surface is heated unevenly. In tropical regions, where the sun's rays strike most directly, the surface is intensely heated. In contrast, at polar regions, where the rays strike obliquely and not at all during the long winter darkness, the ground receives minimal heating (Fig. 10-2).

Wind and Pressure Belts at the Earth's Surface
The uneven heating stirs the lower atmosphere into rising and descending currents which cause the horizontal air movements called wind. Theoretically—if the Earth did not rotate—there would be two global convection cells, one in the Northern Hemisphere and one in the Southern. Heated light

air would be expanding and rising along an *equatorial low-pressure belt*. Cold dense air would be descending into *polar high-pressure centers*. Since air flows from regions of high atmospheric pressure towards regions of low pressure (down the pressure gradient), the Earth would develop two belts of surface winds, each one blowing from the poles to the equator.

The Earth's rotation, however, introduces a complication known as the *Coriolis effect*. This force deflects winds towards the right in the Northern Hemisphere and toward the left in the Southern. Thus in addition to the existing, thermally controlled, equatorial low-pressure and polar high-pressure zones, the rotational force creates two *subtropical high-pressure belts*. One lies at about 30° north latitude and the other at about 30° south. In higher latitudes, the rotation produces two *subpolar low-pressure belts* at about 60° north and 60° south latitudes. As a result, the Northern and the Southern Hemispheres each have three belts of prevailing surface winds. In low latitudes, winds blowing from the subtropical highs towards the equatorial low are deflected eastward by the Coriolis effect to form the steady *trade wind belts*. Air spilling poleward from the subtropical highs towards the subpolar lows is deflected to the right to form the strong *westerly wind belts* in the middle latitudes. In the high latitudes, weak *easterly winds* radiate outward from the polar highs towards the subpolar lows.

The global wind and pressure belts "follow the sun," shifting slowly northward from late December until the end of June and southward during the rest of the year. Although constant over broad expanses of the Pacific and Atlantic oceans, the general pattern of circulation is complicated over the continents by another seasonal phenomenon, the *monsoon effect*, which disrupts broad belts into alternating centers of action (Fig. 10-3). On land, the ground surface both heats up and cools off more rapidly than the water in the oceans. Thus in summer low pressures develop over the rela-

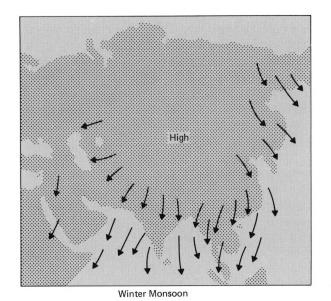

Winter Monsoon

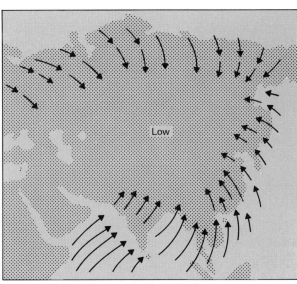

Summer Continent Monsoon

Fig.10-3 Seasonally reversing winds of the monsoons. In summer they move into the warm continent interior, deflected to the right by the Coriolis effect but drawn into the low-pressure center in a modified spiral. In winter cold dense air spills out from the interior.

tively warmer, large landmasses—notably Asia—while relatively high pressures exist over the cooler, more conservative oceans. In this season, moisture-bearing winds blow from the oceans into the continents. In winter, the situation is reversed. Dry winds blow towards the relatively warmer oceans, outward from high-pressure centers over the more rapidly chilled continental interiors.

The Troposphere's Three-Dimensional Circulation
The Earth's complicated three-dimensional circulation can only be shown in greatly simplified models (Fig. 10-4). It is, however, dominated by a great tropical-subtropical belt of warm air, bounded on the north and south by cold air in two circumpolar vortices.

The meeting of the trade winds along the equatorial belt of low pressure creates the intertropical convergence zone (called the I.T.C.) where air flows together and rises. On rising to great heights, this air drifts poleward. Along with the

Coriolis effect, the high-level poleward drifting air moves into higher latitudes where the circumference of the Earth is diminishing. Thus as the volume of the atmosphere decreases and more air drifts poleward, the air "piles up" at about 30° latitude (north and south) to create the subtropical high-pressure belts. Here the excess of air descends into the lower and denser troposphere, where it is heated by compression and becomes drier (the reverse of the process, p. 263, which leads to cooling and precipitation in the hydrologic cycle). At the Earth's surface this hot dry air creates the world's great low-latitude deserts, such as the Sahara in Africa and the Thar in India. Part of the air then blows back towards the equator as the trade winds. This rise of air over the equator, poleward drift at high elevations, descent over the deserts and return to the equator creates a great system of convectional circulation (called the Hadley cell) that dominates the tropical-subtropical belt of warm air.

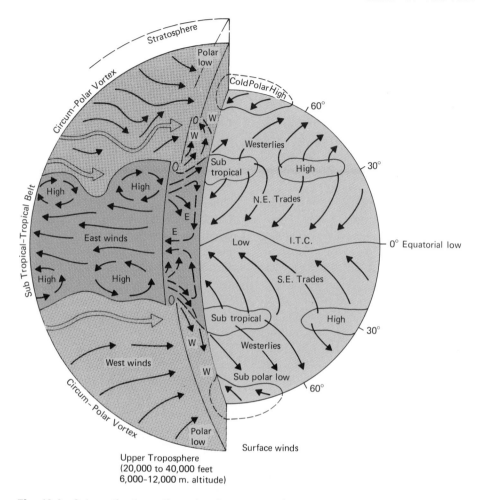

Fig. 10-4 Schematic three-dimensional representation of the atmosphere's circulation in the troposphere. The pattern of wind and pressure belts at the Earth's surface (right side of the diagram) is idealized to show the pattern as it would be if the Earth's surface were homogeneous (not diversified into land and water). In the cross-section (center), where movements in a vertical plane are indicated, e indicates east winds and w indicates west winds. Circulation of the upper troposphere (left side of diagram) is dominated by the two circumpolar vortices and the central belt of tropical-subtropical air. Extended global cooling in the higher latitudes would lead to expansion of cold air in the circumpolar vortices— and a return to glacial conditions at the Earth's surface (Chapters Ten, Fifteen, Seventeen).

Air spilling along the Earth's surface in the trade winds and westerlies is deflected by the Coriolis force, but friction with the ground reduces the effect on the surface winds so that they continue to have poleward and equatorward components of motion. Higher in the troposphere (above about 1000 meters, 328 feet), where friction with the ground has no effect on the moving air, the Coriolis force rapidly converts the winds into due-east winds in the trade-wind region and due-west winds in the westerly-wind region.[3] The Coriolis force is at a minimum at the equator and becomes progressively stronger at increasing latitudes, reaching a maximum in polar regions. Thus poleward of about 30° north and south latitudes, no giant convection cells develop like those dominating the low-latitude atmospheric circulation. Rather, the strengthening Coriolis force dominates the global air circulation and creates the great westward-moving spirals that form the circumpolar vortices.

Studies of the higher part of the troposphere indicate that the changing weather patterns associated with the westerly wind belt develop along the boundaries between the circumpolar vortices and the belt of tropical-subtropical air. In the upper troposphere, the boundaries are offset vertically along *jet streams*, tubelike streams of air that reach velocities of as much as 500 kilometers per hour (300 miles per hour). In the vicinity of the jet streams, the global boundary between the warmer tropical-subtropical air and the air of the cooler circumpolar vortices is thrown into great loops and bends (called *Rossby waves*). Inside the alternate poleward and equatorward bends, warm air moves towards the poles and cold air towards the equator. The bulges of cold air moving towards the equator are eventually pinched off and merge and blend into the warmer air of the subtropical belt.

Close to the ground, where friction with the Earth's surface becomes a factor, the east margins of the great Rossby waves are thrown into smaller waves. They create the cold and warm fronts that characterize much of the changing weather in the middle latitudes. The waves in the lower troposphere are associated with low- and high-pressure centers that generally move from west to east. Moving high-pressure centers, which are associated with equatorward bulges of cold air, develop outward-spiraling wind systems. The high-pressure centers are called *anti-cyclones*. Moving low-pressure centers form in the waves where warm air is moving poleward and is associated with a wind system that spirals inwards, towards the low-pressure center. These storms are called *middle-latitude cyclones* (not to be confused with the far more violent tropical cyclones called hurricanes).

The storms associated with the middle-latitude cyclones develop where air masses having different properties meet along sharply defined boundaries called *fronts*. The air masses acquire their properties from the Earth's surface beneath them at times when the air remains stationary over cold and dry land, cold oceans, hot land, or warm oceans. When the air masses begin to move, set in motion by waves in the upper atmospheric circulation, the warm or cool, moist or dry masses meet along fronts that produce the moving cyclonic storms.[4]

Geologic Work of Wind

Wind directly affects unconsolidated materials where vegetation is sparse, as in dry deserts, or locally in humid regions where vegetation is absent, as on beaches, peripheral to glaciers, on exposed stream bottoms, and—unfortunately—on plowed ground (Fig. 10-5).

Blowing Sand and Dust　Although wind can blow well over 160 kilometers (100 miles) an hour in a hurricane, it usually transports only dust and sand, and, at best, only pebbles, because the density of air is so low, 850 times less than that of water. Hence, wind is a good sorting mechanism; it leaves larger particles behind as *lag gravels* and carries away dust and sand. Blowing sand moves close to the ground, rarely over 2 meters (6 feet) above the

---

[3] They are called geostrophic winds because they blow along the parallels of latitude (*strophic* comes from the Greek word meaning "revolving").

[4] The mechanism of precipitation, rain and snow, is discussed with the hydrologic cycle in Chapter Eleven.

Fig. 10-5 Dust Bowl. Windblown soil trapped by obstruction on a deserted homestead. Soil Conservation Service photo by McLean.

surface. Some desert travelers, caught in sand-storms, have been engulfed in stinging sand to shoulder height while their heads were above the cloud.

Pioneering studies of sand movement and dunes were made in the Sahara during the 1920s and 1930s by Lt. Colonel R. A. Bagnold of the British Army. Later, he supplemented his desert observations, using a home-made wind tunnel. Bagnold found that sand moves mainly by a hopping or skipping motion, called *saltation,* wherein sand grains fly through the air, drop to the surface, and either carom into the air again or hit other particles that take up the motion (Fig. 10-6). Over a pebbly surface, sand grains bounce to a maximum height;

over a sheet of sand, they fly lower because their landings are cushioned by the loose grains. Here, the impact of saltating grains shoves the surface sand along in a slow *surface creep.*

The movement of dust, those particles finer than sand, is a different matter. Dust is transported to great heights and distances by the turbulence of winds and currents. In the drought of the 1930s, material from the Dust Bowl from Oklahoma to Colorado was carried to the New England states where it created a dirty cover on snow (Fig. 10-7). In 1883, when the eruption of Krakatoa pulverized the volcano, dust rode up thousands of meters into the atmosphere on heated air currents to be carried completely around the world.

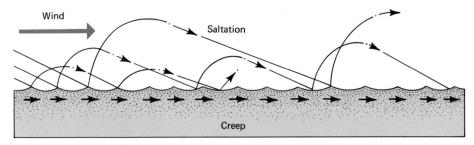

Fig. 10-6   Movement of windblown sand. After R. A. Bagnold.

Fig. 10-7   Advancing front of dust storm in Union County, New Mexico. Soil Conservation photo by Al Carter

Fig. 10-8 House undermined by deflation of sand in Nebraska. U.S. Forest Service photo by C. A. Taylor.

Erosional Landforms Wind erodes by *deflation,* the blowing away of loose materials, and by *abrasion,* the natural sandblasting of rock surfaces (Figs. 10-8, 10-9). Deflation in the Dust Bowl stripped the valuable top soil, exposing the less fertile but better consolidated *B* soil horizon beneath. The principal landforms of deflation are *deflation basins,* wind-scooped depressions ranging in size from a meter to several kilometers in diameter. The Big Hollow just west of Laramie, Wyoming, is an elliptical deflation depression, 14.5 kilometers (9 miles) long, 3.2 kilometers (2 miles) wide, and as much as 36 meters (120 feet) deep.

Desert pavements are layers of closely fitted, polished pebbles often naturally cemented to give a resistant surface. They develop where fine materials are deflated and the coarse pebbles that lag behind are jostled by wind into a tight mosaic. Once formed, such pavements armor desert floors from further extensive deflation (Fig. 10-10).

Windblown sand abrades most effectively 30 or 60 centimeters (a foot or two) above the ground,

a fact demonstrated by wind-carved notches or grooves on fence or telephone poles in sandy deserts. Above 1 meter (3 feet) the erosion is much less because the saltating grains reaching this height are too light to strike with much impact. Wind abrasion—natural sandblasting—forms relatively minor features. Hard rock surfaces may acquire *wind polish* resembling that on highly polished building stones. *Ventifacts,* wind-blasted pebbles and cobbles, have faceted surfaces that may join in sharp edges. *Yardangs* were first recognized in Chinese Turkistan and have since been found in other deserts. They are sharp-crested ridges, or small flat mesas, separated by wind-scoured troughs. Yardangs range from several centimeters to tens of meters (a few inches to tens of feet) in height, and develop in soft rock where wind blasts along the sides of pre-existing ridges (Fig. 10-11).

Depositional Landforms Sand dunes are the most impressive landforms made by wind. Most dunes move slowly downwind, although some, which

Fig. 10-9 Wind-abraded sandstones near Tuba City, Arizona. Photo by Tad Nichols.

purists prefer to distinguish as sand drifts, remain stationary in the lee of fixed obstacles. Characteristically, dunes have a gentle windward slope and steeper leeward side, called the *slip face,* that averages about 33° or 34°. Sand blows up the windward side, across the crest, and spills down the slip face; thus dunes migrate because sand is progressively stripped from their windward face and deposited on the downwind side (Fig. 10-12).

Although dunes are most common in deserts, they can occur wherever vegetation on sandy materials is breached or absent (Fig. 10-13). Sand covers over 1,000,000 square kilometers (400,000 square miles) of Arabia, a sizeable tract, but still only one-third of that country's desert surface. In the Sahara, 777,000 square kilometers (300,000 square miles) are mantled by sand, which is a little over 10 percent of the total desert area. Thus, Hollywood movies to the contrary, deserts are not completely covered by sand. Smaller dune areas appear in humid regions along beaches and in river bottoms of semiarid lands.

The distinctive *barchans* are crescent-shaped dunes whose horns point downwind. They move slowly, from 10 to 20 meters (30 to 60 feet) a year, and may reach 30 meters in height, and, perhaps, 300 meters (almost 1000 feet) in width. Barchans often occur in fields, or clusters, whose

Fig. 10-10 Desert pavement in Death Valley, California. Watch in center gives scale. Photo by John H. Maxson.

individual dunes are moving across bare rock surfaces (Fig. 10-14). *Transverse dunes* form in seas of sand where bedrock is completely buried. They resemble giant ripples or coalescing barchans whose crest lines are at right angles to the prevailing wind (Fig. 10-15). *Longitudinal dunes* are elongate in a downwind direction. They may be drifts in the lee of an obstacle, or true, free-moving dunes. They are locally abundant in the desert of the Little Colorado River and in a broad expanse of central Australia. The Australian "sand ridges" average 15 meters (40 feet) in height and may be hundreds of kilometers long. These dunes, lying about half a kilometer apart, are separated by rocky desert floor.

The *seif dunes*, named from the Arabic word for sword, are large spectacular features, originally described by Bagnold in the Libyan desert. They are complex dunes of irregular shape, as much as 200 meters (650 feet) high, more than 1 kilometer wide, and 100 kilometers (over 60 miles) long. Chains of seif dunes may trail across the desert for several hundred kilometers. Fundamentally, seif dunes are great longitudinal dunes controlled by

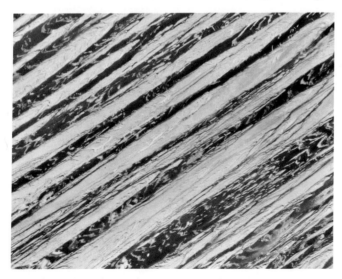

Fig. 10-11 Vertical air view of yardangs that are as much as 30 meters (100 feet) high, 460 meters (1500 feet) wide, and several kilometers long. Courtesy of U.S. Geological Survey.

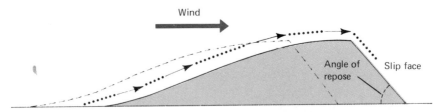

Fig. 10-12 Cross-section illustrating dune migration and position of slip face.

Fig. 10-13 Gleaming gypsum sands form over 56,680 hectares (140,000 acres) of dunes, some 9 meters (30 feet) high, in the White Sands National Monument, New Mexico. Photo by Tad Nichols.

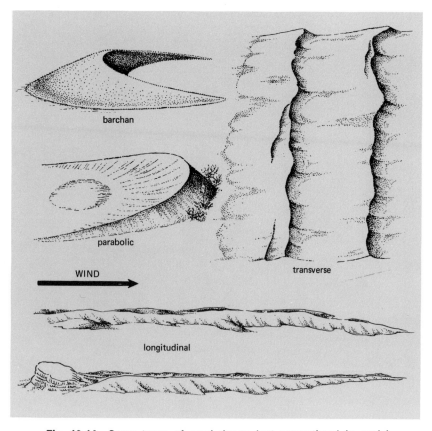

Fig. 10-14 Some types of sand dunes (not proportional in scale).

a prevailing wind but modified by intermittent strong winds from other directions. All the dunes discussed so far are free dunes, little affected by vegetation.

The *parabolic* dunes resemble barchans in being crescentic; however, the horns of parabolic dunes point upwind, and the steep slip face is on the outside of the crescent. These dunes are influenced by vegetation that partially anchors their sand. They originate around blowouts, where deflated sand is trapped by vegetation on the downwind side. Parabolic dunes sometimes acquire hairpin shapes, because the dune funnels the wind so

that sand moves most rapidly over the central part while the sand in the horns lags behind.

A dominance of vegetation stabilizes dunes. The Sand Hills of western Nebraska and adjacent states, covering some 45,000 square kilometers (about 18,000 square miles), are old dunes now anchored by grass. When dunes become stabilized by vegetation, soil eventually forms, and thereafter sheet wash and creep slowly reduce them. Regions such as the Sand Hills indicate a climatic change from more severe conditions associated with shifting sand in free dunes to a more benevolent climate favoring a mantle of vegetation.

Fig. 10-15 Complex transverse dunes near Colorado River Delta, Sonora, Mexico. Photo by Tad Nichols.

If the wind direction is fairly constant, the type of dunes developed is a function of three main factors: wind velocity, sand supply, and vegetation. Longitudinal dunes prevail where moderate to strong winds work on a relatively small sand supply. Barchans develop where both the wind velocity and sand supply are moderate. Transverse dunes form where sand is very abundant. Parabolic dunes reflect the influence of vegetation.

Loess Loess is a deposit of windblown dust. Typically, it is a buff-colored, unlayered deposit of angular silt-sized particles consisting of various common minerals. Long-continued dust storms sweeping from barren source areas have deposited great loess blankets up to several hundred meters (several hundred feet) thick in downwind regions where the vegetation is adequate to hold the settling fine material in place. Some 518,000 square kilometers (200,000 square miles) of the Mississippi Valley and its tributaries are underlain by loess. In eastern Washington, Alaska, southern Germany, Russia, and China, extensive areas of fertile soil have developed on loess.

The Chinese loess is derived from the Gobi Desert and has been carried eastward and deposited in the more fertile parts of the country by great dust storms that continue to the present day. The sources of loess in the Mississippi Valley and Europe were great barren areas traversed by melt-water streams issuing from receding continental glaciers. Such loess deposits reflect a change to more amicable climates during the fluctuating cold periods of the Ice Age.

WAVES AND SHORE PROCESSES

Coastal dwellers need not be convinced that the ocean shapes the edge of the land. Crashing storm waves topple cliffs, alter beaches, and destroy great concrete breakwaters. Even in fair weather the sight and sound of ordinary breakers and the swash and backwash on a beach suggest the timeless gnawing of waves upon the continental margins (Fig. 10-16).

Dynamics

Wind, blowing over standing water in oceans and lakes, initiates the waves responsible for the major sculpturing of the shores. Some waves are created by volcanic explosions and earthquake shocks on the ocean floor, and some currents are caused by tidal changes or different densities in water masses; but by and large, most coastal modification is the work of waves that are generated by wind blowing across open water.

Waves of Oscillation The friction of moving air ripples a glassy sea. Pushed on their backsides and dragged along by air spilling over their crests, the ripples grow into waves. Waves are moving ridges in water. Their height is the vertical distance between a crest and the lowest point in an adjacent trough. *Wave length* is the distance between successive crests. *Forced waves* are actively impelled by wind. They may reach considerable heights during storms and travel hundreds or thousands of kilometers beyond their breeding grounds. *Swells,* breaking along a beach on a windless day, are free waves that may well have originated in a distant storm.

Although the shapes of waves rush across a water surface, the water particles do not. They move in orbits whose paths are shown by a floating cork (Fig. 10-17). With each orbit, the water particles make a slight advance beyond their starting point; that is, they drift downwind. Surface particles have the largest orbits, of a diameter equal to the wave height. Orbital motion dies out rapidly downward, being negligible at a depth equal to one-half of a wave length. This depth, called *wave base,* is the lowest limit at which windblown waves stir sediments on the sea floor.

Most of the geologic work of waves occurs where oscillatory waves move from open waters onto a shoaling bottom. As the water becomes shallower than wave base, the drag along the bottom causes wave lengths to decrease and the waves to steepen. At a critical point, the waves fall forward in breakers, sending masses of water surging onto the shore (Fig. 10-18). Erosion of hard rock results from the crashing impact and hydraulic pressure of the water masses, along with the abrasive grinding by gravel and boulders carried along as tools.

Wave Refraction Rugged, embayed coasts are strongly modified by wave refraction. Advancing waves are first slowed off promontories, where the shallow bottom projects further seaward, while wave fronts continue their unimpeded advance in the deeper waters of adjacent bays. As a result, the waves wrap around the headlands, and rise into powerful breakers, often 10 times higher than those in nearby bays. Thus, wave energy is concentrated in an attack on the headlands, while it is lessened in bays (Fig. 10-19). As a result, wave refraction straightens an initially irregular shoreline.

Fig. 10-16 Sea cliff near Le Havre, France. *The Cliff at Etretat* by Claude Monet. Courtesy of the Metropolitan Museum of Art, bequest of William Church Osborn, 1951.

Beach Drifting Inside the line of breakers on a beach, water surging onto the beach as swash and returning downslope in backwash transports sand parallel to the water's edge when waves strike the shore at an angle. The swash drives sand particles obliquely up the beach, and backwash receding directly downslope carries them straight towards the water's edge (Fig. 10-20). Thus, sand grains migrate laterally, in looping paths, along a beach.

The Longshore Current Outside the line of breakers, the longshore current moves sand along

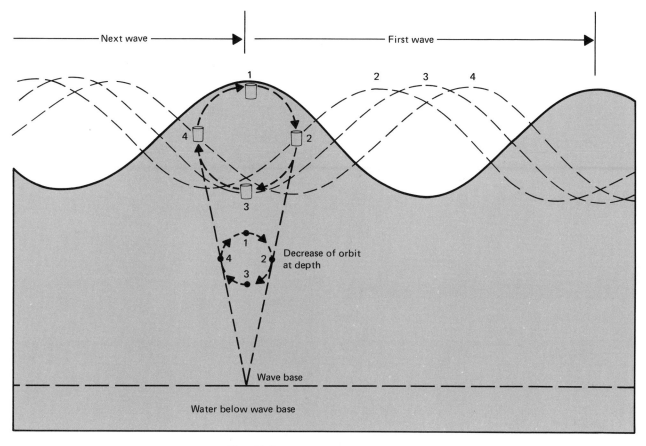

Fig. 10-17 Orbital motion of waves.

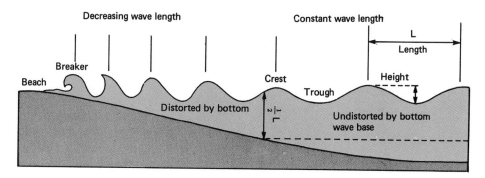

Fig. 10-18 Cross-section of waves breaking on a beach.

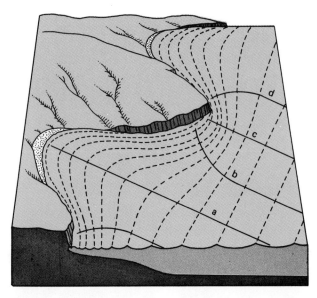

Fig. 10-19 Wave refraction on rugged shore. Dashed lines represent wave crest; solid lines a, b, c, d divide wave fronts into units of equal energy which concentrate on the headlands.

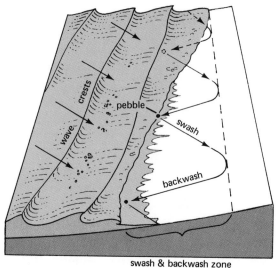

Fig. 10-20 Beach drift.

the ocean bottom (Fig. 10-21). The current originates when winds, blowing obliquely onshore, create a slight "piling up" of water because its shoreward drift is impeded by the shoaling bottom. The excess water escapes by setting up a current parallel to the shore. The combination of beach drifting and the longshore current transports sand laterally along the shore, and thus is an important mechanism for extending beaches that lie

against the land into sand bars which project into open waters.

Coastal Topography

Erosional Landforms   *Sea cliffs* and *abrasion platforms* are the principal landforms cut on a steep coast. They appear on promontories of embayed coasts where wave refraction is active; along straighter coasts they may extend for many miles, broken occasionally by estuaries. The evolution of

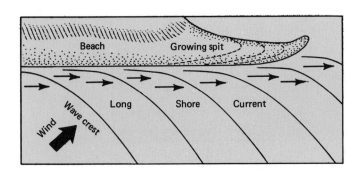

Fig. 10-21 Longshore current.

cliffs and platforms begins when waves cut a small horizontal notch into a sloping shore. As the notch is eroded deeper, the land is undermined, producing a cliff of progressively increasing size. Accompanying the retreat of the cliffs, a platform is cut, at sea level or slightly above, in the zone of strongest wave attack. Wave action sweeps debris, carved from the land, across the expanding platform and deposits it off the lower end, thereby combining the wave-cut platform with a depositional, or wave-built, terrace.

Where the terrace is narrow, beach deposits are absent or temporary, being present when waves are running normally, but stripped away during storms. With widening of the platform by cut and fill, the energy of waves is consumed in friction across the shallow terrace; thus waves reaching the landward margin of the platform are weakened, allowing permanent beaches to develop.

The development of sea cliffs and wave platforms produces a variety of lesser landforms (Fig. 10-22). Hanging valleys form where the wave-attacked cliff recedes more rapidly than streams flowing to the shore can deepen their valleys. Sea caves result where waves pounding on a cliff face

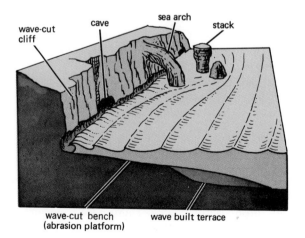

wave-cut cliff

cave

sea arch

stack

wave-cut bench (abrasion platform)

wave built terrace

Fig. 10-22 Erosional features on a rugged cliffed coast.

exploit fractures or other zones of weakness in the rock. Narrow headlands attacked on two sides may develop into sea arches (Fig. 10-16), roughly resembling a flying buttress. Stacks, detached relics marking the former extent of a cliff, are pillarlike islands standing above the wave-cut platform (Fig. 10-23). Some result from the collapse of a sea arch.

Depositional Landforms Beaches, the most characteristic depositional landforms along a coast, consist of sand and, sometimes, wave-worn cobbles (Fig. 10-24). Technically, a beach lies along a coast from the low-tide line to an inland limit marking the highest point reached by storm waves. A variety of sandy reefs extending from promontories into open waters are called spits, bars, and barriers. They are all, essentially, extensions of beaches whose sand has been transported parallel to the coast by beach drifting and the longshore current (Fig. 10-25).

Great offshore sand bars, or barrier islands, characterize the eastern seaboard of the United States, where the near-flat coastal plain slopes gently out beneath the sea (Fig. 10-26). The development has produced extensive protected lagoons and marshes lying between the barrier beaches and the shore. The main source of material for these barrier beaches has long been controversial. Many seem to be composed of deposits drifted from headlands by beach drifting and the longshore current. However, some workers maintain that the material is largely scoured from the bottom, seaward from the barriers. The action of waves breaking well offshore could toss sand into submarine bars that eventually build up above sea level. The barrier beaches off Cape Hatteras, being isolated islands, could have originated from bottom scour. On the other hand, it has been suggested that they, too, were once headland bars, cut off from their original source by violent storm waves or a long-term rise in sea level. Conversely, perhaps they were submarine bars that became exposed when sea level fell.

Fig. 10-23   Stacks and island remnants along United States West Coast. Courtesy of the Field Museum of Natural History, Chicago.

Evidence of Shifting Shores

Once landforms at the water's edge are appreciated, similar features, some of which are high and dry and others drowned beneath the sea, can be recognized.

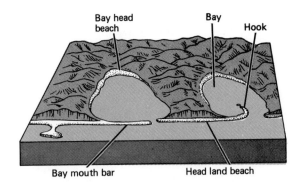

Fig. 10-24   Depositional beach forms.

Emergent Features   The coasts rimming the Pacific Ocean and on its islands are cut into giant steps (Fig. 10-27). The California coast near Santa Barbara, for instance, has 17 terraces, ranging from a few meters to 300 meters (1000 feet) above the present sea level. The lower terraces are clearly former wave-beveled platforms, preserving minor shore features such as stacks. The higher terraces are more stream-dissected, and partly masked with slope wash and stream deposits from higher slopes, but nonetheless of marine origin (Fig. 10-28).

Ancient shorelines are also preserved along the broad, gentle coastal plain of the eastern United States. The coastal terraces here are less obvious than those of the Pacific shores, for they are as much as 65 kilometers (40 miles) wide and bounded by scarplets only a few meters high. Two such scarps, the outer one 7 to 10 meters (20 to 30 feet) above sea level and the one further inland

Fig. 10-25 The great spit with a smaller secondary one forming the tip of Cape Cod around Provincetown, Massachusetts. Photo by John S. Shelton.

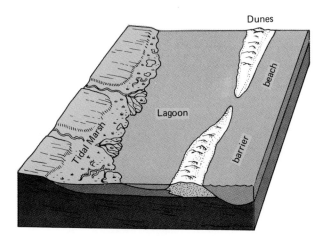

Fig. 10-26 Barrier islands as found along the Atlantic coastal plain.

30 to 35 meters (90 to 100 feet) above sea level, extend from New Jersey to Florida, thence westward along the Gulf Coast.

Abandoned shorelines are also common around inland seas and lakes as, for example, the Baltic Sea in Scandinavia, the Dead Sea in the Holy Land, the large lakes of East Africa, and the Great Lakes and Great Salt Lake in the United States.

Although emergent shore deposits, being largely unconsolidated, are less enduring than wave-cut features, elevated beaches, bars, and deltas are sometimes preserved. For example, Trail Ridge, 6 kilometers (4 miles) wide, 130 kilometers (about 80 miles) long, and 30 meters (100 feet) high, rises prominently above the generally flat plain of southern Georgia. Originating as a barrier island in a former shallow sea, the ridge was built by the northward drift of sand from islands in what is now north-central Florida.

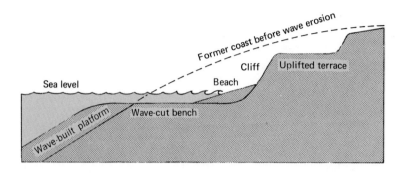

Fig. 10-27 Origin of emergent wave-cut terraces either from sea-level changes or coastal deformation.

Fig. 10-28 Uplifted marine terraces in Palos Verdes, California. Courtesy of Spence Air Photos.

Fig. 10-29 Drowned topography along the east side of Chesapeake Bay, Maryland. Photo by John S. Shelton.

Submergent Features Although drowned landforms are more difficult to locate and may be masked by later deposition, a number have been found. For instance, the Franklin "Shore" is a submerged scarp that can be traced at least 300 kilometers (over 180 miles) along the bottom parallel to the present shoreline of the east coast of the United States. Terrestrial features, those made on dry land, have been drowned as well. Surveys of the ocean floor off Alaska, north of the Aleutian Islands, reveal submerged hills and valleys very similar to those on land. San Francisco Bay on the West Coast and the Chesapeake and Delaware Bays on the East Coast are estuaries, submerged valleys of stream systems that once flowed to more distant shores (Fig. 10-29). Drowning of these main valleys dismembered the stream systems, so that their former tributaries now empty as separate rivers directly into the salt water of the estuaries.

The marked shifting of the shore indicated by both emergent and drowned landforms has two possible causes: either the volume of the oceans and of inland lakes has changed; or the water level has remained the same while the land itself has risen or subsided. In some instances, a combination of both is probable.

Worldwide fluctuations of sea level, called *eustatic changes,* are intimately related to the Pleistocene, late Ice Age. The water, locked in the former great continental ice caps, came originally from the reservoir of the oceans. Sea level fluctuated, falling in periods when the great glaciers expanded, storing water on land, and rising as the glaciers melted and their waters flowed back into the sea. In earthquake areas of active mountain-making, where the coastal rocks are locally rising or being depressed, many shoreline terraces have been cut, then warped. Thus, traced any distance the terraces have markedly different elevations above the present sea level. Around the Baltic Sea and Great Lakes, warped and elevated shorelines reflect the rise of land freed of its glacial load by the melting and disappearance of the continental ice caps.

Thus in late geologic time, whatever the cause, the topographic evidence indicates a continuous fluctuation of the shores. Cross-bedding, fossils, and other marine features of sedimentary rocks also make it clear that, through much of geologic time, the continental interiors have been flooded by migrating seas far more extensive than those of the present day, and that shore processes are responsible for the characteristics of rocks remote from existing seas.

SUGGESTED READINGS

Bagnold, R. A., *The Physics of Blown Sand and Desert Dunes* (reprint, Methuen and Co. Ltd., London, 1942).

Bascom, Willard, *Waves and Beaches,* Garden City, N.Y., Doubleday and Co. Inc., 1964 (paperback, Anchor Books).

Russell, R. J., "Instability of Sea Level," in *American Scientist,* Vol. 45, No. 5, pp. 414–430, 1957.

Sheppard, F. P., *Submarine Geology,* 2nd edition, New York, Harper & Row, 1963.

Drumlins in barrens of the Canadian Shield. R.C.A.F. photo courtesy of Canadian Department of Mines and Technical Surveys.

11

The Hydrologic Cycle

"All streams run to the sea, but the sea is not full. To the place where the streams flow, there they flow again." Whether the author of Ecclesiastes (1:7) envisioned a hydrologic cycle, we do not know. But the concept of a continuous circulation of water from the oceans to the land and back may date back to the Babylonians and certainly dates at least from 650 B.C. At that time Thales, the Greek philosopher, proposed that sea water is driven into rocks by wind and pumped by the pressure of rock into the mountains where, emerging in springs feeding streams, it flows back to the sea. Subterranean passages and various ingenious explanations for desalting the water and forcing it into the mountains were later added to this basic scheme which, in a way, is still with us in the "underground water seams" of water witches.

The problems of salt and lift were avoided by Aristotle (384–322 B.C.) who, observing evaporation and condensation, held that heat from the Sun changes water into air—which would leave salt behind in the sea—and that chilling changes air back into water. He attributed rain to the cooling air and observed that some percolates into the ground and some runs off in streams; however, the main nourishment of streams, he held, results from the conversion of air to water, in cool caverns inside mountains which act like great dripping sponges. Thus, Greek thinkers contributed a cycle, evaporation and condensation; they erred in insisting that rain alone could not feed streams, and (despite the views of Aristotle) that rock is generally impervious to water.

Though the Greek theories dominated educated thought for well over 2000 years, there were dissenters whose views were essentially correct and modern. Marcus Vitruvius in the time of Christ, Leonardo da Vinci (1452–1519), and Bernard Palissy in the sixteenth century all realized that precipitation infiltrates the ground and makes surface streams. Finally, the true path of the global water system was established in the seventeenth century, when grand theorizing was subjected to the test of careful measurement.

Pierre Perrault (1608–1680) determined the amount of water flowing down the Seine over a period of three years. For the same period, he calculated the volume of rain and snow falling on the slopes leading to the river by multiplying the average depth of precipitation by the ground area, which he had obtained from maps. Rain and snow had six times the volume of water discharged by the

Fig. 11-1 Falls of the Iguazu near the Argentine-Paraguayan border rival Niagara Falls. They are 60 meters (200 feet) high and 3.2 kilometers (2 miles) long. Photo by E. A. Carter.

Seine. Edmé Mariotté (1620–1684) verified Perrault's findings and further demonstrated that infiltration feeds springs. He showed that seepage into the cellar of the Paris Observatory corresponded to the rainfall. The English astronomer Edmund Halley[1] (1656–1742), after experimenting with salt solutions, calculated that evaporation from the Mediterranean Sea could easily supply the water returning to it in streams. Although crude by modern standards, these various measurements established the proper mechanism of the hydrologic cycle.

Heated by the Sun, water evaporates from the ocean surface as invisible vapor that mixes with the air above (Fig. 11-2). Vapor-laden air moves inland in currents and winds, rising and expanding in the higher and thinner reaches of the atmosphere.

Expansion causes cooling[2] that decreases the air's ability to hold water vapor. When a certain temperature, the *dew point*, is reached, the air is saturated; further cooling expels water in liquid form, condensing it into countless minute droplets or, if the temperature is below freezing, ice crystals that form the visible clouds. As the cloud particles grow too heavy to be held aloft on rising currents, gravity takes over and they fall as rain or snow.

Of the water reaching the ground, part evaporates or is transpired—used and given off as vapor by plants—returning to the cycle without ever reaching the sea. Part runs overland, in thin sheets that soon concentrate into the linear channels of streams, and part sinks to a level of saturated ground, forming a supply that nourishes springs and streams. This is the essence of the hydrologic cycle.

1 Of Halley's comet fame.

2 Noticeable if the vapor from an aerosol spray touches your hand.

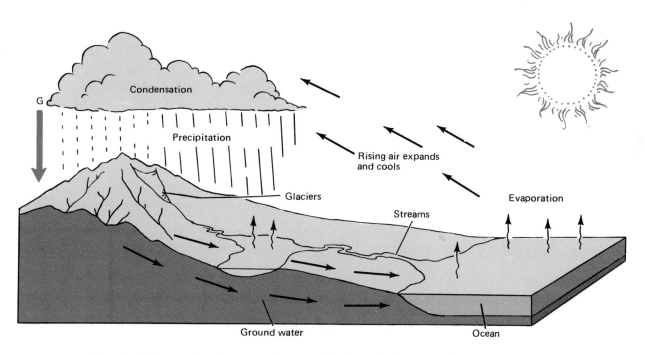

Fig. 11-2 Powered by the sun and gravity (G), the hydrologic cycle keeps water in constant circulation from ocean to land and back to the ocean again.

Although precipitation varies widely from place to place, overall budgets for the hydrologic cycle have now been estimated from a wealth of instrumental readings. The face of the continental United States, for example, receives 75 centimeters (30 inches) of precipitation each year, averaging data from all climatic zones. Of this total, 70 percent is recycled by evaporation and transpiration. The remaining 30 percent feeds streams and replenishes one of our most important natural resources—groundwater.

WATER IN THE GROUND

The economy of groundwater, which we obtain from wells, is like a bank account. What goes in and what comes out determine the balance. In arid regions, such as parts of Arizona and California, with growing populations, swimming pools, air-conditioning units, and extensive irrigation, the groundwater situation is becoming acute. It might seem obvious that more water could be obtained by drilling more wells. But as more wells are drilled, the level at which water is obtained goes deeper. Deeper drilling chases the water table downwards, and with great depths the return decreases. The water account dwindles because replacement comes only from rain and melting snow—a fact not appreciated by the many people of this modern day and age, who are ignorant of the nature of groundwater and the hydrologic cycle.

Infiltration   If water is slowly poured into a bucket already filled with sand, up to a quarter of a bucket of water may disappear into the sand before water spills out. The water is stored in voids, the spaces between sand grains. *Porosity* is a measure of the total void space, and all mantle and bedrock in the upper three kilometers (two miles or so) of the Earth are porous to some degree because of cracks, pores, and other minute channels. Rock porosity ranges from 1 percent or less in granite to well over one-third of the total volume of some clastic rocks. The ease with which groundwater flows through rock, calle*d permeability,* is not a function of porosity alone. The kinds of openings, their size, shape, and interconnections, determine permeability. Shale usually has greater porosity than sandstone; however, it is far less permeable, because its voids are minute and poorly connected. Even dense, massive granite may transmit a fair amount of water if the rock is well fractured.

Groundwater Zones   Studies of wells and other excavations indicate two main zones of groundwater: an upper *zone of aeration* where voids are

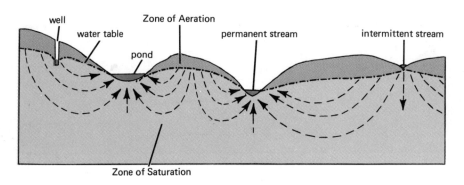

Fig. 11-3 Groundwater zones and movement.

not completely saturated and considerable air is entrapped; and a lower *zone of saturation* whose pores are water-filled. The *water table* separating the two zones is generally a sloping surface, roughly parallel to the ground.

Water is moving beneath the ground, albeit slowly (Fig. 11-3). It normally sinks through the zone of aeration down to the water table, except in times of drought or in desert regions where capillary action may carry water up to the surface where it evaporates. The level of the water table fluctuates, as is evident in wells, in response to additions from infiltration and losses of water to the surface at springs and into stream bottoms. Water is also removed by plants, and pumped from wells.

The movement of water in the zone of saturation has been studied in wells by mapping water levels; i.e., introducing special dyes and waiting for their appearance in adjacent wells or springs. Groundwater flows slowly, a few meters per day, or as little as a few meters per year. The direction of flow beneath the water table is not a simple downslope movement. Water migrates downward from the water table in roughly curving paths that bend and rise into stream bottoms. Thus, deeper waters are replenished and replaced, and do not become stagnant.

Water from the Ground Water seeps out wherever the water table is cut. A well, for instance, is a hole punched through the water table. Lakes and swamps fill basins intercepting the water table. Permanent streams, flowing year round, are nourished by groundwater emerging in their channels, even when rain and surface runoff have long ceased. The channels of intermittent streams, which are periodically dry, lie above the water table. Thus, they carry water only when directly fed by surface runoff from storms or melting snow, or when the water table temporarily rises to the channel level. In arid regions especially, water flowing in the channel may be lost by downward seepage to the water table.

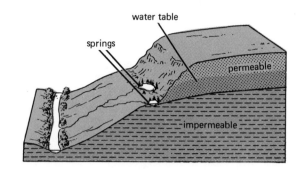

Fig. 11-4 Common type of hillside spring.

Ordinary springs are natural seepages on a hillside. They appear where groundwater is deflected to the surface along impermeable rock barriers or along fractured zones (Fig. 11-4). Pressure forces water to the surface in *artesian springs*, a special sort. They require special conditions: sloping strata containing an aquifer, or permeable rock layer, sandwiched between impermeable beds. Water entering such a system flows downwards in an enclosed system and emerges under pressure, like water from a faucet in city waterpipes fed from an elevated tower. In both cases, the weight of water extending to the higher source gives a hydrostatic head that drives water through the confined system (Fig. 11-6).

Hot springs result where surface water descends to magma at depth and is then recirculated to the surface (Fig. 11-7). *Geysers* are a spectacular variety of hot springs that periodically erupt boiling water and steam. They are common in the recently volcanic regions of New Zealand, Iceland, and Yellowstone Park in the United States. Geysers, according to the Bunsen theory, occur where groundwater fills irregular systems of fissures and cracks in hot rocks. At depth, the higher pressure raises the water's boiling point. On slowly heating, the deeper waters expand and flush water out of the geyser mouth, thereby reducing the weight of the water column. Pressure decreases, until the deep, superheated water flashes into steam, setting off an erup-

Fig. 11-5   Spring issuing from limestone. Courtesy of Wayne M. Sutherland.

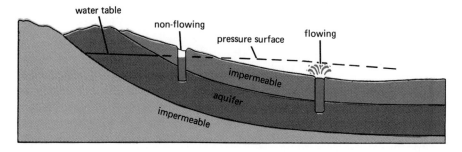

Fig. 11-6   Elements of an artesian system.

Fig. 11-7 Mammoth Hot Springs of the Yellowstone. Baroque depositional forms deposited by hot springs charged with calcium carbonate. Photograph by the author.

tion that empties the fracture system. After the force is spent, water slowly seeps back into the geyser plumbing, and the whole process is eventually repeated.

Subterranean Topography

The subterranean world is a realm of total darkness, containing permanent ice, labyrinths of baroque magnificence with flocks of bats, and underground pools and streams inhabited by blind fish and salamanders. Extensive caverns, with a few exceptions, are restricted to regions of limestone, a structurally strong rock which is, however, strongly susceptible to weathering by carbonation in acidic groundwaters filtering through its ever-present cracks (Fig. 11-8). The Greeks, being familiar with such limestone caves, assumed them typical of the whole underworld, and, as a result, arrived at their erroneous generalizations about the origins of springs and streams.

In humid climates, limestone develops a surface topography known as *karst*, from a typical locality on the Adriatic Sea in Yugoslavia. Karsts, as in the Causses of France and Yucatán in Mexico,

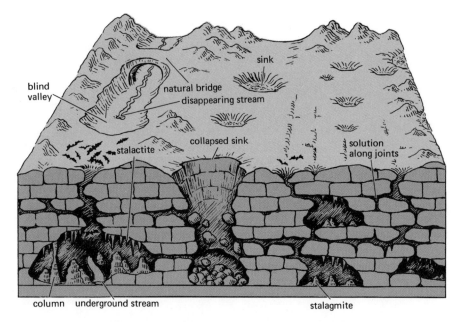

Fig. 11-8 Karst topography and cave features.

contain dry valleys, holes called *sinks*, natural bridges, and disappearing streams that flow into sinks (Fig. 11-9). Downward seepage and water moving slowly beneath the water table corrode out the cavern systems.

Limestone caves are ornamented by a variety of intriguing features, largely evaporation deposits of dripping water highly charged with lime. Most impressive are stalactites, iciclelike limestone pendants from cavern roofs, and stalagmites,[3] mounds built beneath a dripping spot on the cavern floor. Stalagmites and stalactites may grow together, forming grooved and fluted columns (Fig. 11-10).

Because such dripstone features evidently formed in air-filled cavities, it has been proposed that caves are created by underground streams flowing above the water table. However, some caves lined with sharp calcite crystals—resembling

supergeodes—must have been water-filled when such crystals formed. Thus the origin of caves is still in doubt; locally one or the other theory may be correct, but as with many geologic problems, the answer, in many instances, may be a compromise involving elements of both.

RIVERS

Grand Canyon is an awesome gash, 1.6 kilometers (a mile) deep and more than 16 kilometers (10 miles) wide, cut through thick beds of colorful sedimentary rocks, forming cliffs and slopes that recede from a somber inner gorge of crystalline rocks. At the bottom the Colorado River is glimpsed as a tiny brown thread (Fig. 11-11). A park ranger, concluding his lecture to a group on the rim, called Grand Canyon "a monument to erosion by the Colorado River." One tourist asked, incredulously, "Do you really believe that little river cut that great

3 To help remember which is which: stalactite has a *c* for ceiling and stalagmite, a *g* for ground.

Fig. 11-9 Stream disappearing into a sink hole. Courtesy of Wayne M. Sutherland.

big canyon?" The question echoes a debate between scientists in an earlier day. Do stream's make their valleys, or are valleys created separately and later occupied by streams?

In the eighteenth and nineteenth centuries, the conservative Catastrophists believed valleys were formed suddenly, in tremendous floods or violent rendings of the ground. Streams then spilled into the newly formed valleys, which suffered little change thereafter. The opposing uniformitarian view, that most valleys are slowly excavated by their streams, was stated in 1802 by John Playfair in his Law of Accordant Stream Junctions. Its essence is that streams flow in valleys proportionate to their size, and that streams in a drainage network are so nicely adjusted to each other that their junctions are accordant (joining at the same level). This, he held, would be most unlikely, unless each valley was carved by the stream in it.

Gullies, tens of meters deep and hundreds of meters long, forming within a few generations, as well as the load of mud and gravel moving in turbid streams, do indicate the erosive power of running water (Fig. 11-12). But in fairness to the early Catastrophists, such indications are inadequate to account for a major valley like Grand Canyon—if the Earth is only a few thousand years old, the age that theologians had established from biblical research. The uniformitarian philosophy demands an appreciation of the immensity of geologic time.

Fig. 11-10   Stalactites and stalagmites of Carlsbad Caverns, New Mexico. National Park Service photo.

Fig. 11-11 The Grand Canyon of the Colorado River. Photo by J. H. Maxson.

Stream Systems

Rain striking the Earth flows off in thin sheets that soon concentrate into linear rills on the uneven ground. Rills combine as brooks and creeks that feed rivers which, in turn, join in ever-enlarging streams.

Drainage Networks Most drainage networks are *dendritic* having a treelike pattern of a trunk river with branching tributaries. Dendritic patterns characterize either near-horizontal layers or homogene-ous rock of the same resistance to erosion through-out. In the *trellis* pattern, which resembles a lattice, short tributaries are at right angles to long axial streams. This pattern prevails in eroded rock folds, where the main streams follow the axes of weak beds in valleys and the short tributaries flow down the slopes of more resistant beds. The *rectangular* pattern has right-angle junctions like the trellis, but lacks dominant long-trunk streams. Both tributaries and main streams have sharp bends along their courses. The pattern develops where streams are controlled by weak zones along fractured rock. A

Fig. 11-12 In parts of Georgia, gullies such as these in Stewart County have expanded notably in a few generations, threatening roads and farms. Courtesy of Soil Conservation Survey.

radial pattern of drainage characterizes streams flowing outwards from the highpoint on a volcano or structural dome (Fig. 11-13).

Thus the distinctive drainage patterns allow the prediction of bedrock and structure from examination of maps and air photos before visiting a region on the ground. From the broader view of drainage systems, let us turn to the individual streams.

Stream Volume Stream *discharge* is the amount of water flowing through a cross-section of the channel in a given time, and is usually measured in cubic feet per second. Some streams have a constant discharge throughout the year; others are variable, commonly high in spring and low in fall; and some, especially in arid regions, are intermittent, flowing during or immediately after a rain or when fed by melting snow, but at other times being completely dry.

Climate controls the discharge of streams, both directly and indirectly. Expectably, relatively constant rainfall throughout the year tends to give streams a regular discharge, while periodic or seasonal rainfall leads to fluctuating flow. Climate also affects the groundwater nourishment of streams. In humid regions where the infiltration is great, high water tables slope towards and intersect stream channels. Continuous seepage into the channels insures a constant flow, even after many clear bright days. Water tables in arid regions are usually below the channel bottom. Streams flow intermittently when charged by surface runoff, but groundwater adds nothing to sustain the stream between periods of rain. Here, water from the channel seeps down to the water table, diminishing the volume of the stream.

In well-vegetated humid regions, streams have

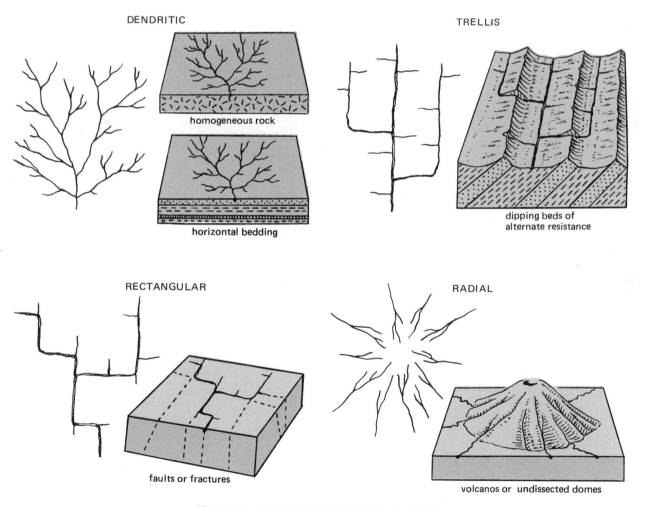

DENDRITIC

homogeneous rock

horizontal bedding

TRELLIS

dipping beds of
alternate resistance

RECTANGULAR

faults or fractures

RADIAL

volcanos or undissected domes

Fig. 11-13   Common types of drainage patterns.

a more constant flow because plants impede sheet flooding and encourage seepage to the water table. Less water from storms or snow-melt runs overland and soaks more slowly into the ground to augment the streams in dry spells. On the barren slopes of desert regions, most rain runs off rapidly on the surface, swelling streams during storms but adding little to the underground reserve (Fig. 11-14).

Rock also influences stream flow. Impervious clays and shales allow little infiltration and cause flashy runoff, so that stream discharge is irregular or intermittent. Streams on permeable rock, such as sandstone, have a more constant flow, since much of the rainfall penetrates the ground.

The Work of Streams The obvious function of streams is to drain the land; in the process, they

Fig. 11-14 Removal of vegetation by overgrazing of California range land caused this gullying, except in the spot protected by the oak tree. Rapid runoff during storms actively erodes the bare slopes. U.S. Forest Service photo by N. W. Talbot.

Fig. 11-15 Hydraulic action during flood in Whatcom County, Washington, destroyed farm land and undermined this barn on the Ray Syre farm. Photograph courtesy U.S.D.A. Soil Conservation Service.

act as the major gradational agent. Waves and currents work only on the continental margins. Wind erosion is limited by the low density of air and by protective covers of vegetation. Glaciers are restricted to cold regions of relatively abundant snowfall. Streams, however, dissect the ground in deserts as well as in humid regions from the arctic to the tropical climate zones. An estimated five million kilometers (over three million miles) of channels, from rills to major rivers, scar the face of the continental United States.

Streams erode in three main ways, the first of which is by the force of moving water alone. Although this *hydraulic action* (Fig. 11-15) is most effective on loose material—sand, gravels, boulders, and clay—it also quarries large fractured blocks from solid bedrock. In *abrasion*, which effectively erodes gorges in solid rock, a stream uses cobbles and boulders as tools to pound and grind the channel deeper (Fig. 11-16). *Corrosion*, the chemical solution of rock, is most important in eroding limestone and other soluble terrains.

The ability of a stream to transport is defined in two ways. *Capacity* refers to the total amount

Fig. 11-16 Boulders in Tantalus Creek corraded its channel. Photo by J. K. Hillers, photographer for Major John Wesley Powell, the first geologist to descend the Colorado River by boat. U.S. Geological Survey, National Archives.

of a given size of material that a stream is capable of transporting. *Competence* relates to the largest particle that a stream can move. Thus a mountain torrent capable of moving great boulders has high competence, yet if the stream is small, its capacity would be far less than a slower-flowing river, carrying a much greater total load.

Material moves along a channel bottom in the *bed load*. Larger fragments move by *traction*, a dragging or rolling along the channel bottom. Some smaller material is transported in *saltation*, a hop-

ping or jumping of particles along the stream's floor. It is the material carried as *suspended load* that gives a muddy look to a river. Fine sands and silts are held in *simple suspension* by the pulsing and swirling, characteristic of the turbulent flow of streams. Clays, the finest clastic particles, travel in *colloidal suspension*, held aloft by the constant jostling of water molecules (Brownian movement) which never ceases even in still, standing waters. The finest particles of all, the invisible ions dissolved from rock, are transported in *solution*.

Bed load and the suspended sands and silts are deposited when the velocity of flow decreases, as when a mountain torrent emerges on a flatter valley floor, or when stream flow is completely checked by flowing into the standing water of ponds, lakes, or the oceans. When a stream recedes from flood, the gradual decrease in velocity tends to sort clastic material, because coarser particles drop out first and progressively finer materials settle in the higher layers.

Colloidal material settles by the process of *flocculation*. Clay particles remain suspended in fresh water because each miniscule particle carries a negative electric charge. Since like charges repel each other, clays remain dispersed. But on entering the sea, they are neutralized by positive charges in the salt water. No longer repelling each other, the clay particles come together, or flocculate, forming clots that sink to the bottom.

Dissolved substances remain in solution, adding to the saltiness of lakes with no outlets and, on a grander scale, to the salt of the sea. However, when water is completely evaporated, ions deposit in a salt crust. Salts may also be precipitated from concentrated solutions by changes in temperature, or other physical-chemical conditions, before evaporation is complete. Some ions are extracted from the water by animals like clams, oysters, and coral, which convert them into relatively insoluble shell materials, a major constituent of many limestones.

The Well-Adjusted Stream To attribute human qualities to nonhuman things is anthropomorphism, which is frowned on in proper scientific circles. Yet, in their dynamic action, streams show one almost human quality. They tend to do their job with the least possible work—a condition of equilibrium characterizing graded streams, whose capacity is nicely balanced with the load they transport. Normally, however, no stream remains in perfect equilibrium. In floods, the increased velocity and discharge give a stream excess energy to scour and deepen its channel, while in low-water stages it deposits along the channel bottom. Thus a graded stream is like a tight-rope walker, out of balance in one direction and then the other, but keeping a general equilibrium.

The long profile of a stream, from mouth to source, has many local irregularities, yet, overall, it approaches a concave curve of increasing steepness upstream (Fig. 11-17). Why should the profile steepen if the stream is generally adjusted along its length? Of the many complex factors involved, three seem most important. The Mississippi, and other rivers as well, shows a progressive downstream decrease in the size of particles transported. Streams have a greater capacity for fine material than coarse; therefore, a lesser gradient should suffice to keep the stream in balance. Also, the discharge of a river increases downstream, as the number of its tributaries grows.

Careful measurements show that downstream velocities remain constant or even increase, which is surprising, for water should logically flow more

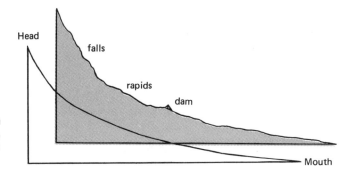

Fig. 11-7 Idealized long profile of a stream from source to mouth, represented by the smooth curve, as compared to the less regular profile found in actual streams.

slowly in a gentler channel gradient. The key to the paradox lies in the number, shape, and size of channels. When small tributaries unite in a single channel, the water surfaces are accordant; however, the single, large channel is deeper.[4] The frictional drag on flowing water is proportionately less, because a single deep channel carries water more efficiently than several small ones. Hence, the combined water volume continues flowing at the same rate, or even accelerating, on the gentler gradient, and its capacity remains adequate for the overall load.

Most graded streams have been disturbed at some time in their history. Of the various possible causes, a change in load is often important. Increased load commonly results from weakening or destruction of vegetation by forest fires, excessive timbering, improper farming, and other geologically short-term factors. Accelerated erosion on the unprotected slopes supplies an overload to streams in the affected part of a basin. The ultimate return of vegetation would reverse the process (Fig. 11-18). The effects of climatic fluctuations, which have been pronounced in the last million years of Earth history, are most important, widespread, and long-lasting. Such changes not only alter vegetational patterns, but may markedly upset streams by increasing or decreasing their discharge.

Overloading of a graded stream starts a sequence of cause and effect that eventually leads to a new and more steeply graded profile. A stream whose capacity is exceeded deposits the extra load, back-filling its channel and the valley bottom. As the channel steepens, the stream flows faster, and its capacity increases. The increasing capacity eventually comes in balance with the load again, in a more steeply graded profile. A decrease in load leaves a once-graded stream with excess energy for downcutting. As the channel deepens, its gradient decreases. As a result, the stream's velocity and

capacity also decrease, until equilibrium with the lesser load is achieved in a new, graded profile, downcut to a less steep gradient.

The dynamic behavior of streams is a matter of practical importance. For example, when the Hoover Dam was completed in 1935, it upset the equilibrium of the Colorado River. Flowing through a desert, the Colorado is a heavily loaded stream that deposits a delta on entering artificial Lake Mead. As was anticipated in the planning, the growing deltaic deposits will ultimately fill the lake —in 400 years by a recent estimate—ending the dam's usefulness. The dam, however, will more than pay its cost through the sale of electrical power. The water returned to the channel below the dam is clear, its sediments trapped in Lake Mead. Freed of its load, the Colorado River deepened its channel from 6/10 of a meter to 2 meters (2 to 6 feet) for a distance of 16 kilometers (10 miles) downstream, within six months of the dam's completion. Thereafter the scouring lessened as a newly graded profile was approached. In other cases, where irrigation water comes directly from a river below a dam, channel deepening has economic consequences: irrigation becomes expensive if pumps must be installed to lift water into a ditch system.

Stream Sculpture

A landscape is a mosaic of smaller forms of two genetic types: erosional features carved in bedrock, and depositional ones built from rock debris.

Erosional Landforms Valleys are cut by streams, yet if this were the only process involved all valleys would be vertical slots, since a stream channel, like a saw, works only in the valley bottom. The familiar retreating slopes of valley walls result from weathering, mass movements, and sheet wash, all of which supply part of a stream's load. Throughout their evolution, streams both deepen and

4 It is apt to be much wider, too, but this does not seem to have such a direct connection.

Fig. 11-18 Photograph shows the erosion protection given by rocks, twigs, and leaves on the ground. Each rock, leaf, and twig is supported by a little pedestal of dirt. The surrounding soil has been washed away by rain. South Portal Canyon, Saugus District, Angeles National Forest, California. Taken by L. E. Berman, August 1954.

widen the valley bottoms. The importance of vertical or lateral cutting varies, however, with a stream's phase of development.

Before reaching the "well-adjusted" condition of equilibrium, the stream has waterfalls and rapids and excess energy that is largely devoted to downcutting. In this phase, valley walls slope, directly to the channel, giving a V-shaped cross profile (Fig. 11-19). On reaching equilibrium, the marked irregularities in the channel disappear; the stream

is graded, its load matching its capacity. The stream may still downcut slowly, but these effects are masked by the now-dominant lateral cutting that undermines the valley walls, creating a flat-floored valley bottom. This low, flat strip, inundated at high water, is a flood plain and is commonly floored by alluvium. The *flood plain* is a narrow strip in its early evolution, but, with time, it characteristically broadens considerably through the retreat of valley walls (Fig. 11-20).

Fig. 11-19 The Canyon of the Yellowstone River shows a youthful stream marked by a steep V-shaped valley, falls, and rapids. Geological Survey photo taken in 1871 by W. H. Jackson, National Archives.

Valley walls in humid regions are smooth slopes because thick soil and waste mantle, anchored by abundant vegetation, mask inequalities in the resistance of underlying rock. The canyon walls of arid regions commonly have steplike outlines, with cliffs of resistant strata, like sandstones and lava flows, that alternate with gentler slopes on softer beds such as shale (Fig. 11-21).

Widespread erosion of near-horizontal rock layers, some hard and some soft, produces *mesas* and *buttes* (Figs. 11-22, 11-23). A mesa[5] is a flat-topped hill, usually capped by a resistant layer that often protects weaker rocks beneath. A butte is simply a smaller residual pinnacle, lacking an extensive flat top (Fig. 11-24). *Hogbacks*, or homo-

5 The word in Spanish means "table."

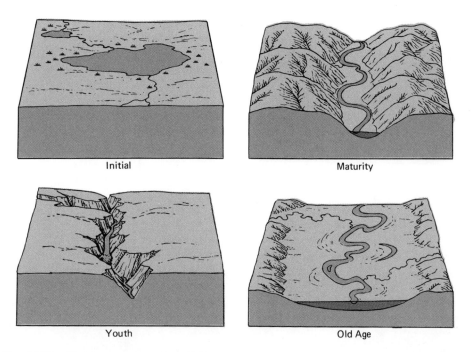

Fig. 11-20 Stream evolution described in stages of a "life" cycle. Initial stage lacks a valley, has swamps and ponds. In youth, downcutting produces a valley and eliminates swamps and ponds as it drains more efficiently. The mature stage has a continuous flood plain indicating the dominance of lateral cutting when stream achieves equilibrium. The stream cycle has reached old age when the flood plain is very broad and marked by abandoned river meanders forming oxbow lakes.

clinal ridges, are the typical erosional landforms developing on tilted strata of alternate resistance (Fig. 11-25). They are common where mountain-building forces have buckled rocks into folded structures. *Water gaps* are notches cut through hogbacks by streams; *wind gaps*, despite their name, have a similar origin, but have been abandoned by their parent stream.

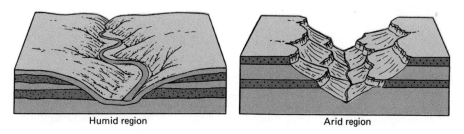

Fig. 11-21 The softer slopes of valleys mantled by soil in humid regions, whatever the underlying strata, contrast with the sharper outlines of arid canyons, especially where rocks have differing resistance to erosion.

Fig. 11-22 Mesas and buttes.

The *pediments* of arid regions are smooth surfaces beveled impartially across hard and soft bedrock. They slope gently away from an angular junction at a mountain foot and are usually blanketed by gravels that are thin, or absent, along the mountain front but thicken considerably into adjacent valleys (Fig. 11-26). Though pediments are conspicuous desert landforms, their origin is still in question. According to one hypothesis, they develop through lateral planation, where streams emerging from a mountain valley are so nicely balanced that they neither downcut nor deposit extensively. Rather, they swing back and forth, carving the pediment by undercutting the mountain front.

Fig. 11-23 1822 lithograph by Carey and Lee showing mesas just east of the Rocky Mountain front. Courtesy of the Library of Congress.

Fig. 11-24 Monument Valley in the Colorado Plateau along the Utah-Arizona state line. Photo courtesy of Tad Nichols.

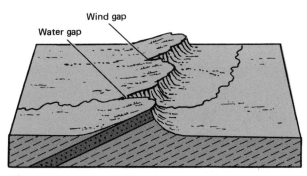

Fig. 11-25 Hogback ridge.

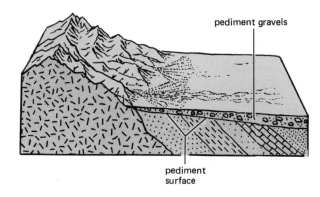

Fig. 11-26 Pediment.

Fig. 11-27 Gemini IV view of the Nile delta, whose area is roughly 1,300,000 square kilometers (500,000 square miles). Upper right are Gulf of Suez (nearer) and Gulf of Aqaba (farther) with Sinai Peninsula between them. The gulfs are developing over spreading axes of tectonic plates. Courtesy of N.A.S.A.

Depositional Landforms   Unlike the erosional features carved into bedrock, depositional landforms are built from unconsolidated materials, the load being dropped by streams when their capacity is reduced.

Where streams flow into standing water, *deltas* of various shapes develop. The delta of the Nile is arcuate, roughly triangular in plan view, suggesting on a grand scale the Greek letter from which all these landforms are named (Fig. 11-27). The Mississippi delta with its projecting fingers exemplifies the bird-foot, or lobate, type. The Tiber delta, a less common variety, is cuspate, suggesting a sharp tooth. Estuarine deltas, such as the one at the mouth of the River Seine, are confined by narrow bays (Fig. 11-28).

The shape of a delta is influenced by many factors, including the configuration of the shore, the material being deposited, and the action of shore processes. Arcuate and cuspate deltas reflect the deposition of loose sandy material which is readily reworked by waves and shore currents. The intri-

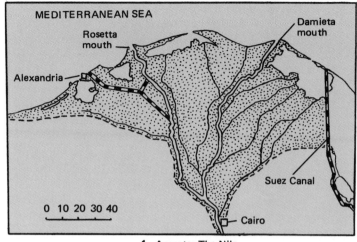

1. Arcuate: The Nile

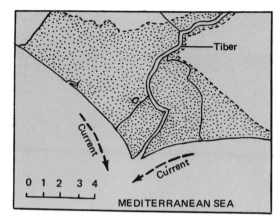

3. Cuspate: The Tiber

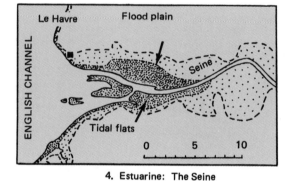

4. Estuarine: The Seine

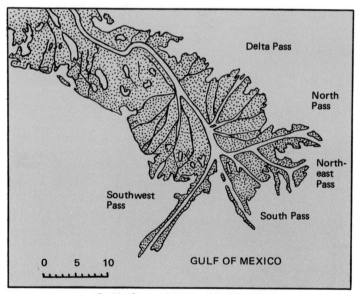

2. Birdfoot or Lobate: The Mississippi

Fig. 11-28 Types of deltas: arcuate, lobate, cuspate, and estuarine. Scales in kilometers.

cate bird-foot deltas usually contain much floccu-lated clay, which is relatively tough and cohesive. Where a river empties into a narrow bay, the out-line of the shore becomes the major factor in pro-ducing the estuarine type of delta.

Alluvial fans are aprons of stream debris laid down where canyons open onto a broader valley floor (Fig. 11-29). Deposition here results from the decrease in a stream's velocity and capacity on reaching a more gently sloping floor. Since the

Fig. 11-29 Alluvial fan.

stream's position is fixed in the valley, but free to migrate on the flat beyond, it swings back and forth, building the fan-shaped deposit of alluvium.

The most characteristic depositional landforms of graded streams, *flood plains,* contain a wealth of associated features. Although most streams are sinuous to some degree, those on flood plains often have meanders, symmetrical bends that shift and grow (Fig. 11-30). Like many other geologic phenomena, the commonplace meandering is not easily explained. It has been described in great detail, without leading to general agreement concerning its cause.

Studies by the Mississippi River Commission, using very large models, suggest that streams mean-

Fig. 11-30 Meandering course of the Laramie River across a flood plain marked by the scars of older abandoned meanders. U.S. Geological Survey photo by J. R. Balsley.

der because they flow in easily eroded material, which would be mostly alluvium. Observations indicate that slight bends in a channel throw the strongest thread of current against the outside bank. Caving along this bank supplies material that moves a short distance down the same side of the channel and is then deposited, thus deflecting the current there against the opposite bank, which is in turn undermined, as the process is repeated down the course of the channel. Once started, the meanders expand laterally by undercutting their outer banks while depositing in the slack water on the inside of the bend, where they build a gradual "slip-off" slope, or point bar.

Reasonable as the mechanism seems, it may not be the whole answer, because streams flowing on glacial ice, where there is no bank caving, also meander. A theory that accounts for such meandering, and perhaps meandering in general, dates from the late nineteenth century. Water has a helical (or corkscrew) flow in bends because the more rapidly flowing surface water is thrown to the outside of a bend by centrifugal force, causing a slightly higher water level there. Water on the channel bottom is displaced towards the inside of the bend. As a result, material eroded in the more turbulent and faster-flowing outer side of the bend moves toward the inside of the bend, where it settles out in the quieter water, thus building the point bar. This mechanism may be a more general explanation, for it accounts for meandering in wide and relatively shallow rivers where the banks are far apart.

Whatever their basic cause, meanders do not grow indefinitely, for as they expand the neck of land projecting into the bend is narrowed. Eventually a short-cut forms, either when flood waters jump the neck forming a channel in which the stream remains, or when the neck is gradually cut through by bank erosion on either side. Once formed, the shorter route through the cutoff gives the channel a steeper gradient. This increases the stream's velocity and, consequently, bank erosion downstream, encouraging the growth of a new meander. *Oxbow lakes* are abandoned meanders whose mouths are gradually sealed off by silt and mud expelled from the faster-flowing water in the new cutoff. Overall, a meandering stream keeps a relatively constant length and a graded profile, because meander growth is balanced by cutoffs (Fig. 11-31).

The Regional Erosion Cycle   Individual landforms are existing details in a broader, and—in the perspective of geologic time—ever-changing scene. The regional erosion cycle, proposed by William Morris Davis (1850–1934) of Harvard, is the most complete presentation of the concept that a landscape gradually evolves through time. A cycle begins with an uplift, caused by mountain-making forces within the Earth, and streams start downcutting. Through time, the affected landmass is worn down through a distinctive sequence of stages culminating in a *peneplain*, a low rolling surface of broad extent near sea level (Fig. 11-32).

Early development, the stage of *youth*, is characterized by high, flat areas, remnants of the uplifted surface between deepening valleys. The region stands high, and much rock must be removed before the cycle is completed. As time goes on, valleys deepen and the landmass becomes more intricately dissected, as the upland flats are gradually destroyed.

A region in the *mature stage* is largely cut into slopes, the original high flat has disappeared, relief is at a maximum, and in general the scenery is most rugged. Landslides and other mass movements are active on steep valley walls. As time progresses, the relief decreases and low flats—flood plains—appear in valley bottoms.

Subdued topography distinguishes the *old-age stage*. Broad valleys are floored with thick alluvium in flood plains whose meandering streams lower the land at exceedingly slow rates. The bedrock in the subdued divides between stream valleys is deeply weathered. Low hills, called *monadnocks* after the mountain in southern New Hampshire, are scattered relics of the once-impressive mountains. Eventually

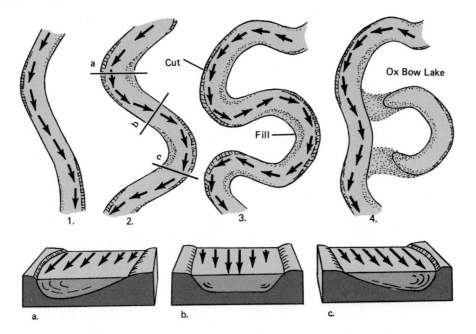

Fig. 11-31 Meander development, ending with the creation of an oxbow lake. Arrows in drawings 1–4 show thread of fastest current. Overall water movement is shown by arrows on cross-sections a, b, and c, locations of which are marked on drawing 2.

the surface becomes a peneplain near sea level, the ultimate base level to which streams can erode the land.

At any stage, renewed uplift may set the cycle back. The result is a complex landscape, marked by landforms of disturbed streams and features of both cycles, until the rejuvenated streams finally obliterate the last vestiges of the earlier cycle.

The concept of the erosion cycle is most valuable in bringing home the gradual evolution of a landmass, in uniformitarian terms, through the constant flow of water in the hydrologic cycle. Moreover, it is very useful in classification, for youth, maturity, and old age are terms that create a strong visual image of topography, without resort to detailed and often tedious empirical descriptions which treat every landscape as if it were somehow different from all others. Yet the Davis concept has been criticized generally in recent times, in part be-

cause it is an oversimplification, a nice device for elementary teaching—which it is—but also because it is hard to apply to actual topography. Most landscapes have had a far more complex history than the scheme suggests. Certainly, as Davis realized, a relatively rapid initial uplift of the Earth's crust, followed by an unbroken period of standstill, is most unlikely, because nowhere is the crust completely stable, and certainly not in regions of mountain-building, which is a prolonged process. Some geologists object to the scheme's pronounced anthropomorphism.

Dynamic Equilibrium A different model of regional development proposed by John T. Hack and others envisions topography as rapidly adjusting to differences in the bedrock of uplifted masses (Fig. 11-33). Thereafter the landscape changes little in form because a dynamic equilibrium exists between

topography develops on shales and other soft rocks whose fine debris can be moved on gentle slopes by streams of low gradient. Unlike the Davis scheme of the continuous gradual change in landscapes, dynamic equilibrium, once established, produces continuous removal of material without gradual evolution of a region through various different stages. Since rugged mountains and broad lowlands in uplifted regions often appear adjusted to the type of bedrock, the dynamic equilibrium theory seems a better model than the regional erosion cycle.

Landforms of Disturbed Streams Many landforms record events in Earth history that have interrupted the steady gradation of the crust by external processes. Uplifted and submerged shore features are of this sort, and so are various erosional and depositional landforms of streams. Some, being carved in hard rock, are relatively long lasting, and many record older events; others in unconsolidated alluvium are less permanent, but are more sensitive indicators of the effects of climatic change on streams.

Terraces in general are steplike surfaces, flat benchlike strips bounded on one side by a steeply rising slope and on the other by a steeply descending slope (Fig. 11-34). They may be developed in bedrock or unconsolidated materials and are common on many coasts, mountain fronts, and valley walls. They have many origins. Some stream-made terraces, such as the cliffs and broader slopes in hard and soft rock layers of barren canyon walls, result from continuous downcutting (Fig. 11-35); but others, not structurally controlled, indicate dynamic changes in streams.

Cyclic terraces, cut in homogeneous resistant rock, give a "valley-in-valley" cross-profile in many mountain regions. They form when a stream, flowing in a broad-floored valley, is rejuvenated and cuts a V-shaped inner valley, above which the flatter terrace tread is a remnant of the original valley floor. Cyclic rock terraces indicate that after a region was uplifted, it remained generally stable long

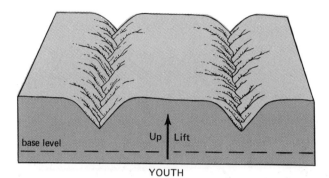

YOUTH

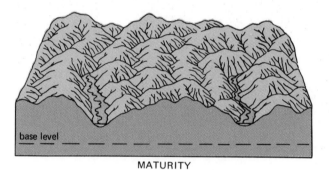

MATURITY

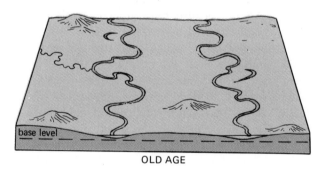

OLD AGE

Fig. 11-32 Diagrams of Davis's regional erosion cycle in an area of homogeneous rock.

material being supplied to streams by weathering and mass movement and the materials being carried away. Thus, a "mature" topography of the Davis cycle develops in resistant rocks whose coarse debris requires steep slopes for its removal. "Old"

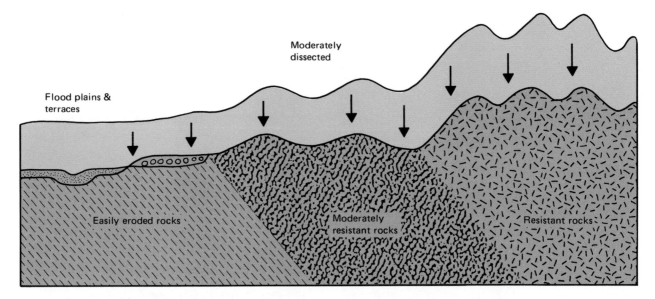

Fig. 11-33 Dynamic equilibrium, Hack's concept of equal downwearing of different rock types to produce diverse topographies.

enough for a stream to reach the equilibrium, or graded, stage, and form a broad-floored valley. Then, later uplift set the stream to downcutting again, which initiated the inner valley (Fig. 11-36).

Entrenched meanders may also indicate rejuvenation wherein a stream at equilibrium, and meandering broadly on a flood plain, has been disturbed. As a result, a winding V-shaped valley is incised along the old meander course (Fig. 11-37).

Alluvial landforms of disturbed streams normally record a history of alternate cutting and filling. Flood plains with thin or moderately thick alluvium are a normal feature of valley widening by graded streams; however, many flood plains, from those of nameless creeks to rivers as large as the Mississippi, have deep alluvial fills recording a history of major upsets in normal stream development. These flood plains indicate periods of aggradation, i.e., back-filling of a valley by overloaded streams. On an aggrading flood plain, streams commonly

have a braided pattern of many small channels that split and join because they are constantly being clogged and diverted by excess debris.

Alluvial terraces usually record at least three main geologic events. First, a stream downcuts to form a valley; next, the stream becomes overloaded and back-fills the valley, forming a thick flood plain; finally, the stream returns to normal downcutting and starts excavating the alluvial fill. Remnants of the thick flood plain now become terrace surfaces on either side of the downcutting stream (Fig. 11-38).

Unpaired terraces, which lie at different levels along opposing valley walls, develop during uninterrupted downcutting by a stream into the alluvium. Each terrace surface represents a former level of the river, as the whole channel migrates back and forth across the valley, while it cuts progressively deeper into the fill. *Paired terraces,* whose surfaces have comparable levels on both sides of a valley, have two general origins. They may represent spas-

Fig. 11-34 Terraces along the Madison River Valley, Montana, just west of Madison Gorge. U.S. Geological Survey photo by J. R. Stacy.

modic downcutting, in which case they are similar to cyclic rock terraces, except for the unconsolidated material in which they are formed (Fig. 11-39).

In other cases, however, paired terraces are formed by several alternations of downcutting and back-filling, a repetition of the three-fold sequence of events producing alluvial terraces. In other words, a stream cuts a valley into its alluvium until its equilibrium is again upset and it commences to back-fill again, creating a younger aggrading flood plain within the valley in alluvium; if the stream once more returns to downcutting, the surface of the younger flood plain becomes a terrace. Many rivers,

especially in the western United States, have a series of such terraces that result from recurrent changes in regional vegetation, stream discharge, and load. These terrace-making changes are in turn the results of a more basic cause—the marked fluctuations of climate during the last million years of Earth history, the time of the most recent Ice Age.

ICE ON LAND

We live in an Ice Age, a mild phase but, nonetheless, an Ice Age. Today 10 percent of the Earth's land surface, almost 15 million square kilometers (6

Fig. 11-35 Structural terraces developed in near-horizontal rocks of alternate resistance. Grand Canyon from Desert View, South Rim. U.S. Forest Service photo by B. W. Muir.

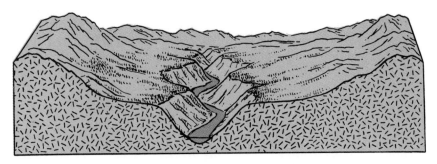

Fig. 11-36 Cyclic terraces cut in resistant rock.

Fig. 11-37 Gooseneck of the San Juan River. Photo of entrenched meander by J. H. Maxson.

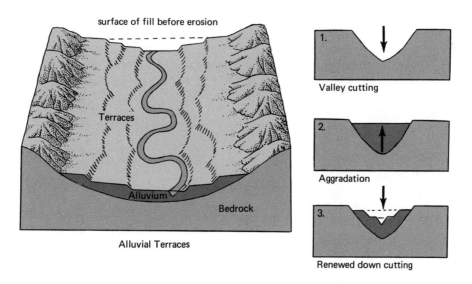

surface of fill before erosion

Terraces

Alluvium

Bedrock

Alluvial Terraces

1. Valley cutting

2. Aggradation

3. Renewed down cutting

Fig. 11-38 General sequence in creating alluvial terraces.

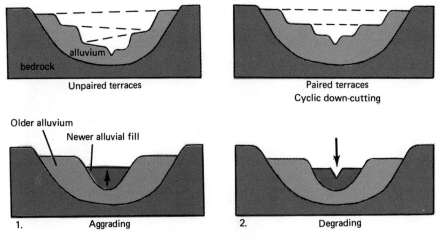

Fig. 11-39 Unpaired terraces and two possible origins for paired terraces.

million square miles), lies under glacial ice. The largest part is in the inhospitable Antarctic, much of the rest is in Greenland, and the remainder is widely scattered in high mountains more noted for their scenery than population (Fig. 11-40). In the not too distant past, in Pleistocene time, 30 percent of the land, about 46 million square kilometers (18 million square miles), was blanketed by creeping ice caps that rasped and gouged bedrock, dumping their debris around their fluctuating margins in ill-sorted bouldery ridges. One great glacier buried North America as far south as the Missouri and Ohio rivers; another spread over most of northern Europe and part of Siberia. Were such glaciers to come again, Moscow, Dublin, New York, Chicago, and many other major cities would be overwhelmed. Far beyond the ice, streams were upset, alternately back-filling and downcutting. Modern deserts were hospitable when the now-populous regions lay under the ice. Sea level fluctuated from 30 meters (100 feet) above to 100 meters (300 feet) below its present elevation, because water moving through the hydrologic cycle was alternately locked in ice

on land, and then released to the ocean as glaciers waxed and waned.

Some Glaciology

A glacier is a large mass of flowing ice that originates on land from recrystallized snow. Thus the ice surrounding the North Pole is not a glacier but, rather, the frozen Arctic Ocean, as the passage of nuclear submarines beneath the pole has demonstrated.

Kinds of Glaciers The shape of a glacier is controlled by the topography over which it flows and the thickness of the ice (Fig. 11-41). *Valley glaciers* are narrow tongues hemmed in by the valley walls of mountainous regions (Fig. 11-42). *Ice sheets* develop where relief is slight, or the ice buries topographic irregularities. They may be *high-level ice sheets,* which spread across plateaulike surfaces until concentrating in marginal valleys. *Piedmont glaciers,* such as the Malaspina glacier in Alaska,

Fig. 11-40 Glacial tongues pouring from the ice-locked heart of Greenland. U.S. Air Force photo, by permission of the Royal Danish Navy.

form where valley glaciers, emerging on broad flats at the foot of mountains, spread and merge in a broad sheet. The two existing *continental glaciers*, the Greenland ice cap, over 1⅓ million square kilometers (½ million square miles) in area, and the much larger Antarctic sheet, of about 13 million square kilometers (5 million square miles), bury all but the highest mountain peaks beneath ice up to 3 kilometers (2 miles) thick (Fig. 11-43).

The Origin of Glacial Ice Glaciers form where winter snowfall exceeds the summer losses from melting and evaporation over a long period of time. As a result, each year a layer of snow survives and is added to that below (Fig. 11-44). Newly fallen snowflakes are delicate, feathery ice crystals. With time the flakes lose their sharp, intricate outlines as water molecules melt from the edges, then migrate towards the center of the crystal and re-

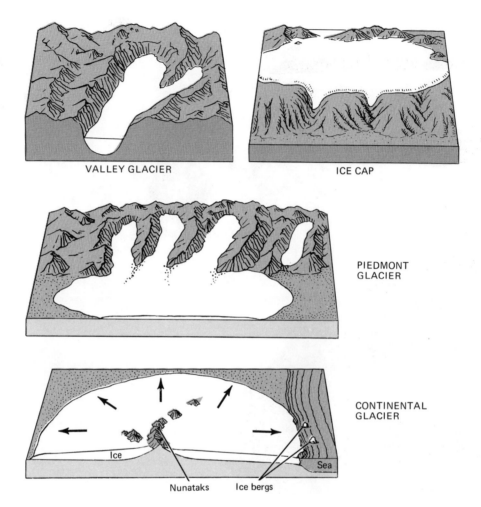

VALLEY GLACIER

ICE CAP

PIEDMONT GLACIER

CONTINENTAL GLACIER

Ice

Sea

Nunataks   Ice bergs

Fig. 11-41 Kinds of glaciers, not to the same scale. The valley glacier might be one and a half to three kilometers (one or two miles) wide whereas the continental glacier would cover hundreds of thousands of square kilometers.

freeze. Eventually, the snowflakes become grains of ice, called *firn* or *névé*. As the mass thickens, the weight of the accumulation compacts the loose firn and expels air to form solid crystalline ice. When the ice reaches a thickness of about 45 meters (150 feet), its static weight squeezes out the lower ice, causing it to flow as a glacier. The upper brittle part of a glacier is broken by large cracks, or crevasses, which close and disappear at the depth of flowage.

Macabre evidence that ice flows was provided by some unlucky mountaineers who fell into crevasses of glaciers in the Alps. Some 40 years later their remains appeared at the glacier's end, several

Fig. 11-42 A main valley glacier joined by two tributary valley glaciers. South Sawyer Glacier, Alaska. U.S. Forest Service photo by L. J. Prater.

kilometers down the valley. As evidence of flowage, however, boulders from distinctive outcrops will do as well. The plastic flowage of solid ice is thought to result from slippage along minute atomic planes within the individual ice crystals, so that the motion is somewhat like the slippage in a tilted deck of cards. In valley glaciers, the flow is gravitational down the sloping floors; in continental glaciers the static weight of overlying ice apparently squeezes the deeper ice out over sometimes flat, sometimes uneven rock floors.

Glacial movement is exceedingly slow, ranging from several centimeters (a few inches) to, at most, several meters (yards) per day. The flow of valley glaciers has been studied by driving a straight line of stakes across the surface. In a few days or weeks, the line bulges downstream, indicating a faster movement in the center, away from the retarding

Fig. 11-43 View toward the interior of Greenland showing the margin of the continental glacier. U.S. Air Force photo, by permission of the Royal Danish Navy.

friction of the valley walls. Movement at depth is shown by driving iron pipes into the glaciers. Pipes in valley glaciers slowly tilt downstream, showing that the surface ice moves faster than that below (Fig. 11-45).

The Glacial Economy Glaciers are dynamic. They expand and shrink, thicken and thin, in response to a delicate balance of nourishment by snow in their upper ends and wastage by melting and evaporation towards their lower ends (Fig. 11-46). The ice

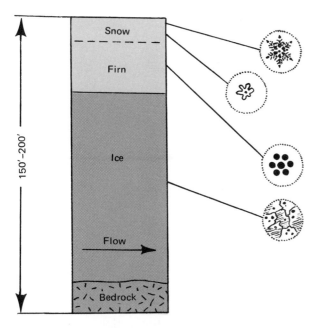

Fig. 11-44 Formation of glacial ice from snow. Thickness of zones is not to scale.

moves forwards whether the glacier's front is advancing, retreating, or relatively fixed. If accumulation of snow exceeds wastage, the front advances; if wastage predominates, the front retreats, because the forward flow of ice does not replace the loss. A perfectly stable front requires a balance of accumulation and wastage that is rare because of variations in snowfall and in average summer temperatures from year to year. Thus glaciers are sensitive indicators of climatic change.

Glacial Topography

Erosional Features   Glaciers erode by *abrasion,* the scraping or rasping action of debris-laden ice, and by *quarrying,* a lifting out of blocks from well-jointed bedrock (Fig. 11-47). Knobs overridden by ice are smoothed by abrasion on the side facing the ice's source, and roughened by quarrying on the lee, or downstream side. When loaded with fine debris, ice may buff down hard rock to a finish resembling highly polished building stone. Sharp cobbles, dragged over bedrock, make striations, or grooves, which are useful in reconstructing the direction of ice movement long after the glaciers have disappeared (Figs. 11-48, 11-49).

The erosional landforms created by continental glaciers are generally unimpressive, because the land, being completely buried by ice, is planed and smoothed as if by a great sanding block. The smaller valley glaciers carve and undermine the mountain masses rising above them, producing a

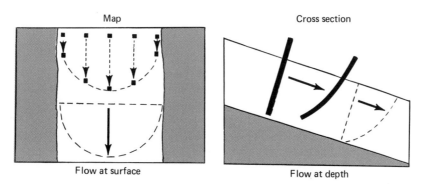

Fig. 11-45 Flow in valley glaciers. Map view shows shifting of points on surface with time. Cross-section shows movement of pipe indicating lesser rate of flow at depth.

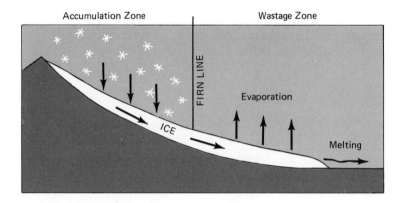

Accumulation Zone   Wastage Zone

FIRN LINE

Evaporation

ICE

Melting

Fig. 11-46 Economy of valley glacier. Firn line is the lower limit of firn (névé) at maximum of summer melting.

rough, spectacular topography that contrasts markedly with the full, round-bodied forms of unglaciated mountains modified by streams, weathering, and mass wasting. *Horns,* classically illustrated by the Swiss Matterhorn, are sharp pyramid-shaped peaks formed where encroaching *cirques,* broad glacial valley heads, converge towards a central summit. *Arêtes* are sharp knifelike ridges between glacial troughs (Fig. 11-50).

Glacial valleys originate in distinctive cirques: broadly rounded basins having an armchair shape. Cirques are actively cut back into mountains, es-

pecially during times of alternate freeze and thaw, such as spring and fall. During the day meltwater coursing down between the ice margin and the adjacent rock wall penetrates the cirque floor, while at night the water freezes, causing intense frost-shattering at the floor's edge that undermines the headwall (Fig. 11-51).

Glaciated valleys are characteristically U-shaped troughs, quite unlike the V-shaped cross-sections of stream valleys, and are produced when ice moves down preexisting stream valleys, gouging out the valley bottom and planing off the ends of

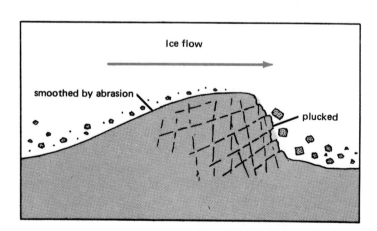

Ice flow

smoothed by abrasion

plucked

Fig. 11-47 Glacial abrasion and quarrying, or plucking, reflect the direction of ice movement.

Fig. 11-48 Outcrop smoothed and striated by glacial action. U.S. Forest Service photo by C. A. Duthie.

interlocking spurs in the process. Actually the U-shaped glacial trough corresponds in function to the infinitely smaller channel of a stream. However, a far larger trough is required to drain an area of creeping ice than of fast-flowing water (Fig. 11-52).

Streams in previously glaciated mountains seem a notable exception to Playfair's law. Tributaries flowing into larger glaciated troughs commonly emerge from *hanging valleys* whose lips create rapids and falls (Fig. 11-53). However, the lack of accordant stream junctions reflects the fact that the streams occupy abandoned glacial troughs, rather than valleys which they have cut. Glaciers widen and greatly deepen earlier stream valleys into the troughs, adjusted to the flow of ice. The ice surface at the junction of glaciers is usually accordant, hence conforming to the law, but the greater volume

of ice in the main glacier requires a larger trough, whose bottom is cut far below that of its tributaries.[6]

U-shaped valleys with less precipitous walls develop beneath continental ice sheets moving down old stream valleys. Such troughs, in upper New York State, are now occupied by the Finger Lakes, including Lakes Seneca and Cayuga. *Fiords,* deep, narrow bays typical of the spectacular glaciated coasts of Alaska, Scandinavia, and New Zealand, are U-shaped valleys that have been flooded by the sea (Fig. 11-54).

Depositional Features A glacier acts like a great conveyor belt. Material scraped from its underlying floor and, in the case of valley glaciers, cascaded

6 Stream channels have the same relations, but they are not so clearly shown because the channels are so much smaller.

Fig. 11-49 Glacial grooves, Flathead National Forest, Montana. U.S. Forest Service photo.

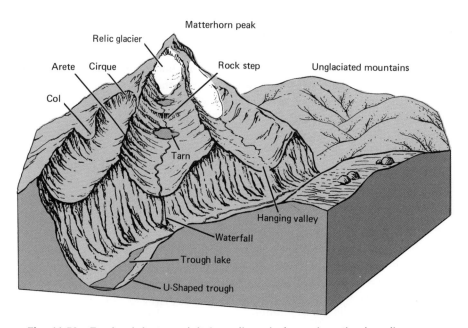

Fig. 11-50 Erosional features left by valley glaciers when the ice disappears.

Fig. 11-51 Steep-walled cirques in the high Wind River Mountains, Wyoming. Photo by Austin S. Post, University of Washington.

onto the ice from precipitous slopes above, is constantly carried to the glacier's end and dropped.

Erratics are individual blocks, some as large as a small house and weighing many tons, that are foreign to the bedrock on which they rest (Fig. 11-55). Trains of erratics stemming from a distinctive bedrock source have been used to reconstruct the direction and pattern of ice movement. The unconsolidated material forming the depositional landforms of glaciers is collectively called *drift*, a term inherited from the earlier misconception that

it was rafted in and dropped by melting icebergs when the land was submerged. Drift includes two sorts of deposits: *till* deposited directly by the ice, and *stratified drift* laid down by meltwater.

Till contains a wide range of particle sizes; for ice, being unselective in its transport, deposits clay, sand, cobbles, and boulders in an unsorted jumble. The cobbles and boulders are often striated, like the bedrock floors, and may also be snubbed, or flattened, where they were dragged across a rock floor. Stratified drift is better sorted, because the

Fig. 11-52 U-shaped glacial trough on south side of Tracey Arm, Alaska. Small glaciers still surround arêtes and horns of adjacent peaks. U.S. Forest Service photo by L. J. Prater.

fine materials are washed out of it, while the boulders and cobbles, too large to be washed along by water, are not carried into it. The washed deposits of stratified drift include those laid down along meltwater streams and those carried into lakes or ponds and the sea.

Moraines are the major landforms composed of till. *Ground moraine,* forming low plains, generally is a thin veneer that accumulates under a

glacier and prevails where a glacier's front was continuously advancing or retreating. *End moraines* are prominent ridges formed around glacial margins and are largest where a glacial front was essentially in equilibrium, so that debris was dropped in the same place.

The topography of end moraines tends to be smoothly rolling if the till is largely clay. A hummocky terrain of knobs and basins, which may con-

Fig. 11-53 Old lithograph of Yosemite Falls cascading into the glaciated trough of the Yosemite Valley. From the Library of Congress Collection.

Fig. 11-54 West arm of Fords Terror Fiord, Alaska. A waterfall spills from a hanging valley into the head of the fiord. Photo by L. J. Prater, U.S. Forest Service.

tain small ponds called kettles, characterizes coarse, gravelly, and bouldery tills, common in valley glaciers (Fig. 11-56). The end moraines of valley glaciers may form a loop a few kilometers across whose curved end is called a *terminal moraine*. It often merges with *lateral moraines* extending back along valley walls (Figs. 11-57, 11-58, 11-59).

The end moraines of continental glaciers form broad sweeping arcs, often traceable for many kilometers, that are usually lower than those of valley glaciers. Moraines of the former great Pleistocene ice sheets in North America and Europe are rarely over 45 meters (150 feet) high, whereas those of valley glaciers may be 150 to 300 meters (500 to 1000 feet) high. The reason for the difference seems to be that glacial tongues in mountains flow rela-

Fig. 11-55 Glacial erratics on Moraine Dome, Yosemite National Park. U.S. Park Service photo by R. H. Anderson.

tively rapidly down steep valley floors; hence they erode more actively and acquire a greater load from their floor, as well as receiving additional material falling from the higher valley walls.

Drumlins are streamlined hills of till, resembling giant inverted teaspoons, that appear in swarms behind an end moraine. Boston's historic Bunker Hill is a drumlin, in a field along the Massachusetts coast. Other drumlin fields occur in upper New York State south of Lake Ontario, and in Wisconsin west of Lake Michigan. Drumlins apparently form beneath glacial ice heavily choked with fine debris. The load is initially deposited on irreg-

ularities in the rock floor and then plastered on in successive crude layers building the hill, which may reach a kilometer or two in length and as much as 60 meters (200 feet) in height (Fig. 11-60).

Outwash plains, the most extensive landforms of stratified drift, usually extend outwards from moraines as broad aprons deposited by loaded meltwater streams issuing from glaciers. The south shore of Long Island, New York, is a flat outwash plain, bounded on the north by hilly moraines of the former continental ice sheet. Valley glaciers also develop distinct but smaller outwash plains. *Valley train* consists of meltwater deposits, fillings that

Fig. 11-56 Knob and basin topography of valley moraine in Jackson Hole, Wyoming. Photo by Herb Pownall.

give flat-floored bottoms to originally U-shaped glacial troughs (Fig. 11-61).

Eskers are long sinuous ridges of washed gravels and sands, frequently providing a natural route through swampy land whose drainage has been disrupted by glacial deposition. Forming behind moraines, eskers may lead into outwash plains. They originate as deposits from streams flowing through tunnels in the glacier; when the ice disap-

pears, the esker remains as a natural cast of the former stream channel.

Ice Ages and Their Causes

Since the early nineteenth century, when geology was first emerging as a science, glaciers have stimulated controversy and speculation. It started with the concept of an Ice Age.

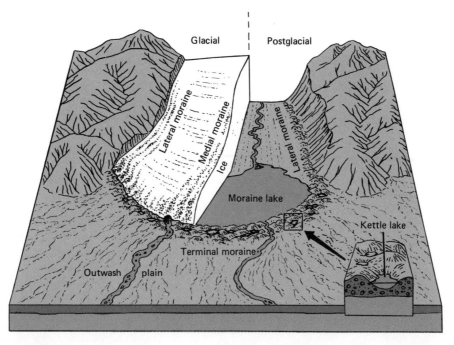

Fig. 11-57 Depositional features of valley glaciers schematically shown with and without ice. Inset shows detail of knob and basin topography. After W. M. Davis.

The Glacial Theory That the valley glaciers of the Alps had once been far more extensive than in recent times has long been recognized by Swiss peasants, whose farms and pastures contain abandoned moraines, scratched and polished bedrock surfaces, snubbed boulders, and other features like those developing around active glaciers. From just such a comparison of landforms and deposits around living glaciers with those now remote from the glacial ice, the geologic concept of an Ice Age became established. The former expansion of valley glaciers had been appreciated by several Swiss naturalists in the seventeenth century and was first formally proposed at a scientific meeting in 1821 by Ignatz Venetz-Sitten, a Swiss civil engineer.

The more startling idea, that vast ice sheets had recently overwhelmed much of Europe, was first suggested by Professor A. Bernhardi, who wrote in 1832 that glaciers had once extended from the polar regions well into Germany, on the basis of the patterns of moraines and erratics there. Louis Agassiz, a colorful Swiss zoologist, is often considered the father of the Ice Age; however, his main contribution—which is important—was to force the existing concept into the mainstream of scientific acceptance. Agassiz, originally a skeptic, became an enthusiastic convert to glacial expansion after touring the Rhone Valley glaciers with a Venetz supporter in 1836. In a paper the next year, he proposed a great period of ice, with a sheet extending from the polar regions to the Mediterranean (a slight overstatement, as the Alps were a southern island of valley glaciers). He attributed this glaciation to climatic change.

Today Ice Ages are acknowledged, and healthy controversy revolves around their causes and the more general problem of past climatic changes. Any satisfactory theory must encompass a number

Fig. 11-58 Looping end moraine of valley glacier confining a lake in Jackson Hole, Wyoming. Photo by Herb Pownall.

of geologic findings, most important being the fluctuations from warm to cold intervals during the late Ice Age and the spasmodic occurrence of glaciation far back through geologic time.

Multiple Glaciation The Pleistocene Ice Age began almost two million years ago and ended about 10,000 years ago. During the Pleistocene, the European and American ice caps had not one, but four, distinct advances, separated by warmer intervals when these glaciers disappeared. As each inter-

glacial time lasted more than 10,000 years, the present could well be an interglacial interval, to be followed by still another advance.

Other Ice Ages have occurred at random through geologic history. About 200,000,000 years ago, in late Paleozoic time, glaciation affected much of the Southern Hemisphere, including India, Africa, Australia, and South America. There is scattered evidence of at least one very ancient glacial period 700 million years ago, in late Precambrian time, on every continent but South America. Moreover, glaci-

Fig. 11-59 Mount McKinley, Alaska. Lateral moraines which become medial moraines within the ice tongue where two glaciers join. National Park Service photo by Lowell Sumner.

ation reflects only the extremes in a continuous pattern of climatic change. For even in long periods between Ice Ages, when the Earth's climate was largely subtropical, there were climatic fluctuations.

Of the many mechanisms proposed, none provides a generally accepted explanation for climatic change. The problem is complicated by the largely circumstantial nature of the evidence, and inter-

pretations require a synthesis of fact and theory from astronomy, archeology, botany, meteorology, zoology, chemistry, and physics, as well as geology.

Solar Variations The Sun being the original source of heat for the atmosphere, variations in solar radiant energy should change air temperatures, so that a period of lesser output might cause an Ice Age.

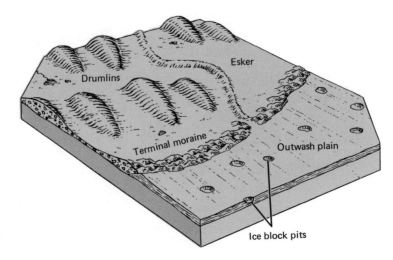

Fig. 11-60 Depositional landforms of continental glaciers.

Perhaps more significant are sun spots, which change the kind of solar radiation. When sun spots are abundant, the weather in the middle latitudes gets colder, the moisture-bearing storm tracks shift southward, and glaciers expand. Periodic variations in sun spots have been recorded since the mid-eighteenth century and, although insufficient to account for an Ice Age, larger, long-term variations might well be an adequate cause. Most solar theories account for glacial periods by the cooling of the atmosphere. Some, however, attribute Ice Ages to increased solar radiation, which would accelerate evaporation from the ocean surfaces, causing increased cloudiness and precipitation. Critics of this scheme point out that the increased precipitation would be largely rain, which does not nourish glaciers.

Planetary Motions Periodic variations in the motions and path of the Earth in space are also considered a cause that shifts the distribution of heat at the Earth's surface. Using known variations in the Earth's orbit, the tilt of its axis of rotation, and other factors, a complicated graph has been constructed showing the changes in solar heat received at various latitudes through time. The curve, most fully developed by the Yugoslav M. Milankovitch, does not provide a cause for Ice Ages, as its variations are regular and cyclic whereas the times of glaciation have been sporadic; however, it may well reflect the causes for lesser climatic variations. The curve has received support from a different line of study. Recent techniques allow the determination of the temperatures at which sediments in the ocean bottoms were deposited. Changes in ocean temperatures, determined from bottom cores, do correspond to the theoretical Milankovitch curve.

Changes in Atmospheric Components Some theorists stress variations in the amounts of carbon dioxide and water vapor, important in atmospheric heating because of their greenhouse effect. If these insulating gases decrease, thus locking carbon dioxide in rocks, as would be the case during periods of extensive limestone deposition, less reradiated heat would be trapped in the atmosphere, leading to climatic refrigeration. Increases and decreases of these gases do occur, but their quantitative importance, based on observed effects of the carbon dioxide released by burning coal and fuels since the

Fig. 11-61 Valley train extending from the ends of two ice tongues issuing from the margin of the Greenland Ice Sheet. U.S. Air Force photo, by permission of the Royal Danish Navy.

Industrial Revolution began, seem totally inadequate to bring on a glacial epoch.

A more geologic hypothesis also involves substances in the atmosphere. Suspended volcanic dust reflects the relatively short incoming waves of solar radiation, but readily allows the escape of longer reradiated heat waves from the ground (which is the reverse of the effect of the insulating gases). For three years after the eruption of Krakatoa in Indonesia, the incoming radiation at a solar observatory as far away as France was decreased by 10 percent. However, there is no geologic evidence suggesting continuous periods of excessive eruptions lasting for the thousands of years needed to create an Ice Age.

Geologic Factors Through geologic history, Ice Ages have been notably absent when continents are low and extensively flooded by the sea; and they do seem to correspond to times of extensive mountain-

Fig. 11-62 Louis Agassiz (1807–1873). Agassiz, originally trained in medicine and zoology, wrote several major publications on fossil fish. He began his work on glaciers and the Ice Age concept in Europe. Later, as a professor at Harvard, he organized the Museum of Comparative Zoology, extended the glacial concept to America, and gave exceedingly popular lectures.

building and general emergence of the land. Such is the case for the latest Ice Age, was probably true in the late Paleozoic glacial period of 200 million years ago, and could well have been true in the earliest glaciation althought its record is scant. The reason seems twofold. High mountains are breeding grounds for glaciers, and extensive emergence of the continents prevents ocean currents from spreading warm water into higher latitudes where it creates milder climates. The effect becomes even more pronounced, once glaciation starts, by the locking of water in ice on land.

The theory of polar wandering, which has had its ups and downs, now seems rather firmly established. Advocates held that glaciation results when the Earth's poles migrate onto land. In the late Paleozoic glaciation, the South Pole apparently lay in Africa; today it lies in the Antarctic continent, whose tremendous volume of glacial ice is largely responsible for the present lowered sea level and the resulting worldwide emergence of the continents.

In their hypothesis, William Donn and Morris Ewing assume the Pleistocene Ice Age began when the North Pole moved into the generally land-locked Arctic Sea. The sea lies in a basin that opens to the Pacific through the narrow Bering Straits and into the Atlantic at Hudson Straits, Baffin Bay, and across a broader expanse around Iceland, all of which are underlain by shallow platforms. Donn and Ewing propose that the Arctic Sea is ice-free during glacial times but freezes over in interglacial periods. This supposition, coupled with the effects of eustatic changes on ocean currents, provides a shutter mechanism, closing and opening the arctic basin to warm waters as the adjacent glaciers wax and wane.

When the Arctic Sea is unfrozen, evaporation from its surface leads to increased snowfall on North America, Europe, and Eurasia, and creates continental glaciers. The growth of the continental ice caps, storing water on land, lowers sea level. As the sea falls, the circulation of warmer waters from the major oceans into the Arctic is eventually cut off by the emergence of the basin rims, with the result that the Arctic Sea cools and freezes over. With the sea no longer a great evaporating pan, snowfall decreases, starving the glaciers, which recede and eventually disappear.

But the waning glaciers then release water that starts sea level rising until warm currents from the Atlantic once more spill into the arctic basin. This melts the frozen lid, opening the Arctic Ocean so that once again glaciers grow and advance. Thus a natural thermostat regulates glacial and interglacial alternations in the Northern Hemisphere, and will probably continue to do so until the Antarctic ice melts, raising sea level to a point where the arctic basin is permanently accessible to warm ocean currents.

Fig. 11-63 Aconcagua, South America's highest peak at 6960 meters (22,834 feet), is a maturely eroded volcano. It displays a wealth of stream and glacial features: dendritic drainage, youthful stream valleys, cirques, glacial troughs, and looping moraines, as well as features of mass-gravity movements. Photo by Wilhelm Mueller.

Neat and plausible though it sounds, the case for the Ewing–Donn hypothesis is not proven. The evidence for an ice-free Arctic Ocean during glacial maxima is scanty: a very few cores from the Arctic sea floor that may indicate warmer water during the maxima, and a few boulders dredged from the bottom that might be erratics deposited by glaciers flowing into an open sea. R. W. Fairbridge suggests that Ewing and Donn have overlooked the tremendous volume of Antarctic ice that could be the deciding factor. Some doubt that the shutter can account for the many minor fluctuations within the main glacial advances. And so it goes. As is so often the case with new hypotheses, more evidence is needed.

Ultimately, the explanation for Ice Ages, and the broader problem of climatic change, may well involve elements of many theories. Today changing climates and Ice Ages remain a challenging scientific puzzle.

SUGGESTED READINGS

Drury, G. H., *The Face of the Earth,* Baltimore, Md., Penguin Books, 1959 (paperback, Pelican Book).

Dyson, J. L., *The World of Ice,* New York, Alfred A. Knopf Inc., 1962.

Leopold, L. B., "Rivers," in *American Scientist,* Vol. 50, No. 4, pp. 511–537, 1962.

Leopold, L. B., and Langbein, W. B., *A Primer on Water,* U.S. Geological Survey, Miscellaneous Report, 1960.

Morisawa, M., *Streams, their Dynamics and Morphology,* New York, McGraw-Hill Co., 175 pp., 1968 (paperback).

Sharp, R. P., *Glaciers,* Condon Lectures Publications, Eugene, Ore., University of Oregon Press, 1960.

Shimer, J. A., *This Sculptured Earth,* New York, Columbia University Press, 1959.

Precambrian History

With some notion of physical geology in mind, we are now ready to tackle Earth history. The remote and hazy past will be first on the agenda. Although the records of the earliest events in the Earth's history were long ago erased from our dynamic planet, they can now be inferred from evidence preserved on the more static faces of our neighboring planets.

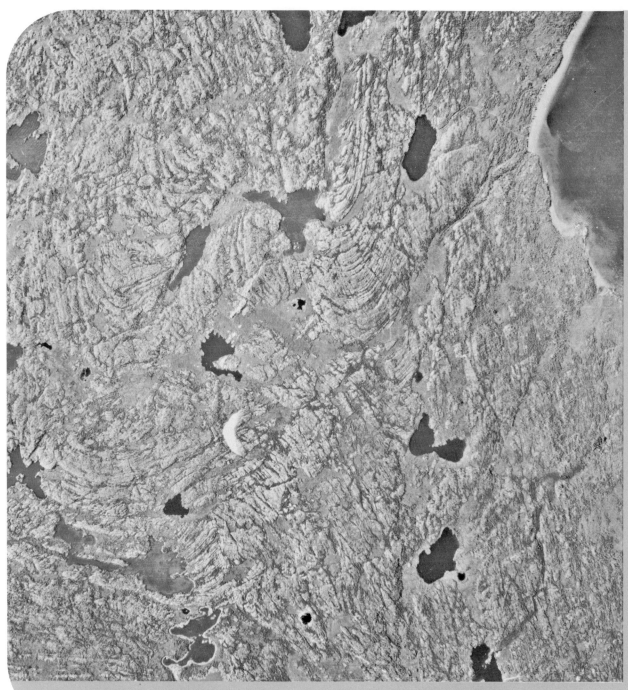

Canadian Shield. R.C.A.F. photo courtesy of Canadian Department of Mines and Technical Surveys.

12

The Precambrian Earth and our Neighboring Planets

The Earth's geologic history—like the history of civilizations—becomes progressively more vague as we search back through time. For while erosion, mountain-building, and igneous phenomena have continuously added new rocks to the geological record, they have also masked and erased the evidence of earlier events. Thanks to fossils, however, a coherent physical history has been compiled for the last 570 million years, back through the Cambrian period. Our reconstructions before the Cambrian period, however, are far less certain because of the difficulties of correlating geological evidence. The Precambrian story, representing three-fourths of geologic time, will remain the least known, even though radiometric dating is helping to clarify this most ancient era of Earth history.

PREGEOLOGIC HISTORY OF THE EARTH

Since the oldest known rocks are something over three billion years old, and the Earth's age is now estimated as over four and a half billion, much of our planet's earliest history is pregeologic, since the events cannot be directly reconstructed from rocks. Thus the birth of our planet, and the rest of the solar system, is in the realm of cosmogony, a speculative branch of astronomy dealing with theories of origin for the universe. Although it is in the broad province of geology, the Earth's early differentiation into core, mantle, and crust, hydrosphere and atmosphere, can only be deduced by applying chemical and physical theory to the existing architecture and conditions of the planet.

In the Beginning

The Origin of the Earth   Perhaps the Earth was born in a great cosmic accident, or perhaps it appeared in the normal evolution of a star called the Sun. These two schools of thought date back to the beginnings of scientific speculation on

Two-body accidental mechanisms	One-body evolutionary concepts

Two-body accidental mechanisms

1. Buffon

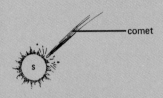

comet

Collison of sun and comet

2. Chamberlin

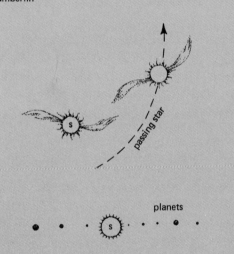

passing star

planets

Near approach of star

3. Hoyle

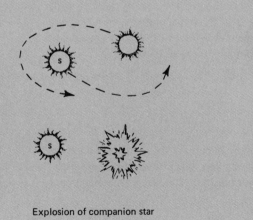

Explosion of companion star

One-body evolutionary concepts

4. Kant–Laplace

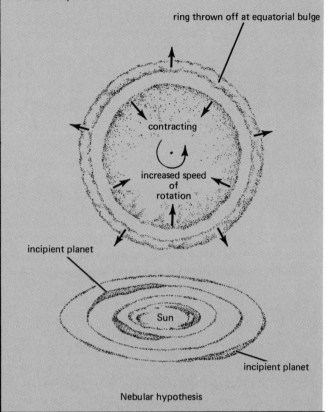

ring thrown off at equatorial bulge

contracting

increased speed
of
rotation

incipient planet

Sun

incipient planet

Nebular hypothesis

5. Von Weiszacker

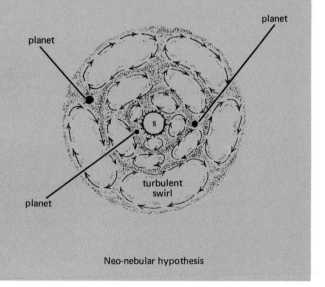

planet

planet

planet

turbulent
swirl

Neo-nebular hypothesis

the origin of the solar system. In 1749 Georges Louis Le Clerc, Comte de Buffon, suggested the first catastrophic explanation of Genesis, wherein a comet striking the Sun sent blobs of matter into space to form the Earth and other planets attending the Sun. About six years later, Immanuel Kant developed the first evolutionary hypothesis: that the Sun and its planets formed simultaneously by the gravitational accretion of particles in a contracting cloud of cosmic gas and dust. With modification, these basic themes have prevailed to the present (Fig. 12-1).

In 1796 Pierre Simon Laplace, apparently unaware of Kant's work, developed an evolutionary hypothesis based on a contracting cloud of hot gas called a nebula. In his thinking, the initial slow rotation of the cloud was accelerated as it cooled and contracted,[1] throwing off rings[2] that gradually coalesced into planets, while the larger central mass formed the luminous Sun. Despite flaws in its special mechanics, the Laplace nebular hypothesis dominated scientific thinking for a century.

About 1900 T. C. Chamberlin and F. R. Moulton of the University of Chicago proposed a new hypothesis, reminiscent of Buffon's comet. A comet, we now know, has far too little mass to disturb the Sun, but Chamberlin and Moulton substituted a near-collision with a passing star whose gravity pulled great tidal blobs from the Sun. On cooling, gas from the armlike[3] blobs condensed into dust, which gradually aggregated at centers that ultimately grew into planets. This hypothesis, with subsequent modifications, was in vogue until about 1950, when it fell victim to insurmountable mechanical objections, the chief one being that incredibly

hot masses pulled out of the Sun would explode into space and never coagulate.

Two presently contending hypotheses were formulated in the early 1940s. Since twin stars and stellar explosions producing novae are not uncommon, the British astronomer Fred Hoyle envisioned the Sun as once accompanied by a nearby star that exploded, with some of its debris coalescing into the planets. About the same time, C. F. Von Weiszacker of Germany, Fred Whipple in America, and others renovated the Kant–Laplace concept. They assumed a collapsing cosmic cloud of cold dust, wherein turbulent eddies eventually coalesced into knots whose gravitational attraction swept in more dust, and ultimately grew into the planets. The larger central mass, heated by tremendous internal compression, became the Sun. So, reinforced by new facts, deductions, and sophisticated calculations, the catastrophic and evolutionary schools survive to the present.

The evolutionary view that Sun and planets formed together has intriguing implications. If it is true, planets should normally accompany stars. Thus the possibility is good that other planets situated like the Earth (not too close and hot, or too far and cold, from a star) could have evolved higher, perhaps intelligent, forms of life. Of the catastrophic hypotheses, Hoyle's explosion of a twin star reduces the likelihood of planets and the possibility of life on other worlds. Chamberlin's hypothesis, considering the incredibly low odds for a chance near-encounter of stars in the vastness of space, makes other planetary systems most improbable, and life on Earth a unique cosmic accident.

The Embryonic Earth

Evolution of the Mantle, Core, and Crust During its pregeological history, the growing Earth must have differentiated into its major architectural

1 As a figure skater can speed up his spin by drawing his arms close to his body.
2 It's not recorded, but Saturn's rings could have influenced Laplace's thinking.
3 The arms may have been suggested by some of the spiral galaxies, which at that time were thought to be gaseous nebulae rather than great clusters of stars.

Fig. 12-1 Hypotheses for Earth's origin (facing page).

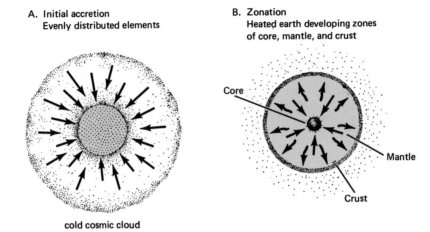

A. Initial accretion
Evenly distributed elements

B. Zonation
Heated earth developing zones
of core, mantle, and crust

cold cosmic cloud

Fig. 12-2 Hypothetical early and later stages in accretion and zonation of the embryonic Earth.

zones. Assuming the now-popular view that the planet developed from a cold cosmic cloud, the chemical elements must have initially been distributed evenly throughout the growing mass (Fig. 12-2). Subsequently, the kinetic energy of particles falling into the proto-planet generated considerable heat (added to by compressional heating and the decay of radioactive elements) and the Earth's mass grew. As a result, the Earth became molten; and as in a blast furnace, the excess of heavy iron and nickel filtered towards the center, forming the core, while the rest, combined with silicon and oxygen, produced the overlying mantle.

The basalt of the crust could represent lighter substances which separated from the molten mantle and floated to the top, like cream on milk; or if the mantle is eclogite, the basalt is a phase change in the outer zone of lower pressure. Whether the Earth was ever completely molten is debatable, but in any case the blast-furnace analogy is a generally acceptable explanation for the zonation into core, mantle, and the simatic part of the crust.

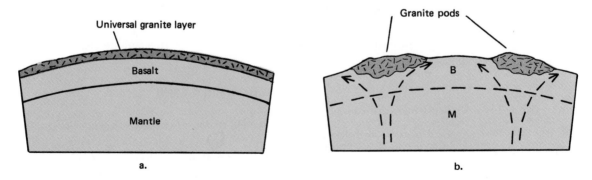

Fig. 12-3 Hypotheses for an initial granite crust: (a) universal layer; (b) isolated pods.

About Continents The origin of the granite forming the sialic layer in the continents involves two opposing viewpoints. In the older and still stoutly defended one, granite is a still lighter material that separated from the basalt, floated to the surface, and hardened (Fig. 12-3). Thus, ideally, granite should form a worldwide layer on top of the basalt. Early students attributed the interrupting ocean basins to the foundering of great crustal blocks. Aside from the difficulty of explaining how lighter sialic blocks could have subsided into the denser sima beneath, it is now known that the major ocean floors are devoid of granite and underlain entirely by basaltic rocks.

To account for the absence of granite from the major ocean basins, Sir George Darwin, son of Charles, suggested that while the Earth was still plastic, tidal forces tore out a great mass, forming the Moon and leaving as its scar the basalt-floored Pacific Ocean basin (Fig. 12-11). Thereafter it can be assumed that the relic patch of granite split asunder and its fragments drifted off until they froze in place, forming the existing continents. Subsequent calculations by Sir Harold Jeffreys proved tidal forces totally inadequate to remove a mass as large as the Moon; so he proposed that while the Earth was molten, convection cells stirring the mantle carried granitic magma to the surface, where it concentrated in superblobs, which later congealed into the continental platforms.

Recently the whole idea that continental masses of sial were derived from basaltic magma during the Earth's pregeologic history has been attacked. Granite, the proponents of the process of granitization believe, is mainly an end product of metamorphism accompanying the continuing action of the rock cycle. To them the absence of sial from the ocean basins, far from being an enigma, is expectable. They assume the initial crust was entirely basaltic, and granite has subsequently been evolving through time, causing an expansion of the continents (Fig. 12-4).

The continents were initiated where upwarps

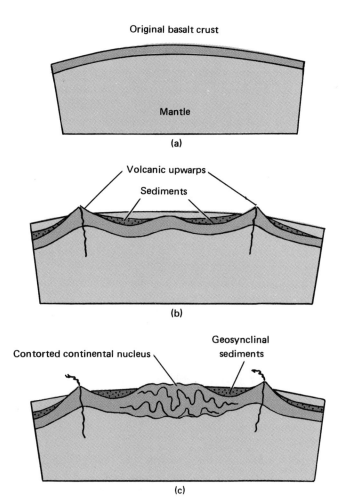

Fig. 12-4 Hypothesis of gradual evolution of continent through orogeny and granitization of sediments: (a) crust originally basaltic; (b) warping and volcanism with deposition in adjacent downwarps; (c) early contorted continental plate with newly formed marginal geosynclines.

of the original basaltic crust produced elongate fracture zones. Here, slightly less basaltic magma could have originated and erupted on the Earth's surface. Thereafter weathering and erosion would winnow certain of the original volcanic constituents and concentrate quartz and clays in the sediments carried into flanking geosynclines. Chemically, quartz and clay, in sandstones and shales, contain about the same elements in the same proportions as granite. During orogenic destruction of the geosyncline, metamorphism of the deeper sandstones and shales would lead to increasingly granitic material. Thus the fusion of geosynclinal sediments into relatively rigid blocks produced the embryonic continents. The orogenic destruction of successive geosynclines forming along the margins of the rigid blocks has led to the expansion of the sialic continental platforms at the expense of the ocean bottoms.

This theory—neat as it sounds—is not above criticism. For instance, to make a proper granite by the weathering and erosion of basalt, large quantities of iron, magnesium, and aluminum would have to be permanently removed from the system. The question is: Where have they gone? The quartz and clay of geosynclinal sediments are highly deficient in calcium and sodium, and low in potassium—all very important components of granitic rocks. Although possible answers have been given for these objections, today we can only say that primary differentiation early in the Earth's history, or the growth of continents with granitization through time, are two basic alternatives for the origin of the continental granites.

Development of the Oceans and Air According to an older view, the newly formed Earth had a dense primitive atmosphere charged with water vapor. No oceans existed initially, however, because rainfall was immediately vaporized by the planet's molten exterior and sent steaming back into the atmosphere. Eventually, when the crust solidified and cooled enough to permit water bodies to accumulate, long-continued downpours—the original "Deluge"—filled the ocean basins, setting the hydrologic cycle in operation. Today there is a new trend of thinking. The original atmosphere, if there ever was one, has been lost, and the existing atmosphere and hydrosphere are secondary features that have slowly emanated from the mantle and crust in volcanic exhalations, hot springs, and the weathering of igneous rocks. The reasoning is based on astronomical, chemical, and geologic data.

Using spectroscopes to determine the composition of the stars, astronomers have calculated the cosmic abundance of the various chemical elements. Assuming the cosmic cloud from which the Earth originated was an average sample of matter in the universe, it would have contained considerable neon, argon, krypton, and some other gases. According to calculations, there should be a million times more of these *noble gases* in the air today if the existing atmosphere had been present from the beginning. The lack of the noble gases is the key to the history of our atmosphere, because if they escaped into space, so did nitrogen, carbon dioxide, methane, or whatever other free gases might have originally been present.

What caused the escape of the original gases? We cannot be sure, but they may have been lost from the contracting cosmic cloud, or, more likely, were expelled when the Earth became molten. Gas molecules are in constant motion, and if their speed was accelerated by heating, the "excited" molecules could have exceeded the escape velocity and broken free of the Earth's gravitational attraction.

Why were not the chemical elements, now forming gaseous molecules, lost with the primitive atmosphere? Apparently they were then locked with the heavier elements. Unlike the *noble* gases, so called because they are inert and do not enter into chemical combination, oxygen, hydrogen, nitrogen, carbon, and the other elements in the present-day oceans and air are reactive. In the cosmic cloud, they were chemically combined in iron oxides, silicates, and other minerals. As the Earth became molten, they remained dissolved in the magmas, while the free gases escaped from the surface. Thus our present atmosphere and hydrosphere have most likely seeped from the interior as volatiles expelled from crystallizing magmas after the Earth's surface had cooled and crusted over. Significantly, analyses do show that volcanic emanations could have produced the present atmosphere and hydrosphere, assuming carbon dioxide has been broken down by photosynthesis to produce the unique abundance of oxygen.

So far, our story of the Earth's physical evolution is speculation. It does represent logical deduction reinforced by mathematical calculations. But the deductions are no better than their basic, and often debatable, assumptions. Turning to the history recorded in Precambrian rocks, we do have observable evidence of the geological events, although here too there are conflicting interpretations.

THE PRECAMBRIAN RECORD

Observations

Precambrian rocks, the first positive records of Earth history, form the foundations of all the continents. In the large expanses where the continental platforms are blanketed by later strata, Precambrian outcrops are limited to a few deep valley bottoms, such as Grand Canyon, and to the cores of deeply eroded mountains, like the Rockies (Fig. 12-5). By far the greatest accessible Precambrian exposures, comprising millions of square kilometers of the Earth's surface, are in the shields of the stable continental interiors.

Precambrian Shields The Baltic-Russian shield is exposed across most of Norway, Sweden, and Finland. The Angaran shield forms the heart of Asia. Most of Africa south of the Sahara Desert, adjacent Arabia, and Madagascar form one of the world's greatest Precambrian expanses. The Amazon River follows a broad, gentle downwarp filled with younger sediments that separate the Guianean shield from the Amazonian shield in South America. The least-known shield underlies most of the eastern part of ice-covered Antarctica. The Canadian shield, of nearly 8 million square kilometers (3 million square miles), centers around Hudson's

Bay and extends into Greenland and the arctic islands, with projections into the Adirondacks and Great Lakes region of the United States. One of the most studied shields, it illustrates some of the difficulties, as well as some tentative conclusions, in interpreting Precambrian history (Fig. 12-6).

Rocks of the Canadian Shield Vast tracts of the surface of Canada are granites and granite-gneisses. In this "sea of granite" are islandlike masses, patches, and infolded pods of metamorphosed sedimentary and volcanic rocks, which are most important in deciphering the history of the Precambrian (Fig. 12-7). From their degree of metamorphism and deformation, three general types can be recognized.

A highly altered and contorted group is characteristically isoclinally folded and intruded. In central and northwestern Canada, a group called the Keewatin–Timiskaming type includes relic graywackes and lavas whose pillow structures indicate eruption under water. In southeastern Canada, northwest of the Saint Lawrence River, equally metamorphosed marbles and quartzites that were originally well-sorted sandstones and limestones represent the Grenville type.

A second distinctive group, infolded in widely spaced belts, includes moderately deformed slates, quartzites, and marbles. Originally marine sediments, all lie unconformably on the more highly deformed sequences; they range from flat-lying to closely folded. These rocks, called the Huronian type, include the ironstone beds which include some of the most productive iron ores in North America. In places, coarse conglomerates are interpreted as Precambrian glacial deposits.

Least extensive is a third group of flat or gently tilted sediments and lavas, lying unconformably across the more deformed groups. The little-altered rocks, known as the Keweenawan type, resemble many Paleozoic and younger rocks in their lack of metamorphism. In the Great Lakes region, where they were first studied, lavas and dark-colored in-

Fig. 12-5 Granite dikes cross-cutting older, contorted Precambrian rocks in Medicine Bow Mountains, Wyoming. Photograph by the author.

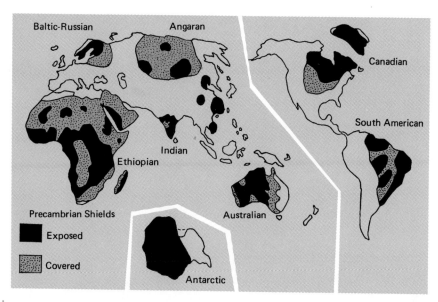

Fig. 12-6 Precambrian shield areas of the world. Dark areas have generally exposed Precambrian rocks. In the stippled areas, shields are generally covered by uncontorted and nearly flat-lying younger strata.

trusive sheets are overlain by arkosic, continental redbeds.

Older Interpretations

Despite the relatively simple summary, the reading of Precambrian rocks is difficult. Many have been so twisted and metamorphosed by successive orogenies that their volcanic or sedimentary origin is hard to determine. The larger part of them has been lost to erosion, not only during their long and eventful Precambrian history, which is marked by unconformities, but also in the long interval since the end of Precambrian time, during which the stable shield has been largely exposed. At best, the sedimentary and volcanic rocks are mere patches in a deeply eroded complex of gneissic and granitic mountain roots.

The Original Crust In the 1840s and 1850s, Sir William Logan pioneered the study of Canada's complicated Precambrian geology. He recognized, mapped, and named many of the important rock groups. Until 1880, however, the widespread granites and gneisses were considered the original crust, inherited from the Earth's molten phase. Then A. C. Lawson of California, working in the shield, made a significant discovery. The granite was intruded into metamorphosed sediments and volcanics. Clearly, they must be older than the granites. Moreover, the most ancient sediments must have been (in part) derived from and along with the volcanics, also lain on still older rocks, but these have been obliterated by invading granites and mountain-building. Further work has nowhere disclosed an original crust; perhaps it will be found under the oceans although recent findings make this unlikely, but in the continents there seem to be no remnants exposed.

The Introduction of Time Units As mapping progressed through the last half of the nineteenth century, confusion arose in attempts to correlate

Fig. 12-7 Early Precambrian metamorphosed sediments (dark grey) intruded by granite (light grey) in large bodies, up to 0.8 kilometers (half a mile) across, and smaller layerlike bodies. Late Precambrian basaltic dikes (dark grey) cut both meta-sediments and granite. R.C.A.F. photo, courtesy of the Canadian Department of Mines and Technical Surveys.

Precambrian rocks, which had been mainly studied for economic reasons in mining districts of the vast shield area. So, in the early 1900s, a committee of Canadian and American geologists was appointed to straighten out the matter. They classified the Precambrian rocks into four divisions, based on the rock types in the Great Lakes region. Shortly thereafter, Chamberlin and Salisbury, in their popular *Textbook of Geology*, proposed that the four divisions represented two great geologic eras. Thus

the Precambrian received time designations, like those previously established for the later part of the geologic calendar.

The Archeozoic era, whose rocks are called Archean, was conceived as a "fiery" era in the Earth's early evolution. Although the dramatic concept has slowly disappeared, the name Archean has persisted for the most highly metamorphosed and largely granitic rocks. The "quiet" Proterozoic era was based on the moderately and little-altered sequences resting unconformably on the Archean rocks.

Based on the four rock divisions in the Great Lakes region, the Archeozoic era was divided into the Keewatin and Timiskamian periods, and the Proterozoic era into the Huronian and Keweenawan periods. Three major orogenies, or geologic revolutions, named from intrusive granites were proposed. The Laurentian orogeny, marked by the Laurentian granite, ended the Keewatin period and was followed by erosion as the Timiskamian period began. The Algoman revolution, starting in late Timiskamian time, closed the Archeozoic era. The Huronian period opened with erosional leveling of the Algoman mountains and ended with local deformation and erosion inaugurating the Keweenawan period. The end of the Precambrian was marked by the Killarney orogeny, associated with granite emplacement. Major erosion followed, producing a widespread surface of low relief upon which the invading Cambrian seas left the first abundantly fossiliferous strata, marking the beginning of the Paleozoic era.

Such a history depends on the basic assumption that the Canadian shield went through three or four major cycles, each affecting the whole of the shield area, and each, in a grand way, representing the events—with the addition of a granite invasion—that produce an unconformity. The events include: (1) the laying down of sedimentary or volcanic rocks; (2) orogeny deforming the surficial rocks and the intrusion of granite; (3) erosional leveling of the resultant mountains followed by the onset of a new cycle. Moreover, in designating the Archeozoic and Proterozoic as eras, and the Keewatin and others as periods, these units were assumed to be worldwide time divisions.

Modern Views

This relatively simple Precambrian history may well represent the relative order of events in the Great Lakes locality, but it is probably not a universal pattern for the Precambrian of all Canada, and the rest of the world, as well. In the late nineteenth century, Lord Kelvin's decree of 40 million years as the age of the Earth, or even the 100 million years assumed by many geologists, made a relatively simple series of Precambrian cycles seem reasonable. Today, however, radiometric dating of rocks indicates a minimum Precambrian time span of about three billion years. From the complicated history of deposition, eruption, mountain-building, and erosion during the 600 million years since the Paleozoic began, we should suspect an even more complex Precambrian history on uniformitarian grounds.

Problems and Reinterpretations The assumption of universal granite intrusion associated with orogenies across the vast expanse of the Canadian shield is not like the well-established patterns of later times. Granite emplacement accompanying the major mountain-building since the Precambrian has been generally restricted to long narrow belts; so why should the Precambrian be different?

For want of fossils (or radiometric dating), early correlations were based on the lithology and degree of metamorphism of Precambrian rocks. In widely separated regions, the highly metamorphosed and intruded sequences were assumed to be time equivalents; moderately metamorphosed groups were equated in age; and so were the little-altered rocks. The pitfalls of far-flung correlation based on similar lithology are demonstrated by the work of Lawson. He correlated a granite over 1600 kilometers (1000 miles) to the west and north with the Laurentian granite in the southeastern part of the shield. Subsequently, radiometric dating gives the true Laurentian an age of 1.1 billion years and Lawson's granite an age of 2.5 billion years, making his an understandable, but magnificent, miscorrelation.

Even without radiometric dating, the dangers of lithologic time correlation are evident. Granites in Idaho that intrude Mesozoic strata are quite like some in the Precambrian. Triassic redbeds in

New Jersey resemble the Keweenawan arkoses. Graywackes in California differ little from some in the Precambrian terrains, and, were it not for fossils, might be correlated lithologically. Moreover, in later and well-established systems, highly metamorphosed rocks subject to intense orogeny are known to be the same age as the very little altered rocks away from the active parts of a geosyncline. Cambrian clays in Russia, for example, are the same age as crystalline metamorphics in the British Isles. Thus in light of recent radiometric age determinations, uniformitarian philosophy, and hindsight, our concepts of Precambrian rocks and history are changing.

Precambrian Provinces   Based on recent investigations, the Canadian shield is now considered to consist of broad belts, called provinces, each distinguished by characteristic rock assemblages and generally similar structural trends (Fig. 12-8). Adjacent provinces have different structural trends and seem to be separated by zones of faulting. Radiometric age determinations on granites yield different ages for the various provinces. The granites, being intrusive, give minimum dates, probably representing times of orogenies that affected the somewhat older sedimentary and volcanic rocks.

In the Canadian shield, the oldest province, called the *Superior*, contains many narrow curving belts of Keewatin-type volcanics and graywackes. These highly metamorphosed rocks of surface origin are surrounded by granites dated radiometrically at 2.5 to 3 billion years. South of the Superior province lies the *Grenville* province, containing vast areas of granite and granite-gneiss, originally named the Laurentian granite by Logan. These rocks have been radiometrically dated as from 1.1 to 0.8 billion years. They are intrusive into strongly metamorphosed sediments that are quite different from the Keewatin and Huronian types of the Superior province. The quartzites, marbles, and schists of the Grenville province must have originated as sandstones, limestones, and shales. They indicate deposition in broad marine basins where they were

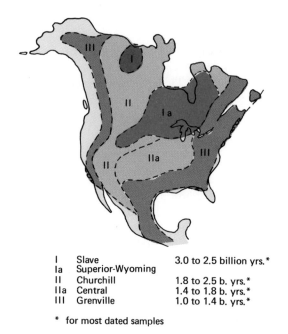

I	Slave	3.0 to 2.5 billion yrs.*
Ia	Superior-Wyoming	
II	Churchill	1.8 to 2.5 b. yrs.*
IIa	Central	1.4 to 1.8 b. yrs.*
III	Grenville	1.0 to 1.4 b. yrs.*

* for most dated samples

Fig. 12-8   Generalized scheme of Precambrian provinces of North America. After A. E. J. Engel, 1963.

reworked and sorted by wave action, in contrast to the rapidly "dumped in" graywackes with their associated submarine volcanics. The *Slave* province, in the northwest part of the shield, is comparable to the Superior province although smaller in area. Between these two provinces lies the *Churchill* province, where ages range between 1.0 and 1.4 billion years, containing less lava and more rocks of sedimentary origin than the older adjacent ones. The *Great Bear* province, northwest of the Slave, is comparable to the Churchill province.

The arrangement of the Precambrian provinces has been taken as evidence for the growth of continents (Fig. 12-9). The Superior and Slave provinces could be continental nuclei, relics of the first-formed parts. The other provinces could represent later additions produced by orogenic destruction of successive marginal geosynclines through Precambrian time. The case for expanding continents is not proven—some geologists argue strongly against it—

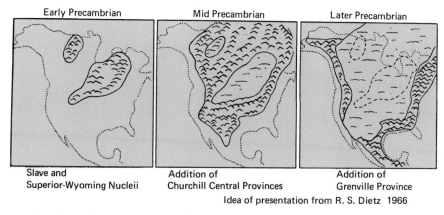

Early Precambrian   Mid Precambrian   Later Precambrian

Slave and   Addition of   Addition of
Superior-Wyoming Nucleii   Churchill Central Provinces   Grenville Province

Idea of presentation from R. S. Dietz 1966

Fig. 12-9 Hypothesis of continental accretion during Precambrian time.

but preliminary work in the Precambrian of Africa, India, and Australia suggests the presence of provinces comparable to those of Canada. Also, in later time, the evolution of the Appalachian mountain system and the Coastal Plain geosyncline, for example, seems to follow the pattern. In any case, the Precambrian shields appear to be deeply planted complexes of mountain roots, forming the stable continental platforms along whose margins occurred the greatest mountain-building activity of Paleozoic, Mesozoic, and Cenozoic time.

That the Precambrian provinces resulted from the global workings of the plate tectonic mechanism is a reasonable hypothesis which fits the evolutionary philosophy of uniformitarianism. But Precambrian rocks are generally much-altered assemblages that reflect a very complicated sequence of events spanning more than three-quarters of the Earth's recorded geologic history. Thus in the present state of our knowledge, any plate tectonic models for these most ancient rocks are best viewed as theoretical speculations; they do, however, provide worthwhile suggestions for future lines of field investigation. It seems probable that by late Precambrian time the plate tectonic mechanism was in operation. On the other hand, the earliest stages of the Earth's development were almost certainly dominated by a catastrophic mechanism—the im-

pact of large meteoroids, possibly asteroids. The evidence is based on recent investigations of our neighboring planetary bodies in the solar system.

ASTROGEOLOGY

Prior to the space program, *astrogeology* (the geologic study of planets and other solid bodies in the solar system) seemed little more than science fiction. Now it is a well-established field, for, supplementing the information obtained from Earth-based telescopes and radar, space probes have sent back physical data and close images of the nearest satellites and planets; astronauts have placed instrument packages on the Moon and collected nearly a ton of rock samples for laboratory studies; and the Viking project landing vehicles are on the surface of Mars.

From their investigations of our neighbors in the solar system, scientists in the space program have obtained important clues to the Earth's early history. Admittedly Venus, which is shrouded by a dense opaque atmosphere, remains largely a geologic mystery, but a wealth of new information is now available for the Moon, Mercury, and Mars. Their relatively static surfaces preserve vestiges of the formative stages of the Sun's inner group of

Fig. 12-10 Apollo 15 landing site on the Moon (1971). Lunar Lander "Falcon" is on the left, Lunar Roving Vehicle with Astronaut James B. Irwin is to the right. The powdery lunar "soil" is marked by footprints and tire tracks. The margin of the Apennine Mountains is in the left background; the crater behind the rover is about 5 kilometers (3 miles) away. Photo by Astronaut David R. Scott, courtesy of N.A.S.A.

smaller rock-cored *terrestrial planets*,[4] one of which —the dynamic, ever-changing Earth—has a markedly different face than the others.

[4] The outer group of the *great planets* includes Jupiter, Saturn, Uranus, and Neptune (Pluto, the outermost planet, is small). Although very massive because of their great size, the great planets may have cores largely composed of liquid or solid hydrogen.

THE MOON

Long before the space age, astronomers had determined many of the Moon's aspects that are fundamental to an understanding of present-day studies of the geology of that planet. Strictly speaking, the Moon is classed as a satellite because it orbits around the Earth. Yet the Moon is unique among the satellites because it is so very large when compared to its controlling planet. Thus many astronomers suggest that the Earth and Moon form a pair, best classed as a "binary planet."

Preliminary Matters

Astronomical Data The distance to the Moon (actually from the Earth's center to the Moon's center) is about 384,600 kilometers (roughly 239,-000 miles), varying from some 28,000 kilometers (about 17,400 miles) nearer than that figure to about 82,300 kilometers (about 51,000 miles) farther because the Moon's orbit around the Earth is elliptical. The lunar diameter of about 3475 kilometers (some 2160 miles) is about one-quarter of the Earth's. Gravity on the Moon's surface is about one-sixth as strong as on Earth. Since the overall lunar density of 3.3 (about 60 percent of the Earth's 5.5) is comparable to that of its basaltic crust, the Moon seems to lack a dense central core.

Only one side of the Moon is visible from the Earth. Formerly, when the Moon spun faster, all its surface may have been exposed to the Earth; now, however, the Moon makes one rotation on its axis while making one orbital revolution around the Earth. This keeping of the same face towards the Earth is attributed to a slowing of the Moon's rotation by tidal friction from the Earth's gravitational attraction. Thus features on the far side were unknown until Russian and American space ships sent back pictures from trips around the Moon.

The Moon's surface is heated and lighted by the Sun; moonlight is reflected sunlight (a small amount of sunlight reflected back from the Earth also reaches the Moon). Temperature changes on the Moon's surface are extreme by earthly standards, and range from over 100° C (some 200° F) above zero on the bright side to 116° C (240° F) below zero (colder than dry ice) on the dark side. These extremes result because the powdery lunar surface permits little conduction and absorption of heat, and because the Moon has no appreciable atmosphere to reflect, absorb, and distribute heat.

Several lines of evidence indicated the absence of a lunar atmosphere. The Moon's strong shadows, clarity of surface forms, and lack of clouds are suggestive. Moreover, stars disappear instantly behind the Moon's edge without the fading or apparent shifting that an atmosphere would cause. Apparently, the Moon's low gravity allows rapidly moving gas molecules to escape into space, in contrast to the Earth, where a much higher escape velocity is needed to overcome the stronger gravitational attraction. Because the Moon lacks an atmosphere and hydrosphere, its surface has not been constantly reworked and altered like that of the Earth. Thus in contrast to that of the Earth, the lunar landscape still retains features that were created several billion years ago.

Possible Origins of the Moon Three sorts of speculative hypotheses have been proposed for the creation of the Moon. The hypothesis that the Moon is a large mass of terrestrial material thrown out by centrifugal force long ago when the Earth spun much faster explains the relatively low density of the Moon as compared to the Earth by assuming that only crustal material and some mantle were expelled. However, theoretical calculations based on the energies involved do not support the concept (Fig. 12-11).

Perhaps the Moon formed in the same way and time as the Earth—from a separate clot of matter that grew by accretion in a cloud of gas and dust—and later went into orbit around the Earth. By this second hypothesis, the Moon should have about the same materials and density as the Earth; however, the Moon seems to lack a heavy central core.

In a third suggestion, the Moon was once a separate planet orbiting around the Sun. During a close approach of the two planetary orbits, the Moon was captured and became a satellite revolving around the Earth. This hypothesis accounts for

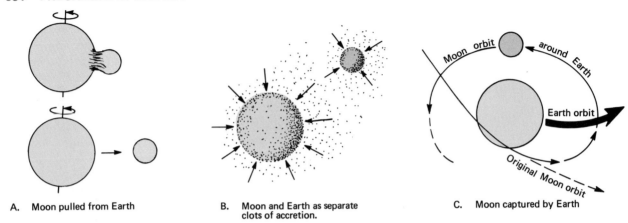

A. Moon pulled from Earth   B. Moon and Earth as separate clots of accretion.   C. Moon captured by Earth

Fig. 12-11 Three possible origins of the Moon.

the large size of the Moon relative to the Earth. Theoretical calculations indicate that such a capture is possible, but it requires very special conditions. Since the Earth's origin is still debated, little wonder that the creation of the Moon is a subject mainly for speculation.

Lunar Topography and Geologic Mapping

Even to the unaided eyes of ancient observers, the lunar surface displayed two contrasting irregular divisions: darker regions (forming "the man in the Moon") and lighter, more reflective ones.

Maria and Terrae Appreciation of the true diversity of the surface began in 1610 when Galileo turned his primitive telescope on the Moon. He named the darker regions *maria* (Latin for "seas") —in the mistaken belief that they were large bodies of water—and the lighter-colored regions he called *terrae* (lands). He saw mountains around some larger maria, and he discovered the craters of the Moon. Since then astronomers have studied and mapped the Moon with telescopes of increasing power and resolution. Now orbiting spacecraft have sent back extremely detailed views, and astronauts have taken photographs on the lunar surface.

More than three-quarters of the lunar surface consists of terrae, rough, densely cratered uplands— an unknown fact until orbiting spacecraft sent back images from the Moon's previously invisible far side. Maria, which form almost half of the near side, are plains. Most of the 14 recorded maria are connected. Mare Imbrium, some 1100 kilometers across (about 700 miles) and the largest "sea," is an off-shoot of Oceanus Procellarum. Although relatively smooth when compared to the highly cratered terrae, the maria surfaces are not featureless. They contain long trenchlike features, called rills; large linear scarps, such as the Straight Wall; elongate welts, designated maria ridges; low irregular domes; and many craters which, however, are notably less abundant than in the uplands (Fig. 12-12).

Circular Structures Most distinctive in the Moon's topography—in terrae and maria alike—is the dominance of circular patterns. They range in size from immense basins, such as Mare Imbrium, through rimmed craters like Copernicus (64 kilometers, or about 40 miles, in diameter), to small pits about 10 centimeters (4 inches) in diameter (Figs. 12-13, 12-14).

The vast basin containing Mare Imbrium is partially encircled by the "Mountains of the Moon," whose principal peaks rise 7900 meters (about

Fig. 12-12 Craters and rills are prominent in this Apollo 8 photograph (December 1968) of the lunar surface. Courtesy of N.A.S.A.

26,000 feet) above the mare plain. Although they have been given such names as Jura, Alps, and Caucasus, the analogy with the Earth's global belts of folded mountain ranges is misleading. The lunar mountains are parts of a gigantic circular ridge.

The Orientale Basin, an impressive feature some 900 kilometers (560 miles) in diameter, is classed as a *multi-ringed basin*. Its bull's-eye pattern of concentric ridges is exceptionally clear because only the central part contains maria materials.

nicus is a *rayed crater* characterized by bright streaks that radiate outwards, like giant splash-marks, for thousands of kilometers. Other craters are not as fresh-looking as Copernicus. Some, although well defined, lack ray systems, and many in the pitted terrae have notably less sharp rims.

Establishing a Lunar Time Scale By carefully mapping surface features, scientists of the U.S.

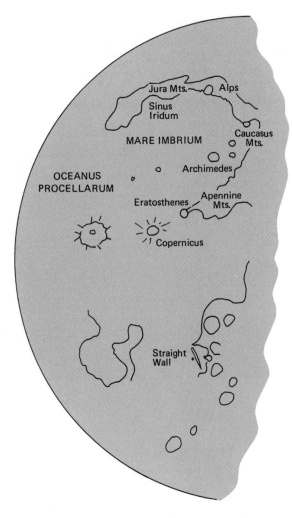

Fig. 12-13 Last-quarter photograph of the Moon showing the features mentioned in the text. Mount Wilson Observatory photograph.

Copernicus is a fresh-looking, well-preserved crater having infacing walls as high as 360 meters (roughly 12,000 feet) which are marked by concentric steplike terraces. The relatively flat crater floor is broken by a cluster of central peaks. The outer flank, beyond the crater rim, has a light-colored hummocky topography. Beyond this blanket of ejecta, the lunar surface is scarred by shallower, smoother-sided, often elongate depressions. Coper-

Fig. 12-14 Diagram showing the location of features in the preceding photograph.

Geological Survey established a geological time scale for the Moon. The first geologic maps (largely of possible Apollo landing sites) were made from telescopic observations and interpretation of telescopic photos; later ones are based on detailed views of the Moon sent back by space vehicles.

The time scale was constructed by adopting classical methods for determining a relative sequence of geologic events. The criteria included: intersection (younger features cut across older ones) and superposition (younger strata lie on older), as well as the density of cratering (greater on older surfaces) and relative "freshness" (older features less sharply defined).

The lunar time scale was determined from events recorded in the rocks around Mare Imbrium (Fig. 12-15). The *Pre-Imbrian period*, the oldest time division, was based on observations of the

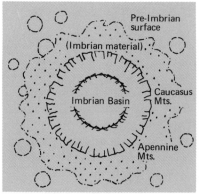

A. Imbrian Period, early events
Creation of Imbrian basin, and
surrounding mountains.

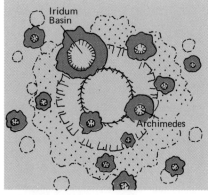

B. Imbrian Period, later events
Later craters form in and around
Imbrian depression. Archimedes is
typical crater

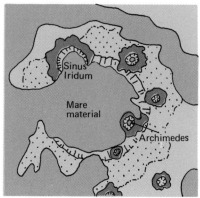

C. Latest Imbrian Period
Mare material floods region.

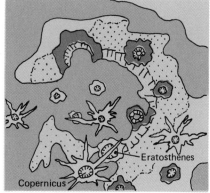

D. Post mare Periods
Older Eratosthenian Period craters
now lack rays. The younger Copernican
Period craters have rays which sometimes cross the older craters.

Fig. 12-15 Schematic diagram of events in the Mare Imbrium vicinity, based on Shoemaker.

highly cratered uplands, or terrae, south of Mare Imbrium. It seems to have been a period of intense bombardment of ancient lunar crust by large meteoroids. The ensuing *Imbrian period* began when the vast Imbrian basin was created. The Imbrium event probably represents a major episode when gigantic planetoid-sized bodies crashed into the Moon's Pre-Imbrian surface, blasting out great quantities of debris and upturning crustal rocks to form the elevated rim. Later in the period, many craters and associated deposits were formed on the Imbrian Basin rim. The crater Archimedes and the curved scarp around Sinus Iridum are representative of this episode. The last major event of the Imbrian period was a vast flooding, or partial drowning, of the Imbrian Basin and some craters by mare materials. These are now known to have originated as vast eruptions of highly fluid volcanic lavas.

Post-Imbrian time is divided into the *Eratosthenian period*, recorded by craters that lack bright rays and light-colored ejecta blankets; and the *Copernican period*, the latest time division, whose fresh-looking craters are characterized by conspicuous ray systems. Originally all craters probably had bright rays, but it is thought that the materials have altered and darkened with increasing age. That the rate of cratering had decreased markedly in Post-Imbrian time seems evident from our present knowledge of the age of the maria surfaces and the density of cratering they display. Although they are well cratered, the crater density on these surfaces is very much less than that in the older highlands.

The geologic time scale determined in the Imbrian region has, in its broad form, provided a valid scheme which has been successfully applied to other regions of the Moon. Establishment of a mapping program and determination of a relative time scale were essential aspects in the program of lunar study, but they could not in themselves provide answers to some major questions about the nature and dynamics of the Moon. The answers required "ground truth," which could only be obtained by actual landings to collect rock samples and to place recording instruments on the surface of the Moon (Fig. 12-16).

Major Geologic Processes

The Nature of the Moon's Interior The early years of the space program were marked by a healthy controversy among scientists whose conflicting theories for the origin of the maria and lunar craters led to markedly different models for the Moon's interior. One group of geologists—philosophical descendants of Dr. Hutton and the Plutonists—formed the "hot Moon" school. Envisioning considerable igneous activity within the Moon, they considered the maria to be vast lava flows, with perhaps some included volcanic ash. The craters of the Moon they considered mainly volcanic features: Hawaiian or Krakatoa type calderas that grew to far greater diameters than their earthly counterparts because of the lesser gravity on the Moon. The "cold Moon" school of geologists—whose concepts have vestiges of nineteenth-century geological catastrophism—rejected any major igneous activity at the surface or in the outer zones of the Moon. They considered the maria to be underlain by a great thickness of dust and larger fragments of debris. They attributed the lunar basins and craters to impacts by giant meteoroids. The "warm Moon" school were compromisers. They attributed the large crater forms to impact, but admitted the possibility of

Fig. 12-16 Apollo 15 landing area (Hadley North was selected) lay at the foot of the Apennine Mountains. Hadley C Crater and the Hadley Rill are in the plains. Unlike many straight rills that seem bounded by faults, the sinuous Hadley Rill suggested to some scientists the work of running water. This mechanism, however, has been ruled out on the Moon; this rill may represent a collapsed tube of once-molten lava. Courtesy of N.A.S.A.

NASA-S-70-6053-V

○ HADLEY NORTH

Hadley δ

HADLEY ○

HADLEY

RIMA

○ HADLEY SOUTH

0 2 4 6 8
KM

some volcanic eruptions and of lava flows beneath a surface blanket of dust.

The early space vehicles visiting the Moon (Ranger, Luna, and Surveyor) sent back much valuable data, but some of the major theoretical problems were not resolved until the Apollo landings. During the Apollo program, manned landing vehicles visited maria plains, a probable Copernican ray, ejecta blankets from two vast basins, the foot of the Apennine Mountains, and the lunar terrae highlands. Among the contending schools of thought, the compromisers were generally, but not completely, correct. The hot Moon school was right about the maria, but not much else. The cold Moon school—the catastrophists—were closest to the present-day interpretations of the Moon's interior, and right about the origin of the great lunar basins and craters. Explosive impact seems, by all odds, the major process that shaped the face of the Moon.

Impact Mechanisms   That the vast basins and craters of the Moon are impact scars caused by collisions with gigantic meteorites was proposed by the distinguished American geologist G. K. Gilbert in 1893. The hypothesis went virtually unnoticed until the 1940s, when it began to attract some attention.

Theoretically, if a meteoroid weighing thousands of metric tons plunges into a planet's surface at 150,000 kilometers per hour (roughly 100,000 miles per hour), the kinetic energy (of motion) is converted into heat almost instantaneously. The meteoroid vaporizes and creates a shower of fragments in an explosion having the violence of multiple H-bombs. The explosion generates a tremendous compressional (shock) wave in the planetary crust. Immediately the rock in the resulting crater rebounds elastically in a rarefaction (expansion) wave that flings up adjacent bedrock into an elevated rim mantled with a surge of blast debris.

In recent times, the mechanism of explosive impact cratering has been studied in various ways. In the laboratory, high-speed sequential photographs have been taken of drops falling into water and of small projectiles shot into sand boxes. Detailed studies of the immediate effects and aftermath of underground atomic and nuclear bomb tests have been most instructive. Where these explosions are set off at shallow depths, they produce large craters in the Earth's surface which strongly resemble the lunar craters. Such manmade craters have steep infacing walls, are surrounded by a hummocky topography of debris blasted from the crater, and develop rays of fine ejecta.

Field studies of the Earth's known meteoroid craters are most relevant. Of the eight circular features accepted as impact structures—because of abundant particles of meteoric iron in their vicinities—the most thoroughly investigated is Meteor Crater, Arizona (also called Barringer Crater).[5] It is a nearly circular depression, about 1200 meters (4,000 feet) across and some 180 meters (600 feet) deep (Fig. 12-17). The encircling rim, which rises some 46 meters (150 feet) above the surrounding plateau, contains upwarped limestone strata. Descending slopes outside the rim are underlain by debris with a hummocky topography. Except for the lack of rays—destroyed by erosion since this "prehistoric" feature formed—Meteor Crater duplicates what we observe in many youthful-looking lunar craters.

Multi-Ringed Basins   The Moon's Orientale Basin may well preserve characteristic features of the most gigantic impact craters (Fig. 12-18). In others, such as Mare Imbrium, the multi-ringed nature may be obscured by later volcanic floodings of basalts. The multiple rings in the vast lunar basins have been explained in several ways. They might be relics of "frozen" shock waves generated by the tremendous explosive impact of great planetoid-sized bodies. In this case, compressional waves are interpreted as so violent that they permanently exceeded the solid planetary rock's elastic limit (abil-

5 The site was used for geologic training of astronauts.

Fig. 12-17 Barringer (Meteor) Crater between Flagstaff and Winslow, Arizona, shows an upturned rim mantled with hummocks of debris blasted from the crater. It strongly resembles a lunar crater. This site was used as a study area by astronauts training for the Apollo program of lunar landings. Photo by the author.

ity to rebound to an original shape)—thereby permanently deforming the lunar surface into the gigantic bull's-eye pattern.[6] In a different concept, the multiple rings are interpreted as gigantic slump blocks. They might have formed in minutes or in days when the initial crater spread outwards because of gravitational collapse of its steep high walls.

6 As if the rings made by a stone thrown into still water were instantaneously frozen.

Impact Craters Craters such as Eratosthenes and Copernicus represent impact structures of a lesser—but still impressive—magnitude than the multiringed basins (Fig. 12-19). The concentric steplike terraces which are common on the infacing walls of craters exceeding 20 kilometers (12.5 miles) in diameter are generally interpreted as great slump blocks. The central peaks, standing above the flat floors of many large craters, have several possible origins. Perhaps they result from the rebound of rock that was depressed at the center of the im-

Fig. 12-18 The nature of the Orientale Basin near the edge of the lunar nearside was not appreciated until vertical images were returned by Lunar Orbiter 4 (1967). The crater is a large multi-ringed basin whose encircling mountains reach some 6100 meters (20,000 feet) above the adjacent surfaces. Radiating valleys in the rings may have been gouged out by ejecta. Mare material in the center and smaller younger craters are visible; the flank (right side) consists of a blanket of ejecta. Courtesy of N.A.S.A.

Fig. 12-19 High-resolution oblique view of the crater Copernicus photographed by Lunar Orbiter II. The north wall (view is to lunar north) is 3000 meters (10,000 feet) high. The large terraces on the wall are interpreted as giant slump blocks. The central peaks in the floor of the crater are 300–900 meters (1000–3000 feet) high. The Carpathian Mountains are on the horizon. Courtesy of N.A.S.A.

pact crater. Perhaps they represent jumbled central piles of debris which spilled rapidly inwards off the initially formed crater walls. The hummocky topography extending outwards from the crater rim is considered a debris blanket blasted out of the central crater. Pits on the lunar surface beyond the blankets are interpreted as secondary craters gouged out by large rock chunks thrown from the crater. Since rays are best seen under the direct light of a full Moon, and they disappear and cast no shadows under oblique light, they have long been considered as very thin streaks of fine debris expelled great distances from the craters.

Moon Rocks

The rock samples brought back by astronauts have provided much critical evidence—some of it surprising—for the geology and history of the Moon.

Fig. 12-20 This view from the orbiting Apollo 16 service module (1972) shows much of the highly cratered terrae surface that characterizes the farside of the Moon. Circular mare on the upper right horizon (Mare Crisium) and some of the other darker areas are on the lunar nearside, where the maria are prominent. Note different generations of craters indicated by contrasting sharpness of their walls and cross-cutting relations. Close examination shows some central peaks and slump steps on inner walls of larger craters. Courtesy of N.A.S.A.

Briefly, the investigations of the specimens in many laboratories indicate two main groups of lunar rocks: crystalline igneous types; and clastic (fragmental) types containing fused and glassy components.

Terrae Anorthosites   Bedrock in the terrae, the ancient uplands, is thought to be mainly *anorthosite*, a special variety of the gabbro family, consisting largely of calcium-rich plagioclase feldspar with only minor amounts of ferromagnesian minerals. Samples collected by astronauts commonly contained fragments of anorthosite which were apparently blasted out of the Moon's ancient crust. Further evidence for the nature of highland bedrock was obtained in orbiting space vehicles carrying remote sensing devices designed to record characteristic X rays emitted by elements in lunar rocks under bombardment by solar radiations. The instruments detected much aluminum and silicon (which would be abundant in anorthosite), but little magnesium, which would indicate ferromagnesian minerals characterizing such rocks as basalt. Rock samples from the Moon's far side (Fig. 12-20) would have been desirable to confirm the composition of the terrae, but landings there were beyond the capabilities of existing space technology.

Since anorthosite is an igneous rock (crystallized from magma), the Moon's outer part must, at one time, have been molten. The intense heat required for melting was probably generated by constant explosive impacts of large meteoroids during a period of intense bombardment of the lunar surface. Additional heat may have been provided by the accumulation of radioactive elements and increasing internal compression as the Moon grew by accretion to its final size. When cooling began in the deep ocean of magma enveloping the Moon, fractional crystallization probably took place. During crystallization, the denser iron-rich crystals would tend to sink to the bottom of the partially molten magma to produce a deep-seated zone of mafic rock, such as peridotite; while the relatively less dense crystals of aluminum-rich calcium-plagioclase were concentrated near the surface to form the outer lunar crust of anorthosite.

Lunar Basalts   The maria are vast plains of basalt. Like the Earth's Columbia, Deccan, and other great lava plateaus, the maria originated from vast eruptions of highly fluid flood basalts. The convincing evidence includes close views of lobate lava-flow margins from orbiting space vehicles; astronaut observations of a series of lava flows separated by layers of breccia and ash, exposed in the nearly vertical wall of a lunar valley (the Hadley Rill); and laboratory analyses of lunar basalts which, like those on Earth, consist of plagioclase feldspar, pyroxene, and olivine. The mineralogy of the Moon's basalts is unique, however, in containing appreciable amounts of titanium and chromium. These elements are highly refractory (resistant to heat) and would only occur in very hot magmas, hotter than those producing the Earth's basalts. Moreover, the minerals in lunar basalts have less of the relatively volatile elements (such as sodium, potassium, and sulfur) than their counterparts on Earth, and—significantly—no chemically combined water. Perhaps the Moon's original composition differed somewhat from the Earth's, or else the more readily vaporized elements escaped into space from very hot magmas.

Since the vast eruptions of maria basalts followed the creation of the great lunar basins, these major events in the Imbrian period seem related. When the giant planetoids collided with the Moon, the tremendous explosive impacts must have shattered the lunar crust to a depth far below the immense craters. Thus the maria eruptions might have been triggered when deep-reaching fractures tapped a still-molten or partially molten zone beneath the Moon's crusted-over exterior. On the other hand, the Moon's former molten surface layer (from which an anorthosite crust differentiated) might have cooled and solidified relatively rapidly to form a thick lithosphere. In this case, the generation of the basaltic magma might have resulted from the partial melting of deep-seated, solid peridotite. The generation of heat by a series of tremendous explosive impacts might have triggered a thermal event that caused partial melting in the lithosphere. Basaltic magma could well result from partial melting of low-temperature components of a mafic zone of rock, such as peridotite.[7]

Clastic Rock Materials   The *lunar breccias*, like those on Earth, are characterized by angular fragments. Unlike most terrestrial breccias, however, the Moon rocks contain appreciable quantities of glass particles and globules, and they are welded into cohesive masses. Lunar breccias come from ejecta blankets whose materials were welded together by the shock waves and the intense heat generated by explosive meteoroid impacts.

The Moon's surface is mantled with a powdery *regolith*, ranging in thickness from about 10 centimeters (around 4 inches) on ejecta from the youngest craters to about 8.5 meters (just under 30 feet) on older lunar features. Although called "soil" (in the engineering sense of unconsolidated material), the regolith is certainly not produced by the chemical weathering and organic processes that create terrestrial soils. Instead, the lunar regolith is a churned mixture of crystals, rock fragments, and glassy particles that originates from fine impact debris, infalling meteoric dust, and cosmic ray bombardment.

Radiometric Dating   Absolute dating of the astronauts' rock samples—from a half dozen localities —provided critical points for calibrating the lunar time scale in years. The project was a major scientific accomplishment because the significance of the sophisticated laboratory measurements, which were not always numerically consistent, required

careful scientific evaluation. For example, some minerals, as in meteoroids, contain radioactive elements and decay products whose ratios record the time since the solar system originated—when the "radiometric clock" was started. If minerals are melted, however—as might have occurred when the lunar surface became molten or during the Imbrian impact events—the clock is "reset" to record time since new-formed minerals crystallized from magmas (p. 111). Moreover, ancient rocks subjected to several episodes of intense fracturing and reheating could lose volatile elements that are critical to ratios required for accurate dating. Thus dates obtained for similar samples by using different radiometric methods (for example, uranium-lead and rubidium-strontium) do not always precisely agree. Despite such problems, an acceptable framework for absolute dating has been established, and it indicates that the Moon's surface features are very old.

A few samples from terrae soils (if they have been properly dated and interpreted) indicate an age of some 4.6 billion years for the upland materials—a date favored because it corresponds to independent estimates for the age of the solar system based on evidence from meteoroids collected on Earth and on astronomic calculations. Anorthosite, in a breccia interpreted as material blasted from the crust during the violent episode of Imbrian explosive basin creation, provides an age of about 4.15 billion years. Because this date seems young in comparison to those of maria basalts, it possibly reflects the Imbrian impacts rather than the time of origin of the highland anorthosite crust. A basalt from Mare Tranquillitatis yielded an age of about 3.65 billion years, and a specimen from Oceanus Procellarum was dated at about 3.26 billion years. Thus the Imbrian period, during which the vast lunar basins and their ejecta were created and then flooded with maria basalt, was probably an extended episode which may have lasted a billion years. A specimen collected from a supposed ray of Copernicus yielded an age of 0.9 to 0.8 billion

7 Such speculations are part of the game of science, wherein —armed with few facts and an imagination, along with a certain amount of chemical and physical theory—one can construct all sorts of models.

years. It seems that this crater, which represents the last period of major changes on the Moon, was created in Precambrian time of the Earth's time scale.

The Little-Changing Moon

The face of the Moon, seen through telescopes or in space-vehicle photos, comes close to fitting the poetic image of "eternal mountains and everlasting hills." Based on existing evidence, the Moon has no major internal and external mechanisms like those whose continuing interplay alters the Earth's surface features.

The lack of internal processes is indicated by the absence of ongoing volcanic activity and faulting as well as the total lack of folded mountain chains. Seismic records from instrumental packages placed on the lunar surface by astronauts do, however, indicate moonquakes. But their weak vibrations (all less than magnitude three on the Richter scale) suggest a very thick, strong lithosphere, about 1000 kilometers (some 620 miles) in thickness. Since moonquakes occur when the Moon, in its orbit, is closest to the Earth, they are thought to be related to rock tides generated by the Earth's gravitational attraction and not to dynamic internal operations.

External geologic processes acting on the lunar surface are very limited. Since the Moon has no atmosphere or hydrosphere, its surface is not constantly reworked by streams, glaciers, wind, and waves—features related to the Earth's hydrologic cycle—nor by chemical weathering processes, which require the presence of moisture. The rate of deposition of lunar regolith, from measurements of its thickness on features of known age, is incredibly slow—about one millimeter (0.04 inch) in a million years. Weathering and erosion of rocks forming the lunar landscape are probably restricted to marked temperature changes, gravitational flows and slides, and the impact of micrometeorites. Un-

like the Earth, where small meteoroids flash (forming meteorites) and are burned up by friction in the protective atmosphere, the weak impact of these sand-sized particles may very slowly modify the Moon's unprotected surface features. Overall, however, the Moon seems a long-static planet whose scarred surface preserves abundant evidence of an early catastrophic history several billion years ago.

THE INNERMOST PLANETS

Venus

In size, mass, density, and distance from the Sun, Venus is the Earth's closest counterpart in the solar system. It has a diameter of 12,194 kilometers (7579 miles). But Venus remains the least known of the terrestrial planets because its face is masked by a dense cloudy atmosphere consisting of carbon dioxide (90 percent or more), some oxygen, and traces of water vapor. Two Soviet spacecraft landed on the planet in 1975 and sent back data for about an hour each until their instruments failed as a result of the intense atmospheric heat. They recorded temperatures of 465° C (about 870° F), which are hot enough to melt lead, and an atmospheric pressure about 90 times greater than on the Earth's surface. The very high temperatures probably reflect an extreme greenhouse effect (p. 236) in the dense atmosphere of carbon dioxide.

Earth-based radar stations, which have bounced their pulses through the dense atmosphere on Venus, have determined that the planet rotates in a direction opposite to that of the Earth, and that the surface of Venus has mountains, craters, and plains whose local relief is comparable to any on Earth. Photos from the Soviet spacecraft indicated some angular rocks and some flat-topped rock masses but no loose sand or dust. Despite the recent increase in knowledge, however, Venus remains a little-known planet posing many interesting

Fig. 12-21 Cratered surface of Mercury as recorded by Mariner 10 on September 23, 1974. Courtesy of N.A.S.A.

theoretical problems but providing little evidence of its geologic history.

Mercury

Mercury, the closest planet to the Sun, has a diameter of about 4900 kilometers (3000 miles), a density of about 5.0, and a mass about 6 percent of the Earth's. Because of the resulting low gravitational attraction, about 27 percent as strong as the Earth's, Mercury retains no atmosphere. Surface temperatures range from about 370° C (700° F) on the bright day-side of the planet to −180° C (about −300° F) on the dark night-side.

The surface of Mercury, as recorded by instruments aboard space vehicle Mariner 10 in 1974, is dominated by crater-forms having a density of distribution and range of sizes comparable to those on the Moon (Fig. 12-21). The largest features are multi-ringed basins, as much as 1250 kilometers (780 miles) in diameter, which are associated with apparent floods of marialike material. Rayed and unrayed craters, averaging 40 kilometers (25 miles) across, commonly have central crater peaks. Compared to lunar craters, those on Mercury seem shallower, the extent of their ejecta blankets seems less far-flung, and secondary craters are concentrated closer to primary crater rims. Despite these differences—which may reflect a stronger field of gravity—the surface of Mercury seems to preserve a record of events in the formative stages of the inner planets of the solar system which is very similar to that of the Moon.

MARS

Unlike the Moon and Mercury, whose faces seem ancient relics little changed in the several billion years since their early catastrophic histories, Mars remains a geologically dynamic planet. It has a diameter of 6760 kilometers (roughly 4200 miles), an average density of 3.9, and a mass equaling 11 percent of the Earth's. The gravity field, about a third that of the Earth's, is enough to hold a gaseous envelope. This atmosphere—as recorded by instruments of the Viking 1 landing vehicle—consists of some 95% carbon dioxide, 2%–3% nitrogen, 1%–2% argon, 0.3% oxygen, detectable water vapor, and minute traces of the noble gases krypton and xenon. At the surface of Mars, the atmosphere is a bit less than 1/100th as dense as that at sea level on the Earth. Despite the low atmospheric pressure of less than ⅛ kilogram per square centimeter (less than ½ pound per square inch), the Martian surface is swept by violent winds, reaching velocities of 270 kilometers per hour (around 170 miles per hour).

Polar Regions

Seasonal Polar Caps   Mars has winter and summer seasons—like the Earth—which are marked by the expansion and contraction, alternately in northern and southern hemispheres, of white polar caps (Fig. 12-22). In winter they extend to about 45° of latitude, about half way to the Martian equator. In summer the caps recede to within 5° latitude from the poles.[8]

The temperatures at the Martian surface are colder and the range is far greater than on Earth (because of Mars's greater distance from the Sun), and far more extreme (because of the thin atmo-

8 A degree of Martian latitude equals 59 kilometers, or 36.7 statute miles.

Fig. 12-22 Mars, showing white south polar cap. Olympus Mons volcano is the "pimple" a bit left and above center; light-colored area just right of center is the Tharsis Ridge region. Mariner 7 photo, August 1969, from a distance of 472,000 kilometers (293,200 miles). Courtesy of N.A.S.A.

sphere); however, the temperature variations do not rival those on the Moon and Mercury. In equatorial regions, Martian temperatures may rise to about 30° C (85° F) on a sumer day and fall to −80° C (−112° F) at night. In polar regions, temperatures during the long winter night probably drop well below −125° C (−190° F)—the freezing point of dry ice (solid carbon dioxide).

Whether the polar caps were mainly ice (frozen water) or dry ice (solid carbon dioxide) was a much-debated question. It was settled by data from Viking orbiting vehicles in 1976. The caps are ice. The seasonally fluctuating parts of the white caps are probably just a cover of frost, at most a few meters thick. However, Viking photographic images do suggest that in places the ice is, or has been, thick enough to flow like a glacier and erode the ground. The polar caps are associated with two distinct terrains that are exposed in summer when the caps recede.

Terrains   The "etch-pit" terrain is characterized by many depressions, which might be deflation hollows in a windswept landscape. The "laminated" terrain, which lies poleward of 80° latitude, con-

Fig. 12-23 Phobus, the larger of Mars's two satellites, is about 25 kilometers long and 21 kilometers wide (15.5 by 12.5 miles). In the Viking Orbiter photo only about half of the surface facing the Sun was illuminated. The large crater (right end) is about 5 kilometers (3 miles) across, and craters as small as a few hundred meters are visible. Possibly this irregular-shaped satellite is a captured meteoroid of the sort that blasted out the great craters on the inner planets and the Moon during their early history. Courtesy of N.A.S.A.

tains flat-lying alternating layers of lighter and darker materials. As many as 50 such layers, having a total thickness of about a kilometer (more than 3280 feet), have been observed. Laminated terrains are intrepreted as sedimentary deposits, with dark layers dominantly composed of windblown dust and light-colored layers of ice.

Geologic Processes

Impact Structures The oldest regions on Mars, comprising about half the planet's surface, have densely cratered landscapes resembling the face of the Moon.[9] Mar's largest circular structure, the

9 By chance, the paths of the first Mariner space vehicles

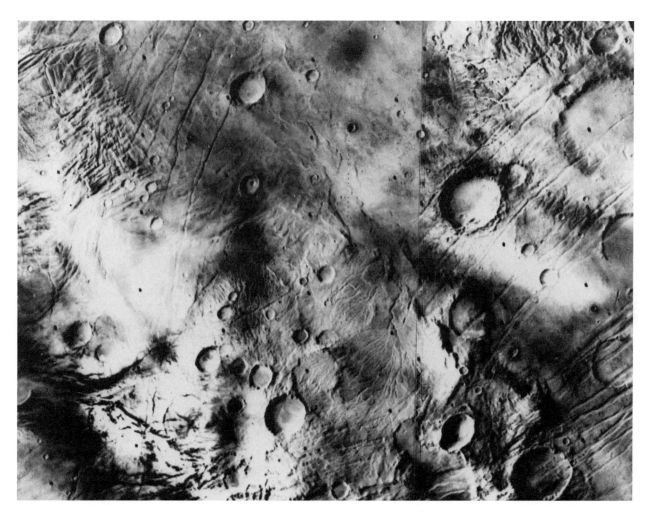

Fig. 12-24 Mosaic photo from Viking Orbiter (1976) shows the southern highlands of Mars. The prominent impact craters and plains are offset by many faults indicated by the linear trends. The visible dendritic patterns (just right, below center) strongly suggest the work of running water at some past time. Bright areas are parts of the South Polar frost cap. Courtesy of N.A.S.A.

Hellas Basin, has a diameter of some 2000 kilometers (1240 miles). This feature, twice the size of

the Imbrian Basin on the Moon, might reflect the proximity of Mars's orbit to planetoid missiles in the adjacent asteroid belt (Fig. 12-23). That the largest basins on Mars originated as multi-ringed features, like those on the Moon and Mercury, is suggested by the relics of concentric ridges in several of the basins. Admittedly, no concentric rings are visible

mainly crossed cratered regions. That the surface of Mars was more varied than the Moon's was not fully appreciated until 1972 when Mariner 9 transmitted over 7000 excellent images during its useful life of nearly a year in orbit around the planet.

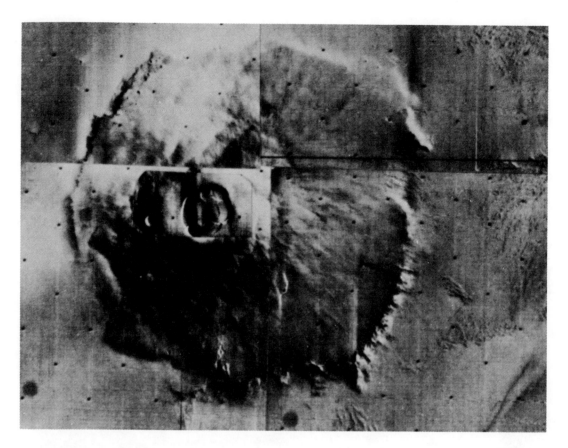

Fig. 12-25 Olympus Mons, the solar system's greatest known volcano, as recorded by Mariner 9. Courtesy of N.A.S.A.

in the Hellas Basin, but its smooth floor seems a vast plain developed by basin-fill which may have buried internal rings.

The many impact craters pitting the presumedly older terrains on Mars mainly range from 100 to 20 kilometers (62 to 12 miles) in diameter. Although clearly comparable in origin to those on the Moon and Mercury, Martian craters are somewhat different. Their density on the ground surface is somewhat less, central peaks exist but most craters are flat-floored, their rims are topographically more subdued, ejecta blankets are less sharply defined, and crater rays are conspicuously absent. These contrasts with the craters of the Moon and Mercury could well result from the erosional and depositional activity of strong Martian winds.

Volcanism Besides vast lava plains that resemble lunar maria and the far younger lava plateaus on the Earth's surface, Mars has tremendous shield volcanos, some of which—based on the youthful patterns of their radial flows and vaporous clouds—may still be erupting. The largest known volcano in the solar system is Olympus Mons, one in a group of four•aligned shield volcanos, each of which is larger than any on Earth (Fig. 12-25).

Olympus Mons has a basal diameter of some 500 kilometers (over 300 miles), rises 25 kilometers (15 miles) above its surrounding plains, and has a crater (or caldera) 65 kilometers (40 miles) wide. It is three times as high as Mount Everest. Although dwarfing the island of Hawaii, greatest volcanic pile on the Earth's crust, it seems significant that the volume of Olympus Mons about equals the total volume of basalts composing all the volcanos in the Hawaiian island chain. The volcanic piles forming the Hawaiian Islands probably represent successive eruptions through a drifting sea-floor plate that is gliding across a long-lived mantle plume. On Mars—which apparently lacks any horizontally drifting plates of lithosphere—far greater volcanic piles could erupt upon unmoving crustal plates standing over fixed mantle plumes.

Tectonic Features   The dynamic nature of Mars's interior is indicated by its great crustal fracture systems. Many bound straight-sided valleys 1 to 5 kilometers (0.6 to 3 miles) wide and several hundred kilometers (at least 1000 miles) long. Interpreted as grabens, they indicate tension (stretching) of the Martian crust. One of the most impressive features on Mars is the Coprates Canyon, 120 kilometers (75 miles) wide, as much as 6 kilometers (3.7 miles) deep, and 5000 kilometers (3100 miles)

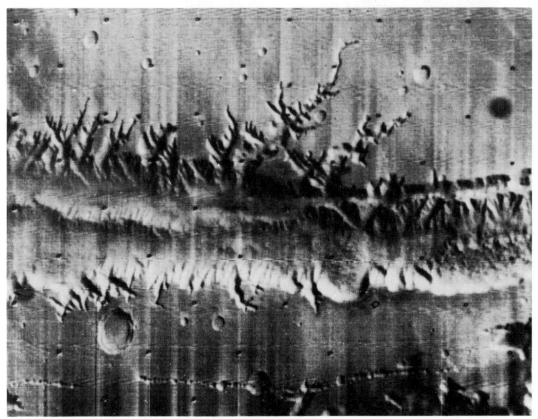

Fig. 12-26   The great Coprates Canyon is probably a graben structure. Tributaries appear joint or fracture controlled and may have been eroded by strong Martian winds; however, similarity to some stream valleys on Earth is interesting. Canyon area shown is roughly 270 by 480 kilometers (235 by 300 miles). Courtesy of N.A.S.A.

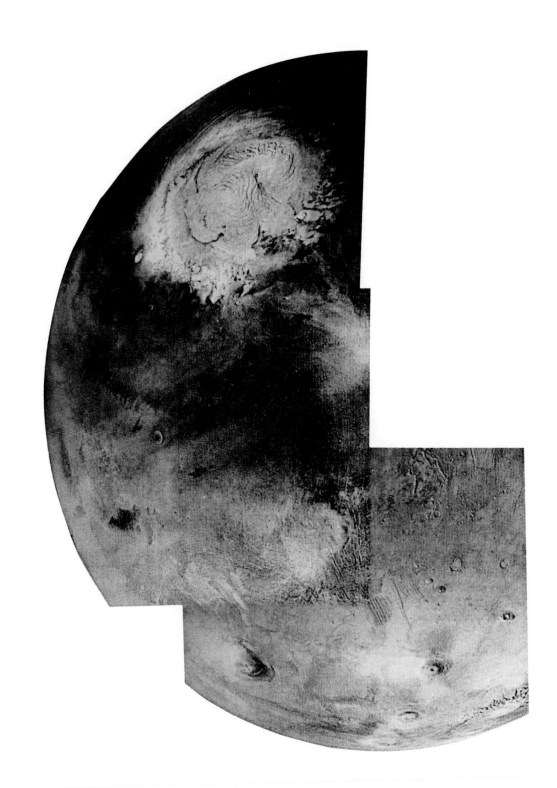

long (Fig. 12-26). This "Grand Canyon of Mars," which would stretch across the continental United States, is four times as deep and six times as wide as its namesake on Earth. It is interpreted as a complex of grabens.

Some of the fracture systems are closely related to great volcanos and impact basins; however, many other systems seem related to a vast regional pattern. From a center at the Tharsis Ridge, near Mars's equator, the fracture systems radiate outwards through the region of volcanos across almost a quarter of the planet. Based on knowledge of the Earth's crustal deformation, it can be speculated that the planetary pattern reflects tensional faulting of a vast domal uplift. Whatever the deep-seated cause of Mars's crustal deformation, it is not directly comparable to the Earth's global mechanism of plate tectonics.

Perhaps, as some scientists have speculated, Mars is in a "youthful" stage of planetary tectonic development. In one of several tentative hypotheses, the Coprates graben complex has been compared to an axis where plates are spreading, like the Earth's mid-oceanic ridges. Perhaps this feature, or an even vaster dome centering on the Tharsis region, reflects rising mantle material from within the planet. But Mars lacks one of the Earth's most obvious global features: nowhere is there any evidence of appreciable crustal compression—Mars has no great chains of folded mountains like the Alps, Andes, or Himalayas.

Atmospheric Sculpturing Wind seems the major atmospheric agent now shaping the Martian landscape. When Mariner 9 arrived and began orbiting Mars, its photographic mission was delayed because a great dust storm obscured the face of the planet. Although such violent wind storms may be rare, occurring about once in a century, there is ample evidence for the geologic work of wind

(Fig. 12-28). A crater west of the Hellas Basin contains a large field of transverse dunes. Streaks, best interpreted as eolian deposits, extend well "downwind" of many Martian craters. Deflation hollows may form the pits on polar terrains. Where sand-sized material is available for abrasion, erosional eolian landforms are likely on Martian bedrock.

Only traces of water have been detected in the Martian atmosphere. Nothing resembling the Earth's surface waters such as streams, lakes, or seas exists on Mars.[10] Yet the Martian surface has dry dendritic drainage patterns, meandering and braided dry channels and other features whose only counterparts on Earth are the work of running water (Figs. 12-24, 12-26). Moreover, some Martian canyons closely resemble the distinctive canyons on Earth that were scoured out by the surging floods of tremendous volumes of water (when the collapse of glacial-ice dams released the water in deep lakes; see Chapter 15).

All explanations, so far, for the stream-like features on Mars are merely working hypotheses, scientific guesses. The water on Mars could have originated—like the Earth's atmosphere and hydrosphere—from volcanic "degassing" of the mantle (p. 324). Today, however, all water on Mars seems locked in ice in the polar regions: in the ice caps, the light-colored layers of laminated terrains, and perhaps within pores of permanently-frozen ground like that of the Earth's arctic tundras. That Mars once had a hydrosphere of sorts but later lost it is a deceptively simple explanation. Carrying this reasoning a step further, however, leads to a major problem: Why, after a hydrosphere and an atmosphere formed, was only the water lost? Perhaps more acceptable is the hypothesis that Mars has

10 Space vehicles have sent back no evidence whatsoever of the "canals of Mars"—a long-standing topic in science-fiction stories and in some scientific speculations.

Fig. 12-27 Major geologic features of Mars. Shows Coprates Canyon (lower right), Olympus Mons (lower left), Tharsis region (lower one-third), fracture zones (lower center), and laminated polar terrains (upper left). Photo Mozaic from North Pole to just south of the equator. Courtesy of N.A.S.A. (facing page).

had pronounced climatic changes (like the Earth). Today Mars's water could be mainly locked in laminated terrains or frozen ground. But if the polar regions warmed appreciably—perhaps because of volcanic eruptions, or perhaps because the planet's axis of rotation became more steeply tilted towards the Sun—melting ice might send water flooding across the Martian surface. Such speculations reflect our present inadequate knowledge—for the stream-like features are only one of several major problems in Martian geology.

CONCLUSIONS

As regards the changing Earth—whose primeval rocks were long ago obliterated by the workings of the plate tectonic machine—modern geologists can only agree with James Hutton, who wrote in 1795: "I see no vestiges of a beginning." Yet thanks to technology and the space program, we can now infer the Earth's "pregeologic" history from our neighbors in space. Their little-changed faces may well preserve vestiges of the formative stages of the Sun's terrestrial planets. Perhaps Dr. Hutton would accept recent theorizing, since he also wrote: "A theory is nothing but the generalization of particular facts, and in a theory of the Earth, these facts must be taken from observations of natural history." Now geologic laboratories are so deluged with new facts about the Moon, Mercury, and Mars that any theorizing needs constant updating. Nevertheless, from the welter of detail some broad generalizations seem reasonable.

Initial Accretion

Conceived as clots of matter in a swirling cosmic cloud, the embryonic planets grew as their increasing gravitational fields swept in dust and smaller accreted bodies. Heated by compression as their masses increased, by collisions with impacting bodies, and by the energy from accumulating radioactive elements, the planets became molten masses in which lighter materials floated to the surface and crystallized into primordial crusts. This part of the history of planetary growth remains pure speculation, but has a reasonable scientific basis.

Catastrophic Impacts and Volcanic Floodings

The final phases of catastrophic accretion, when the planets were essentially full-grown, can be reconstructed from observable features on the surface of the Moon and of our neighboring planets. The highly cratered lunar terrae and comparable regions on Mercury and Mars seem relics of an intense period of meteoroid bombardment, from about 4.65 to 4 billion years ago. During the next major historical episode, from about four billion to perhaps three billion years ago, giant planetoids crashed into the older highly cratered terrains and blasted out vast impact basins. These events probably triggered the subsequent eruptions of great floods of maria-type basalts.

Later History

During the last three billion years, the face of the Moon seems to have been largely static. Meteoroid impacts continued but at a greatly decreasing rate. The rayed crater Copernicus, one of the Moon's least altered and most youthful-appearing craters, was created during the Precambrian era of the Earth's time scale. Mercury, although possibly affected by later volcanism,

Fig. 12-28 The surface of Mars from the landing site of Viking I. The bouldery surface (foreground) is mantled with dunes, indicating the action of winds on the planet. (The arms on the landing vehicle record meteorological data: air pressure, temperatures, wind directions and velocities). Courtesy of N.A.S.A. (facing page).

seems to record a comparable history. Mars preserves vestiges of an ancient catastrophic history, but it has remained a geologically active planet. The Martian surface reflects continuing volcanism and tectonic processes as well as surface sculpturing by wind and, probably, running water. Quite probably the Earth's earliest history was marked by events like those recorded on the Moon and neighboring planets. But the evidence here was obliterated by the ongoing internal and external processes that constantly destroy old rocks and create new ones on the surface of our more dynamic planet.

Aside from a continuing physical evolution resulting from such mechanisms as plate tectonics and the hydrologic cycle, the Earth is unique among the Sun's family of planets in producing a vast array of complicated and changing biological organisms. Mars might have developed primitive living things.[11] But the possibility of life on other planets in the solar system is most unlikely because of their exceedingly harsh surface environments. Unlike the hospitable Earth, our neighbors in space are characterized by extreme temperature variations coupled with either the lack of any protective atmospheric blanket, as on the Moon and Mercury, or suffocating atmospheres, as on Venus. Thus the changing panorama of life through time is a topic limited exclusively to the later chapters of Earth history.

[11] Experiments at the two Viking landing sites neither confirmed nor denied the possibility of very primitive life on Mars. (Certainly no Martians walked by the T.V. cameras.)

SUGGESTED READINGS

Bowker, D. E., and Hughs, J. K., *Lunar Orbiter Photographic Atlas of the Moon,* NASA SP-206, Washington, D.C., National Aeronautics and Space Administration, 1971.

Engel, A. E. J., "Geologic Evolution of North America," in *Science,* Vol. 140, No. 3563, pp. 143–152, 1963.

Jones, H. S., *Life on Other Worlds,* New York, New American Library, 1956 (paperback).

Murray, B. C., "Mercury," in *Scientific American,* Vol. 233, No. 3, pp. 58–68, 1975.

Mutch, T. A., *Geology of the Moon,* Princeton, N.J., Princeton University Press, 1970.

Pollack, J. B., "Mars," in *Scientific American,* Vol. 233, No. 3, pp. 106–117, 1975.

Siever, Raymond, "The Earth," in *Scientific American,* Vol. 233, No. 3, pp. 82–90, 1975.

Wood, J. A., "The Moon," in *Scientific American,* Vol. 233, No. 3, pp. 92–102, 1975.

On the History of Life

Sometime during the long Precambrian history, life began on Earth. We will now consider the nature of living things and the changes in past life as determined from the fossil record. Fossils—aside from their intrinsic interest as indicators of the past floras and faunas that inhabited the Earth—are important because they provided the timepieces needed to construct the geologic time scale. The time scale was essential for unraveling the Earth's past sequence of complicated physical events.

Hand specimen of fossiliferous limestone. Photo by Tad Nichols.

13

"Dramatis Personae" in the History of Life

BIOLOGICAL INTRODUCTION TO THE FOSSIL RECORD

Plants, Animals, and the Chemical Cycle of Life

Just as the physical world is unified by a great geochemical cycle, so is the world of life unified by chemistry. From a purely chemical point of view, life is part of a long-continued circulation of six main elements (C, H, O, N, P, S)[1] from the water, soil, and air into complicated organic molecules and back again. The most critical role in this great life cycle is played by plants. They freshen the air with oxygen and manufacture the food on which the animal kingdom depends; where the cycle is retarded, they may furnish coal and some of the volume of oil. Most important in this grand scheme of life is the remarkable process of photosynthesis, wherein the green cells of plants use solar energy to convert carbon dioxide gas and water into sugars and starches with the liberation of free oxygen.

Because of the abundant oxygen in its atmosphere, the Earth is unique among planets of the solar system. If the present atmosphere is derived from volcanic exhalations—the presently favored scientific view—the free oxygen cannot be an original constituent, for this reactive element is absent from volcanic gases, being combined with hydrogen and carbon in H_2O and CO_2. Some oxygen might have resulted from the splitting of water vapor when the Earth was still very hot, and some may be continuously formed by the solar radiation bombarding and breaking down water vapor in the outermost atmosphere. But the largest proportion of oxygen is best attributed to photosynthesis, which began in geologically remote time with the appearance of the first primitive green plants.

In a sense, all animals are parasites, for they cannot make food from the Earth's raw materials. A host of predators eat other animals, but towards the

1 Carbon, hydrogen, oxygen, nitrogen, phosphorus, sulfur.

bottom of any food pyramid are vegetarians living off plants. The starches and sugars produced by photosynthesis are carbohydrates, energy producers essential for the active lives of animals. No less important, plants also manufacture proteins, the foods for growth and replacement of body tissues. Plants take nitrogen from compounds in soil and water and combine it with the carbon, hydrogen, and oxygen of carbohydrates to synthesize amino acids, the "building blocks" of protein. Then, by adding sulfur and phosphorus, they convert the amino acids into the extremely complex and varied molecules called proteins. These, mixed with considerable water, form protoplasm, which makes up the cells of all living things.

Protoplasm is the essential stuff of life. Its constant physical and chemical changes, collectively called metabolism, enable it to build new protoplasm exactly like itself. It oxidizes organic foods to liberate the energy required for vital processes, producing waste materials of CO_2 and H_2O and, in the case of proteins, nitrogenous urea as well. On the death of an organism, decay bacteria break down its remains into the simpler substances from which it was constructed and into ammonia, the end product of protein decomposition.

The simpler products of metabolism and decay return to the air, water, and soil, where they are recycled. Certain bacteria convert ammonia into nitrate compounds, fertilizers that can be assimilated and used again by plants. In special cases where bacterial decay is inhibited, coal may result from the partial breakdown of woody plant material, and oil from incomplete decomposition of simple plants and animals. When burned, the free carbon in coal and the hydrocarbons of oil are rapidly oxidized to CO_2 and other simple products, ending this interruption of the grand cycle. Perhaps the most striking aspect of the geochemical cycle, through geologic time, is the remarkable structural plasticity of protoplasm that has allowed the evolution of the myriad forms of plants and animals.

The Concept of Organic Evolution As the physical world has gradually changed through geologic time, so has the world of life, for converging lines of evidence clearly indicate that more advanced plants and animals developed from simpler ancestors. This is the doctrine of organic evolution which, like physical uniformitarianism, has triumphed over catastrophic explanations. Baron Cuvier, for instance, who was one of the first to recognize faunal succession, was a whole-hearted Catastrophist, convinced that differing populations were specially created in their final form and later exterminated wholesale, to be eventually replaced by newly created and different forms. Today, no competent scientists dealing with life and its history doubt that organic evolution has occurred; the uncertainties now revolve around its causes.

Biological Evidence of Evolution

Several lines of evidence for evolution are biological, being based on the study of living plants and animals.

Experimentation Man has experimentally accelerated changes in certain groups by selective breeding. All dogs from Chihuahua to Saint Bernard have been produced from wolflike forebears. The highly productive types of hybrid corn result from selective breeding. Unfortunately, so do bacteria immune to antibiotic drugs. The disease germs are an expanded population descended from a few ancestors who managed to survive the application of the drugs, and, unlike the other examples, are a case of unplanned experimentation. However, despite their seeming differences, all dogs belong to a single species, or kind, of animal, *Canus familiaris,* because they can all interbreed. Moreover, general experimentation has not produced a new genus. So such cases, although suggestive, are not in themselves proof of evolution.

Comparative Anatomy The anatomy of living forms is best explained by evolution. For example, the skeletal plan of all legged vertebrates, including such diverse animals as lizards, dogs, monkeys, and men can be matched bone for bone. In examining front limbs, the flipper of a seal and the wing of a bat, although used for quite different purposes, have corresponding bones which can also be matched in the arm of a man or the front legs of a crocodile. Each has a single upper bone (called the humerus), two bones below the elbow (the radius and ulna), a cluster of wrist bones (carpals), and groups of finger or toe bones (phalanges). The bones do differ in shape and size, but this reflects their special uses. The soft anatomy, including muscles, nerves, the circulatory systems, and internal organs, are also remarkably similar. Moreover, structural resemblances can be demonstrated in groups of invertebrates and plants. Such anatomical similarities are most easily explained if groups of animals and plants are related, rather than products of special creation.

The biological case is further strengthened by *vestigial structures*. The appendix in man has no function although in other animals, rabbits for instance, it is a functional part of the digestive system. Pelvic girdles in some snakes, wings of ostriches, and splint bones of horses are all useless structures corresponding to functional parts of other creatures. All such remnants suggest a degeneration of features once useful in some ancestral form—hence evolution.

Embryology "Ontogeny recapitulates phylogeny" is an impressive statement attributed to the German zoologist Haeckel, which is known as the Biogenetic Law. It means that as an individual animal develops from a fertilized single cell through the early stages before acquiring its adult form (this individual history being ontogeny), the animal retraces the history of its entire race (phylogeny). In man, for example, human embryos pass through stages when they have nonfunctional gills (suggesting an aquatic life) and tails, and in a late prenatal stage are covered with hair which is lost before birth. Support for the parallelism of individual and racial development is given in the paleontological record of many invertebrate stocks where juvenile forms show stages like the adult forms of earlier geologic periods. Admittedly, living forms do not follow the sequence perfectly because many phylogenetic stages may be omitted, the sequence may be partly juggled, and sometimes new features appear that do not characterize an adult ancestor.[2] Thus the biogenetic "law" is really not a law. In general, however, animals which are thought to be closely related from other lines of evidence have similar embryonic development; differentiation and specialization come about at maturity. Thus the similarity of earlier stages strongly suggests a common descent.

Biochemical Evidence Human blood yields little precipitate when mixed with the blood serum of a reptile, somewhat more with that of a monkey, and large amounts when mixed with serum from anthropoid apes. In short, the closer the assumed relationship of animals, the more similar their blood. Any one of these "facts of life" is hardly proof, yet they do give converging lines of evidence. This biological case for organic evolution is strong because it deals with the complete spectrum of living organisms whose whole anatomy, both hard and soft, can be studied in great detail, and which can be manipulated or experimented with to some extent. Moreover, the work of Alfred R. Wallace and especially of Charles Darwin that established the concept was largely based on living forms. But biology treats only the end products, plants and animals as they now exist, so its case rests entirely on circumstantial evidence.

2 The umbilical cord is the most obvious example.

PALEONTOLOGY

Paleontologists

For the record of life through the ages, we are indebted to paleontologists—hybrids, part geologist and part biologist. In defining their work, some purists talk of paleobiology which includes *paleobotany*, the study of fossil plants, and *paleozoology*, the study of fossil animals. Commonly, however, those who deal with animals call themselves paleontologists: *invertebrate paleontologists* deal with spineless creatures, such as worms, sponges, clams, and crabs; *vertebrate paleontologists* study animals with a spinal cord; *micro-paleontologists* squint through a microscope at tiny noncellular organisms or minute invertebrates. Whatever the label, they must be well grounded in biology, for in paleontology the present is as much the key to the past as in other geologic fields.

Since fossils are defined as evidence of past life, the question arises, how far past? The long-extinct dinosaurs are clearly fossils; the remains interred in the local cemetery are not. "If it still smells take it to a zoologist, if not, to a paleontologist" is an oversimplification. The beginning of written history is used by many paleontologists as a convenient dividing line between present life, treated by biologists, and "prehistoric" life, or fossils. But some would quibble, so perhaps we can only say that fossils are those past remains that paleontologists and paleobotanists study. In any case, without fossils, and paleontologists too, there would be no satisfactory history of the Earth.

The Uses of Paleontology

Besides being timepieces and environmental clues, fossils are—by definition—signs of past life; so paleontologists with a more biological bent use them to reconstruct the particular plants and animals they represent and study them as communities of individuals interacting with each other, and, perhaps most important, use them to gain some insight into organic evolution.

Reconstructing Extinct Organisms Restorations of animals are largely based on hard parts. External skeletons, or shells, are often found intact in the case of clams and other small invertebrates living in marine environments where quick burial and preservation are common. Vertebrate skeletons, however, are usually found as dismembered pieces, for unless those on land were mired in a swamp, overwhelmed by an ash fall, or caught in some other comparable situation, their burial takes time, leaving them prey for scavengers. Yet, from a knowledge of the skeletal anatomy of modern animals, skillful piecing together of fragments, and those few lucky finds of complete skeletons, paleontologists have a good idea of the bony structure of most fossil vertebrates.

Clothing skeletons with muscles, fat, hair, and skin, or filling in the fleshy anatomy of invertebrates, requires, at best, a certain amount of guesswork, for corruptible flesh is rarely preserved, and largely known only in such late fossil forms as the Pleistocene mammoths frozen in the arctic tundras. Mummified remains are even more rare, although in Wyoming two dinosaurs were dried out before fossilization, thereby preserving details of their hides. Such things as the color of extinct animals, vertebrate or invertebrate, may never be surely known. For instance, dinosaurs may have been either gaily colored or somber; the modern reptiles, to which they may be compared, are found with both types of coloration.

But bone and meat are functionally related, so museum mounts, pictures, and the like showing the living shape of extinct animals are not pure guesswork. Based on the anatomy of living animals and the mechanical workings of their various parts, musculature can be inferred from the size and shape of bones and the scars where muscles were attached.

Fatty structures, however, are less predictable. Whether fossil camels had humps is as impossible to determine from skeletons as whether a man had a paunch or a woman was curvaceous. Yet reconstructions do represent the admirable urge of paleontologists to present their scraps of stone as living things, and, overall, most reconstructions are not concoctions of unbridled imagination. For instance, lobe-finned fish, a group from which amphibians probably arose, were thought to have been extinct since Cretaceous time. Since 1938, however, representatives of the lobe-fins have been taken from deep waters off Africa. They are remarkably similar to reconstructions of a kind of Cretaceous lobe-fin previously known only from reconstructions based on fossils.

Environments and Ecology   Once fossils are visualized as living plants and animals, they allow the reconstruction of past environments. Any animal, including fossil forms, reflects the sum total of complexly interwoven physical and biological elements of the world in which it lived. The basic elements of environments, past and present, are heat, light, and the surrounding medium. So, to prosper, organisms must be adjusted to temperature, pressure, and movements of their native air or water—to deep or shallow, muddy or clear water in oceans, streams, and ponds; to humid temperate climates, tundras, rainforests, or deserts—in short, to the many elements of their surroundings. In an earlier section we mentioned how different forms that reflect different conditions have caused problems in correlation, as well as helping in determining environments of depositions.

Many paleontologists today are much interested in ecology—the study of organisms in relation to their biological communities and to their physical surroundings. It is the study of total environment or, put another way, "the balance of nature." For instance, in a modern Rocky Mountain population, deer are browsers on leaves and twigs and do not really prosper on hay, and mountain lions prefer a diet of venison. Squirrels live on nuts and seeds and are, in turn, preyed on by agile martens who are no threat to deer; and so it goes. Each animal has a somewhat different role or ecological niche to fill, and should two groups with similar demands come together, one of them must change its habits, depart, or die out. These relations are not completely understood in living populations, so the study of paleoecology based on fossils is even more challenging.

Evidence of Evolution   The proof of evolution has come from the fossil record, which yields the only direct—that is, observable—evidence of gradually changing forms of plants and animals. Faunal succession, seen in rocks stacked one on the other, adds the needed dimension unavailable to biologists—the geological dimension of time. There are certain drawbacks to working with fossils. As previously mentioned, the reconstruction of individuals as they were in life is often difficult and tentative in varying degrees. More important, the fossil record is definitely biased towards certain groups, so that very rarely is anything like a complete population represented as it must have been in life. Marine invertebrates such as clams, coral, and the like are preserved out of all proportion to their relative importance in the original living population, because they had resistant hard parts and lived in an environment where rapid burial was common. Yet there is no doubt that even in the seas multitudes of delicately armored and soft-bodied shell-less creatures lived with them. The Burgess Shale of Cambrian age in the Canadian Rockies has a locality where, because of unique conditions, a multitude of soft-bodied invertebrates are preserved. All other Cambrian fossil localities would leave the erroneous impression that the early Paleozoic seas were populated almost entirely by hard-shelled forms. On land the chances of preservation, in general, are poor because of the dominance of erosion, but a few places such as the Cenozoic Florissant Lake beds of Colorado do give rare, but significant, glimpses of the true diversity of past floras and faunas.

The genealogical record of the commonly preserved organisms is not as complete as we would like it. Because of unconformities, changes in conditions suitable for preservation, and migration of populations, there are gaps in the vertical record of changing faunas through time—recall that the master geologic rock column had to be pieced together from many different localities. However, despite the admitted imperfections of the fossil record, hundreds of thousands of fossils have been collected and studied, and, even if they represent a miniscule fraction of former life, they quite adequately demonstrate certain basic facts. Life began in the remote past as simple organisms and has been continuous, and more complex forms have evolved from simpler ones, which is the essence of the geological account of life through the ages.

Organizing the Record

If minerals seem difficult to comprehend because of their variety, the animal and plant kingdoms are an even greater challenge. As the living and fossil organisms so far described number over a million and a quarter, a system of grouping and naming is absolutely necessary to make any sense of them.

Classification   The Swedish botanist Carl Von Linne (1707–1778), better known by the Latinized name Linnaeus, set up the system of biological classification that, with modifications, is in use today. Linnaeus did not invent taxonomy, the science of plant and animal classification, but in his *Systemus Naturae* published in 1758 he organized previous attempts into a generally workable system. The Linnean classification, which is used by paleontologists, sorts living things into a hierarchy of progressively smaller groups.

All living things belong either to the plant kingdom or the animal kingdom,[3] each of which is subdivided into major branches called *phyla*. These are, in turn, subdivided into smaller groups called *classes*, the classes into *orders*, orders into *families*, families into *genera*, and genera into *species*, which represent the smallest groups and are composed of *individuals*. As an example, let us classify man according to this hierarchy: kingdom animal; phylum chordata (animals having backbones or cartilaginous columns); class mammalia (warm-blooded animals that suckle their young); order primate (who live in trees or walk on their hind legs, have five fingers and five toes with flat nails); family hominidae (tool-makers); genus *Homo* (human beings); species *sapiens* (including all living groups of men, to which you as an individual belong). Precise classification requires still further grouping by adding "sub-" or "super-" to certain of the original divisions. The Linnean system is not a simple classification scheme, but nobody has proposed a more useful one.

The designation of species has given paleontologists considerable trouble, for they deal with extinct life forms. Biologists define a species to include animals that interbreed. This is fine for biologists, because they can examine the sexual preferences of their subjects; but almost by definition paleontologists cannot, and must define their species by structural similarities which may well reflect interbreeding, but this cannot be positively proven. Moreover, in dealing with countless generations, paleontologists must contend with gradually changing organisms, which makes it difficult to draw hard lines between ancestral and descendant species. Paleontologically, a species may be only a snapshot of changing life through time.

Nomenclature   To name a plant or animal it is not necessary to list each group it belongs to in descending order of the Linnean hierarchy; a dual system

3 Modern usage sometimes includes a third kingdom, the protista, unicellular organisms which are hard to classify as either plant or animal.

of nomenclature, also established by Linnaeus, suffices. It uses two names, one for the genus and one for the species, rather like the family name and given names used for people. But naming organisms is not as easy as you might think. For scientific purposes, modern languages will not do because they would bring the confusion of many tongues and because there just are not enough common names to go around. Moreover, the same common name is often applied to quite different organisms in different places. A gopher, for instance, is a turtle in Florida, a ground squirrel in the High Plains, and a pocket gopher in the Rocky Mountains. So names are made up using ancient Greek and Latin which are international, or once were, among educated people, and, being "dead" languages, their word meanings do not change with usage. Many such scientific names are quite descriptive, if one knows Greek and Latin; *pachycephalus*, for instance, means "thick head," to coin an uncomplimentary name; *Felis domestica* translates as "cat, domestic," a species including all house cats; *Tyrannosaurus rex*, the "tyrant king," was the largest of the flesh-eating dinosaurs.

Originally, biological classification was strictly empirical, a descriptive "pigeonholing" of plants and animals by their similarities and differences. Animals with backbones, for instance, form a major group further divisible according to whether they have scales, feathers, hair, and various other structures. When the concept of evolution came into vogue and was reinforced by accumulated evidence from paleontology, the classification was seen to be genetic—organisms were related, the more complex being descendants of simpler ones. Classification had an integrated framework whose elegance is further evidence of evolution.

THE PLANT KINGDOM

Plants are broadly divisible into two types: those with a vascular system of vessels and ducts for piping dissolved nutrients and food; and those without. The more primitive nonvascular group, the lower plants, are largely adapted for life in water, where the kingdom originally evolved. The higher, vascular plants are characterized by roots, stiffened stems, and leaves, suiting them for life on land.

Nonvascular Plants[4]

Thallophyta A simple body, the thallus, characterizes the thallophytes (*phyte* means "plant"). Despite a great diversity of sizes and shapes meriting their division into a number of separate phyla, two main groups are recognized: the *algae*, having chlorophyl, the green protein used for photosynthesis, and the *fungi*, which lack it.

Most algae live in the sea, some in fresh water, and a few on wet ground, environments where they can absorb water directly into their cells without elaborate piping systems. They include pond scums, simple green algae, and numerous brown marine types, ranging from one-celled micro-organisms to large seaweeds and kelps 91 meters (300 feet) long. Although by far the oldest group and ancestral to all higher plants, only certain algae types are common in the fossil record.

Common fossil algae include various marine species that secrete calcium carbonate within and around their cells to build plates, cylinders, or encrustations with distinctive micro-structures. Such algae still live today, and fossil forms are found in rocks as old as Ordovician. In other cases, communities of many algal species form filamentous mats that, judging by modern ones, do not precipitate much solid material in their cells. However, they do trap and bind sediment particles to form dome-shaped, laminated structures from a few centi-

---

4 Plant classification being in something of a turmoil, it seems best not to designate subkingdoms, phyla, classes, or the like, but simply to refer to groups.

meters to a meter or so in diameter and height. Their swirling bunlike masses are relatively common in some Precambrian rocks, but unfortunately of no value in dating and correlation, for modern representatives are almost identical to the most ancient (Fig. 13-1).

Some 15,000 species of diatoms secrete an array of intricate siliceous skeletons. These microscopic plants live in water, mainly in the sea, where vast numbers floating as plankton are the foundation of the food chains there. On settling to the ocean floor, their hard remains form diatom ooze—es-pecially in cool waters—and eventually the rock diatomite (diatomaceous earth), some deposits being up to 914 meters (3000 feet) thick. Diatoms are common Cenozoic fossils, occurring in Cretaceous deposits, and are possibly present in Jurassic rocks. Their absence from older deposits may reflect a late evolution of hard parts, or long-term solution of the colloidal, less stable variety of silica in their tests (shells) by penetrating waters.

The fungi, which include bacteria, molds, rusts, toadstools, and such, are probably algae descendants who lost their chlorophyl. In any case, many

Fig. 13-1 Fossil algae of Cambrian age from upper New York State. Courtesy of the American Museum of Natural History.

live on land, though usually in moist environments, and all are parasitic on living plants, animals, or organic matter. Through geologic time, the role of bacteria has certainly been of prime importance, else the world of life would long ago have been overwhelmed by its own wastes and dead tissues. There is direct evidence of large fungi growing on trees of the Carboniferous coal forests. However, the geologic record of this group is insignificant.

Bryophyta   *Liverworts* and *mosses* are bryophytes, a minor group of lower plants that are mainly adapted for life in wet places. They differ little from thallophytes in lacking a vascular system and probably evolved from algae. Liverworts have a flat thallus body attached to the ground by threadlike structures called rhizoids. The more familiar mosses, some of which are hardy and adapted to very cold or dry climates, do have very primitive stems and leaves but rhizoids instead of roots, and no vascular system. In common with the higher plants, and unlike thallophytes, the fertilized ovules, or eggs, of bryophytes are retained and protected in the plant's tissues for a time after fertilization, giving them a better chance of survival.

Although mosses are peat-formers, the fossil record of bryophytes in general, which extends back to the Carboniferous, is too poor to be enlightening. They are interesting as a primitive attempt of plants to adapt to land life; however, they seem a separate line of thallophyte descendants, giving rise to no higher types. The "missing link" between the simpler water-dwelling algae and the vascular land plants is still missing. It was probably a seaweed, like the large *Nematophyton*, that evolved a tube-filled body for conducting water.

Vascular Plants

Long ago the algae invaded the swamps and ponds of the wet terrestrial environments, but they lacked the necessary equipment for life on land. In adapting to land life, their progeny, the higher plants, have developed the following:

1. An outer coat, or epidermis, that prevents drying out in air, yet has pores allowing exchange of gases involved in photosynthesis.
2. A fibro-vascular system that supports plants deprived of the buoyancy of water and serves as a circulatory system.
3. Anchoring roots that tap groundwater and mineral nutrients.
4. Leaves for increasing the area of photosynthesis.
5. In the highest types, reproductive systems freed of the need for surrounding water and producing a protected embryo.

From the fossil record, and the comparative anatomy of primitive survivors and more advanced plants, we can see how plants evolved and met terrestrial requirements.

Psilopsida   The fossil record of plants in the early Paleozoic when the land plants were appearing is very poor; yet through a quirk of preservation, Devonian strata of a silicified peat bog in Scotland contain excellent specimens of *Rhynia* which, even if not the actual progenitor of the higher types, is certainly a vascular plant of the most primitive sort. This small, rushlike plant, one-third to two-thirds of a meter (a foot or two) high, was little more than a creeping stalk with spore cases at the tips; leafless, hence photosynthesis must have occurred in its stem; and rootless, being anchored by a creeping root stock bearing threadlike absorbing hairs (*rhizoids*) instead of true roots. However, the stem had a fibro-vascular system. Other fossil Psilopsida include slightly more complicated types, whose stems bore scales that foreshadowed the evolution of leaves. Fossil Psilopsida are all of Silurian and Devonian age. No later ones have been found, although two interesting little plants living today in the tropics seem relics of this archaic group, whose fossils provide a good transition to the three main groups of higher vascular plants.

Lycopsida   Today the club mosses are insignificant creeping plants of tropic and temperate zones, exemplified by the ground pine. They seem descendants of the scaly Lycopsida, because their solid stems are clothed by many spirally arranged leaves. A step up in the evolutionary ladder, they possess true roots which, being adventitious, are essentially downward branches from a root stock. Reproductively, ground pines bear small cones filled with a single type of spore. However, one living Lycopsida produces two kinds of spores that develop into male and female plants; and some of the large fossil forms had even evolved a seed habit, a parallel but separate development to that in the higher modern plants.

Questionable remains have been reported in Siberian rocks of Cambrian age, but the heyday of the Lycopsida was in the Carboniferous when they rivaled the size of many large modern trees. These great scale trees, abundantly preserved in coal deposits, were lycopsids which shed leaves from their trunks, leaving prominent scales on the bark. The trunks were straight, some varieties having a few branches towards the top. Specimens of *Lepidodendron*, as much as 30 meters (100 feet) high, were scale trees with roots, and almost as large was *Sigillaria*. Why some of these large trees which had evolved a seed habit of reproduction did not survive is something of a mystery.

Sphenopsida   The horsetails, called scouring rushes because their siliceous stems were once used to clean pots and pans, are modern representatives of the Sphenopsida. Horsetails have spore-bearing cones, stout root stocks, and some roots. They are distinguished by vertically ribbed, hollow stems that have bamboolike[5] joints from which swirls or clusters of leaves radiate. Although a bit more advanced

than the club moss group, the Sphenopsida also arose from the Psilopsida stock, had their climax in late Paleozoic time, and are now relatively unimportant in the modern flora. The oldest known representatives occur in Devonian strata; however, the giant of the stock, *Calamites*, a 24-meter (80-foot) tall hollow tree, grew in thickets in the Carboniferous coal swamps.

Pteropsida   Club mosses and horsetails are the remnants of two lineages of Psilopsida descendants whose late Paleozoic day of grandeur is long past. The third collateral line, the Pteropsida ("fernlike" plants) now dominate the landscape, clothing it in green. The incredibly diverse and prolific Pteropsida are divisible into three groups: *ferns,* the parent stock; and two groups of seed-bearing descendants, the *gymnosperms,* which have naked seeds and in turn gave rise to the *angiosperms,* whose seeds are inside a case that is contained in a flower. The Pteropsida in general are set apart from other vascular plants by the nature of their leaves.

The *ferns* (technically *Filicinae*), which first appear in Devonian rocks, are the most primitive Pteropsida (Fig. 13-2). Although 12-meter (40-foot) *tree ferns* still live in New Zealand, the abundant and familiar ones are all relatively small. Their stems are typically inconspicuous, subterranean rhizomes with adventitious roots. The feathery, lacelike leaves are prominent, and from an evolutionary point of view most important. Unlike the horsetail and club moss groups, whose generally small and simple leaves are modified scales, fern leaves evolved from branches. Some of the ancestral Psilopsida have slightly flattened branch tips, foreshadowing things to come. In fossil ferns, transitional stages indicate that the true, or modern, leaf evolved by the continued flattening of small branches covered with simple scale leaves. The simple leaves ultimately fused into a single sheet, a superior light-catching device for photosynthesis, which forms the most advanced type of leaf.

---

5 Bamboo, however, is a more advanced plant related to the grass family.

Fig. 13-2 Fossil fern of Pennsylvanian age from Illinois. Courtesy of the Smithsonian Institution.

Ferns, however, despite their efficient leaves, are restricted to shady moist environments, for their reproductive methods are little improved over those evolved in the water-dwelling algae. So let us review plant reproduction, which needed some overhauling—specifically the development of the seed habit—before pteridophytes could complete the conquest of dry land. From a strictly geologic view, *spore* and *pollen,* the reproductive cells that plants shed in prodigious quantities, are important fossils, especially useful for dating and environmental studies in younger terrestrial sediments. Pollen is produced by gymnosperms and angiosperms, whereas spores are reproductive structures of ferns and other seedless plants.

Many spores are asexual single cells, each of which can grow into a new plant under proper surrounding conditions. Algae broadcast theirs in water, and terrestrial plants theirs through the air. So far, the method seems simple and easily adapted to a thoroughgoing land life. But plant reproduction was complicated in the remote past when some ancestral marine algae "discovered" sex. Sexual reproduction involves two kinds of reproductive cells, the gametes, called sperm and egg, that fuse into one which develops into a new individual animal or plant. As a result plant reproduction is more complicated than in animals, for plants lead a double life as alternate sexual and asexual generations. An asexual plant, the sporophyte, sheds only spores that develop into sexual plants, called gametophytes. The gametophytes are sperm and egg producers whose offspring are new sporophyte plants. The typical fern, for instance, is the sporophyte; the gametophyte is an inconspicuous different-looking plant, a thin green sheet attached to the ground by rhizoids. Sperm are loosened on the underside of a fern gametophyte, where they swim through a film of water to fertilize the eggs. Because sperm cannot stand drying out, they are the weakest link in the reproductive cycle of spore-bearing terrestrial plants.

The gymnosperms were the first Pteropsida with a seed habit. In seed plants a minute female gametophyte remains and functions inside the reproductive apparatus of the sporophyte. Grains of pollen (a kind of micro-spore resistant to drying) are transported through the air in great quantities. Those reaching a female element generate sperm that fertilize the egg. When fertilized, the egg develops into an embryonic plant surrounded by food housed inside a tough cover—this whole structure constitutes a seed. Thus the evolution of pollen freed terrestrial plants from the necessity of wet ground; and the seed, being an embryonic plant provided with stored food, has a good start on life when conditions favor germination.

The descent of seed-bearing Pteropsida from ferns is documented by clearly transitional fossils, the extinct *seed ferns* found in Devonian to Jurassic rocks. They had both treelike and sprawling habits, and are so like the true, or spore-bearing, ferns that unless associated seeds are found, the two groups are indistinguishable. Having seeds attached to their leaves, however, makes seed ferns the simplest of gymnosperms, a group characterized by exposed seeds.[6]

Some late Paleozoic leaves indicate that *cycads*, palmlike gymnosperms,[7] evolved from seed ferns. Some of the cycads, an extinct group abundant in the Mesozoic flora, had stubby, oval trunks crowned with palmlike fronds that bore seed cones. As this group also had primitive flowers, it is thought by some paleobotanists—but not all—to have been ancestral to the angiosperms, the flowering plants. The living cycads are distinguished by cylindrical tree trunks. A rather primitive gymnosperm stock, they are first found in Triassic rocks, and possibly in late Paleozoic.

The ginkgos, which may be found on many college campuses, are reported living wild in western China. They date back to the late Paleozoic, and were especially common in early and middle Mesozoic time. Superficially resembling angiosperms, they are in fact rather primitive, and their one remaining species is called a "living fossil." Cordaites, a late Paleozoic tree, was common in the Carboniferous coal swamps, but died out before the Mesozoic. Although bearing straplike leaves as much as a meter long, it seems closely related to conifers in most other features.

6 *Gymno* means "naked."
7 The true palm is an angiosperm.

Fig. 13-3 Miocene leaf from Colorado. Courtesy of the Smithsonian Institution.

The *conifers* include most of the familiar evergreens, such as pine, spruce, fir, and those all-time giants of the plant world, the redwoods, reaching 90 meters (300 feet) high and 12 meters (40 feet) through the trunk. Seeds are borne underneath the scales of the characteristic cones, from which the group is named, and the leaves have evolved into needles. Conifers first appeared in late Paleozoic rocks, reached their peak in late Mesozoic time, and are still common.

The angiosperms are a tremendously diversified and specialized collection—there are some

135,000 living species—including trees, like oaks, elms, maples, willows, palms, and many others; some smaller woody plants, or shrubs; grasses; garden vegetables; and the flowers. Unlike the gymnosperms, which are all trees or woody plants, the angiosperms have evolved into the rapidly growing soft-bodied forms called herbs. The distinguishing feature of all angiosperms is their reproductive apparatus, the flower. These complicated structures enclose the female elements and, after fertilization, the seed, inside a chamber called the ovary. The enclosed angiosperm seeds are largely insect-pollinated by bees attracted to the flowers, in contrast to the dominantly wind-pollinated exposed seeds of the gymnosperms.

The oldest fossils of angiosperms appear in Triassic rocks. As these first representatives are distinct and well evolved, the group may well have originated from a gymnosperm representative in the late Paleozoic. Unfortunately, a clearly transitional type is yet to be found. Angiosperms have abounded since Cretaceous time, when many forms appeared that are quite like those of the present. Geologically, some have been very useful indicators of climatic changes, such as fossil magnolias in northerly climes. The evolution of horses, cows, and other grazing mammals was made possible by the great expansion of the grasses in mid-Cenozoic time. Today angiosperms are the most important food source for animals and man, by all odds the dominant flora, and the culmination of plant evolution.

THE ANIMAL KINGDOM

The million or more animal species so far described comprise about 30 phyla,[8] of which only a dozen are common in the fossil record. Arbitrarily, we shall include these phyla in two general, and most unequal, groups based on the kinds of paleon-

tologists who study them, rather than on taxonomic procedures. Invertebrates, animals without backbones, make up all phyla save one. Vertebrates, having backbones or something similar, include the most advanced animal types.

Invertebrates

The geologic time scale—the framework for Earth history—is largely based on relative dating using invertebrates. Because of the prevalence of marine rocks, they are by all odds the most common fossils, and their evolution, especially in three phyla, Protozoa, Mollusca, and Arthropoda, makes some of them highly diagnostic index fossils.

There is no direct evidence as to how the invertebrate phyla are related to each other, for the main groups were all distinct and well evolved before fossils became abundant in earliest Paleozoic time. However, their branching from the "Tree of Life" can be logically inferred from the embryology, comparative anatomy, and degree of complexity of living invertebrates. It may seem pointless to discuss soft anatomy, which is largely internal and rarely preserved in fossils, yet it represents the vital parts of animals, and in a large measure is the basis for arraying the phyla in increasing order of complexity.

Phylum Protozoa The progenitors of all other animals were protozoa, one-celled animals of great abundance.[9] However, as fossils they are virtually unknown, with two notable exceptions: the orders *Foraminifera* and *Radiolaria* of the class *Sarcodina*.

The foraminifera, living mainly in salt water, have hard tests, or exoskeletons (Fig. 13-5). Most "forams" secrete calcite, a few silica, and some cement foreign grains together. The tremendous variety of tests makes them very useful to micro-

---

8 Taxonomists cannot quite agree on the exact number.

9 Classed with one-celled plants, and one-celled things that are hard to class as either plants or animals, in the Kingdom Protista of some classifications.

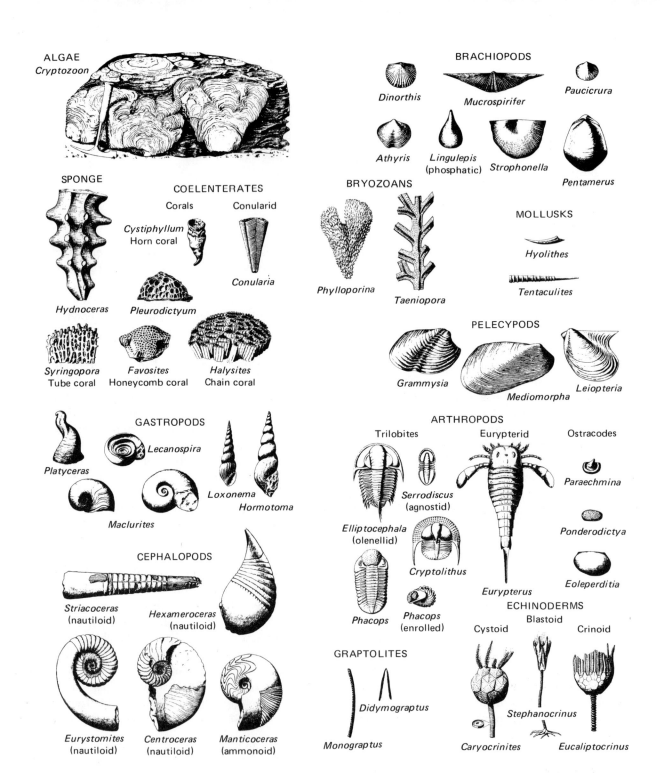

ALGAE
Cryptozoon

BRACHIOPODS
Dinorthis
Mucrospirifer
Paucicrura
Athyris
Lingulepis (phosphatic)
Strophonella
Pentamerus

SPONGE
Hydnoceras

COELENTERATES
Corals
Cystiphyllum Horn coral
Conularid
Conularia
Pleurodictyum
Syringopora Tube coral
Favosites Honeycomb coral
Halysites Chain coral

BRYOZOANS
Phylloporina
Taeniopora

MOLLUSKS
Hyolithes
Tentaculites

PELECYPODS
Grammysia
Mediomorpha
Leiopteria

GASTROPODS
Platyceras
Lecanospira
Loxonema
Hormotoma
Maclurites

ARTHROPODS
Trilobites
Serrodiscus (agnostid)
Eurypterid
Ostracodes
Paraechmina
Elliptocephala (olenellid)
Cryptolithus
Ponderodictya
Phacops
Phacops (enrolled)
Eurypterus
Eoleperditia

CEPHALOPODS
Striacoceras (nautiloid)
Hexameroceras (nautiloid)
Eurystomites (nautiloid)
Centroceras (nautiloid)
Manticoceras (ammonoid)

GRAPTOLITES
Didymograptus
Monograptus

ECHINODERMS
Cystoid
Blastoid
Crinoid
Caryocrinites
Stephanocrinus
Eucaliptocrinus

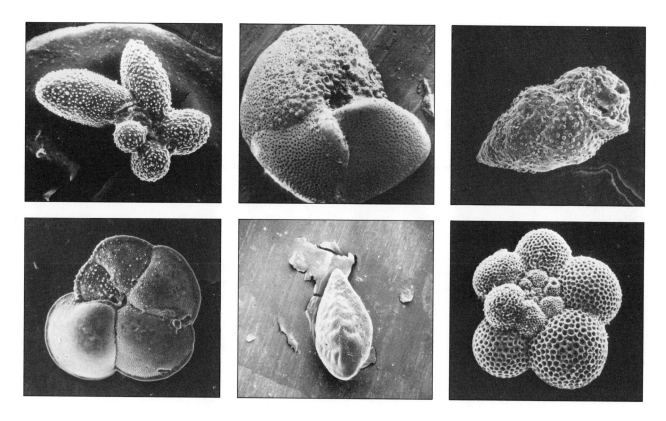

Fig. 13-5 Some types of foraminifers. Scanning electron microscope images (70 to 90 times magnifications). Courtesy W. E. Frerichs.

paleontologists, especially for subsurface rock correlation from drill-hole data. Some foraminifera are important rock formers; for instance *Globigerina,* the shells of which make calcareous oozes. Forams are typically small, about the size of a wheat grain or less. However, some, such as *Nummulites,* were coinlike forms several inches across, examples of which can be seen in the limestones of the Egyptian Pyramids. The radiolarians are saltwater dwellers with siliceous exoskeletons, loosely built

and often of a netlike form. Their tests, being extremely small as well as fragile, are not the best guide fossils; however, unlike the calcareous foram remains that dissolve in cold waters, they form characteristic radiolarian oozes and cherts that can exist in cold deep water.

Phylum Porifera For some time taxonomists were not sure whether sponges were plants or animals, as they live attached to the ocean floor and show

Fig. 13-4 Representatives of major invertebrate fossils. Courtesy Geology N.Y. State, Albany, (facing page).

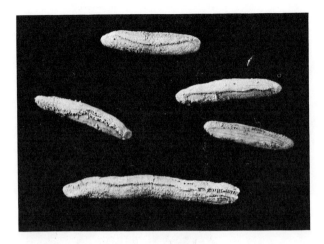

Fig. 13-6 Large foraminifers from Permian of west Texas. Courtesy of Dr. G. A. Cooper of the Smithsonian Institution.

little reaction to outside stimulation. They are animals, however, for they feed on organic materials. Despite an exceedingly low level of bodily organization, sponges, being multi-celled animals, are a step above the protozoa. The sacklike sponge body contains needlelike skeletal parts, spicules often preserved as fossils (Fig. 13-7). Calcareous spicules are common, and siliceous ones are probably the source for some beds of chert. The *archaeocyathids*, a group of extinct reef-builders, which may or may not be truly porifera, are important sponge-like Cambrian fossils. In general, however, sponges are a rather dismal group for paleontological purposes.

Phylum Coelenterata Somewhat higher in the evolutionary picture are the coelenterates, jellyfish

Fig. 13-7 A slab of Paleozoic sponges. Courtesy of the Field Museum of Natural History, Chicago.

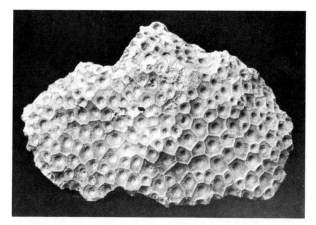

Fig. 13-9 Hexacoral compressed cups. Courtesy of the Field Museum of Natural History, Chicago.

Fig. 13-8 Solitary Paleozoic "horn" coral, a tetracoral. Courtesy of the Smithsonian Institution.

others added in between. The great variety in coral shapes, crowding or loose arrangement of colonial individuals, as well as the arrangement and number of septa, has made them useful guide fossils (Fig. 13-10). Colonial corals, aided by other organisms

and the like. They have a digestive cavity—a not particularly attractive one—that ingests food and expels waste through the same opening. This mouth is usually surrounded by tentacles that direct food towards it and are armed with stinging cells.

Geologically, the most important coelenterates are the corals because they secrete distinctive calcareous skeletal structures. Some corals are solitary individuals (Fig. 13-8); others grow by budding, to produce coral heads or colonies, fused masses of small coral cups each occupied by a tiny tentacled animal (Fig. 13-9). The cups of many species are braced by vertical plates, called septa, that converge inwards from the outer wall. Paleozoic septate corals have four basic septa, and most have secondary additions to the pattern. Mesozoic and Cenozoic corals commonly have six septa and

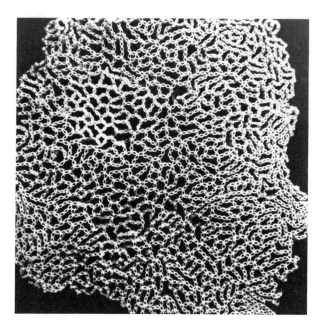

Fig. 13-10 Silurian "chain" coral from Kentucky. Courtesy of the Smithsonian Institution.

(notably algae), have produced great reefs in tropical seas, past and present. Solitary types, in contrast, are not good indicators of water temperature.

Graptolites   As different paleontologists have classed these extinct animals with Coelenterata, Bryozoa, and Hemichordata, it seems safest to treat them separately. Graptolites look like pencil marks on rock (in Greek, *graptos* means "written" and *lithos*, "stone"), for they are usually flattened impressions on shale bedding. Their detailed anatomy, which has been largely reconstructed from less common, undistorted specimens in limestones, consists of a central element, surrounded by branches composed of chitin,[10] which bear many small cups housing the individual animals. They apparently grew by budding.

Some graptolites were shrublike forms attached to the sea floor. Many others were floaters, suspended by a chitinous thread from floating seaweed or a gas-filled sack. Graptolites are important index fossils in Ordovician and Silurian dark shales (Figs. 13-11, 13-12). The floating species were

---

[10] Organic material somewhat like your fingernails.

Fig. 13-11 Graptolite specimen form Virginia. Courtesy of the Smithsonian Institution.

Fig. 13-12 Graptolite from Quebec. Courtesy of the Smithsonian Institution.

widespread; their abundance in dark shales apparently reflects a reducing environment more conducive to their preservation than the better-aerated sea floors of limestone deposition. The evolutionary trends making graptolites useful for dating include: reduction in the number of branches, changes in the angle of branches from straight down through straight up, and changes in the shapes of the individual cups.

Worms   Several phyla can be treated together as "worms." Their burrows and trails, though generally of little diagnostic value, are preserved in rocks as old as Precambrian, but actual worm remains are rare. If they were important in giving rise to higher phyla, paleontological evidence is lacking.

Phylum Bryozoa   The unimpressive bryozoans are relatively advanced invertebrates and useful fossils. These "moss animals" are microscopic water-dwelling individuals that build lacelike, twiglike, or mounded colonies of calcium carbonate and chitinous materials (Fig. 13-13). They often appear as

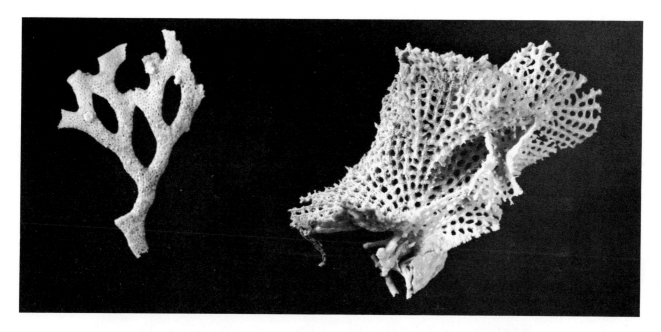

Fig. 13-13 Permian bryozoa from Glass Mountain, Texas. Courtesy of the Smithsonian Institution.

encrusting mats on shells and rocks. The bryozoa have rudimentary nerves, a muscle system, and a digestive system that takes in material at one end and expels it at the other—an improvement over coelenterates although the gut, being U-shaped, has anal and oral openings adjacent. Bryozoans range from Ordovician, or possibly Cambrian, to recent time. The group as a whole has been abundant, and some forms having a short time range and wide distribution are good guide fossils. However, the individual fossils require microscopic techniques for study, which may be why they have not received the attention they deserve from paleontologists.

Phylum Brachiopoda The brachiopods are bivalves, superficially resembling clams, which are actually quite different. The two brachiopod shells are usually of different shapes, and one covers the bottom of the animal, the other the top (Figs. 13-

14, 13-15). Clam shells, in contrast, are generally mirror images and arranged side-by-side with a hinge at the top; moreover, many clams are mobile.

Fig. 13-14 Early Paleozoic brachiopods from Ohio. Courtesy of the Smithsonian Institution.

Fig. 13-15 Early Paleozoic brachiopods from New York State. Courtesy of the Smithsonian Institution.

Only the larvae of brachiopods migrate; the adults are all sedentary, commonly attached to the ocean bottom by a fleshy stalk.

Brachiopods, which have calcite, phosphatic, or chitinous-phosphatic shells, are very abundant fossils. Based on comparison with living ones and muscle scars in fossil shells, the soft anatomy of brachiopods was relatively complex with digestive, circulatory, nervous, and muscle systems. Inside their shells, they had fleshy loops covered by minute tentacles that swept microscopic food particles into their mouths. Some "brachs" developed hard supports for the fleshy loops.

Being conservative beasts who never abandoned a sedentary life, brachiopods are all broadly similar in basic plan. They did, however, develop marked differences in shape and ornamentation of their shells. Some are smooth, except for growth lines, but others are strengthened and ornamented with a wide variety of ribs, spines, and such (Fig. 13-16). Thus their details are diverse enough to allow the recognition of 200 living and some 3000 fossil species. Among the brachiopods, *Lingula* is a "living fossil" (Fig. 5-3); however, many others of wide distribution and limited time ranges are useful in dating. Unlike many other bottom dwellers, some genera were tolerant to different condi-

tions on the sea floor, hence useful for facies correlation. Overall, the brachs have been a most important group. They were the key to much of the Paleozoic history of North America.

Phylum Echinodermata Echinoderms are different. Their five-fold radial symmetry, exemplified by the starfish, suggests an aberrant branch on the Tree of Life; yet they may have stemmed from the same ancestor as the vertebrates. The radial symmetry, while reminiscent of the coelenterates, is a secondary development. Echinoderms are far more advanced in their internal workings, with well-developed digestive and nervous systems and a unique circulatory system. They pump water through ducts or vessels which, in some representatives, have many minute projections called tube feet. This water vascular system serves for food gathering and respiration; in some cases it also serves for locomotion and even grasping, for the tube feet operating by changes in water pressure can be used as suction cups. *Echinodermata* translates as "hedgehog skin" which describes their leathery body walls, reinforced by loose calcareous spicules or plates which are often intricately fitted.

Two subphyla are recognized: attached forms

Fig. 13-16 Permian spiney brachiopod from Glass Mountain, Texas. Courtesy of Dr. G. A. Cooper of the Smithsonian Institution.

Fig. 13-17 Mississippian crinoids from Iowa. Courtesy of the Smithsonian Institution.

and free-moving ones. Of the attached echinoderms, the *crinoids* are most important (Fig. 13-17). These so-called "sea lilies" have been diverse and plentiful, surviving from the Cambrian to the present day, and customarily live in brightly colored swarms suggesting submarine flower gardens. Typical crinoids have a body with a mouth at the top, a set of flexible branching arms, and a flexible stalk. The bulblike body, a few centimeters or less across, protects the vital parts. The arms have furrows throughout their length along which microscopic food particles are wafted to the animal's mouth. The supple stem, usually about ⅗ meter (2 feet) long, but some as long as 15 meters (50 feet), consists of buttonlike discs bound together by animal tissue. At its bottom is a rootlike hold-fast, or in some cases coiled or grappling hook arrangements, to anchor the animal to the sea bottom.

In death the binding tissue decays, scattering the body plates and stem buttons over the sea floor. The stem discs, being the most abundant parts, are the main components of some late Paleozoic limestones. The various arrangements of the five or six rows of body plates have allowed paleontologists to describe some 750 genera, and many more species, of crinoids.

The free-moving echinoderms include the common *starfish,* which have five stout arms and move along on their many tube feet, and the *brittle stars,* which move by wriggling their narrow arms—which they can shed—by means of muscles. Although fossils of these still-living stars date back to the Paleozoic, well-preserved specimens are rare, more curiosities than useful fossils. The *echinoids,* whose name is easily confused with that of the whole phylum, are free-moving, armless echinoderms represented by sea urchins and sand dollars. Their globular, disc, or heart-shaped tests (shells) consist of many well-fitted plates in a radial arrangement. They are protected by sharp spines, also used for locomotion along with their tube feet. Known since the Ordovician, the echinoids have become quite varied and abundant from the Cretaceous to the present day. They would be good guide fossils if they were not so limited in their environmental preferences.

Phylum Mollusca For spineless creatures, molluscs have been a very progressive group. Clams, snails, and squids are modern representatives of this paleontologically most significant phylum. Anatomically they are more complex than any of the animals so far discussed; only arthropods and vertebrates surpass them. Represented in various members of the phylum are circulatory systems including a chambered heart; digestive systems replete with livers and kidneys; sensory systems with nerves, rudimentary brains, and well-developed eyes. Distinctive molluscan features include bilateral symmetry; a fleshy body covering; the mantle that secretes calcareous shells; a rasplike tongue, the radula, which is a ribbon of many tiny hard teeth for obtaining food, sometimes by drilling holes in other mollusc shells. They also have a single muscular foot, designed originally for crawling but now adapted to other types of locomotion.

Snails and their kin, class *Gastropoda,* are successful and complex animals (Fig. 13-18). Producing some 60,000 species since early Cambrian

Fig. 13-18 Devonian snail collected in Michigan. Courtesy of the Smithsonian Institution.

time, they now inhabit the ocean bottom from 5500 meters (18,000 feet) below sea level to the tidal zone, ponds, and streams; some have even invaded the dry land. Most crawl about on a large muscular foot—albeit at a snail's pace—and some, the pteropods, whose remains make considerable ooze on the ocean floor, can swim. Most are vegetarians; some, with gamier tastes, are scavengers; and a few are voracious carnivores, drilling the shells of other molluscs with their filelike radulae. All gastropods have a distinct head with a pair of eyes and one or two tentacles, and most breathe with gills, although land dwellers have evolved lung sacs.

Except for naked land slugs and marine *nudibranches*, snails are protected by a single, coiled shell. Its single opening can be closed with a horny or shelly trap door, in some cases. The shells are commonly screwlike spirals, most being right-handed, or clockwise, although a few spiral to the left. Shell forms range from high and pointed to low, nearly flat spires, and some are flat coils with no spire point at all. In some gastropods, abalone and limpets, for example, the single shell is a simple, open saucer or cone having no twisting passage. The surfaces of shells are variously ornamented with lines, grooves, ridges, spines, or knobs.

Although many groups are long ranged and many distinguishing features are not reflected in their hard anatomy, snails are useful fossils. Their variety of form and ornamentation gives them value, especially in fresh water and terrestrial deposits where other fossils are often lacking.

The clams and oysters—a generally tasty lot—have been called "bivalvia" by Germans, "lamellibranches" by the British, and *pelecypods* by Americans. Translating the names in order gives a fair description of these molluscs as two-shelled, having leaflike gills, and with a hatchet foot. Most move by their wedge-shaped muscular foot, have a shell-secreting mantle, and breathe by fairly complicated gills. Besides extracting oxygen from water, the gills sweep microscopic food particles to the animals' mouths. In contrast to snails, the pelecypods lack a well-defined head, eyes, tentacles, and radula.

Clams are typically bilaterally symmetrical, with each of their two shells a mirror image covering each side of the animal (Figs. 13-19, 13-20). Internal muscles close the shells, and elastic ligaments along a hinge line at the top open them. The shells are calcareous, usually lined on the inside with "mother of pearl," and protected on the outside by a chitinous film preventing solution by the salt and fresh waters of their environments. Some pelecypods have lost the mirror symmetry of equal-sized shells, notably the sedentary oysters who cement themselves to the sea floor, and the scallops, who swim by clapping their shells. However, most clams are equivalved, and crawl or burrow. Razor clams are surprisingly rapid diggers. In size clams range from a fraction of a centimeter long to one 2 meters (over 6 feet) long, the 180-kilogram (400-pound) South Sea *Tridacna*.

As fossils, clams are rare in Cambrian rocks and relatively common since the Ordovician. Most

Fig. 13-19 Devonian clams from New York. Courtesy of the Smithsonian Institution.

remains are casts or molds because clam shells are largely composed of aragonite, a mineral like calcite in composition but a less stable form more readily dissolved. Despite a wide variety of shell forms differing in shape, ornamentation, and detailed hinge structure, the pelecypods are a rather disappointing fossil group for rock correlation—more useful in defining large time units than small ones.

The 17-meter (55-foot) giant squid is a living "sea monster," the greatest invertebrate of all time, and a member of the most progressive molluscan group, the *cephalopods*. Modern cephalopods are active swimming carnivores with keen senses. They breathe with gills, have a radula, two beaklike, horny jaws, eight, ten, or more tentacles, and eyes the equal of our own. Naked cephalopods, one of three subclasses, are represented in modern seas by the octopus, which has no skeleton whatever, and by squids and cuttlefish which have much-reduced internal skeletons (cuttlebone is seen in canary cages). These unarmored cephalopods are unimportant fossils except for one extinct group, the *belemnoids*. They resembled living cuttlefish and had stout, cigar-shaped, internal skeletons. Once considered actual thunderbolts, the skeletons are found in late Mississippian through Cretaceous rocks, and are most plentiful in certain Jurassic and Cretaceous formations.

Shelled cephalopods abounded in the Jurassic strata of Britain, where William Smith pioneered the art of stratigraphic correlation. Yet, of the bygone hosts of this tremendously important fossil group (10,000 fossil species), only the pearly nautilus survives. It swims in a flat-coiled shell, buoyed up by gas and partitioned into chambers by crossbulkheads called septa. The last, or living, chamber,

Fig. 13-20 Cretaceous clam from Tennessee. Courtesy of the Smithsonian Institution.

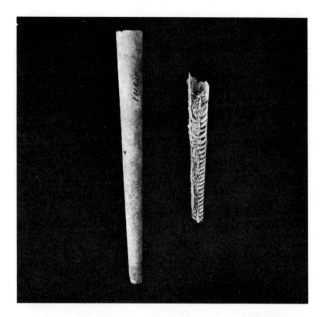

Fig. 13-21 Late Paleozoic straight nautiloid cephalopod from Glass Mountain, Texas. Courtesy of the Smithsonian Institution.

The *ammonoids,* which appeared in the Silurian and became extinct at the end of the Mesozoic, are distinguished by increasingly more complicated septa through time. Most Paleozoic forms had rather simple septa (called goniatitic) whose sutures appear as simple curved and angled lines (Fig. 13-22). In late Paleozoic and Triassic time, a tribe appeared whose more complex partitions were warped into crenulated lobes and rounded saddles (ceratitic type), producing a suture like the teeth of a coarse bucksaw. A third group living from Permian through Mesozoic time had wrinkles on both the saddles and lobes (ammonitic), creating sutures like scribbled handwriting (Fig. 13-23). Most ammonoid shells are flat coils (Ammon was the Egyptian ram-headed god), little different from their nautiloid kin; however, a few spiraled, like large snail shells, and towards the end of their history some straight forms evolved with only a small coil in the first-formed juvenile chambers.

which is occupied by the animal, connects by a calcareous tube, the siphuncle, to the smaller empty chambers that were vacated as the animal grew. The partitions are visible only when the outer shell is stripped, or in fossils after the chambers have filled with mud and the shell has dissolved away. The junctions of the septa with the outer shell form lines called sutures, whose tracelike handwriting identifies species. From the intricacies of the septa, and their resultant suture lines, two groups of armored cephalopods are recognized.

The *nautiloids,* of which the pearly nautilus is a member, are characterized by simple, saucer-shaped septa that make smooth or gently curving sutures. Nautiloids, which first appear in Cambrian rocks, abounded in the time interval from Ordovician to Devonian (Fig. 13-21). The earliest ones were uncoiled; coiled shells are a later evolutionary development. Most had smooth exteriors, although some were bedecked with ribs and spines.

Fig. 13-22 Late Paleozoic coiled cephalopod (goniatite) with relatively simple sutures. Indiana. Courtesy of the Smithsonian Institution.

Fig. 13-23 Cretaceous coiled cephalopod (ammonite) with complicated suture. Courtesy of the Smithsonian Institution.

Ammonoids are index fossils *par excellence.* They are large and easy to study, had a worldwide distribution—reflecting their swimming habit—were abundant and readily preserved, and had a long history and geologically rapid evolution. Thus they are ideal for relative dating and correlation of both large and small time units, as well as invaluable evolutionary materials. They are delightful fossils.

Phylum Arthropoda   Three-quarters of the living species in the animal kingdom are insects, centipedes, crabs, spiders, and their close relatives. So, if numbers are the criterion, the arthropods are the most successful of all animal groups. Adapted to all possible habitats, they fly, swim, walk, crawl, burrow, or in the case of barnacles lead a sedentary life. Many are complex animals with complicated respiratory, circulatory, reproductive, excretory, and sensory systems including brains and eyes. They are exceedingly prolific; the record egg-producers are arthropods.

The most diagnostic arthropod feature is jointed appendages forming legs, pincers, and in some cases antennae (*arthro* means "joint"; *pod* means "foot"). Arthropods have a chitinous exoskeleton, usually segmented for flexibility, that is strong, waterproof, and resilient—witness the grasshopper which strikes a windshield and survives. Yet a suit of armor presents a problem if the occupant is to grow. The arthropod solution is molting, whereby they periodically shed their suits and replace them with larger ones. From Cambrian onwards, arthropods have been a dominant part of the Earth's population, through one branch or another.

Most arthropods are insects: flies, mosquitoes, ants, moths, and beetles; they alone comprise two-thirds of the living animal species. Because insect life spans are brief—some live less than a year—many generations develop in a short time. Thus insects have had a greater opportunity to evolve into diverse and very specialized forms than most other creatures. Wingless insects are known as far back as Devonian time, and flying varieties appear in the Pennsylvanian, a time of dragonflies with wing spreads of ⅗ meter (2 feet) and cockroaches 10 centimeters (4 inches) long. Unfortunately, fragile insect corpses are not particularly amenable to preservation, so their use in geologic dating and detailed evolutionary studies is almost nil.

The *trilobites,* another group of arthropods, were lords of the Cambrian seas, where they represented 60 percent of the known population (Figs. 13-24, 13-25). They thrived in the Ordovician, but thereafter gradually declined to their extinction at the end of Paleozoic time. Most trilobites were small, from 2 to 10 centimeters (1 to 4 inches) long; a giant of the tribe, *Paradoxides,* grew to 45 centimeters (18 inches). The name *trilobite* reflects the division of their chitinous shell, or carapace, into three lengthwise lobes. The central one usually stood in highest relief, set off from the marginal lobes by grooves. The animals also had three transverse parts, from front to back: a head, or cephalon;

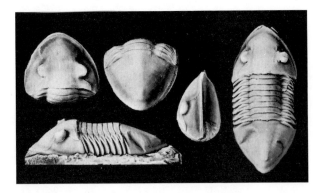

Fig. 13-24 Ordovician trilobites from New York and Ohio. One of the species is flat; the other is enrolled. Courtesy of the Smithsonian Institution.

a body, the thorax, of from 2 to 29 unfused cross-segments; and a tail, the pygidium, of several fused segments. Some had compound eyes, like the fly,

and paired, jointed appendages. The common preservation of their carapaces results because their chitinous shells were impregnated with calcium carbonate.

Swimming, crawling, and burrowing in shallow parts of the sea floor, they apparently grubbed in the mud for small animals or organic debris. Though now extinct, hence evolutionary failures, they left 1000 fossil species, exhibiting interesting adaptations to their way of life, and, as we have previously mentioned, are important guide fossils for the early Paleozoic (Fig. 5-2).

The *eurypterids* were most common in Silurian and Devonian time, although species have been found in rocks from Ordovician to Permian age (Fig. 13-26). These extinct "sea scorpions" are the only important fossils in the group of arthropods whose living members include scorpions, spiders, and a "living fossil," the horseshoe crab. All have a fused head and thorax and lack antennae. The

Fig. 13-25 Ordovician trilobite from Oklahoma. Courtesy of the Smithsonian Institution.

Fig. 13-26 Silurian eurypterid from western New York State. Courtesy of the Smithsonian Institution.

Fig. 13-27 Mass of Tertiary vertebrate bones from Nebraska. Courtesy of the Field Museum of Natural History, Chicago.

eurypterids had poison glands, stingers, and, in some representatives, large claws. Thus they were probably active predators. Especially fearsome ones, the largest arthropods known, were almost 3 meters (10 feet) long. Their general plan suggests that the eurypterids may have been ancestral to the scorpions, who were the first animal invaders of the land.

Vertebrates

The last major group to emerge on the Tree of Life is the *Phylum Chordata*—creatures supported with an internal flexible rod. Worms, arthropods, and molluscs have been proposed as their immediate forebears; however, good evidence suggests a seemingly unlikely group, the echinoderms of the five-fold symmetry. Their larval forms are bilaterally symmetrical and resemble the larvae of acorn worms, a lowly group of peculiar chordates. Moreover, blood serum tests and the chemistry of muscles suggest the affinity of echinoderms and chordates. Apparently, echinoderms evolved along two lines; one, acquiring a secondary radial symmetry, led to the starfish, echinoids, and sea lilies; the other gave rise to the chordates.

Chordates include two broad groups: the primitive *Acraniata*, lacking definite heads, brains, or paired appendages; and the *Craniata*, or vertebrates. The former group had several insignificant subphyla—sea squirts, the acorn worms, and the sea lancelets. The lancelet, *Amphioxus*, occupying an important evolutionary spot, has a front end, but no real head, eyes, ears, bones, or fins; but it does have a jelly-filled rod stiffening its back, with a nerve chord above and a simple digestive tube below, as well as fishlike gills. Amphioxus is a living prototype—there is no fossil record—of the ancestral vertebrate. For, with certain improvements, amphioxus could become a very primitive fish.

All vertebrates have an internal skeleton with a spinal column of jointed cartilage or bone segments, called vertebrae; a skull housing a brain and various sense organs; appendages, usually paired, forming fins, feet, arms, or wings supported by skeletal girdles; and usually sets of ribs. The sub-

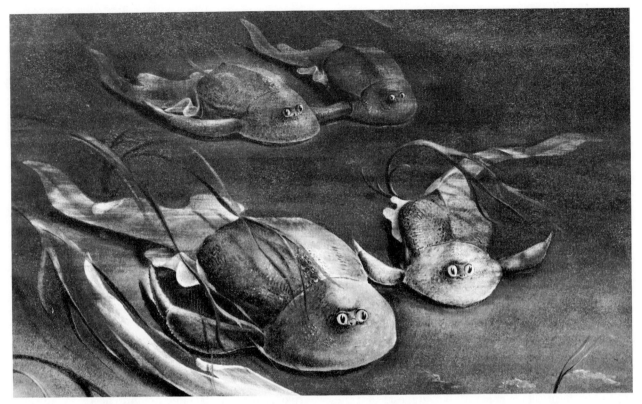

Fig. 13-28 Devonian placoderms. Courtesy of the Field Museum of Natural History, Chicago.

phylum *Vertebrata* is customarily divided into two superclasses: *Pisces,* meaning "fish," and the *Tetrapoda,* including the amphibians, reptiles, birds, and mammals.

Primitive Fish Scattered bony flakes, first found in Colorado and Wyoming, tell us that in Ordovician time vertebrates existed, but little more. Better fossils in late Silurian and Devonian rocks allow reconstruction of the first vertebrate animals. They were *ostracoderms,* which were jawless, and, with one exception, finless, bone-headed fish of an archaic sort. The heads were heavily armored, perhaps as protection from eurypterids, but the rest of the skeleton was cartilaginous. As their mouths were mere holes or slits under the heads, and the animals lacked stabilizing fins, ostracoderms were probably bottom-dwelling mud-grubbers.

The first improvements in vertebrate design occur in *placoderms,* a rather diverse group of armored fish first appearing in late Silurian and Devonian strata (Fig. 13-28). The placoderms included flat-bodied, full-bodied, and eellike forms. Their revolutionary development was moveable jaws, evolved from gill arches, as well as paired fins. While placoderm history ended in the blind alley of extinction at the end of Paleozoic time, some of its early members, as yet unknown, spawned a new lineage whose progeny are the living fish.

Higher Fish The higher fish are of two types: those with cartilaginous skeletons, the sharks and

Fig. 13-29 A 4.2 meter (14-foot) Cretaceous fish (Xiphactinus) with his last meal, a 1.8 meter (6-foot) fish (Gillicus) inside. Collected and prepared by George F. Sternberg. Photo by E. C. Almquist, courtesy of the Sternberg Memorial Museum, Fort Hays, Kansas, State College.

rays; and the bony fish. *Sharks* are mainly marine fish, aggressive members of the saltwater community whose record extends back into Devonian time. Having cartilaginous skeletons, their fossil remains are mostly teeth and spines, which are notably abundant in some Mesozoic and Cenozoic formations.

The *bony fish,* now dominating both fresh and salt water, include two groups. The first consists of the ray-finned fish whose primitive members are represented by sturgeons and gars and whose advanced representatives include such various types as eels, catfish, salmon, sailfish, tuna, and a host of others. These fish, by far the most abundant group of vertebrates, are the culmination of animal adaptation to aquatic life (Fig. 13-30).

Fig. 13-30 Cast of a coelocanth, a still-extant lobe-fin fish. Courtesy of the American Museum of Natural History.

The other branch of the bony fish includes the lungfish and lobe-finned fish which, although subordinate in numbers, species, and adaptation to aquatic living, have a most significant position in the Tree of Life, for some ancient member of the group begat the first land-dwelling vertebrates. Lungfish are today limited to the southern continents of Australia, Africa, and South America. Using their lungs to breathe air directly, they can survive in stagnant pools fatal to gill-breathing fish, and inside burrows in dried-out river bottoms.

Lobe fins, or *crossopterygian* fish, were abundant in Devonian time, as were the lungfish, but were believed extinct in the Cretaceous. However, in 1938 a 1.5-meter (5-foot) representative was netted in deep water off East Africa, and more have been caught since. The Devonian crossopterygians may well have been reluctant invaders of the land, literally fish out of water. Equipped with lungs, they could survive in air as they struggled to find new water holes in drought-seared river bottoms. In many ways, the lobe fins resemble the first land vertebrates. They have lungs, nostril openings, similar teeth and skull elements, and a rather rugged skeleton. Their fins extend from fleshy stalks braced by a single upper and double lower bone, a structure analogous to the limbs of terrestrial vertebrates.

Class Amphibia The high point of animal evolution is reached in the *tetrapods*, the second broad subclass of vertebrates. Their main stream of evolution involves progressive improvement for life on land. Some of them have gone further and taken to the air, and some have reverted to water-dwelling, but in such cases the structural adaptations are superimposed on a terrestrial reptilian or mammalian body plan. Like the plants, which preceded them, animals faced formidable problems in moving from water to land. They, too, needed skins that would hold in their body fluids, special systems of respiration, new reproductive methods, and stiffer support to counter the increased effect of gravity.

Moreover, the strengthening of their skeletons had to include a better design for overland travel.

Though the *amphibians* were the first vertebrate land dwellers, they never quite succeeded as terrestrial animals. Amphibians are not completely divorced from life in water. The juveniles are tadpoles, gilled and legless swimmers. The adults must breed in water, except for a few specialized types who use other damp places, and except for toads, periodic swims are a necessity to prevent desiccation. Of existing amphibians, frogs and toads are specialized forms. The salamanders are closer to the basic stock.

Skeletons and tracks of the earliest amphibians, the *labyrinthodonts*, occur in late Devonian rocks of east Greenland. These primitive types were ungainly creatures whose broad flattened skulls of heavy bone insured a better fossil record than most later amphibians which were more lightly built (Fig. 13-31). Abundant in the later Paleozoic, the labyrinthodonts died out in the Triassic. The name comes from the labyrinthlike pattern of their tortuously enfolded tooth enamel, a characteristic of many lobe-finned fish. Their sprawling, stubby legs were most inefficient, but the skeletal structure was stouter than in their fishy forebears and adequate for overland travel unsupported by the buoyancy of water. If little more than modified fish, these Devonian amphibians were nonetheless the precursors of all tetrapods.

Class Reptilia Until a suitable means of reproduction evolved, the vertebrates could not become thoroughly terrestrial. It was the development of the amniote egg—a hen's egg is an example—that freed reptiles from life in water. Such an egg allows an embryonic animal to develop in a moist environment, surrounded by food inside a protective shell, until it can survive in air. To complete their adaptation to land living, the reptiles also evolved scaly or platy skins that can stand air indefinitely, improved hearts and circulatory systems for active life on land, and improved legs.

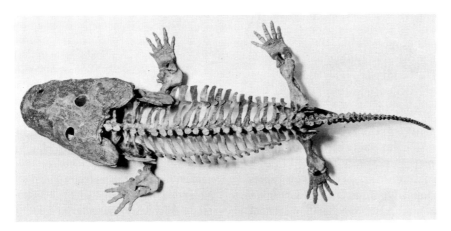

Fig. 13-31   Skeleton of Eryops, a Permian labyrinthodont amphibian, attaining 1.8 meters (6 feet) in length. Courtesy of the Field Museum of Natural History, Chicago.

The *cotylosaurs*, ancestors to all other reptiles, abounded from Pennsylvanian to the end of the Paleozoic and lingered on into Triassic time. *Seymouria*, a one-meter-long representative of this group, had the teeth and skull of a labyrinthodont, but the rest of the skeleton was of a more advanced and reptilian design. If Seymouria laid shelled eggs, it was a reptile; if not, an amphibian. As the first known fossil egg is Permian, and reptile skeletons are known from the Pennsylvanian, the exact classification of Seymouria is unsettled. In any case, Seymouria is an ideal connecting link between amphibians and reptiles.

Turtles, snakes and lizards, crocodiles, and rhynchocephalians (whose sole survivor is the small Tuatara, a primitive lizardlike inhabitant of New Zealand) are the living remnants of Mesozoic reptilian hordes that ruled land, sea, and air. The now extinct *thecodonts*, which evolved from the cotylosaurs by early Mesozoic time, are progenitors of the living crocodiles and also birds, as well as flying reptiles and the spectacular dinosaurs that left no descendants after holding the Mesozoic limelight for well over 100 million years. One group of thecodonts were four-footed carnivores called *phytosaurs*. They closely resembled crocodiles except

that their nostrils were just in front of their eyes. Despite the resemblance, however, crocodiles are not descendants of the phytosaurs who, although creatures of similar habit, became extinct at the end of Triassic time.

Other thecodonts, called *pseudosuchians*, were bipedal reptiles walking on strong hind legs whose forelimbs were arms for grasping, which had long narrow skulls. *Ornithosuchus,* found in upper Triassic rocks of Scotland, typified this basic stock that produced the dinosaurs. They had birdlike hind legs, each with three-clawed toes, and the sturdy pelvis essential for animals balancing on two legs. The skeleton and skull were of relatively light construction containing many hollow bones, and a long tail counterbalanced the forward-leaning body. The pseudosuchians, like the phytosaurs, died out at the end of Triassic time but several lineages evolving from them retained basic family resemblances, notably the dinosaurs. Although some of the dinosaurs were to revert to a quadruped (four-footed) existence, the hind legs were always notably bigger than the front, reflecting their bipedal ancestors.

The intriguing *dinosaurs* are the subject of some popular misconceptions. Their name translated means "terrible lizard"; however, they were not lizards, and if some were unquestionably terrible—the largest terrestrial carnivores ever—many were unaggressive vegetarians. Nor were they all giants, another misconception, for they ranged from 26-meter, 36,000-kilogram (85-foot, 40-ton) monsters down to adults no larger than a plucked chicken (which they superficially resembled).

The name was given to the lot of them before it was recognized that the dinosaurs include two quite different reptilian orders, markedly different in their hip and pelvic structure. Most higher vertebrates, including you and me, have three main bones in the pelvic girdle: the ilium, ischium, and pubis. In *saurischian* dinosaurs, the downward-projecting ischium and pubis diverge from the ilium above, as they do in most other reptiles; hence these are the "reptile-hipped" group. In the *ornithischian* order, the two downward-projecting bones are parallel, adjacent, and directed towards the back as in birds; hence they are the "bird-hipped" dinosaurs.

The saurischians produced two lines: the only carnivorous dinosaurs, bipedal forms culminating in the great *Tyrannosaurus rex*, and four-footed, unarmored vegetarians including the well-known *Brontosaurus* (Fig. 13-32). The ornithischians were all herbivorous and of four main types. One group

Fig. 13-32   Skeleton of Brontosaur excavated near Sheep Creek, Wyoming. Mount by S. H. Knight (in foreground). University of Wyoming photograph by H. Pownall.

Fig. 13-33 Duck-billed Trachodon and crested dinosaurs with a squat "reptilian tank" in center. Mural by C. R. Knight. Copyright the Field Museum of Natural History, Chicago.

of largely unarmored bipeds included the duckbills, like *Trachodon*, some of which developed bizarre head structures (Fig. 13-33). The other "bird-hipped" dinosaurs were all quadrupeds. The *stegosaurs* were distinguished by erect bony plates rising from their backs (Fig. 13-34). The *ankylosaurians* were "reptilian tanks," heavily armored with overlapping bony plates, and wielded a macelike tail studded with spikes. The toothless, beaked *ceratopsians* had large heads protected by a bone neck frill projecting from the backs of their skulls, and usually long spearlike horns projecting forwards (Fig. 13-35).

Besides the thecodonts which gave rise to the dinosaurs and several other reptile groups, the primitive cotylosaurs spawned *mammallike reptiles*, a lineage of great interest in subsequent evolutionary developments. The remains of the "sail-backed" *pelycosaurs* (Fig. 13-36) have been found in late Pennsylvanian and Permian rocks of North America. These flamboyant, lizardlike beasts, up to 4 meters (12 feet) long, sported a great flap braced by long spines rising from the vertebrae. The function of the "sail" is not known, but it may have helped control the body temperature of this cold-blooded group. The differentiation of teeth shows that some pelycosaurs were predators and others plant eaters.

Sometime before the appearance of the more bizarre pelycosaurs, offshoots from this lineage gave rise to the *therapsids*, whose fossils occur in middle Permian and Triassic rocks on all continents. Although the early therapsids differed little from the ancestral cotylosaurs, later representatives developed distinctly mammalian features. *Cynognathus*, an early Triassic therapsid as big as a large dog, stood well off the ground (Fig. 13-37). Although its body extended well into the tail in a reptilian manner, its legs, vertebrae, skull, and jaws had mammallike structure. Significantly, the dental array has incisors, canines, and cheek teeth in contrast to the relatively undifferentiated sharp pegs forming the teeth of most reptiles. *Cynognathus* neatly links the reptiles and mammals.

Class Aves There are and have been flightless birds, such as ostriches and penguins, but anatomically the great majority of modern birds are su-

Fig. 13-34 *Stegosaurus*. Mural by C. R. Knight. Copyright the Field Museum of Natural History, Chicago.

Fig. 13-35 *Triceratops* and *Tyrannosaurus*. Mural by C. R. Knight. Copyright the Field Museum of Natural History, Chicago.

Fig. 13-36 A "sail-backed" lizard from the Permian of Texas. Courtesy of the Smithsonian Institution.

Fig. 13-37 Several representatives of *Cynognathus,* a carnivorous mammallike reptile, looking hungrily at a vegetarian relative. Mural by C. R. Knight. Copyright the Field Museum of Natural History, Chicago.

perbly adapted to the rigorous demands of flight. Feathers, which are highly modified scales, distinguish birds from all other animals. These structures are excellent body insulation and provide wing surfaces that—unlike the membranes of bats and flying reptiles—are not easily damaged. Birds are warm-blooded animals with a very efficient heart and high metabolism appropriate for their energetic way of life. Their skeletons are evolutionary masterpieces, combining light weight with superior strength and rigidity needed to anchor their strong flight muscles. The pelvis is solidly fused and the breast bone, which anchors the wing muscles, is large.

Despite their specialized structure, birds are clearly reptilian descendants. Their legs are strong, rather like those of the reptile ornithosuchys. They have scales on their legs (note chickens) and around their mouths. Though all modern birds have beaks, there are traces of teeth in embryonic ducks,

and birds lay amniote eggs like reptiles. Because their bones are light, porous, and often hollow, bird fossils are rare. However, birds are linked to pseudosuchian reptiles by two excellent specimens. These represent the first known bird, called *Archaeopteryx* (Figs. 13-38, 13-39).

The first fossil was discovered in 1861, in a fine-grained limestone[11] laid down as a calcareous mud in a Jurassic lagoon near Solenhofen, Germany. Had the skeleton not been completely fringed by the impressions of long feathers, it would almost certainly have been classed as a small reptile. The bones were solid, the neck was long, the hind legs and pelvis were reptilian, the vertebrae extended into a long bony tail, and the front limbs bore three clawed fingers. All in all, a most unbirdlike skeleton, but the feathers meant a warm-blooded animal, and their distribution along the front limbs indicated

11 Quarried in those days for lithographic printing.

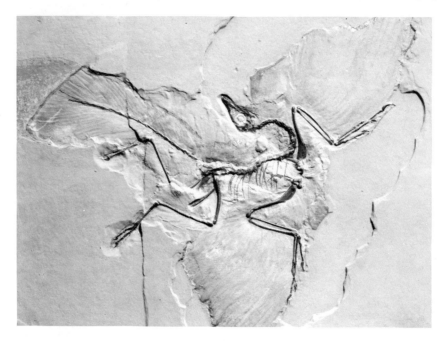

Fig. 13-38 Cast of fossil *Archaeopteryx*. Courtesy of the American Museum of Natural History.

Fig. 13-39 Restoration of *Archaeopteryx*. Courtesy of the American Museum of Natural History.

wings. By Cretaceous time, birds were essentially modern except for the presence of teeth, and since the early Cenozoic when they had evolved horny beaks, birds have shown virtually no basic structural changes (Fig. 13-40).

Class Mammalia Today the lands are ruled by mammals. Their success is largely a matter of superior physical endowment. Like birds they are warm-blooded, maintaining a constant body temperature that allows a more energetic life as well as relatively constant activity, despite daily or seasonal temperature changes. Cold-blooded animals, like reptiles and amphibians, are handicapped because they go torpid when the temperature is low, and must, on the other hand, seek shade on hot days to avoid sunstroke. The warm-blooded mammals

usually have protective coats that help control their body temperature. This insulation may be subcutaneous fat; more commonly it is hair, the mammalian trademark, just as feathers are for birds.

Mammalian reproduction gives their young a better chance of survival. Unlike the embryos of reptiles and birds which often fall prey to egg-loving creatures, those of mammals are carried inside the mother until born. And mammals are superior parents, suckling their infants with milk (*mamma* is Latin for "breast") and protecting and training their offspring until they can fend for themselves.

Paleontologists, of course, usually deal only with skeletons, but here, too, the mammals are distinctive. The mammals' well-constructed legs are under their bodies so the bones carry most of the

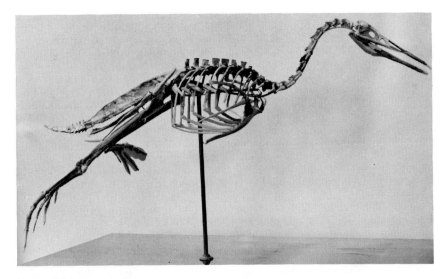

Fig. 13-40 Skeleton of Cretaceous, toothed, swimming bird (*Hesperornis*) from Kansas. Courtesy of the Smithsonian Institution.

weight, and little muscular energy is wasted in merely supporting the body. Most distinctive is the skull. The larger brain of mammals, marked by a larger brain case in fossil types, gives them a decided intellectual advantage over the other vertebrates. Also, the skull has two round knobs called *condyles*—typical reptiles have only one—that ride on the uppermost vertebrae. The mouth has a secondary bony palate, missing in reptiles, that separates the food and air intake systems. Whereas reptiles have seven bones in their lower jaws, mammals have only one, and it is hinged differently to the rest of the skull.

But of all the hard parts, mammal teeth are most distinctive. Often a vertebrate paleontologist can tell the type and age of mammal remains from the teeth alone.[12] Reptiles and lower vertebrates develop an indefinite number of teeth during a lifetime, replacing those shed or lost, and the teeth are all roughly similar. But mammals normally have

only two sets, the milk and adult teeth, and there is considerable differentiation. Their basic dentition includes incisors, or nipping teeth, followed by a set of canines, or tusks, behind which are diverse cheek teeth, premolars and molars, designed for grinding or cutting. This basic pattern shows many specializations. Note how different are the teeth of elephants, horses, beavers, rats, and men, to name but a few.

The incredibly diverse mammals, living and fossil, defy brief summary. Some 34 orders are lumped into various larger cohorts, superorders, and subclasses; and on the other hand divided into smaller suborders, superfamilies, and families. All mammals are, however, divisible into three basic, but numerically disproportionate groups: monotremes, marsupials, and placentals.

Monotremes The only known monotremes are the duck-billed platypus and two genera of spiny anteaters in Australia and New Guinea. Being egg-layers with other reptilian features, they were once considered survivors of the transitional stock that

---

12 Which is not to say he could reconstruct an unknown animal from one tooth.

produced all mammals. Now monotremes are generally thought to be a completely separate warm-blooded, hairy, milk-producing lineage from the reptiles. Monotremes have no fossil record prior to the Pleistocene.

Marsupials   Opossums are "living fossils," very like the ancestral marsupials but a bit larger, whose reproduction characterizes the group. Born in a very immature state, marsupial young climb into their mother's pouch, where the more fortunate ones attach themselves to teats and grow until ready for life outside. Skeletally, marsupials are rather like placental mammals; however, paleontologists can identify their skeletons from their special pouch-bearing bones, smaller brain cases, and other features. In the Cretaceous they competed successfully with the placentals. In the Cenozoic, except for the opossums, they were largely restricted to an isolated community in South America and another in Australia, where such marsupials as kangaroos, koala "bears," wombats, and carnivorous Tasmanian "wolves" still survive.

Placentals   Some 95 percent of Cenozoic mammals are placentals, the most advanced, intelligent, and successful vertebrates. They range from A to Z (aardvark to zoril), from pygmy shrews no heavier than a dime to 30-meter (100-foot) blue whales weighing 135,000 kilograms (150 tons), from flying bats to burrowing moles and a host of other living and fossil types, including man.

Small-brained and primitive in dentition, the *insectivores*, exemplified by the mouselike shrews, moles, and hedgehogs, represent the ancestral placental stock. Since their late Cretaceous appearance, insectivores have remained small [the giant of them was about ⅔ meter (2 feet) long], and lived on a diet of insects and worms. Bats, edentates, and primates form a supergroup, or cohort, of mammalian orders whose descent from insectivores can be traced.

Bats are essentially flying insectivores whose front limbs are highly modified as wings in which the greatly elongated fingers support a thin membrane. They had mastered true flight by the Eocene, the only mammals ever to do so.[13]

The *primates* are apparently tree shrew descendants whose specializations reflect an active arboreal life (Fig. 13-41). They have five digits with flat nails on feet or hands adapted for grasping, keen eyes with good depth perception, and a general emphasis on brains. Man, a primate along with lemurs, monkeys, and apes, is the acme of organic evolution in terms of intelligence, but he is certainly not in physical structure. His teeth are degenerate, and his body primitive and obscenely naked—by mammalian standards.

The *edentates,* a peculiar group including sloths, anteaters, and armadillos, may be toothless, as the name implies, or more commonly have simplified teeth devoid of enamel. Although edentates are highly specialized, as exemplified by anteaters, armor-plated living armadillos, and extinct glyptodonts, their descent from insectivores can be paleontologically demonstrated. The same is not true of the remaining mammal groups; the first known cetaceans, rodents, carnivores, and ungulates were already specialized and distinct.

Whales, dolphins, and porpoises are *Cetacea,* mammals that, like some reptiles before them, returned to a fishlike existence. Their bodies are well streamlined, their front limbs have become flippers, and the hind limbs are vestigial with no external expression. The first-known whales, represented by the fossil *Zeuglodon* of the Eocene, were already highly specialized marine mammals, giving no clues as to their original derivation. Supported by water, whales have evolved the giants of all time, far larger than the brontosaurs.

The prolific *rodents* have more species than all other mammals combined. Rats, mice, gophers, porcupines, beavers, and guinea pigs are among the

---

13 Flying squirrels merely glide on skin flaps stretched between their legs.

Fig. 13-41 *Notharctus,* an early (Eocene) primate. Courtesy of the American Museum of Natural History.

best-known representatives. Rabbits, once classed as rodents, now seem a separate line that evolved rodentlike teeth. Rodents are gnawers, distinguished by two pairs of long, chisellike, self-sharpening incisors that grow throughout life as they are continually worn down. Otherwise, rodents are rather unspecialized, having long skulls with low brain cases and flexible clawed feet bearing five well-developed toes. Fossil rodents, being generally small, have attracted little paleontological attention.

The *carnivores* and hoofed animals, or ungulates, seem to be descendants from a common ancestor in the late Cretaceous, because their first-known and most primitive members of earliest Cenozoic time have a strong "family" resemblance. The carnivores, specializing in agility and intelligence, claw and fang, have preyed on their hoofed cousins. Carnivore bodies are relatively primitive; they are most specialized in teeth. Typically (examine your pet dog or cat) the strong incisors are for tearing and ripping; the canines are daggers for stabbing and killing; and the cheek teeth are efficient bone crushers or meat shears, except in bears, who have reverted to a more general diet. Terrestrial carnivores include all dogs, cats, bears, badgers, and raccoons. Seals, sea lions, and walruses are aquatic carnivores, for despite external appearances, their skulls are very like those of their land-living cousins.

The *ungulates* are hoofed vegetarians. Unlike the carnivores, which are included in a single order, the very diverse ungulates comprise about 16 different orders, some only distantly related. They have specialized in feet and teeth. Browsing ungulates, eaters of leaves and twigs, such as deer, have low-crowned cheek teeth; whereas grazers, the eaters of silicous abrasive grasses, for example horses, have high-crowned molars suited for much wear. Their cheek teeth, in general, have evolved towards square-topped grinding mills with distinctive ridges and wrinkled surfaces. Ungulate feet are modified for walking on "tip-toe" with the heel well off the ground (in contrast to "flat-footed" creatures like possums, rodents, and men).

Perissodactyls are the odd-toed ungulates, such as the rhinoceros, which have three toes, and the horses, which having only one functional toe represent the ultimate foot evolution in the line. The functional toes are greatly lengthened and strength-

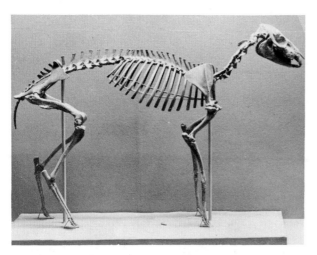

Fig. 13-42 Skeleton of a fossil three-toed horse (*Meso-hippus*) from Wyoming. Photo by Herb Pownall, courtesy of the University of Wyoming.

ened; the useless side toes (which are not counted in describing an animal as even- or odd-toed), no longer touching the ground, are reduced or lost. *Artiodactyls*, those of the "cloven hoof," are the even-toed ungulates like cows and deer. The vertical axis of their foot passes between two central toes. Pigs represent primitive artiodactyls in having a simple stomach and eating almost anything. Advanced artiodactyls are ruminants, vegetarians who can quickly crop and gulp food into the first section of

their complicated stomach for storage, later to be regurgitated for further chewing, before passing it back to other stomach compartments for final processing. This stomach mechanism has great advantages for vegetarians harassed by carnivores, and may well be the reason artiodactyls are still thriving, while the perissodactyls, with their simple stomach, are not. Today only domestic horses are at all abundant; wild members of the once-great hosts of odd-toed ungulates are perilously close to extinction.

Ponderous ungulates, such as elephants, have evolved pillarlike legs whose toes are pulled together creating a short, broad stump of a foot that is usually "shod" with a shock-absorbing pad. Elephants are *proboscidians*, so called because of their proboscis, or nose, which with the upper lip forms the distinctive trunk. Until near the end of the Cenozoic, the proboscidians were an amazingly abundant and varied stock. They have some rather exotic close kin whose relationship is established from fossil evidence. Sea cows, dugongs, and manatees are aquatic vegetarians that evolved from members of this ungulate group in early Cenozoic time; and so are the conies, primitive rodentlike ungulates mentioned in the Bible. With the ungulates, we conclude the "Dramatis Personae" of some of the important participants in the geologic story of life.

SUGGESTED READINGS

Matthews, W. H., *Fossils*, New York, Barnes and Noble, Inc., 1962 (paperback).

Romer, A. S., *Man and the Vertebrates, Volume I*, Baltimore, Md., Penguin Books, 1954 (paperback, Pelican Books).

Simpson, G. G., *Life of the Past*, New Haven, Conn., Yale University Press, 1953 (paperback).

Cretaceous stemless Crinoids (*Uintacrinus*). Courtesy of the Field Museum of Natural History, Chicago.

14

A Greater Genealogy

Having reviewed the different plants and animals important in the fossil record, we now examine them in floral and faunal groups. As such, they are the basis for relative dating of later rocks and the geologic time scale, as well as glimpses of the changing panoramas of life through the ages.

PRECAMBRIAN LIFE

Better than Pooh-Bah, in Gilbert and Sullivan's *Mikado,* who could trace his ancestry back to a piece of amoebic protoplasm, modern science traces our origin to a virus, a far simpler form—little more than a self-reproducing crystal.

On Origins

Speculations Of the various speculations about the birth of life, *creation by a Divine Being* is a theological concept, not demonstrable by fact and observation; hence beyond the realm of science. Among naturalistic explanations, the *extraterrestrial hypothesis* requires that the original protoplasm on Earth came from somewhere else in space. It has been suggested that carbon compounds recently found in some meteorites are organic. However, the extraterrestrial hypothesis is not generally accepted, in part because it avoids the origin of protoplasm, relegating it to some other planet. The hypothesis of *spontaneous generation*—long outdated—states that life is continuously arising in dead matter. Scientists once argued the idea, because of the appearance of flies and micro-organisms in "sterile" solutions or decaying meat. Some time ago, it was demonstrated that the flies came from minute, maggot-producing eggs, and the idea was finally negated when Louis Pasteur and others in the mid-1800s showed that airborne germs caused the "spontaneous generation" of decay bacteria. Today the prevailing scientific view is that life did indeed arise spontaneously—but under different chemical conditions prevailing in the geologically remote past. This hypothesis is at least possible in view of two recent lines of biological investigation.

Some Evidence First, amino acids, the building blocks of protein, have been synthesized in the laboratory. In 1952 Stanley Miller, a University of Chicago graduate student, mixed water vapor, hydrogen, ammonia, and methane (marsh gas), compounds theoretically present in the Earth's primitive atmosphere. He then subjected these compounds to electrical sparks, simulating lightning discharges or solar radiation. After a week, the mixture contained traces of amino acids where none were present before. The experiment certainly did not manufacture life from nonliving substances, but in the amino acids it did produce some of the basic materials.

Secondly, the nature of viruses—those disease agents far smaller than bacteria—has finally been established thanks to the electron microscope, which provides magnifications of some 300,000 times. Viruses unite crystalline substances of the mineral realm with the world of life. They consist of carbon, hydrogen, oxygen, phosphorus, and sulfur, the elements of the geochemical life cycle found in protoplasm. The much-studied tobacco mosaic virus has a dual behavior. In the cells of tobacco plants, it functions as a living disease organism; yet when isolated it forms a minute crystalline structure, like a mineral substance. Admittedly, known viruses require surrounding protoplasm of higher organisms to function as living entities—a possibility denied to the first-formed life. To date no one has created a virus by laboratory experiment; moreover, the evolutionary advance from a virus to the simplest one-celled organism, such as the amoeba, is very great. But the evidence, from viruses coupled with amino-acid synthesis, does suggest that several billions of years ago some chance combination of the Earth's existing chemicals created a supermolecule with the attributes of life.

Possible Early Evolution

Lacking a fossil record, we have no direct evidence of how the earliest life progressively evolved in the Precambrian seas. However, armed with a knowledge of evolution gleaned from later fossiliferous rocks, and by comparing the living organisms from simple to more complex as arrayed in the Tree of Life, we can establish the likely pattern.

Before the Fossil Record The first miniscule specks of protoplasm must have been plantlike, resembling the existing sulfur bacteria in their ability to use inorganic chemicals for food, without relying on photosynthesis. In the first great evolutionary step, protoplasm formed single cells with a nucleus and other special parts. Those cells acquiring a predatory habit, like the amoeba, begat the animal kingdom, and those developing chlorophyl for photosynthesis were the progenitors of typical plants.

The next great advance was the development of multi-celled organisms, probably as colonies of similar cells joined together. Later, colonial masses acquired specialized cells in a "division of labor" for carrying out different bodily functions. Thereafter tremendous evolutionary changes produced the very heterogeneous thallophytes and most of the invertebrate stocks by the end of Precambrian time. So far the story is logical speculation; what is the actual record from fossil evidence?

Precambrian Fossils There are sparse but significant indications of well-established life, long before the Paleozoic. Bunlike masses of crinkly laminae in Precambrian limestones are virtually identical to some living calcareous algae. In the ancient rocks of Ontario's Gunflint Formation, black cherts radiometrically dated at about two billion years, microscopic algae, and fungi have been authenticated (Fig. 14-1). Although less conclusive, graphite and metamorphosed carbonaceous shales are common in some Precambrian terrains. In later rocks, the carbon in dark shales and graphite deposits is largely organic, so the carbon in Precambrian rocks may be the same.

Fossils of multi-celled animals are periodically reported in Precambrian rocks; however, uncon-

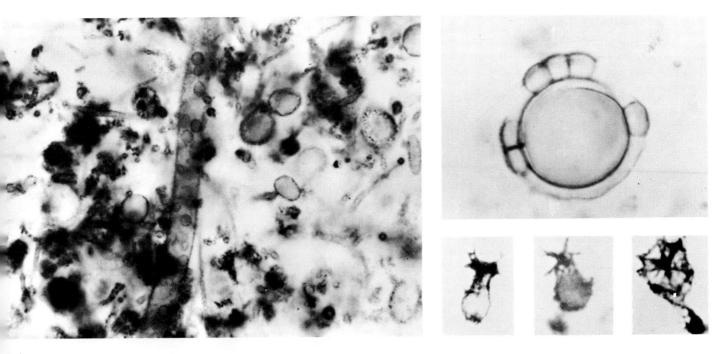

Fig. 14-1 Microorganisms from the Gunflint Chert. Photomicrographs taken in transmitted light through rock thin sections (magnifications: left, 1289×; top right, 1389×; bottom right, 1823×). Courtesy of E. S. Barghoorn.

tested finds are exceedingly rare. Burrows and trails of worms were the only generally accepted ones until recently, when Precambrian rocks in Australia disclosed fossils of jellyfish, worms, corallike animals, and some previously unknown invertebrates. At best, however, the Precambrian fossil list is very unimpressive, considering that life was probably quite profuse.

Thus until the Earth was at least four billion years old, its rocks were virtually devoid of fossils. Then suddenly, geologically speaking, abundant fossils appear worldwide, marking the beginning of Paleozoic time. Most explanations for the delayed and dramatic appearance of fossils are unconvincing. Was life far less abundant in Precambrian times than later? It seems unlikely because of the complexity and diversity of Cambrian faunas. Many

Precambrian rocks are highly metamorphosed and intruded; so has the fossil record been destroyed? Possibly, yet fossils of later eras have been found in metamorphic rocks, including high-grade schists. Moreover, some Precambrian rocks are little-altered sandstones and limestones, virtually indistinguishable from younger fossiliferous strata.

That Precambrian animals lacked hard parts is a reasonable explanation. Most fossils record shell or bone, corruptible flesh being rarely fossilized. Unfortunately, the hypothesis leads to other considerations, as to why hard parts became abundant almost simultaneously. Also, how did muscles evolve without solid areas of attachment? Trilobites, for instance, have complex appendages; how could the muscles controlling them have evolved without a carapace (shell) for anchoring them? Whatever the answers to these questions, for practical purposes the meager Precambrian fossil record is useless for dating and correlation.

ANCIENT (PALEOZOIC) LIFE

Early Paleozoic Life

At the dawn of the Paleozoic, all life as far as we know was in the seas, vertebrates were absent, and the lands were barren. Yet the Cambrian fauna of invertebrates had an advanced development far removed from the original primitive one-celled organisms. The early Paleozoic, encompassing the Cambrian, Ordovician, and Silurian, is rightfully called the "Age of Invertebrates." Beginning in early Cambrian as a rather small bottom-dwelling group of coelenterates, brachiopods, sponges, gastropods, and arthropods, the invertebrates expanded greatly through the Silurian. Thereafter, although still prospering, they had to compete with vertebrate

Fig. 14-2 Diorama of mid-Cambrian sea floor showing algae "organ-pipe" sponges, worms, and trilobites. Courtesy of the Field Museum of Natural History, Chicago.

populations which arose in the seas and eventually followed the plants onto the lands.

The Cambrian All the animal phyla may have been present in Cambrian time; the absence of bryozoans and chordates as fossils could reflect either a lack of hard parts or the vagaries of preservation (Fig. 14-2). Trilobites rapidly reached their all-time peak in varieties and numbers in the Cambrian, for which they are the outstanding guide fossils, then gradually declined to their extinction at the end of the Paleozoic. Brachiopods of a rather primitive sort were second to the trilobites in abundance. Also common were Archaeocyatha, a sponge-like animal forming clusters of tubes on calcareous ocean floors. This particular group is known only in Cambrian rocks.

Minor elements of the population were snails, uncoiled cephalopods, and very rare clams. Graptolites, which were soon to flourish, appeared towards the end of the period. That the fossil sample gives a biased picture of the Cambrian community is illustrated in the dark Burgess Shale of the remarkable Canadian locality where many soft-bodied animals, such as jellyfish and worms, are preserved in abundance.

Moundlike calcareous structures reflecting algal activity are the most common evidence of plants. The record of terrestrial plants is exceedingly poor—explicable in North America because known Cambrian strata are almost wholly marine—although possible lycopod impressions have been found in Siberia. Thus higher plants could have been tentatively invading the wet fringes of the land. But, overall, the continents must have been desolate, eroded badlands.

The Ordovician Marine invertebrates from Cambrian stocks were widespread in the Ordovician seas (Fig. 14-3). Trilobites and brachiopods inhabited limy bottoms of the "shelly" facies of shallow-water deposits, and clams now became relatively common. Nautiloid cephalopods reached their all-time peak, both straight and curved-shelled varieties being present. The giant of the Ordovician was a straight nautiloid, 3½ meters (15 feet) in length. After the Ordovician, nautiloids progressively declined. Above all, however, this was the "Age of Graptolites." Preserved largely in black shale facies, they are among the most important guide fossils for the Ordovician, and following Silurian, as well.

New members of the Ordovician community were the corals, both solitary and colonial tetracorals; the bryozoans; and among the echinoderms, crinoids and starfish appeared. The oldest known eurypterid was found in black shales of this period in New York State. Perhaps most significant are the scattered bony plates heralding the first vertebrates. The generally widespread nature of Ordovician seas makes seaweed and other algae the principal plant fossil; although they may have been present somewhere, we have as yet no record of Ordovician terrestrial plants.

The Silurian Compared to the Cambrian and Ordovician, the Silurian period was relatively short, and its fauna shows few major changes (Fig. 14-4). Brachiopods, molluscs, and corals abounded; graptolites, if less flourishing, produced important guide fossils, until the end of the period, when floating types died out. Trilobites were definitely on the decline. They may well have been meals for nautiloids as well as placoderms, the first jawed vertebrates, for the progressive decline of the trilobites seems to bear a "sinister" correlation to the rise and expansion of fish.

The Silurian was the "Age of Eurypterids," the largest arthropods of all time, which became abundant towards the end of the period (Fig. 14-5). The remains of their close relatives, the scorpions, first found in upper Silurian rocks, are of special interest as the first air-breathing animals potentially capable of venturing onto the land. Doubtless, most of the lands were still barren; however, plants were established in swampy lowlands, for the first good

Fig. 14-3 Ordovician sea floor of shell facies near present site of Cincinnati, Ohio. Algae, straight-shelled nautiloid cephalopods, snails, colonial and solitary "horn" corals, bryozoans, brachiopods, and trilobites are represented. Courtesy of the Field Museum of Natural History, Chicago.

record of Psilopsida and Lycopsida is found in Silurian rocks.

Late Paleozoic Life

The Devonian, Carboniferous, and Permian periods comprise the late Paleozoic, an interval of profound developments in life. Fish came to dominate the waters, and vascular plants and vertebrates spread over the continents.

The Devonian   The Devonian was a most important period in the history of life. Among the established invertebrates, corals, bryozoans, pelecypods, and various echinoderms flourished. Brachiopods, reaching their peak, were to serve as most important guide fossils. Some eurypterids still persisted; trilo-

Fig. 14-4 Silurian sea floor of coral, sand, and patch reef bottom showing algae, chain coral (foreground), honeycomb coral, organ-pipe coral (right), lacy bryozoans, early relations of stalked crinoids (cystoids), brachiopods, nautiloids, and trilobites. Photo of exhibit in the Smithsonian Institution.

bites were definitely on the decline. Important new animals appeared on the Devonian scene. The ammonoid cephalopods, which were to become the most paleontologically significant invertebrates, were in the seas (Fig. 14-6). Insects of a wingless type were well established on land.

Most important was the establishment of several major vertebrate lines. Although the primitive

ostracoderms became extinct in the period, the Devonian was the "Age of Fish." Placoderms flourished, producing, among others, the great 10-meter (30-foot) armor-headed arthrodires. Sharks and bony fish were present; among the latter, the lobe-finned crossopterygians were of greatest evolutionary significance. Towards the end of the Devonian, the first amphibians appeared. In upper Devonian

Fig. 14-5 Silurian brackish water environment dominated by eurypterids, also shrimplike animals and snails. Courtesy of the Field Museum of Natural History, Chicago.

rocks of Greenland, remains have been found that are essentially like crossopterygian fish, except that these animals had very primitive legs instead of fins.

Vascular plants made considerable progress. Early in Devonian time there is evidence of only small primitive plants, mainly Psilopsida, seemingly restricted to wet coastal marshes and swamps. By the middle of the period, however, the Earth's first true forest certainly existed. At Gilboa, New York, for example, stumps up to ⅔ meter (2 feet) in diameter are included in a fossil flora of ferns, seed ferns, scale trees, calamites, and the forerunner of the conifiers, *Cordaites*. Thus by the end of the Devonian, terrestrial vegetation was prominent, and amphibians held a vertebrate beachead (Fig. 14-7).

The Carboniferous   The North American Carboniferous is divided into the Mississippian and Pennsylvanian periods; in European usage, however, the Carboniferous is considered a single period with an upper and a lower division (corresponding to Mississippian and Pennsylvanian in approximate

Fig. 14-6 Devonian sea floor showing a variety of corals, straight and coiled cephalopods, simple and bizarre trilobites, snails, and brachiopods. Courtesy of the Field Museum of Natural History, Chicago.

time). The name refers to the characteristic coal strata, the carbonized remains of vascular plants forming true "forests primeval." Ferns, seed ferns, thickets of large calamites, and the giant scale trees *Lepidodendron* and *Sigillaria* thrived in lowlands of the great coal swamps. Besides this collection, true conifers, precursors of the trees that would dominate later landscapes, appeared by the late Carboniferous.

The coal forests, like modern swamps, were infested with insects (Fig. 14-8). Although mostly of extinct primitive orders, Pennsylvania insects were well evolved, suggesting they had been common in Mississippian and Devonian time as well, but their fragile carcasses were not preserved. In any case, coal-swamp insects include, among others, a dragonfly whose wingspread was almost a meter and some thousands of different species of cockroaches with representatives up to 10 centimeters (4 inches) long.

Sharks and bony fish continued to dominate the seas. In Mississippian time, amphibians must

Fig. 14-7 Devonian forest as painted by C. R. Knight. Copyright the Field Museum of Natural History, Chicago.

have ruled the lands for want of competition; however, their fossils are few because rocks of this period are mainly marine. Pennsylvanian strata have exposed a varied host of peculiar-looking Labyrinthodonts, some being 3½ meters (15 feet) long, who held their own with the first reptiles appearing in late Carboniferous time. These ancestral reptiles, the cotylosaurs, included *Seymouria*, a transitional form from the amphibians.

Among marine invertebrates, the crinoids were extremely prolific, their plates making up a major part of some Mississippian strata (Fig. 14-9). Brachiopods continued to thrive with the appearance of some new spine-covered types. The *fusulinids*, a

most important group of Foraminifera for dating purposes, appeared in early Carboniferous time and were exceedingly common in the Pennsylvanian and disappeared at the end of the Permian.

The Permian The end of the Paleozoic era brought a crisis wherein some plants and animals declined, some died out completely, and others, meeting the challenge, prospered. The decline and fall of many groups is an interesting enigma. It may have related to a pronounced shift in the land-sea ratio. The Permian was a time of widespread continental emergence marked by increasingly extreme climates, and a complementary reduction in area of

Fig. 14-8 Restoration of Pennsylvanian coal swamp. Scale trees and firns (left), giant horsetails (right). Large cockroach on tree trunk (left). Dragonfly (right center). Copyright the Field Museum of Natural History, Chicago.

shallow shelf seas. Thus changing environments may have exerted adaptive pressures on inhabitants of both land and sea.

Crinoids and bryozoans declined markedly. The Permian was the end of the line for two once-great Paleozoic groups: the tetracorals, later to be replaced by modern six-septaed types; and the trilobites, which had once reigned supreme. The ammonoid cephalopods were reduced to one small group. On land the dominant trees of the earlier coal forests approached extinction, but other groups profited. True conifers became abundant, perhaps being better adapted to the cooler climates. The

Glossopteris flora, a plant group found almost exclusively in the Southern Hemisphere, appeared in the Permian and lasted well into the Triassic. They were "tongue-leaf" plants related to seed ferns, or perhaps angiosperms, that lived in temperate and cold climates prevailing then on the southern continents.

The amphibians, including the large, ungainly *Eryops* and the wedge-headed *Diplocaulus*, were still common in the Permian, but amphibians were giving way to reptiles (Fig. 14-10). Notable were the flamboyant "sail-backed" lizards, the pelycosaurs represented in the carnivorous *Dimetrodon*,

Fig. 14-9 Mississippian "garden" of sea lilies and their kin on sandy mud sea bottom. Courtesy of the Field Museum of Natural History, Chicago.

and the less common, herbivorous *Edaphosaurus.* The mammallike therapsids such as *Mosochops,* a large South African vegetarian, had appeared in abundance. The late Paleozoic expansion of reptiles foreshadowed their coming days of grandeur.

MEDIEVAL (MESOZOIC) LIFE

The Mesozoic, era of "medieval life," was the heyday of reptiles on land, in the sea, and in the air. The ammonoids, recovering from their late Paleozoic depression, once again flourished. By the end of the era, the flora had a distinctly modern look.

Early and Middle Mesozoic

The Triassic Widespread desert, semiarid, and cool conditions, accompanying the general continental emergence, continued into the Triassic. In

Fig. 14-10 Permian landscape in Texas showing primitive lizard-like reptiles, sailback, and (lower right) some wedge-headed amphibians. C. R. Knight painting. Copyright the Field Museum of Natural History, Chicago.

terms of life, the period was transitional between the late Paleozoic and the highly characteristic floras and faunas of the subsequent Mesozoic.

The beginning of Triassic time is marked, in marine rocks, by a resurgence of the ammonoids rebounding from their near extinction at the close of the Paleozoic. The generally ceratitic sutures of Triassic ammonoids contrast with the simpler goniatitic sutures prevailing in the late Paleozoic forms. By the close of Triassic time, when ammonitic sutures, the most complex kind, were well evolved,[1] these cephalopods again faltered. This low

in their racial history marking the period's end was followed—as at the Paleozoic-Mesozoic time line—by another expansion in the ensuing Jurassic.

Among other invertebrates, the belemnoid cephalopods were also widespread in Triassic seas; brachiopods declined in most regions; echinoids, starfish, and various crustaceans including lobsters and shrimp were also present. Corals, seemingly extinct in the early Triassic, reappeared as the new six-septaed types. Although vegetation was rarely preserved in the widespread Triassic redbed environments, the conifers apparently flourished; cycads, ferns, and scouring rushes were not uncommon, but seed ferns and scale trees were rare.

1 They had appeared in the Permian.

Of land vertebrates, the Labyrinthodont amphibians had a final surge, then passed from the scene at the end of the Triassic. The cotylosaurs and various mammallike reptiles, other late Paleozoic holdovers, also persisted until the close of the period. Most significant among the reptiles were the newly appeared thecodonts. Of these, the phytosaurs were the crocodile equivalents of the Triassic; the bipedal branch of pseudosuchians, such as Ornithosuchus, were late arrivals during the period. Although 3-meter (10-foot) long running reptiles would be most impressive if living today, the pseudosuchians hardly rivaled their Jurassic and Cretaceous descendants, the dinosaurs (Fig. 14-11).

Still other reptiles readapted a tetrapod structure to aquatic life. These "sea serpents" ruled

Fig. 14-11 Triassic landscape showing a small dinosaur ancestor (lower right), a mammallike reptile (*Cynognathus,* lower center), and larger ancestral forms of the sauropods (middle). Peabody Museum diorama, Yale University.

Fig. 14-12   Marine reptiles. Plesiosaurs and (leaping) ichthyosaurs. Mural by C. R. Knight. Copyright the Field Museum of Natural History, Chicago.

the seas, prevailing over the fish until the end of the Mesozoic era. Turtles appeared in the Triassic seas along with a number of specialized marine reptiles that lived only in the Mesozoic. The *ichthyosaurs* resembled sharks and the later mammalian porpoises externally. However, the fish-shaped ichthyosaurs had a reptilian skeletal structure. The *nothosaurs* had rather stout bodies, limbs and feet adapted as paddles, and long, darting necks for snatching fish. They used their fins as oars, unlike the streamlined ichthyosaurs whose appendages functioned as stabilizers as they swam like fish, using body oscillations and sculling with their long tails.

The Jurassic   The marine invertebrates would not look unfamiliar except for the widespread abun-

dance of belemnoids and ammonoids. After their second racial depression, near the end of Triassic time, the ammonoids once again proliferated. They showed a remarkable evolution in shells and suture complications, allowing exceedingly detailed dating of marine rocks for the rest of the Mesozoic.

In the sea, the fishlike ichthyosaurs were thriving, and nothosaurs had given rise to *plesiosaurs*, "sea monsters" looking like miscegenation of turtle and snake, though not a member of either group (Fig. 14-12). Their shell-less, paddle-driven bodies extended into snakelike necks, some relatively short and others fantastically long. The *geosaurs*, sea-going crocodiles with fishlike tails, appeared and died out, all in the Jurassic. There were dragons, flying reptiles called *pterosaurs* that were apparently soaring carnivores. They ranged from the size

Fig. 14-13 Jurassic landscape showing the first known bird (Archaeopteryx). Mural by C. R. Knight. Copyright the Field Museum of Natural History, Chicago.

of a sparrow to those such as *Rhamphorhynchus*, having long, toothed skulls, light bones, and a tail like a child's kite. Pterosaurs were a separate lineage, not ancestral to birds, because they lacked feathers, had weak legs, and their wings consisted of a flight membrane stretched over a greatly elongated "little" finger. The Jurassic is notable for the first appearance of birds, *Archaeopteryx*, that probably flew among the conifers and cycads prominent in the landscape (Fig. 14-13). Another Jurassic first were the true mammals. No complete skeletons have been found; their presence is indicated largely by jaws and by teeth differentiated into molars, canines, and incisors. These animals were an unimpressive lot about the size of squirrels or rats, a furtive group relegated to insignificance by the dinosaurs.

The "Golden Age" of dinosaurs began. The world's greatest land animals were a group of rather similar dinosaurs, collectively called *brontosaurs*. These saurischian quadrupeds, as much as 24 meters (80 feet) long and weighing an estimated 45,000 kilograms (50 tons) or more, may have spent much of their time immersed in swamps and ponds to support their great bulk. The most formidable of Jurassic carnivores was *Allosaurus*, a 12-meter (40-foot) bipedal saurischian. The ornithischian line produced *Stegosaurus*, a vegetarian quadruped 9000 kilograms (10 tons), whose back sported erect triangular plates and terminated in a tail sprouting formidable defensive spikes. *Stegosaurus* is popularly believed to have had two brains. The true one, in the head, was the size of a walnut and 20 times smaller than the "second brain," an enlargement of the spinal cord controlling the movements of the hind legs and tail (Fig. 14-14).

Late Mesozoic Times

The Cretaceous During Cretaceous time the Earth's floras became modern in most aspects. The coming of the angiosperms, our familiar trees and flowering plants, led to a partial displacement of

Fig. 14-14 Jurassic dinosaurs. From left, *Brontosaurus, Stegosaurus,* the large carnivore *Allosaurus,* and (lower right) *Camptosaurus,* an early "bird-hipped" herbivorous type. Courtesy of the Peabody Museum, Yale University.

the cycads and conifers so prevalent in the earlier Mesozoic.

In the seas, belemnoids persisted, and ammonoids continued to evolve in great variety, producing, among others, the giant of the tribe, almost 3 meters (10 feet) in diameter (Fig. 14-15). Ichthyosaurs, plesiosaurs, and other marine reptiles abounded, and a new group, the mosasaurs, appeared. These fierce-looking 9-meter (30-foot) marine lizards swam with their long lithe bodies, using paddlelike limbs as stabilizers (Fig. 14-16). Though now extinct, the mosasaurs have close living relatives in the split-tongued varanid, or monitor, lizards of the East Indies.

The Cretaceous *Pteranodon,* with an 8-meter (27-foot) wingspread, was the largest flying vertebrate of all time. Unlike the Jurassic flying reptiles, *Pteranodon* was tailless, had a long cantilever projection from the back of its skull, and a toothless beak.

Cretaceous birds are known mainly from fossils of *Hesperornis,* an aquatic bird almost 2 meters (about 6 feet) long having vestigial wings and paddle feet, and the small gull-like *Ichthyornis.* They were skeletally modern except for the presence of teeth.

Dinosaurs were still in ascendency. Among the giant saurischian quadrupeds, the brontosaur group lingered on. The bipedal carnivores culminated in *Tyrannosaurus rex.* This "tyrant king" was the largest flesheater the world has ever seen: 12 meters (40 feet) long, 6 meters (20 feet) high, and weighing an estimated 72,500 kilograms (80 tons). His stunted forelimbs were apparently useless, but the great head and jaws, armed with daggerlike teeth, were adequate for assaulting the largest dinosaurs (Fig. 14-17).

The ornithiscian vegetarians produced several interesting types. The bipedal trachodons, or "duck-bill" dinosaurs, so called from their flattened lower

Fig. 14-15 Cretaceous sea floor showing algae, numerous types of clams and snails, straight and coiled ammonites, and belemnites (swimming). Courtesy of the Smithsonian Institution.

jaws, had had Jurassic ancestors, but the group reached its climax in the Cretaceous. They were mainly aquatic dinosaurs, some, at least, having webbed feet, and their jaws bore as many as 2000 teeth for grinding swamp plants. Some members of this group evolved bizarre skull structures, capping crests, ridges, and thick domes of bone. Armored ornithiscians were the ankylosaurs. These squat-bodied "reptile tanks" were protected by heavy, overlapping bony plates and fringed with protective spikes. The tail was a spiked mace, a fearsome defensive weapon.

The last dinosaurs to evolve, appearing in the late Cretaceous, were the *ceratopsians.* These mas-sive, four-footed ornithiscians achieved lengths of 7½ meters (25 feet) and weighed up to 7000 kilo-grams (8 tons). They had large, toothless beaks, and were characterized by a protective bone neck-frill projecting backwards from their specialized skulls. Most later ceratopsians sprouted horns from the front of their skulls, from a single one, as in *Sty-racosaurus,* whose neck-frill was studded with spikes, to as many as three in *Triceratops.*

The Cretaceous period ended with mass exter-minations. Dinosaurs, flying reptiles, and most marine reptiles died out completely; invertebrates were not immune, for the hosts of belemnites (some may have lingered into the Eocene) and

Fig. 14-16 Some Cretaceous marine vertebrates, a mosasaur, and giant turtle with flying *Pteranodon*. C. R. Knight painting. Copyright the Field Museum of Natural History, Chicago.

ammonoids died out as well. Such mass extinctions are, as yet, impossible to explain. All the Mesozoic victims had been diverse, numerous, and very successful animals. Whatever the cause, the disappearance of these many groups produced a distinct fossil boundary between Mesozoic and Cenozoic rocks.

MODERN (CENOZOIC) LIFE

The collapse of the reptilian dynasty must have left the landscape with a vacant look. This was deceptive, however, for insignificant, furtive mammals who had bided their time for 100 million years were on the scene. Their unprecedented radiation is the major life theme of the Cenozoic, the "Age of Mammals." Abundant and widely dispersed into every environment, these rapidly evolving vertebrates left excellent guide fossils in terrestrial deposits which, because of their relative recency,

are generally less eroded than earlier terrestrial strata.

The Early Tertiary

The early Cenozoic includes the Paleocene, Eocene, and Oligocene epochs. In the Old World, it is called the *Nummulitic Age* because the "giant," coin-shaped foraminifera proliferated in a seaway across southern Eurasia. Otherwise, the marine invertebrates were mainly foraminifera, clams, and snails, not too different from today. In the mild climates of the early Cenozoic, tropical and subtropical floras spread far northward. Magnolias, figs, and palms, for instance, graced the now treeless plains of Wyoming, Montana, and Canada.

The Paleocene and Eocene At the beginning of the Paleocene, terrestrial mammals were all small quadrupeds, at most about a meter (a few feet)

Fig. 14-17 Cretaceous dinosaurs. From left, *Triceratops, Tyrannosaurus, Ankylosaurus,* and a duck-billed *Anatosaurus.* Upper center, the flying reptile *Pteranodon.* Courtesy of the Peabody Museum, Yale University.

high, and of rather similar and primitive structure (Fig. 14-18). They had flat, five-toed feet, and long narrow heads bearing 44 relatively unspecialized teeth. Marsupials, insectivores, and the soon-to-be-extinct multituberculates remained from the Cretaceous fauna. Gradually, new forms appeared: primates, rodents, carnivores, and ungulates. Certain groups evolved rapidly, dominating the early Tertiary in an archaic mammalian fauna, some of whose members achieved a considerable size.

Of the archaic groups, *creodonts* were the earliest abundant carnivores. They were, initially, small animals like weasels having slender bodies and rather short legs. Evolving rapidly, the creodonts diversified into types resembling wolves, lions, and bears. One, with a meter-long skull, represents the largest known predatory mammal. Creodonts are classed as archaic because their teeth were primitive adaptations of insectivore dentition. By Oligocene time, most of these flesheaters were extinct, replaced by the more intelligent ancestors of modern carnivores.

Several orders of archaic ungulates reached a climax in the early Tertiary, then, like the creodonts, gave way to forerunners of modern hoofed lineages. The giant amblypods included *Uintatherium,* which resembled a present-day elephant in size and pillar-like leg structure. Its defensive weapons were impressive: three sets of horns on the top of its head and two long canine teeth set in the upper jaw (Fig.

Fig. 14-18 Paleocene landscape in Rocky Mountain region, where a subtropical flora of modern-type plants existed. With the great dinosaurs extinct, small primitive mammals took over. Peabody Museum diorama, Yale University.

14-19). *Coryphodon* was as large as a living hippopotamus, hornless, but armed with fierce canines. The failure of these animals to survive may relate to their low-crowned teeth, suitable only for soft vegetation, or, more probably, to their small brains and lack of intelligence.

Condylarths, appearing in earliest Tertiary time, were the first and most primitive of ungulates. The sheep-size *Phenacodus* resembled the early prototypes of later carnivores in having a long skull and tail, conspicuous canine teeth, short legs, and five-toed feet. Although too large and too late to be the direct ancestor of modern ungulates, Phe-

nacodus displayed significant preliminary changes. Its molars were square-topped, and its toes were tipped with individual small hooves instead of claws. Also present in Eocene time were small, relatively unspecialized ancestors to oreodonts, titanotheres, rhinoceroses, camels, and horses.

Famed *Eohippus*[2] was a slim, relatively long-legged and long-headed lowbrow about the size of a fox—an unlikely-looking forebear of noble *Equus*, the modern horse. Significantly, however, the "Dawn Horse" of the Eocene ran on the tips of its

2 Technically and more properly called *Hyracotherium*.

Fig. 14-19 Mural of Eocene flora and fauna in western Wyoming. Large animal is *Uintatherium*. Early members of many mammal lines are shown. Courtesy of the Smithsonian Institution.

toes, each capped with a little hoof. There were four useful toes on each front foot (and a fifth nonfunctional toe), and three on each hind foot. Dentally, small incisors and canines were present, and the molars, although low-crowned, were already square-topped. The ancestors of the other future great lines of perissodactyls were as humble as, and almost indistinguishable from, Eohippus in Eocene time.

The Oligocene By Oligocene time, the large archaic mammals had given way to more modern lineages (Fig. 14-20). Small carnivores could be

distinguished as dogs and cats. Miniature mastodonts, the first elephant stock, roamed the Old World continents. The perissodactyls, odd-toed ungulates, were, generally, the dominant vegetarians. Horses were represented by the three-toed, sheep-size *Mesohippus*. The rhinoceroses had diversified into three lines from hornless, slim-legged, fleet Eocene ancestors only a bit larger than Eohippus.

Oligocene rhinos included the following types: larger, running members whose habits differed little from the contemporary horses; large, stout, short-legged amphibious forms outwardly resembling hip-

Fig. 14-20 Oligocene mammals and flora of North Dakota. Included are titanotheres, entelodonts (fighting), ancestral horses (left), running rhinos, and many other vegetarians and carnivores. Courtesy of the Smithsonian Institution.

popotamuses; and forebears of "true" rhinoceroses. The running and amphibious tribes became extinct about the end of the Oligocene, while the main line continued. *Baluchitherium*, a hornless rhinoceros of the Oligocene and early Miocene, qualifies as the largest known terrestrial mammal. Although not as large as the greatest dinosaurs, it was 7½ meters (25 feet) long and over 5 meters (18 feet) high at the shoulders. Having a rather long neck, *Baluchitherium* was apparently a browser of tree leaves, and, superficially, like a ponderous giraffe.

The perissodactyls had striking offshoots in the now-defunct *titanotheres*. Theirs was a short, spectacular career. From small Eocene forebears, differing little from Eohippus, they evolved bulky giants with large bone-cored horns on their skulls. Flourishing in the early Oligocene, they were gone by the middle of the epoch. Low intelligence may have contributed to their downfall, but a more fundamental flaw was probably dental. Their low-crowned teeth may have worn out rapidly in chewing the harder, siliceous grasses that in mid-Tertiary time were replacing the softer earlier vegetation.

The *chalicotheres*, unique as the only ungulates with clawed feet, were probably diggers, feeding on soft roots along streams. They were "composite" perissodactyls, having three functional toes on each foot, a head like a horse, teeth like a titanothere, and a giraffelike body with longer legs in front than behind. Never numerous, they were, however, a long-lived group, surviving from their Eocene appearance till late Pleistocene.

In early Cenozoic, the artiodactyls of the cloven hoof had not come into their own. Piglike types, from whose members the diversity of later advanced ruminants probably evolved, did produce an impressive offshoot in the now extinct "giant hogs," better called *entelodonts*. Despite a swinish look, entelodonts differ from existing pigs in being adapted for running. They had a straight back, long legs, and, in later forms, just two toes. Ugly in looks and most probably in disposition, they bore knobs or flanges on their cheeks and impressive canine tusks. By Oligocene time, some were twice the size of a bear, and they were to become larger still.

Oreodonts were thriving North American artiodactyls from late Eocene until their Pliocene extinction. Judging by their prolific remains in certain Oligocene beds, herds of oreodonts must have

swarmed over the western regions at that time. They are, perhaps, best classed as "ruminating swine," for their teeth suggest cud chewers, whereas their legs were short with four-toed feet and their bodies were piglike. Some were quite small, many were the size of a sheep, and the largest were comparable to a hog.

The Late Tertiary

The Miocene   Starting in Miocene time, climates on the great landmasses of the Northern Hemisphere became progressively cooler and drier. Grasslands expanded at the expense of forested areas, and, as a seeming result, the hoofed grazing animals proliferated. This was the "Golden Age" of mammals, whose rich fauna, if differing somewhat from that of the present, had a generally familiar look. Although impoverished by the disappearance of the

Fig. 14-21 Miocene menagerie, including *Dinohyus*, Chalicotheres, horses (*Merychippus*), camels, antelope, and others. Courtesy of the Smithsonian Institution.

titanotheres and some other lines, the perissodactyls were still varied and prominent until the end of late Tertiary when they began a final decline. Their cousins, the artiodactyls, ever diversifying and expanding, became the dominant hoofed vegetarians, a position they continue to hold to the present time.

In the Miocene (Fig. 14-21), horses included several types, one of which was the size of a small pony and ran on a single functional toe. Rhinoceroses of several kinds were still common. The chalicotheres were represented by the relatively large *Moropus*. The entelodonts culminated in the piglike *Dinohyus*, as large as an American bison and having a meter-long skull. The brains of this particular animal group were never large, which may account for their mid-Miocene extinction. Oreodonts were still present. Camels were a varied and humpless stock; some were relatively small and designed for fast running, while others became large with long necks, such as *Alticamelus*, who could browse on leaves 3 meters (10 feet) off the ground. Giraffes first appeared in the Miocene as a branch from the deer family. The earliest forms were short-necked; the elongation of neck and legs

Fig. 14-22 Pliocene diorama showing large bear, dog, horse (*Pliohippus*), an amphibious rhino, a "shovel-tusk" mastodon, camels, pronghorn, and the unicornlike ruminant *Synthetoceras* with a branched horn above its muzzle. Courtesy of Peabody Museum, Yale University.

characterizing the living giraffes is a late evolutionary development. Cattle and their close relatives, the antelope, sheep, goats, and pronghorns (American "antelope"), were new arrivals on the Miocene scene. Mastodonts, the elephant stock with low-crowned teeth, some with four tusks, diversified and spread across the Northern Hemisphere. To complete the mammalian array, varied carnivores of the dog and cat families preyed on the multitudes of ungulates

The Pliocene In the Pliocene, the "autumn" of the Cenozoic preceding the Pleistocene "winter," mammals still prospered, and some increased in size (Fig. 14-22). Oreodonts became extinct; however, cattle and their kindred ruminants were expanding.

Mastodons, an Old World stock, were well established in North America, and the true elephants, whose stout, complexly ridged molars adapted them to grazing, appeared. Rhinoceroses and camels abounded and several horse types were present. The ubiquitous carnivores included various dogs, wolves, cats, and bears. A notable development in Africa was the appearance of apelike creatures whose skulls and bodily structure foreshadowed the advent of man.

The Quaternary

Mammals The last two million years of Earth history, the Pleistocene and Recent (or Holocene), has

Fig. 14-23 The Pleistocene was the era of giant mammals. Mammoths, mastodons, giant bison and sloths, large beaver and wolves, saber-toothed tigers, the modern horse (Equus), as well as the "mammalian tanks," the Glyptodonts. Courtesy of Peabody Museum, Yale University.

been a time of climax and crisis for mammals. As the great ice caps waxed and waned, Eurasia and North America were still populated by a greater diversity of mammals than exists today, including the giants of certain lines (Fig. 14-23). *Castoroides,* a now-extinct beaver, was a gigantic rodent as large as a black bear. Buffalo—technically bison—roaming North America included some species having a 2-meter (6-foot) horn spread, far larger animals than existing ones. Most impressive were the proboscidians, the browsing mastodons, and true grazing elephants represented by woolly mammoths. The greatest of these was the *imperial mammoth,* a hairy giant over 4 meters high at the shoulder and bearing tusks almost 4 meters (13 feet) long. The abundant mastodonts and mammoths left hundreds of skeletons in peat bogs of the United States. In Siberia and Alaska thousands of mammoth tusks have been collected for the ivory trade, and some complete specimens, with preserved flesh and hair, have been recovered in permanently frozen ground.

Horses, including the modern Equus, abounded, and rhinoceroses included a woolly type roaming far north in Eurasia. In North America, carnivores included *Smilodon,* greatest of a now-extinct line of saber-tooth cats capable of attacking the largest mammals; lions as large as those living today in Africa; and the giant *dire wolf,* larger than any modern ones. Living beside the giant extinct mammals were many of today's familiar types.

At various times in the Cenozoic, mammals had migrated between Africa, Eurasia, and North America. The mixing was especially pronounced while Pleistocene ice caps lowered sea level, exposing intercontinental land connections, such as at the Bering Straits. Thus, by the late Cenozoic, emigrants from Africa, such as the proboscidians, ranged over North America; on the other hand, horses and camels, native North American stocks, had spread as far as Africa. South America and Australia, in contrast, have been isolated continents, where rather exotic faunas evolved. In Australia

the marsupials, free of competition from the more aggressive placentals, have prospered until today, when man is upsetting their natural balance.

South America long had a provincial population of marsupials and archaic placentals whose ancestors reached the continent in early Cenozoic time. In isolation, they evolved a variety of herbivores and carnivores which, if exotic compared to animals in the rest of the world, took on many of the habits and filled many of the ecological niches occupied elsewhere by more progressive placentals. In the Pleistocene, however, a migration route became exposed along the Panamanian Isthmus. Thus the Pleistocene fauna of North America was enriched by South American armadillos, porcupines, and some now-extinct forms: the giant ground sloth *Megatherium,* the size of a small elephant, and *glyptodonts,* unique armored mammals, with macelike tails, which resembled the "reptilian tanks" of the Mesozoic. North American placentals invaded South America, with disastrous results to the natives there, who were largely exterminated and replaced.

Extermination is a major theme of the Pleistocene, for towards the end of the epoch, most of the giant mammals and some others became extinct on the northern continents. Unlike the decimation of the South American animals, which is explicable through competition with advanced placentals, the decline of the great mammals is as mysterious as the end of the dinosaurs. Changing climates with failure to adapt, competition from man (the most devastating predator of all time), and other hypotheses proposed are not entirely satisfactory. Many of the victims were large, powerful, extremely well-adjusted animals that had survived the marked climatic fluctuations of the Pleistocene.

Whatever the cause, the herds of mastodons and mammoths, the largest bison, saber-toothed cats, sloths, glyptodonts, and other impressive mammals all disappeared from the northern continents. If horses and camels, which evolved in North America, had not spread into Eurasia and Africa in earlier times, they too would be totally extinct.[3] Had south Asia and Africa not provided a sanctuary, many splendid mammals would be known only as fossils. Africa—where elephants, rhinos, hippos, great cats, and antelope herds roam the grassy savannas—presents a last glimpse of the "Golden Age" of mammals. Even here, they are on the decline. Thus the existing populations of large mammals are the impoverished survivors of the great dying, in late Quaternary time.

Man   The human population, now at its greatest number and still rapidly expanding, makes latest Cenozoic time the "Age of Man." Structurally, humans are primates, a group having a scanty fossil record, markedly so for the direct ancestors of man. Most primates were arboreal, living in forests where conditions for fossilization are poor; and being generally intelligent, they were not apt to be trapped in bogs, tar pits, or other environments of quick burial. Thus, for want of evidence, the early descent of man is conjectural.

A few jaw fragments and distinctive teeth, found in India and Africa, have been assigned to a genus called *Ramapithecus*. These small primates, which had the potential of evolving into the genus *Homo*, man, had become different from primitive ancestors of the present-day apes some 15 million years ago—by Miocene or early Pliocene time.

The genus *Australopithecus*, a Pliocene and early Pleistocene group of hominids, is well established from fossil discoveries during the last few decades. In South Africa, Olduvai Gorge in Tanzania, the shore of Lake Rudolph in Kenya, and the Afar region of Ethiopia, australopithecine remains have been recovered from sediments that have been dated (by potassium-argon techniques) from interlayered lava flows. These hominids, who lived from about five million to some two million years ago, were apelike in having projecting jaws

3 Our modern horses were reintroduced into North America by man, starting with the Spanish Conquistadores.

and low sloping foreheads. Skull measurements, however, indicate brain cases that were larger than those of any living apes (although averaging only half the size of that in modern man). Moreover, their teeth were of a human, rather than apelike, pattern; their pelvic structure indicates a good, but not perfect, adaptation to walking upright on two legs; and—significantly—they manufactured crude tools from stones. At least two species of Australopithecines coexisted for several million years: the larger and heavy-boned *Australopithecus robustus*, and the smaller and lighter-boned *Australopithecus africans*. Fossils of the robust species indicate no progressive increase in size of the cranial cavities through time. The more delicate-boned *africans* species, however, apparently did evolve a larger brain. Thus some anthropologists believe it produced the ancestors of the genus *Homo*, or man.

Our ancestors, however, might have still another lineage. A three-million-year-old skull unearthed in 1972 near Lake Rudolph had a brain capacity of some 750 cubic centimeters—notably larger than those of *Australopithecus africans* skulls from deposits of comparable age. Thereafter a number of jaw fragments containing teeth that experts considered more like man's than ape's or Australopithecine's were found and dated as from three and a half to three million years old. In 1974 a remarkable skeleton at least three million years old was discovered in northern Ethiopia. The most complete hominid specimen of Pliocene or early Pleistocene age yet pieced together (almost half the skeleton was recovered), it represents a one-meter-tall—but full-grown—female, based on the pelvic structure which, with the leg bones, also clearly indicates that she walked fully erect. Whether the skeleton, nicknamed "Lucy," should be classified in the genus *Homo*, true man, is still being debated by specialists. But she does seem distinct from her contemporary Australopithecines. This specimen could well represent the ancient population from which mankind evolved.

The first fossil remains accepted by scientists

Fig. 14-24 Diorama of a Neanderthal family by sculptor Frederick Blaschke. Copyright the Field Museum of Natural History, Chicago.

as true man are of early or perhaps middle Pleistocene age—one and a half to one million years old. They are classified as *Homo erectus*, represented by Java and Peking man, who had low, sloping brows and about two-thirds our brain size. Yet their chins and teeth were notably human, and they walked in a fully erect position.

By the third interglacial age—some 100,000 years ago—the modern genus and species of man, *Homo sapiens*, was well represented by the *Neanderthals* (Fig. 14-24). Numerous skeletal remains show them as short, stocky people having sloping foreheads and more rugged features than we possess. Early Stone Age man made tools from chipped flints and had a relatively complex culture.

They were hunters in a world of mammoths and giant cave bears as well as wolves and other nonextinct species.

The modern type of *Homo sapiens* appeared some 40,000 years ago during the last major expansion of great continental glaciers. They were the *Cro-Magnon* people, whose physical appearance was completely modern and whose brains were as large as ours. The Cro-Magnons developed several rather complex Stone-Age cultures and were fine artists whose cave paintings still rank as masterpieces.

Human evolution is an Old World phenomenon. Not until the end of the Pleistocene did modern man reach the Americas by way of the

Bering Straits. No incontrovertible skeletons of these first Americans are known; their record is in distinctive, well-fashioned artifacts, exemplified by the Folsom-type points with which they hunted the last of the giant bison and mammoths. Although these first "discoverers" of America are called paleo-Indians, it is not surely known whether they were ancestors of the Indians living here when the first Europeans arrived.

For geological purposes man's ancestors are poor fossils; moreover, their study has been taken over by the anthropologists. Yet the emergence of *Homo sapiens* makes an appropriate ending for an account of life through the ages. After all, until this species evolved the geologic story of the Earth remained untold.

SUGGESTED READINGS

Carrington, Richard, *A Guide to Earth History,* New York, New American Library, 1956 (paperback).

Colbert, E. H., *Evolution of the Vertebrates,* New York, John Wiley & Sons, 1955.

Edey, M. A., *The Missing Link,* New York, Time-Life Books, 1972.

Rush, J. H., *The Dawn of Life,* New York, New American Library, 1957 (paperback).

Paleozoic and
Later Physical History

The evolution of continents since the early Cambrian comes into clearer focus. Paleozoic and later rocks are generally less complicated by intrusion and metamorphism, less eroded, and less covered than their Precambrian counterparts. Above all, in the Paleozoic and later eras, the abundant fossil record is an invaluable tool for the relative dating and correlation of rock strata. Thus structurally jumbled and far-flung outcrops can more easily be fitted to broad regional patterns.

We now present the ever-changing paleogeographies of North America; and then, in a wider perspective, consider the changing geologic explanations for the evolution of oceans, continents, and mountain ranges.

Red Rocks Amphitheatre, Denver, Colorado. Courtesy of U.S. Forest Service.

15

North America After the Precambrian

Since "the present is the key to the past," we briefly review the existing landscape and the general ages and kinds of rocks now exposed in North America before presenting the ephemeral paleogeographies of the past.

INTRODUCTION

North America Today

North America is rather nicely symmetrical. Its heart is the broad Precambrian shield, a deeply planated, basement complex. Surrounding the shield are the Interior Plains of younger, flat or gently dipping sedimentary beds, covering the shield's margins. The strata, although up to several thousand meters thick in places, are thin compared to the sequences exposed in the highlands beyond.

The Appalachian mountain system, extending from Newfoundland through eastern Canada to Alabama, is broadly divisible into north-south belts. Farthest inland is the Appalachian Plateau. Its little-deformed sedimentary rocks extend eastward into the Valley and Ridge province of relatively open folds to the north and thrust faults and tighter folds to the south. The "crystalline" Appalachians extend from the the Blue Ridge Mountains and Piedmont in the south to Maritime Canada. Their rocks are strongly metamorphosed and intruded by granite as far north as New England, but become less so in Canada.

Beyond Alabama, the mountain structures veer westward, and, though largely buried by Coastal Plain sediments, do emerge in the Ouachita Mountains of Arkansas and Oklahoma, and in the Marathon Mountains of west Texas.

The Coastal Plain is the emergent part of the broad, gently sloping continental shelf, the largely drowned eastern margin of North America. Its relatively unconsolidated, gently dipping sediments form a continuous outer lowland from Mexico to New Jersey, with emergent patches on Long Island and Cape Cod. No comparable feature borders the Pacific shore, where coastal ranges reach the ocean with only a narrow continental shelf beyond.

The complex Cordilleran region west of the Interior Lowland is far broader and more diverse than the Appalachian system. The Cordillera extends from Alaska to Mexico and, ultimately, down the length of South America as the Andean chain. The Rocky Mountains, forming a western bastion to the Interior Plains, include broad anticlinal ranges with broad intervening basins from central Wyoming southward, and more complexly faulted masses with narrower valleys from northeastern Utah through Canada into Alaska.

The Rockies are separated from the Pacific Coast mountain system by broad, often rugged, intermontane provinces including the Columbia lava plateau, fault-block mountains and valleys of the Basin and Range country, and the Colorado Plateau of largely flat sedimentary rocks.

The Pacific mountain system includes the Sierra Nevada, a great tilted block, mostly of granite, which merges northward with the volcanic Cascades, a broad, even-crested upwarp topped by volcanic cones. Separated by broad valleys from the Sierra-Cascade chain are the Pacific Coast ranges, complicated mountains including rocks as young as Cenozoic. With this barest outline of the present situation in mind, let us return to the earliest Paleozoic.

Early Paleozoic Symmetry

North America was also symmetrical in the early Paleozoic, but lacked the flanking Appalachian and Cordilleran mountain systems whose evolution is the major theme of Paleozoic and later history.

Platforms and Geosynclines As the Paleozoic dawned, North America was an emergent continental platform, a subdued or broadly rolling landscape, eroded across the generally contorted and crystalline rocks recording a long and eventful Precambrian history (Fig. 15-1). The platform, which included the present-day shield, stable interior, and some areas beyond, was bordered by seas where mountain belts lie today. Towards the end of Precambrian time, both the eastern and western continental margins had begun to subside

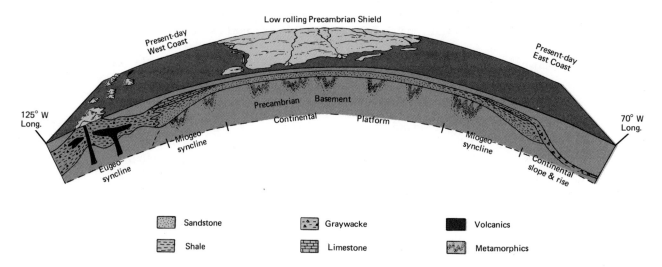

Fig. 15-1 Diagram of symmetrical North America developing through Cambrian and early Ordovician time. Data mainly from Marshall Kay.

and collect sediments in elongate belts—the initial Appalachian and Cordilleran geosynclines.

Each geosyncline had an inner part, the miogeosyncline, and outer part, the eugeosyncline. Next to the continental platform, the deepening troughs accumulated sandstones, shales, and limestones—all well-washed deposits reworked by wave action in shallow water on a stable, or slowly subsiding, sea floor. The eugeosyncline, farther offshore, is characterized by graywackes and other poorly sorted clastics, as well as cherts, often thickly bedded.

An Early Paleozoic Flood   Early Cambrian seas crept along the axes of both the Appalachian and Cordilleran geosynclines. In time the seas expanded inland, leaving a transgressive sequence of basal sandstones, overlying shales, and finally limestones indicating clear waters, with younger rocks progressively lapping over the older. Most of the sediments laid down in the advancing seas seem derived from the continental interior where streams carried material out from weathered Precambrian terrains. The advance of the seas continued unbroken from Cambrian into Ordovician time leaving a continuous rock sequence, so that the time change from one period to the other is determined purely from fossil evidence. Two facies of differing fossil assemblages were well established by the Ordovician: a shelly facies characterized by trilobites and brachiopods in the better sorted rocks nearer shore; and a graptolite facies in dark shales farther offshore. By mid-Ordovician, shallow seas had inundated much of the continental platform, leaving only the heart of the shield exposed.

So far we have treated the continent as a whole, for until mid-Ordovician, when the flood reached its peak and then receded, North America retained its roughly mirror image of a central platform bordered by geosynclines from which the seas advanced and merged across the continental interior. Hereafter we shall discuss the Appalachian and Cordilleran regions separately, for from

mid-Ordovician onwards the events in each were not concurrent.

THE APPALACHIAN STORY

Cambrian and Ordovician rocks preserved in North America are almost entirely marine. Starting in later Ordovician, however, continental deposits appear that reflect the rise and fall of mountain systems along the eastern and southern borders of the continent.

The Collapsing Geosyncline

From the Ordovician to the close of the Paleozoic, the Appalachian eugeosyncline was affected by spasmodic orogenies during which its rocks were deformed and metamorphosed, replaced, and intruded by granite. Mountainous islands arose in the geosyncline and were worn down, shedding clastic debris westward across the more stable miogeosyncline belt and onto the continental interior. Finally, the whole geosyncline was destroyed when its inner part was folded and faulted, near the end of the Paleozoic.

Taconic Mountain Building   Pulsations of Taconian orogeny commenced in mid-Ordovician, reached a peak at the end of the period, and died out by the Silurian. The orogeny was strongest in the northern Appalachian region and New England. In easternmost New York and adjacent regions, the present-day Taconic Mountains, which are eroded stumps of the Paleozoic ranges, contain black graptolitic shales of the outer volcanic belt that some geologists believe have been thrust 65 kilometers (about 40 miles) westward onto rocks of the nonvolcanic geosynclinal belt.[1] Unconformi-

---

1 The extent of the thrusts is debated, but they do seem prominent in the north while dying out to the south.

ties in the Hudson Valley give good evidence of the Taconian orogeny where folded Ordovician rocks are beveled and overlain by relatively underformed Silurian and Devonian strata. Farther east, in the highly deformed rocks of New Hampshire's White Mountains, the orogeny is indicated by an unconformity, as well as granites that invaded Ordovician rocks, but not Silurian.

Doubtless, the Taconian ranges were majestic mountains, perhaps comparable to the modern Rockies or Alps, for they shed tremendous volumes of sediments westward into a new trough, subsiding where the older geosyncline had lain. The clastic flood from the rising Taconian ranges is called the *Queenston Delta*, although strictly speaking it was a compound of many deltas, flood plains, and alluvial fans. Their continental sediments grade into marine deposits that extended far westward on the sea floor (Fig. 15-2).

Near the rising highlands, continental sandstones and red beds predominate, with darker shales beyond representing muds swirled westward and deposited in shallow seas; extensive limestones accumulated on the clear waters of the stable platform. These facies record a retreating sea wherein continental and near-shore deposits expanded everfarther westward during the accelerating orogeny of later Ordovician time. The spread of the continental sediments fluctuated as seas occasionally advanced slightly eastward, depositing marine shales on red beds, before renewed deltaic expansion drove the shore westward again. When the Taconian orogeny reached its climax, at the end of Ordovician time, terrestrial red beds extended to the present-day location of Niagara Falls, almost onto the stable continental interior.

Return of the Seas The Taconian ranges were leveled during Silurian time, while the seas spread back towards the east (Fig. 15-3). That the mountains were still high at the beginning of the period is shown by the lower Silurian deposits which grade from conglomerates on the east into widespread sandstones thinning westward and being replaced by shales on the stable continental platform. But by mid-Silurian the mountains were much reduced, and sluggish west-flowing streams deposited finegrained red beds and shales. With time the region of deltaic clastic deposition contracted, until, by the end of the period, marine limestone was being laid down as far east as the present-day Hudson Valley region. Together, the Ordovician and Silurian clastics form a great horizontal wedge recording the rise and decline of the Taconian Mountains.

In New England and Maritime Canada, geosynclinal remnants persisted through the Silurian. These locally subsiding basins trapped sediments

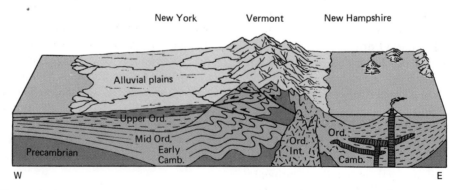

Fig. 15-2 Diagram of the late Ordovician Taconian Mountains (center) and Queenston Delta (west). Based on Marshall Kay and other sources.

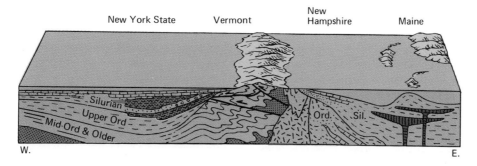

Fig. 15-3 Diagram of the eastern United States in late Silurian time when seas encroached on the eroded Taconian Mountains. Based on Kay and other sources.

and volcanics over three kilometers (1.9 mile) thick, which lie unconformably on the edges of rocks contorted in the Taconian orogeny.

The Acadian Orogeny The northern part of the Appalachian geosyncline, as far east as the beveled Taconian ranges, remained a quiet area of widespread limestone deposition from the late Silurian until early Devonian time. Then the Acadian orogeny began, which affected much of the region already deformed in the Taconian. Thus the already contorted and intruded Ordovician strata, as well as previously unaffected Silurian and Devonian layers, were deformed and invaded by granitic masses to produce such complicated crystalline

rocks as are now exposed in the White Mountains of New Hampshire. By the end of Devonian time, the volcanic belt of the Appalachian geosyncline had been destroyed in the northeastern United States and eastern Canada. Thereafter its crumpled rocks formed a solid addition to the continental platform.

The rising Acadian mountains shed a second clastic flood towards the continental interior during middle and late Devonian time (Fig. 15-4). This clastic wedge, called the *Catskill Delta*, buried the early Paleozoic geosyncline and more recently deposited Silurian and Devonian rocks. As in the preceding Queenston Delta, conglomeratic and sandy red beds, of continental origin, grade into

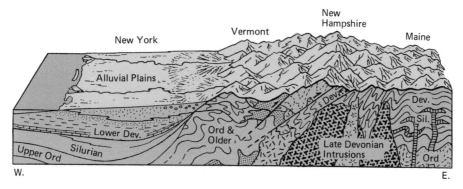

Fig. 15-4 Diagram of conditions at end of Devonian time. Acadian ranges on the east and the Catskill Delta on the west. Data from G. H. Chadwick, M. P. Billings, and others.

marine sands and shales in the shallow sea west of the mountains. Strata of the Catskill Delta also coarsen upwards, reflecting the spread of debris as the sea was driven westward.

Carboniferous Aftermath   Erosion leveled the Acadian ranges of New England and eastern Canada during the Mississippian. Then, in a typical aftermath of orogeny, the region was epeirogenically uplifted with block faulting. Terrestrial sandstones, conglomerates, and shales derived from the new highlands were laid down in local basins along with some interspersed volcanics. Shallow Pennsylvanian seas encroached upon the northern part of the area and left some limestone and evaporites before withdrawing. In eastern Massachusetts and Rhode Island, some terrestrial deposits have the look of glacial tills, and others were coal swamp deposits.[2]

The western Appalachian region, during Carboniferous time, remained a subsiding belt collecting debris eroded from highlands to the east. In Pennsylvania and Virginia where they are still preserved, lower Mississippian conglomeratic rocks grade into extensive sandstones towards the west. The upper Mississippian rocks are largely deltaic red beds of sandstone and shale. Although some in Virginia are coal-bearing, the overlying Pennsylvanian strata have the principal coal deposits of the Appalachian region. The coals, formed in forested swamps, are interbedded with terrestrial clastics swept from uplands to the east. Almost a kilometer (0.6 mile) thick in Pennsylvania, the coal measures become much thicker southward into Alabama, indicating deeper subsidence of the western trough with the creation of a clastic wedge. It, like the older clastic floods to the north, is suggestive of another orogeny, a Pennsylvanian spasm in the Piedmont–Blue Ridge region.

The Allegheny Orogeny[3]   The western Appalachian region remained undeformed until near the end of the Paleozoic. Beds of the early Paleozoic nonvolcanic geosyncline and the overlying clastic deltas are essentially parallel, with no angular unconformities. The end of geosynclinal conditions in the Appalachians came in the Allegheny orogeny, a final spasm of mountain building that created the folded ranges of the sedimentary Appalachians in the Permian. During this orogeny, the crystalline Appalachians were again invaded by granite and parts were thrust westward, overriding and contorting hitherto undeformed strata into open folds in Pennsylvania and tighter folds and thrust sheets southward into Alabama (Figs. 15-5, 15-6).

We have dwelt largely on the classic story of the Appalachians. Their extension in the Ouachita belt, from Alabama to west Texas, had a generally comparable Paleozoic history. Originating in the Ordovician, the Ouachita geosyncline was a relatively quiet site of subsidence and deposition until Mississippian time. Thereafter the destruction of this geosyncline, though telescoped in time, followed the Appalachian pattern. In the early Carboniferous, pulses of orogeny began in the outer volcanic belt where great thicknesses of graywacke and shale accumulated, while floods of clastic debris were shed northward towards the continental interior. In Pennsylvanian time, the eugeosyncline, which had been largely consolidated, was rammed northward, contorting clastic wedges and older miogeosynclinal rocks of a longer-lasting, inner depositional belt. Orogeny had ended by the Permian.

Thus at the end of the Paleozoic era, North America had lost its earlier symmetry, for the Cordilleran geosyncline still remained in the west, while mountains rimmed the continent from Texas through Alabama and northward to Canada (Fig. 15-7).

---

2 Whether these conglomerates are tills is now in doubt; some coals have since been metamorphosed to graphite, suitable for pencil lead but not for burning.

3 Once called the "Appalachian Revolution," when geologists considered it the main mountain-building event. It is now recognized as only the final spasm of several that affected the Appalachian belt.

Fig. 15-5 Present-day view of crystalline Appalachians in the Great Smoky Mountains. Courtesy of the Department of Conservation and Development, Raleigh, North Carolina.

Later Appalachian History

When the mobile geosynclines in the Appalachian region were finally destroyed, their rocks became a relatively rigid extension to the continental platform. But if folding, thrusting, and regional metamorphism had ended, the area was still restive.

Mesozoic Aftermath   The Appalachian ranges were steadily eroded during the early Mesozoic. Debris from the uplands was apparently swept out of the region, which lacked major structural basins to trap sediments, and deposited it in the continental interior and the depths of the Atlantic. As the Triassic progressed, erosion reduced the Appalachians to the low rolling surface of a peneplain (Fig. 15-8).

Late in the Triassic, the subdued erosion surface was broadly upwarped with associated large-scale normal faulting and volcanism, an episode known as the *Palisades disturbance* (Fig. 15-9). Rising blocks shed sediments into a series of grabens, following the earlier mountain trends. The

Fig. 15-6 Z-shaped folds formed in the Allegheny orogeny as seen today. Photo by John S. Shelton.

fault valleys are prominent today along the Bay of Fundy region in Nova Scotia, the Connecticut Valley, and other Triassic lowlands from New Jersey to North Carolina. Drilling through Coastal Plain sediments has revealed buried Triassic grabens as far south as Florida and in the Gulf Coast region of Texas. The infaulted Triassic rocks include arkosic conglomerates, sandstones, and shales that are largely red beds in the northern valleys and dark gray with occasional coal beds in the southern ones. These terrestrial sediments, nearly 6½ kilometers (4 miles) thick in places, represent coarse

alluvial fans near valley walls, with finer sediments of flood plains and lake bottoms beyond. Basaltic volcanism, accompanying the normal faulting, produced lava flows and shallow intrusive sills that are interbedded with the sediments. A striking example is the 300-meter (1000-foot) thick Palisades sill across the present-day Hudson River from New York City.

Following the Triassic epeirogenic rejuvenation of the Appalachian belt, the region stabilized and was eroded during the Jurassic. By early Cretaceous time, a relatively featureless surface called

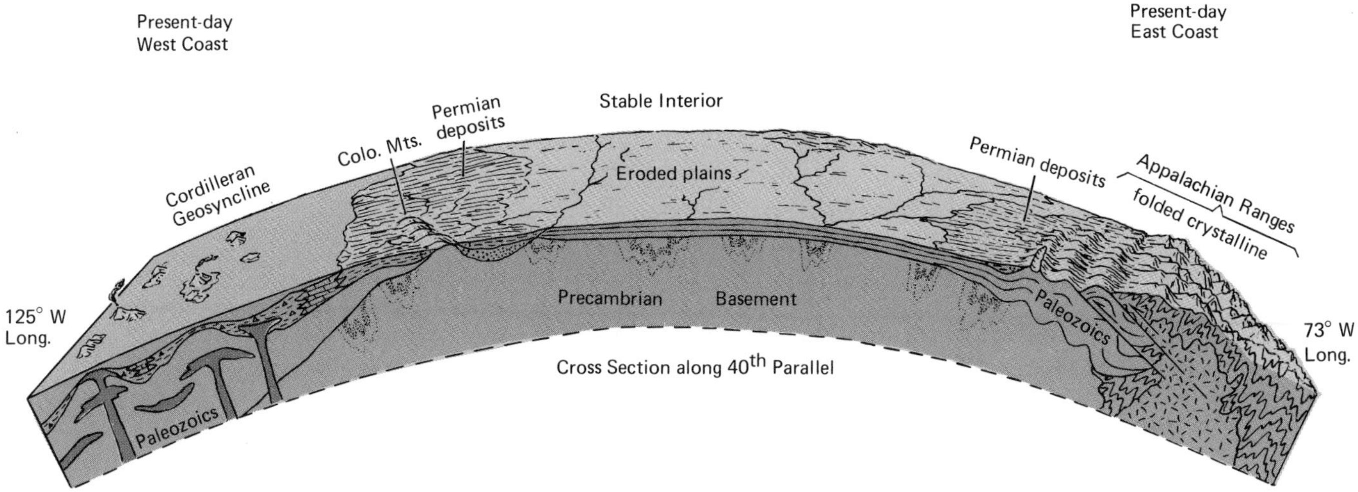

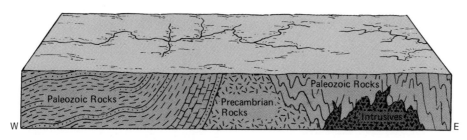

Fig. 15-7 Asymmetrical North America at the end of Paleozoic time.

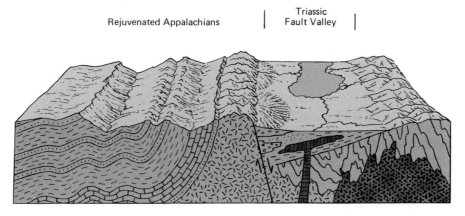

Fig. 15-8 Later history of the Appalachians. Triassic erosion surface beveling Appalachians. Based on D. W. Johnson, 1931.

Fig. 15-9 Diagram of late Triassic uplift (Palisades disturbance) with associated block faulting and volcanism. After D. W. Johnson.

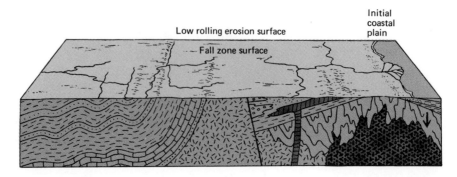

Fig. 15-10 Early Cretaceous erosion surface. Fall zone peneplain.

the Fall Zone peneplain was beveled across the Paleozoic mountain roots and the downfaulted Triassic blocks (Fig. 15-10).

Cenozoic Finale The Fall Zone peneplain was epeirogenically arched and eroded, except where it was protected by flanking sediments of the Coastal Plain to the east. Thus, in the early Cenozoic, another broad erosion surface, called the Schooley peneplain, which is well preserved in New Jersey, was carved across the uplift during an interval of relative stability (Fig. 15-11). This surface forms extensive plateaulike uplands on crystalline rocks from New England to the Blue Ridge, and is responsible for the even crests of the folded Appala-

chians (Fig. 15-12). The Schooley surface was, in turn, broadly uplifted. Rejuvenated streams incised it until an interval of mid-Cenozoic stability resulted in still another subdued erosion surface. It is developed on less-resistant rocks in the Appalachians, and is called Harrisburg surface from its prominence around the capital of Pennsylvania.

The Harrisburg surface was also uplifted and eroded, producing another still less extensive surface on the weakest rocks in the region. Called the Somerville surface, from Somerville, New Jersey, it is now being incised by present-day streams (Fig. 15-13). Thus Appalachian evolution after the orogenic phases of intense folding that destroyed the Paleozoic geosyncline has involved a series of

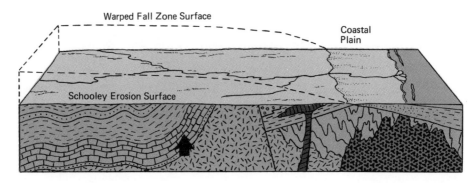

Fig. 15-11 Early Tertiary arching and erosion producing the Schooley erosion surface. After D. W. Johnson.

Fig. 15-12 View of early Cenozoic erosion surface, now uplifted and dissected in north-western Massachusetts. Photo taken from Whitcomb Summit on the Mohawk Trail. Courtesy of the U.S. Forest Service, photo by L. J. Prater.

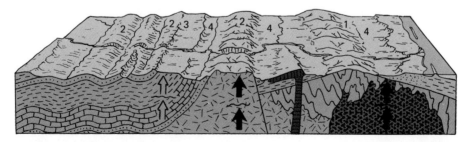

Fig. 15-13 Dissection of the Schooley surface and cutting of lower surfaces to produce present-day topography. (1) Fall zone surface; (2) Schooley surface; (3) Harrisburg surface; (4) Somerville surface. After D. W. Johnson.

broad epeirogenic uplifts with interspersed periods of erosion and stability—a history recorded in the Triassic grabens and cyclic erosion surfaces.

The Coastal Plain

While the Appalachian saga trailed off into an aftermath of gentle upwarpings and erosion, a new geosyncline was evolving as the Coastal Plain. It first appeared when the subdued mid-Mesozoic surface, beveling the Appalachian and Ouachita belts, began to subside along the continental margins. Thereafter ever-thickening sediments collecting in the downwarps built a great terrace whose emerged surface is the Coastal Plain.

Along the Gulf Coast—where subsurface relations are well known from intensive drilling and geophysical prospecting for oil—the basement has subsided to produce a trough filled with sediments as much as 16 kilometers (10 miles) thick beneath the present shoreline (Fig. 15-14). Geophysical evidence also indicates a basement trough beneath the Atlantic seaboard, where sediments are 3 to 5 kilometers (2 to 3 miles) thick. Despite the depth to the basement, none of the sediments in the troughs indicates especially deep water. Older continental clastics and shallow-water deposits (penetrated by deep oil wells) now lie buried thousands of meters below sea level. Thus this modern geosyncline, like its nonvolcanic counterparts in the ancient Appalachian and Cordilleran regions, has subsided slowly as its sedimentary wedges thickened.

Paleogeography At any given time in its history, the Coastal Plain surface had broadly similar geographic belts, as recorded in its sedimentary facies. Flood plains, swampy lowlands, and marshes lay

Fig. 15-14 Surface of the Gulf coastal plain, east of Sabine Pass, Louisiana. Old beach ridges form stripes parallel to the shore. Photo by John S. Shelton.

inland behind deltas and beaches along the shore. Offshore, beach and bottom sands merge into muds of progressively deepening waters. Sands, silts, and clays deposited in these environments were carried from the continental interior by existing streams and their predecessors, chief among them being the Mississippi. In clear waters, well away from turbid river mouths, limy muds, shell banks, and chalks were deposited.

The deposits reflect events in the headwaters of streams draining to the Coastal Plain. In the eastern Gulf Coast, for example, Mesozoic rocks are largely sands and other clastics, indicating strong erosion of the Appalachians; whereas Cenozoic rocks are dominantly limestones. In the western part of the Gulf Coast, in Texas, limestones are the most abundant Mesozoic rocks, and the overlying deposits are largely sandstones and shales, reflecting the active rise and erosion of the Rocky Mountains during Cenozoic time. Coastal Plain sediments also reflect fluctuating shorelines, as the sea flooded and ebbed across the continental margins during Mesozoic and Cenozoic time.

Shifting Seas   The oldest Coastal Plain strata are subsurface Jurassic rocks known only from drilling along the Texas-Louisiana coast. These red beds and salt deposits lie unconformably over the eroded surface of the Ouachita orogenic belt. The thick salt beds are thought to have precipitated from highly concentrated saline waters, periodically renewed by flooding into enclosed basins along a desert coast. The salt has a special interest, for it eventually punched upwards through great thicknesses of younger sediments to form *salt domes,* great subterranean columns thousands of meters in vertical extent. The salt domes, aside from providing a commercial source of salt and associated sulfur, have created some very productive oil-bearing structures.

Late in the Jurassic, seas from the Caribbean flooded over the red bed sequence in the first of some eight advances, each complicated by lesser fluctuations. The Jurassic sea spread as far north as Arkansas before retreating. In early Cretaceous time, the seas returned, spreading as far inland as Kansas and southeastern Colorado. Their withdrawal left a marked erosion surface preserved in an unconformity with late Cretaceous rocks. The late Cretaceous transgression was one of the greatest floodings of North America. It spread Coastal Plain sediments far across the continental margins and created a great north-south embayment across the stable interior as well.

After the Cretaceous deluge ebbed, the Paleocene-Eocene flooding was the most extensive. Middle and later Cenozoic transgressions of the Coastal Plain each fell short of its predecessor. But while the extent of the marine transgressions declined, the sites of thickest deposition shifted southward, building the continental shelf progressively outwards into the Gulf of Mexico.

Summary   We have concentrated mostly on the Gulf Coast region. The Atlantic Coastal Plain differs mainly in having a thinner sedimentary wedge, no salt deposits, no known strata older than early Cretaceous, and—influenced by the warpings of the Appalachian axis to the east—strata that extend less far inland and contain more terrestrial clastics. Overall, however, the Atlantic and Gulf Coastal Plains and their underwater extensions, the continental shelves, are a modern geosyncline. But, unlike its Appalachian and Cordilleran predecessors, the Coastal Plain geosyncline lacks an outer belt of volcanic islands and deeply subsiding basins where graywackes, lavas, and pyroclastics are accumulating. This may merely mean that there are geosynclines and geosynclines.

CORDILLERAN EVOLUTION

To present the highlights of Cordilleran evolution, we must go back in space and time. Towards the end of the Paleozoic, North America as a whole had lost its earlier symmetry. The Appalachian and Ouachita geosynclines had been crumpled into mountain belts east and south of the stable interior. Western North America in the late Paleozoic was still occupied by seas and geosynclines—although here too the paleogeography had changed considerably from that of the early Paleozoic. If differing in many aspects and in the timing of events, the final fate of the Cordilleran geosyncline broadly mirrors the Appalachian pattern.

Devonian to Mid-Triassic

Until late Devonian time, the Cordilleran geosyncline remained, as it had been in the Cambrian and Ordovician, a generally subsiding trough with two north-south trending belts. Well-washed sandstones, shales, and limestones were deposited in the eastern miogeosyncline and adjacent continental platform. Graywackes and other clastics, interspersed with basaltic flows, accumulated in the restless eugeosyncline farther west. Here, in what is now the Pacific border region, basins subsided deeply among rising volcanic archipelagos.

Deformation in the Eugeosynclinal Belt Starting in late Devonian time and lasting into the Mississippian, the western part of the Cordilleran geosyncline was deformed during the *Antler orogeny* (Fig. 15-15). Structurally, this event, which may

have affected most of western North America, is best known from studies in central Nevada where eugeosynclinal rocks were thrust-faulted eastward over miogeosynclinal rocks. This orogeny probably produced a tectonically active island chain whose members periodically rose above sea level and were eroded; eventually this orogenic belt was eroded to a platform beneath shallow seas. The history of the Antler event is recorded in unconformities and coarse clastic debris shed eastward into waters of the adjacent nonvolcanic miogeosyncline. West of the island arc, graywackes and volcanics continued to accumulate in a eugeosynclinal belt.

After an interval of relative quiescence, the western part of the Cordilleran region again became tectonically active during the *Sonoma orogeny*. During this episode of Permian and early Triassic time, recorded in rocks in western Nevada and California, eugeosynclinal deposits along the west side of the Antler orogenic belt were deformed and the region became emergent. Thus, in early Mesozoic time, the Cordilleran region still contained a broad miogeosyncline flanking the continental platform; however, the Antler and Sonoma orogenies in the outer part of the Paleozoic geosyncline represented preludes to the coming eventful history of the region.

The Colorado Mountains East of the region of thick geosynclinal deposition, the western United States underwent an episode of mountain-building

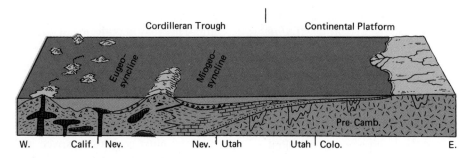

Fig. 15-15 Early stages in Antler orogeny, arch emerging in late Devonian time from the Cordilleran geosyncline.

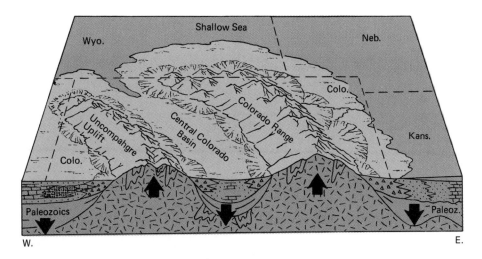

Fig. 15-16 The Colorado Mountains, later destroyed by erosion, orginated in a late Paleozoic orogeny in the general area where, much later, the southern Rocky Mountains would develop.

that centered in Colorado, mainly in Pennsylvanian time. The orogeny may have been linked to the contemporaneous mountain-building in the Ouachita geosyncline. However, unlike the Appalachian-Ouachita and subsequent Cordilleran mountains—which involved more typical thrusting and folding of thick geosynclinal rocks—the late Paleozoic Colorado Mountains arose from the sialic continental platform where Paleozoic rocks were relatively thin or absent. Thus Precambrian basement was soon exposed in these linear uplifts which shed aprons of terrestrial debris into intervening and adjacent downwarps (Fig. 15-16). The resulting sediments—represented by the Fountain Formation in the Red Rocks west of Denver—formed arkose, a sandstone rich in feldspar as well as quartz fragments, indicating rapid erosion of a rugged granitic terrain and rapid deposition, so that the feldspars were not weathered to clay.

These late Paleozoic mountains have sometimes been called the "Ancestral Rockies," which is a bit misleading because the present-day Southern Rockies, although in the same region, are much later features. After the main uplift in Pennsylvanian

time, leveling of the Colorado Mountains is indicated by fine-grained red beds that progressively buried the uplifts and flanking arkoses. Thereafter Mesozoic seas left thick marine deposits over the site of the Paleozoic ranges, before the existing Rockies emerged during the final orogenic phase in the destruction of the Cordilleran geosyncline.

Expanding Orogeny

Cordilleran geography began to change markedly in early Jurassic or, possibly, late Triassic time. Initially, broad warping sent a long, low peninsula northward from lands in southern Arizona and New Mexico, through Nevada and Idaho, and well into British Columbia. Except for northern Canada, the Cordilleran geosyncline was split in two. Thereafter the structural evolution of the western United States involved a western trough, called the *Pacific Coast geosyncline;* an eastern downwarp, the *Rocky Mountain geosyncline;* and the intervening barrier of land and mountains, the *Cordilleran geanticline.* We will return to the complicated story of the Pa-

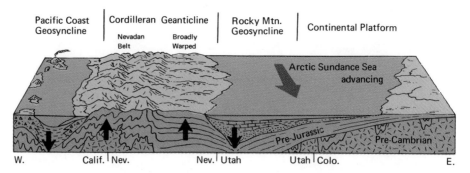

Fig. 15-17 Late Jurassic orogeny divided the earlier Cordilleran geosyncline into two separate troughs.

cific Coast geosyncline after discussing the evolution of the Cordillera to its east (Fig. 15-17).

The word *geanticline* originally meant a broad upward flexure of the Earth's crust wrinkled by minor subsidiary folds—a complementary feature to a geosyncline. At certain times and places the Cordilleran geanticline may have been such, but at others it was intensely folded and faulted. It is now generally considered a complex deformational belt whose broad warps and contorted mountains were notable sources of sediments during active uplift.

Although differing in details, the fate of the Cordilleran geosyncline broadly mirrors the Appalachian pattern. Starting well offshore, the great sedimentary trough collapsed during a protracted spell of mountain-building pulses that progressively enlarged the geanticlinal region as they gradually migrated to the continental platform.

Although some geologists consider particular orogenies as including all types of deformation and mountain-building during certain times and in certain regions, we shall take a different and perhaps simpler approach. The Mesozoic orogenies that converted the Cordillera into a broad and complex mountain region can be separated into three phases, based on the style of deformation. Although they overlap somewhat in time, a generally earlier group of spasms in the western part of the Cordillera can

be collectively called the *Nevadan orogeny.* Pulses farther east, across what was to become the Basin and Range country of Utah and adjacent states, are considered the *Sevier orogeny.* Farthest east, in Colorado, Wyoming, and some places in Montana, the style of deformation characterizes the *Laramide orogeny.*

Nevadan and Sevier Developments Nevadan orogeny began in Jurassic time and strongly affected western Nevada and adjacent California. Jurassic and older rocks in what had been the eugeosynclinal part of the Cordilleran geosyncline were intensely folded, faulted, metamorphosed, and intruded. Today such rocks are exposed in the Klamath and Sierra Nevada ranges.[4]

Although called Nevadan, this orogeny was no local affair. From Alaska to Mexico, and as far east as Idaho, the older Cordilleran geosyncline crumpled into mountain ranges, invaded at depth by extensive granite batholiths. Nor was the Nevadan orogeny a simultaneous spasm of short duration. From Jurassic until about the middle of Cretaceous time, shifting sites of deformation migrated generally eastward through Nevada to western Utah and across Idaho into western Montana.

4 The great tilted fault block of the Sierras is a much later development, but the rocks within it are deformed by the late Mesozoic orogeny.

The Sevier orogeny was under way in easternmost Nevada and western Utah before the granitic intrusions associated with Nevadan orogeny had ceased. Sevier orogeny began in late Jurassic time (perhaps earlier) in western Utah and adjacent regions. Throughout Cretaceous time, and into the earliest Cenozoic, Sevier-type deformation migrated eastward. It affected the thick sequences of miogeosynclinal rocks across the Cordilleran region. The sandstones, shales, and limestones were compressed and broken into great thrust sheets. These structures are great multiple slabs containing intricately folded strata in which older rocks have generally been thrust over younger along low-angle faults. Such *imbricate* structures, which are found in many of the world's folded mountain belts, are now exposed in an almost unbroken mountainous terrain extending southward from Alaska through the Canadian Rockies, Montana, westernmost Wyoming, and adjacent parts of Idaho and Utah (Fig. 15-18).

The Rocky Mountain Geosyncline In early and middle Jurassic time, when North America was largely emergent, a narrow arctic sea crept southward through Canada east of the geanticlinal belt. By the beginning of late Jurassic time, while the Nevadan and Sevier orogenies were active far to the west, waters of this *Sundance Sea* advanced far southward into the subsiding Rocky Mountain geosyncline and widely flooded the western margins of the continental platform. Sandstones, shales, and

limestones of the Sundance Sea are 3000 meters (10,000 feet) thick in Utah where the geosyncline subsided deeply, but thin towards their shoreline margin on the continental platform (Fig. 15-17).

Later in the Jurassic, the sea withdrew. The geosyncline and adjacent platform emerged as a vast swampy lowland traversed by streams heading in the active geanticlinal belt to the west, and in highlands rising across Arizona and New Mexico to the south. The sands, silts, and clays deposited across the lowland comprise the *Morrison Formation*—a world-renowned burial ground for Jurassic reptiles, including the great brontosaurs, and a host rock for major uranium deposits as well.

Conditions of latest Jurassic time continued into the earliest Cretaceous, but then seas reinvaded the Rocky Mountain geosyncline, depositing especially thick sediments in deeply subsiding pockets marginal to the continental platform, in Utah and western Wyoming. These early Cretaceous seas eventually withdrew, northward into the Arctic and southward towards the Gulf Coast.

During the late Cretaceous, the seas returned in the greatest inundation of North America since the Ordovician deluge. They flooded the Rocky Mountain geosyncline and spread broadly across the stable interior, dividing North America into a broad low island in the east where parts of the Canadian shield and eroded Appalachians remained above water, and the mountainous geanticlinal belt to the west. While mountains rose in the expanding Sevier belt, the Rocky Mountain geosyncline warped downwards, notably along the margin of

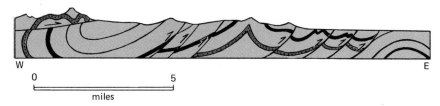

Fig. 15-18 West-to-east cross-section a few miles south of Frank, Alberta, showing multiple thrust faulting and folding, typical of Rocky Mountains, developing from deeper parts of geosyncline. Generalized from B. R. MacKay, Canadian Geological Survey, 1932.

the continental platform, and on the platform itself in south-central Wyoming. Late Cretaceous sediments, deposited in the unevenly subsiding basins, have a rather constant facies pattern at any given time.

Along the foot of the mountains rising to the west lay coarse bouldery fans that grade into fine-grained sands, silts, and clays of a swampy forested lowland (Fig. 15-19). Farther east, the continental clastics are replaced by marine deposits: beach and near-shore sands of the transitional zone which, in turn, grade into muds of the sea floor. At times limestone accumulated in clear waters towards the low-lying interior of the continent.

These facies form a great clastic wedge shed by the expanding Sevier mountain belt to the west. While sediments collected in the subsiding Rocky Mountain geosyncline, the fluctuating shore retreated eastward, so that continental deposits progressively overlap marine beds towards the continental interior. As the Mesozoic era drew to a close, the final collapse of the Rocky Mountain geosyncline was imminent, for to the west continuing orogeny had largely converted the site of the Paleozoic Cordilleran geosyncline into an uplifted mountainous belt.

The Laramide Culmination   Geosynclinal deposition ended east of the Pacific Border region towards the end of Cretaceous time, when the Rocky Mountains arose in the final phases represented by the Laramide orogeny. Laramide structures developed where the sialic continental platform was deformed. They are characterized by vertically uplifted, broad, flat-topped anticlines that are separated by broad, downdropped, synclinal basins. These ranges—rather unique as mountain structures—extend from southern Montana through central and eastern Wyoming into Colorado, and as far south as Santa Fe, New Mexico. Let us focus on the history of these ranges in Wyoming, as a sample.

The broad folds first appeared as islands, emerging from the late Cretaceous seas some 70 million years ago. They divided the earlier wide area of Cretaceous deposition in the Rocky Mountain geosyncline into local basins receiving sediments eroded from the rising folds. Close to the upwarps, coarse clastics and unconformities marked the spasmodic rise, but in the centers of the basins deposition was continuous. Thus no all-embracing unconformity separates Mesozoic and Cenozoic rocks, and the time change from one era to another can be determined only from fossils.

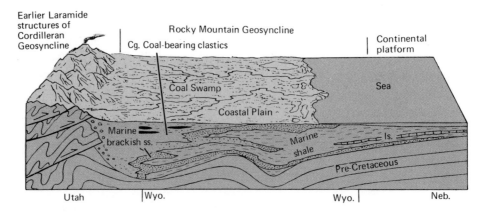

Fig. 15-19 Diagram of late Cretaceous conditions with the subsiding Rocky Mountain geosyncline bounded on the west by Sevier mountains.

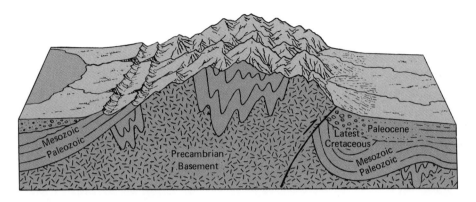

Fig. 15-20 Diagram of the type of early Cenozoic (Paleocene) mountain that emerged in the eastern part of the Rockies from Montana southward through Wyoming and Colorado. After diagrams by S. H. Knight.

As the eastern ranges grew, they were eroded. Thus the youngest Cretaceous rocks of the Rocky Mountain geosyncline were the first to be stripped from the anticlinal crests and were redeposited in the newly formed local basins to form the latest Cretaceous clastics. Thereafter older Mesozoics and Paleozoics were successively exposed in the uplifts until, finally, abundant granitic and metamorphic fragments of earliest Cenozoic age show that the ranges had been breached to their Precambrian cores (Fig. 15-20). As a rule then, the oldest rock fragments made up the youngest conglomerates.

The last vestiges of the seas disappeared from the Rocky Mountain region sometime in the earliest Cenozoic. Marine deposition was wholly replaced by alluvial fans, flood plains, coal swamps, and (in western Wyoming and northern Colorado) some exceedingly large interior lakes. Though the mountains then were probably as rugged as the present-day Rockies, the region as a whole lay near sea level, for early Cenozoic plant and animal fossils document a wet subtropical climate like that of the Gulf Coast region today. The present-day dry Rocky Mountain climate reflects the later evolution of the Cordillera.

Orogenic Aftermath

North America regained its symmetry during the Nevadan, Sevier, and Laramide orogenies, which destroyed the earlier western geosynclines. The continental platform had expanded by consolidation of the formerly mobile geosynclinal belts, but the present-day Cordilleran provinces did not emerge until well into Cenozoic time. Then, in the orogenic aftermath when seas were gone, epeirogenic deformation involving broad upwarping with high-angle faulting and volcanism produced the Rocky Mountains, Colorado Plateau, Columbia Plateau, and the Basin and Range. With this introduction to later evolution of the Cordillera, let us continue the eastern Rocky Mountain story, and then discuss the other regions.

The Rocky Mountains Laramide folding and thrust faulting died out towards the end of the Eocene. By then the Rockies had a rather modern look. Existing mountain ranges were established, and local relief—from valley floors to mountain peaks—was impressive. But the region remained close to sea level and had a humid climate. The

Fig. 15-21 Water-laid volcanic conglomerates and tuffs of Eocene age in Absaroka plateau southwest of Yellowstone Park. Photo by Herb Pownall, courtesy of J. D. Love.

final shaping of the region during an epeirogenic phase of uplift was yet to come.

Through Oligocene time especially, great clouds of ash from prolonged eruptions to the west drifted into the Rocky Mountains and Great Plains. The exact sources of the ash are not everywhere known, but Cenozoic volcanism occurred at various places in the Rockies—a notable example being in the Yellowstone Park region of northwestern Wy-

oming and adjacent states (Fig. 15-21). Although its eruptions seem finished, the Yellowstone is still hot below the surface as indicated by its renowned geysers and other hot-water phenomena. From Eocene until Recent geologic time, however, active eruptions in the Yellowstone region built great lava plateaus, and could well have provided enormous amounts of windblown ash during the Oligocene. Whatever its source, the ash clogged streams, caus-

Fig. 15-22 Air view of the Granite Mountain region in central Wyoming bears similarity to the late Tertiary landscape in the region. Courtesy of Paige Jenkins and J. D. Love.

ing them to aggrade and raise the levels of basin floors while back-filling mountain valleys.

The Rocky Mountains and Great Plains region began to rise gradually sometime in the mid-Cenozoic. Unlike the preceding Laramide deformation, involving the sharp folding and thrust faulting characteristic of orogeny, the mid-to-late Cenozoic uplift was epeirogenic—a gentle regional arching accompanied, in the Rockies, by some normal faulting. In places cobbles and boulders incorporated in a matrix of reworked ash show that the highlands of the Rockies had been washed clean and were being eroded.

Towards the close of the Tertiary period, the Rocky Mountain region as a whole had risen well above sea level, and the rugged local relief of earlier times was gone (Fig. 15-22). Only scattered hills rose above a subdued and rolling landscape, partly cut across resistant rocks in Laramide folds and partly built up on thick basin fills. The surface across the Rockies merged with the Great Plains, a vast eastern apron of volcanic ash and debris eroded from the mountains (Fig. 15-23).

At the beginning of Quaternary time, or perhaps a bit before, the long Tertiary episode of basin-filling ended. Streams were rejuvenated, possibly because the cooler, wetter Pleistocene climates swelled stream volumes. In any event, streams scoured out the broad basins, reexposing the once-buried mountain fronts, and leaving the late Terti-

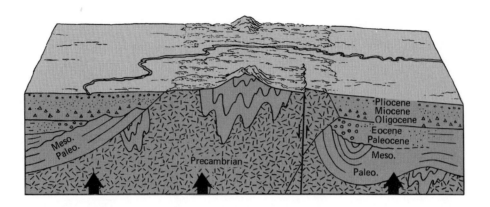

Fig. 15-23 Diagram of late Tertiary (Pliocene) conditions in central and southeastern Wyoming with surface of low relief reflecting filling of intermontane basins and reduction of mountains. After S. H. Knight.

ary surface preserved on resistant rocks as a plateaulike upland with scattered higher peaks (Fig. 15-24).

Normally, streams head near divides and flow outwards to adjacent lowlands, but many streams in the Rockies are unique in crossing major mountains by way of deep canyons. Notable examples are the Flaming Gorge, cut by the Green River through the Uinta uplift; the canyon of the Wind River through the Owl Creek Mountains; and the Royal Gorge of the Arkansas in the Colorado Front Range (Fig. 15-25). Such special cases represent *superposed* streams whose transverse courses reflect the later Cenozoic history of the region. Streams on the subdued late Tertiary surface wandered through sediment-filled divides on buried Laramide structures. Then, during the Quaternary excavation of the basins, the streams were "let down" across the exhumed folds and maintained their earlier courses by carving the transverse canyons.

Thus the Rocky Mountains of today are essentially Laramide structures exhumed from a Tertiary burial—but with additions. The normal faults reflect the last broad uplift of the region. High-level erosion surfaces and superposed streams are inherited from the great interval of cut-and-fill (Fig. 15-26).

The Colorado Plateau   The Colorado Plateau has long been a conservative block of the continental platform. Never a geosynclinal region after the Precambrian, its sedimentary veneer is relatively thin—a few thousand meters (or yards) thick.

Through the Paleozoic era, the region was washed by shallow seas migrating from the Cordilleran trough to the west, periodically emergent and eroded, and at times mantled by terrestrial red beds. In the Mesozoic era, red beds and dune sands of Triassic and Jurassic deserts were succeeded by somber-colored Cretaceous sandstones and shales, signaling the rise and erosion of mountains in the surrounding Cordillera.

During the time of the Laramide orogeny, the Plateau developed long, flat-topped uplifts which are separated by locally sharp, monoclinal flexings from broad intervening basins (Fig. 15-27). In structure as well as age, these great upwarps resemble the broad-backed anticlines in the Southern

Fig. 15-24 Hogback ridges and Rocky Mountain front in the Pikes Peak region of Colorado, as seen from the lowland excavated east of the mountains in Quaternary time. An 1866 lithograph from the Library of Congress Collection.

Rockies. However, the deformation and uplift in the Plateau having been less intense, subsequent erosion has not exposed Precambrian basement (except in the deep inner gorge of Grand Canyon).

Prolonged denudation after the monoclinal flexing stripped weaker Mesozoic rocks from the uplifts. The debris, mingled with that from the Rockies rising to the north and east, was laid down as ter-

restrial basin deposits, unconformably overlapping the warped margins of the uplifts. As in the Rockies, fossil plants and animals indicate a low, humid region until mid-Cenozoic time; then an arid climate developed as the region was epeirogenically uplifted.

High-angle faulting broke the Plateau during its final rise (Fig. 15-28). In places volcanic fields

Fig. 15-25 Canyon cut by superposition of the Wind River across the Owl Creek Mountains of Wyoming. Photo by Paige Jenkins.

erupted onto the surface. A notable example is near Flagstaff, Arizona, where great eroded strato-volcanos of the San Francisco Peaks are surrounded by a swarm of smaller cinder cones, ash beds, and extensive lava flows. Streams, rejuvenated in the general uplift, excavated basin deposits and cut many deep canyons, including the Grand Canyon of the Colorado across the Kaibab monocline. Jagged canyons, great retreating escarpments, relic buttes, and mesas, all carved from colorful strata, as well as stocks and laccoliths, volcanos, and lava flows of dark rock, today make the generally desolate Colorado Plateau a geologic showplace (Fig. 15-29).

The Basin and Range Province Meanwhile, back where it all began, the protracted storm of Nevadan-Laramide orogeny had subsided. West of the Rockies and Colorado Plateau, a generally mountainous region of crumpled geosynclinal rock was being eroded (Fig. 15-30). Relic patches of latest Mesozoic and earliest Cenozoic deposits are largely scattered valley fills, indicating that most of the mountain debris was flushed from the region by streams flowing to the sea.

During the early Cenozoic, the mountains were generally worn low. A subdued upland surface was beveled across the deformed and intruded Nevadan

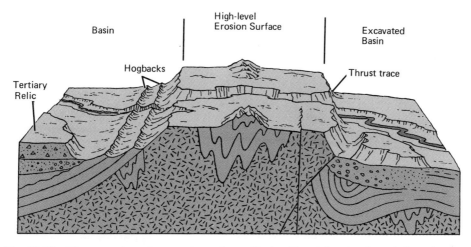

Fig. 15-26 Diagram of eastern and southern Rocky Mountain structure today showing excavated basins, relic erosion surface on mountain core, and superposed stream. After S. H. Knight.

Fig. 15-27 Monoclinal flexure in Pennsylvanian rocks of the Colorado Plateau near Mexican Hat, Utah. Photo by Tad Nichols.

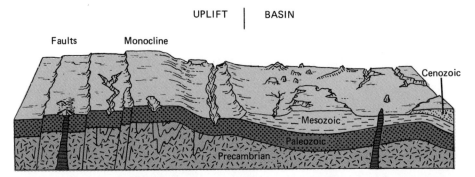

UPLIFT | BASIN

Faults   Monocline

Cenozoic

Mesozoic

Paleozoic

Precambrian

Fig. 15-28 Diagram of Colorado Plateau features. After Powell, 1876.

rocks to the west (Fig. 15-31). Tuffs, lavas, and agglomerates erupted upon this surface in Miocene time. To the east, flood plain, swamp, and lake deposits accumulated near the actively rising Rockies of early Cenozoic time.

In late Cenozoic, mid-Miocene, or perhaps

Fig. 15-29 John Wesley Powell (1834–1902). A one-armed Union veteran, Powell was a leading figure in the heroic period of American geology after the Civil War. His two boat trips down the Colorado River and field work in Utah and Arizona contributed greatly to geologic knowledge of the region and geologic processes in general.

earlier time, normal faulting began blocking out the characteristic Basin and Range structures. Although the Miocene horsts are now leveled by erosion, their terrestrial debris of thick bouldery fans overlain by clays and silts with salt and gypsum resembled deposits in modern grabens. Active block faulting of the region has shifted from place to place with time, so that the present-day mountains show varying stages of erosion, ranging from fresh fault blocks to low relic hills (Fig. 15-32). Grabens between the horsts form broad basins largely filled with coarse clastics shed from the mountains. Where the uplifted blocks have long been stable, the basin margins are broad pediment surfaces eroded at the expense of the mountains.

The Columbia Plateau   Tremendous lava floods erupted in the Cordillera north of the Basin and Range from the Miocene onwards. Flow after flow of fluid basalt emerging from narrow fissures inundated all but the highest peaks of a rugged topography having up to 750 meters (2500 feet) of relief and cut across granite and metamorphics of the Nevadan orogeny (Fig. 7-13). The Blue Mountains of northeastern Oregon are Nevadan islands, partly upwarped, which stand above the volcanics, as do the Steptoe Buttes, much smaller masses of granite marginal to the Northern Rockies in eastern Washington.

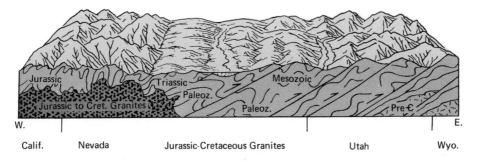

Fig. 15-30 Diagram of late Mesozoic mountainous region of the Nevedan and Sevier deformation where the Basin and Range Province was later developed.

Fig. 15-31 Reduced topography during early Cenozoic time in Basin and Range Province, accompanied by extensive andesitic volcanism.

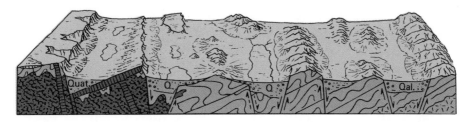

Fig. 15-32 Diagram of late Cenozoic block faulting producing the present Basin and Range topography.

Clay, sand, and gravel beds are sandwiched with the lavas. Between spasmodic eruptions, streams flowing across the surface were dammed by lava tongues to create lakes that persisted until buried by renewed lava outpourings.

Although largely flat-lying, the plateau's strata were warped after most of the lavas had erupted. Parts of the region broadly subsided, perhaps from loss of mass erupted from the depths, or perhaps under the load piled on the crust. However, other

Fig. 15-33 Basin and Range structure with sand dunes in foreground near Last Chance Range, Nevada. Photo by John H. Maxson.

parts rose, forming broad flexures towards the east and sharper anticlines near the Cascade Ranges, along the western margin of the Plateau province.

Youthful streams, in places cascading over retreating Niagara-type falls, flow through many spectacular canyons whose walls expose the plateau's structure. By far the deepest is Hell's Canyon, an 1800-meter (6000-foot) chasm cut by the Snake River through somber volcanics and deep into the underlying granite. As in the Rockies, stream valleys sometimes cut across uplifted structures. However, such streams in the Columbia Plateau are probably *antecedent*. That is, they were established before the plateau surface was warped and have maintained their courses by cutting canyons across the actively rising folds.

Today erosion prevails in the Columbia Plateau, and the igneous episode seems at an end—although the "Craters of the Moon," a line of prehistoric cinder cones with associated lava flows on the youthful Snake River Plateau of Idaho, seem fresh enough to have erupted yesterday. Other major developments in the plateau are discussed later.

West Coast Evolution

Let us go back to the region west of the Cordilleran geanticline.

The Pacific Coast Geosyncline   While late Jurassic and subsequent events were shaping the inner Cordillera, a restive volcanic geosyncline subsided unevenly along the Pacific border. Thus eastern and western North America were comparable, from the Jurassic onwards, in having new marginal downwarps, but contrasted in that the Pacific Coast geosyncline was a less stable volcanic belt.

From late Jurassic through early Cenozoic time, clastic sediments shed from the Nevadan ranges, along with cherts and submarine volcanics, built strata tens of kilometers (or miles) thick on the subsiding floor of the Pacific Coast geosyncline. The region can be imagined as a coastal plain, building from Nevadan highlands into a broad seaway that merged with the Pacific Ocean beyond. The seaway remained a site of shifting volcanic archipelagos and rapidly subsiding basins whose sediments and volcanics, with interspersed unconformities, attest an unstable geosyncline.

Later evolution of the Pacific Border is most easily treated as northern and southern parts.

The Pacific Mountain System (Oregon and Northward)   During early Cenozoic time, the subsiding geosyncline collected thousands of meters of rock, including deltaic deposits to the east and marine clastics and volcanics to the west. In Miocene time, basalts of the Columbia Plateau spread westward, interfingering with sediments in the future site of the Coast Ranges. The modern ranges emerged in the late Cenozoic. A late Miocene phase of deformation folded the geosynclinal rocks into anticlines, trending northward through Oregon and northwest to west in northern Oregon and Washington. These folds were then eroded to a low rolling surface during the Pliocene.

The existing north–south grain of the region appeared in late Pliocene or early Pleistocene time. The earlier erosion surface was broadly arched to form the Cascade and Coast ranges, and, with an intervening downwarp, the Puget Lowland. The Cascade erosion surface attained its present elevation by the early Pleistocene. Upon this surface, andesitic eruptions built such towering strato-volcanos as Mounts Baker, Rainier, St. Helen, and Hood, as well as Mount Mazama, whose decapitation formed the scenic caldera of Crater Lake (Figs. 15-34, 15-35). In the Klamaths the Nevadan basement was broadly upwarped in late Cenozoic time, fencing off the northern downwarp from the interior valley in California.

The Pacific Mountain System (California)   Broad-gauged reconstructions of the paleogeography and geologic evolution of California are—to say the least—controversial. We start where the story seems straightforward.

The Sierra Nevada region has shed sediments into the subsiding Pacific Coast geosyncline since the Nevadan orogeny. The Nevadan orogeny was followed by a late Cretaceous mountain-building episode in the Sierras, when much of the Jurassic granite was digested by renewed intrusions into the mountain roots. During early Cenozoic time, the once-lofty ranges were eroded to a low rolling upland, exposing metamorphic pods and granitic masses formed, at depth, in the preceding orogenies.

In the later Cenozoic, the Sierra region was spasmodically upwarped, so that streams carved several generations of cyclic valleys into the earlier upland surface. However, the Sierras were not yet impressively high; floras and faunas in the valley deposits indicate a lowland environment. The climactic uplift began in the Pliocene and continued through the Quaternary. In this phase the Sierras have risen as a great west-tilted fault block, now reaching 4419 meters (14,497 feet) in the pinnacle of Mount Whitney.

West of the Sierras, in the Great Valley region, a floor of Nevadan basement subsided unevenly

Fig. 15-34 Mount Rainier, a dissected Cascade volcano with extensive active glaciers. Courtesy of the U.S. National Park Service.

during Cenozoic time. Local basins trapped deep-water sediments, notably during the Miocene and Pliocene. Late Tertiary sediments, which reflect shallower water, with time give way to Quaternary terrestrial deposits, in places 2700 meters (9000 feet) thick. Instability of the region during deposition is shown by abundant unconformities, culminating in a mid-Pleistocene episode of folding after which nearly horizontal deposits lapped across the eroded edges of upturned strata. The evolution of the Great Valley is closely related to that of the adjacent Coast Ranges.

The problems of interpreting California's geologic history come to a head in the Coast Ranges.

Unlike their counterparts in Oregon and Washington, which are made up largely of Cenozoic rocks, the California Coast Ranges expose considerable metamorphic and intrusive igneous rock. This basement includes two distinct assemblages. One is a crystalline group of highly metamorphosed schists, quartzites, gneisses, and marbles intruded by granite bodies. Lithologically, the metamorphics resemble rocks in southern California, rather than those to the east across the Great Valley in the Sierras. From potassium-argon dating, granites near San Francisco are late Cretaceous—indicating an orogeny contemporaneous with the one producing much of the Sierran granite (Fig. 15-36).

Fig. 15-35 Crater Lake, showing Wizard Island, a cinder cone, with caldera rim in background. U.S. Forest Service photo by L. J. Prater.

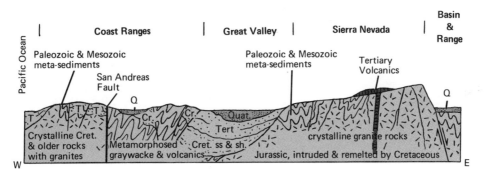

Fig. 15-36 Diagrammatic cross-section through California south of San Francisco Bay. Based on Reed and Hollister, 1936; Hoots, Bear, and Kleinpell, 1954; Taliaferro, 1951; and Jenkins's California Geological Map, 1937.

The other prominent basement group consists of moderately metamorphosed, although strong, deformed graywackes, shales, and cherts, along with interbedded basaltic flows—all of which are characteristic of the eugeosyncline. Although largely unfossiliferous, hence difficult to date, this rock assemblage apparently includes Paleozoic and Mesozoic strata; so some at least are older than the crystalline basement group. Yet younger granites nowhere intrude the graywacke-basalt group. Where adjacent, the two groups are separated by faults.

Sedimentary strata associated with the basement rocks may change markedly in thickness and facies in a few miles. The rocks in neighboring fault blocks may be markedly different, indicating seemingly unrelated histories of uplift and subsidence, erosion and deposition. Structurally, the Coast

Fig. 15-37 Branch of San Andreas fault (looking northwest) near Indio, California. The low hills (lower left) are structurally different from the main hills (right of fault) and came from the south by horizontal fault movement. Courtesy of Spence Air Photos.

Ranges are a jumble. Complex folds, normal and reverse faults abound. Most impressive is the strike-slip fault system dominated by the San Andreas rift. This major crustal fracture slices 800 kilometers (500 miles) southeastward from Tomales Bay north of San Francisco, through the Coast Ranges, and thence into the Gulf of California (Fig. 15-37).

Until quite recently, most geologists assumed that the Coast Ranges evolved where they are today and that their deformation involved largely vertical movements. To some workers, the crystalline basement group represented the roots of pre-Jurassic landmasses extending into California as islands, or peninsulas, from the Pacific. Independent geosynclines between the land masses received

Fig. 15-38 Movement on San Andreas fault which carried splinters of Sierra block northward. Based on Curtis, Evernden, and Lipson, 1958.

sediments both from the western lands and the Sierra region to the east.

By another interpretation, the geosynclinal group of volcanics and graywackes was deposited on subsiding basement of the crystalline group in a continuous sea west of the Nevadan orogenic belt. The local uplifts and basins, in this theory, did not originate until the late Cretaceous orogeny. Thereafter the eugeosynclinal rocks were stripped from the uplifts, exposing the crystalline basement group. In light of later evidence, all such interpretations have two flaws. The relative ages of the two basement groups are wrong, and the San Andreas and kindred faults are assumed late developments of relatively unimportant horizontal displacement.

Recent thinking emphasizes extensive slippage along the strike-slip faults. Some California geologists now suggest that during the late Cretaceous orogeny, the crystalline basement group actually originated in the southern part of the Nevadan belt, now extending through the Sierras. Then, 80 million years ago (still in the late Cretaceous), movement began on the San Andreas rift, and basement rocks south of the present Sierra block gradually slid some 480 kilometers (300 miles) to the vicinity of San Francisco.[5]

Thus the Coast Range crystalline rocks are splintered mountain roots jammed far northwestward into volcanic geosynclinal deposits (Fig. 15-38). The suggestion—staggering at first thought—actually simplifies California paleogeography, for by it the late Mesozoic orogenies would have been restricted to a single north-south belt of the present Sierra axis, which was then bordered by a single continuous geosyncline on the west.

Not all geologists believe in such a great horizontal displacement, but if the known rate of shift on the San Andreas fault in historic time, averaging 6 meters (20 feet) per century, is projected back 80 million years, a 480-kilometer (300-mile)

5 If movement continues, Los Angeles may eventually come abreast of San Francisco.

displacement is quite possible. Most geologists would accept horizontal movement of at least 50 kilometers (30 miles) since Miocene time, for the problem of the markedly different Tertiary rocks in now-adjacent fault blocks can be neatly solved by mentally sliding blocks back. The similar rock types and ancient shore lines (which make especially nice reference lines) can be matched across the strike-slip faults. As you might suspect, progressively older deposits do require increased amounts of "backsliding" for a match.

The complicated Cenozoic geology of coastal California seems to reflect the rising and subsiding, warping and breaking of blocks jostled by continuing movement along one of the world's great strike-slip fault zones. Cenozoic deformation had two peaks: one in the late Miocene when the present pattern of the Coast Ranges was established; the other at the end of Tertiary time when shifting seas drained into the Pacific, leaving California emergent. Today the region, shaken by periodic earthquakes and marked by continued shifting of the ground, remains as it has been since the late Mesozoic—a restless orogenic belt where the North American continent is still actively evolving.

THE PLEISTOCENE FINALE

While the basic continental structure has been several billion years in the making, the face of the land as we know it today was largely etched out in Quaternary time by weathering and mass wasting, streams, waves, wind, and the trademark of the Pleistocene, glacial ice (Fig. 15-39).

The Ice

During the Pleistocene, valley glaciers born in the highlands of Labrador grew into piedmont glaciers that expanded into continental ice sheets, overwhelming Canada and the United States north of the Ohio and Missouri Rivers (Fig. 15-40).

Continental Glaciers Movements of the ice that disappeared 10,000 years ago are reconstructed from the same sort of evidence that originally led to the concept of the Ice Age: striated bedrock, cobbles, and boulders trailed out from distinctive outcrops, great looping moraines, roche moutonée, drumlins, and other landforms. Careful mapping of the moraines has shown that the continental glaciers advanced along preexisting lowlands as a series of broad lobes that eventually coalesced over most of the upper Mississippi Valley (Fig. 15-41).

After Agassiz established the Ice Age concept, the next big advance in unraveling the Pleistocene was the discovery of *multiple glaciation*. The spread of the ice, it was discovered, was no single episode but, rather, four separate expansions separated by warm intervals of glacial retreat. The glacial maxima are called *Nebraskan, Kansan, Illinoian,* and *Wisconsin* from relic drifts in these type localities.

The surfaces of the much-eroded pre-Wisconsin glacial deposits lack morainic topography, and have well-integrated drainage systems. The latest stage, the Wisconsin, is the best known because its deposits are least weathered and eroded and form youthful moraines. Moreover, the moraines and drift indicate that the Wisconsin stage was no single waxing and waning of the ice either, but included as many as seven alternations of cold and warm spells. Although the evidence is gone, the earlier glacial stages were doubtless just as complicated.

The *interglacial intervals,* recorded by deeply weathered zones on tills, are called *Aftonian, Yarmouth,* and *Sangamon* from type localities in the upper Mississippi Valley. Weathered zones on the pre-Wisconsin drifts reach depths of 3 meters (3 yards) and form compact clayey *B* soil horizons.[6] In places these soils are buried under younger glacial deposits, indicating that after a warm soil-form-

---

6 Formerly called "gumbotills," an apt description of their nature when wet.

Fig. 15-39 At times during the Pleistocene, ice like that in this Antarctic glacier overwhelmed North America as far south as the Missouri and Ohio rivers. Walcott Glacier (person on ice-mantled moraine gives scale). Photo by Wayne M. Sutherland.

ing interval the glaciers advanced again. The long time required to produce such soils, along with incorporated pollen, peat, and other plant remains, as well as animal fossils, indicates that the interglacial intervals were both longer and, at times, much warmer than the time span since the Wisconsin.

This conclusion is based on the observation that modern soils—formed after the last retreat of the continental glaciers—are less well developed than the ancient ones.

Yet extensive as the great ice sheets were, not all of the higher latitudes were overwhelmed, for

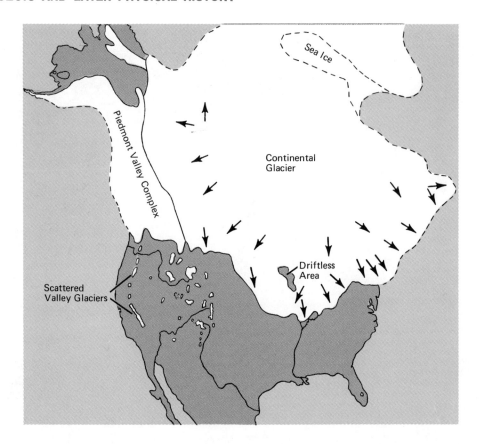

Fig. 15-40   Extent and types of glaciers in North America during maximum Pleistocene expansions of glacial ice. From Glacial Map of North America of the Geological Society of America, 1945.

glaciers require adequate snowfall as well as intense cold. Thus parts of Alaska and vast tracts of Siberia, regions of low precipitation, remained tundras—unglaciated, treeless arctic wastes where intense cold produced deep zones of permanently frozen ground. The Driftless Area of Wisconsin, an island of some 25,000 square kilometers (about 10,000 square miles) surrounded by glaciers, also remained ice-free throughout most of the Pleistocene—but for a different reason. Ice lobes were channeled around this area.

Cordilleran Glaciation   Alaska lacked a continental ice sheet, being occupied instead by extensive

piedmont glaciers fed by a multitude of valley glaciers issuing from mountains and uplands culminating in the great backbone of the Alaska Range (Fig. 15-42).

Most of the Cordillera through Canada and southward had more numerous and extensive valley glaciers by far than now exist. They were responsible for the cirques, horns, and U-shaped troughs creating a spectacularly rugged topography as far south as New Mexico in the Rockies and the San Francisco Peaks in the Colorado Plateau of central Arizona. Some high mountains in the Basin and Range country of Nevada were glaciated. Valley glaciers in the Sierras carved out, among

Fig. 15-41   Moraine topography in northeastern South Dakota. Photo by John S. Shelton.

others, the spectacular trough of the Yosemite. The Olympic Range in the northern coastal region of Washington was strongly molded by ice, while the Cascades, which still bear hundreds of glaciers, provided ice for piedmont glaciers that merged in a thick sheet, filling the Puget Lowland.

Presumably the Cordilleran glaciers waxed and waned with the great continental ice caps, but correlation with Nebraskan, Kansan, and Illinoian stages of the interior lowland is difficult because Wisconsin valley glaciers and strong erosion in the mountain regions have removed all but a few patches of earlier till. Continental and mountain sequences do interfinger from the Canadian–United States border northward, but they only show the relations of Wisconsin-aged deposits.

Beyond the Ice

Although glaciation was its most striking aspect, the imprint of Pleistocene climatic changes extends far beyond the glaciated regions.

Shorelines   The Earth's coastal regions are scarred by the worldwide fluctuations of Pleistocene sea levels. The oceans fell and rose, as water of the hydrologic cycle was alternately locked on land in glacial ice, then released during melting to swell the oceans.

Wave-cut terraces standing high and dry along many of the world's coasts bear marine fossils indicating warm interglacial oceans. So it might seem that the maximum heights of Quaternary sea

Fig. 15-42 Alaska's Mount McKinley, highest mountain in North America at 6,081 meters (20,269 feet), still carries relics of formerly more extensive Cordilleran glaciers. Courtesy of the Alaska Railroad.

levels could be easily determined. Unfortunately, many uplifted shore lines—especially around the Pacific—have been warped and tilted by crustal movements after they were cut. Even in nonorogenic regions, the Gulf Coast, for example, inland terraces may warp down and disappear beneath later sediments on approaching the shore. Some workers maintain that most coastal margins have been tectonically upwarped during the Quaternary; hence sea level may well be near its maximum

now. Yet, because the interglacial times were undoubtedly warmer than today, water released by the shrunken Greenland and Antarctic ice masses should have raised the ocean surface appreciably. The most likely estimate, from evidence in stable crustal regions, places sea level as at least 30 meters (90 feet) higher during past interglacial times.

At the glacial maxima, former sea floors were bared and streams drained to shores farther out on the continental margins. Thus submerged stream valleys, drowned wave-cut terraces, and terrestrial deposits dredged from beneath the sea have been used to estimate the low stages of Pleistocene sea levels. A lowering of sea level by 150 meters (500 feet), more or less, seems likely for the Illinoian stage, and 120 meters (400 feet) for the less extensive Wisconsin glaciation.

Wind The work of wind has been especially prominent in the Quaternary. Loess deposits, wind-laid dust sometimes reworked in water, are widely distributed throughout the Mississippi Valley, for beyond the glacial margins, meltwater streams left broad flood plains of sand and silt. Clouds of dust, winnowed from these barren windswept flats, were carried away to settle in thickening deposits reaching a hundred meters deep in places.

Sand dunes were more widespread than today in various climatic phases of the Pleistocene. For example, the Sand Hills region of western Nebraska, over 60,000 square kilometers (about 24,000 square miles) of now-stabilized grass-covered dunes, was a sea of shifting sand earlier in the Quaternary.

Deflation hollows were actively eroded during warm, dry interglacial times when the western plains and Cordillera were even more a desert than they are now. The blowouts range from the 11-kilometer (7-mile) long Big Hollow of the Laramie Plains in southeastern Wyoming to many smaller depressions, some now occupied by permanent or ephemeral ponds.

Streams That the Ohio and Missouri rivers mark the approximate southern limit of continental glaciation is not coincidence. When the ice buried trunk streams of major drainage systems, north-flowing tributaries ponded along the glacial fronts. Eventually, the marginal lakes overflowed to each other, cutting the channels of the Ohio and Missouri rivers, which remained the major tributaries of the Mississippi after the ice withdrew. Abandoned and drift-filled preglacial valleys clearly show that many present-day streams flowed to trunk streams north of the Ohio and Missouri. In New York, Pennsylvania, and Ohio, tributaries once drained to major streams and emptied into the sea either through the Saint Lawrence Valley or westward through the former Teays River into the Mississippi. Likewise, many tributaries of the Missouri once flowed farther north to a preglacial trunk stream.

Stream regimens were upset during the Pleistocene, causing the alternate back-filling and downcutting responsible for extensive alluvial terraces. In the upper Mississippi Valley, and marginal to Cordilleran valley glaciers, streams beyond advancing glacial fronts were overloaded with debris, thereby aggrading their valleys with alluvium. Wasting glaciers, in turn, swelled streams with meltwater, causing partial excavation of the fills. In the lower Mississippi Valley, however, geologists believe instead that downcutting accompanied glacial maxima. They reason that channel gradients steepened when sea level fell, thus giving streams greater cutting power, so they deepened their valleys. During the interglacial intervals of rising sea level, streams deposited in slack waters of drowning valleys to cause back-filling. Whatever the mechanism—perhaps each is correct in the regions for which it was proposed—thick alluvium and terraces are notable Pleistocene products.

Lakes The Pleistocene spawned myriad lakes, some still existent, and others now disappeared but recorded in lacustrine (lake) deposits, abandoned shorelines, and outlets (Fig. 15-43). Much of Canada, as well as Maine and Minnesota, are "lands of a thousand lakes" because uneven glacial scouring and deposition of hummocky moraines left many closed depressions.

Of the large lakes formed along the glacial margins, the Great Lakes remain as North America's largest existing freshwater bodies. Their basins were carved from preglacial stream valleys by the continental ice sheets, and the lakes appeared as the Wisconsin ice front withdrew. The history of

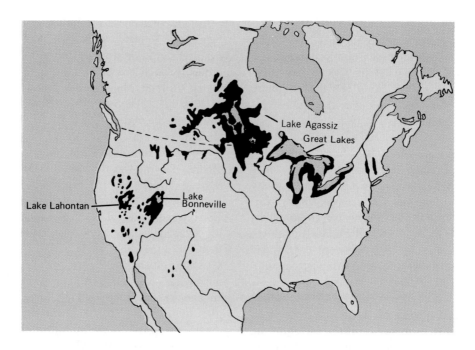

Fig. 15-43 Extent of Pleistocene North American lakes, now greatly reduced or gone. From the Geological Society of America Glacial Map, 1945.

the Great Lakes makes a very complicated story of expansions and contractions read from lake deposits, abandoned shorelines, and outlets. In essence, however, these lakes originally drained to the Mississippi through now-abandoned channels.[7] Lower escape routes emerged as the ice receded northward, first through the Mohawk and Hudson valleys, then, as today, through the Saint Lawrence River (Figs. 15-44, 15-45, 15-46).

The abandoned shores of the Great Lakes have been uplifted, indicating a rising domal uplift where the thickest ice once lay. Apparently, the tremendous load of the Pleistocene glaciers depressed the Earth's crust, while their wastage has allowed its gradual rebound. In a similar, and, perhaps, better documented case involving recovery from the north European ice cap, tide gauge records

7 Such as the former outlet now followed by Chicago's sewage canal.

coupled with fossil evidence on warped shorelines indicate a rebound reaching a maximum of about one meter (3.28 feet) per century in the Baltic region, associated with a broad dome rising across Scandinavia.

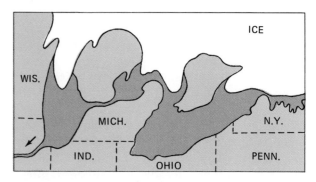

Fig. 15-44 Earlier episode in Great Lakes history when ice forced drainage into the Mississippi River.

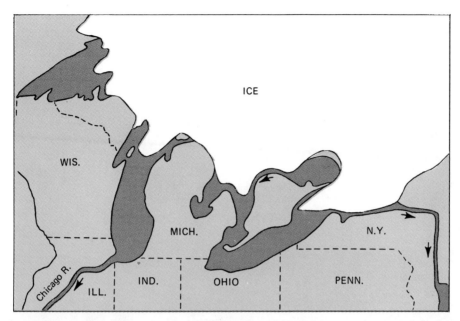

Fig. 15-45 Later episode when ice retreat expanded lakes and opened a new outlet through the Hudson Valley of New York.

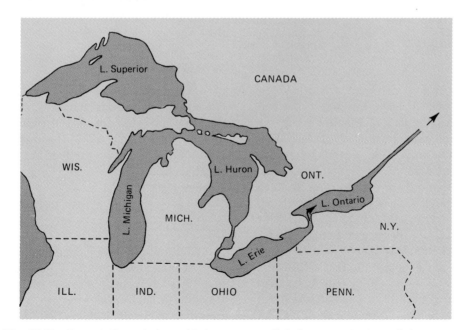

Fig. 15-46 Present Great Lakes with ice gone and drainage northeastward through the St. Lawrence River.

Fig. 15-47   Wave cut and built terraces marking former Pleistocene levels of Lake Bonneville. From classic work by G. K. Gilbert in *Monograph Number 1* of U.S. Geological Survey. Drawing by W. H. Holmes, courtesy of U.S.G.S.

Lake Agassiz, most widespread of North America's former Pleistocene lakes but never particularly deep, covered the flats of Manitoba and the Dakotas. Its shrunken relics include Lakes Winnipeg and Manitoba in Canada. Lake Agassiz also spilled into the Mississippi while the continental glacier dammed its northern margin, then drained off northward through a lower outlet into Hudson's Bay when the ice retreated.

The grabens of the Basin and Range country provided closed depressions that contained many Pleistocene lakes. Lake Bonneville in Utah, whose supersaline remnant is the Great Salt Lake, and Lake Lahontan, whose relics include Nevada's Pyramid Lake, were the largest. Such lakes, far removed from the continental glaciers, reflect either lower evaporation of cooler climates or increased precipitation (or both) accompanying glacial maxima. Lake Bonneville's complex history of deeper and shallower stages is recorded in lake deposits and abandoned shorelines reaching 300 meters (1000 feet) above the existing Great Salt Lake (Fig. 15-47). Similar features indicate a comparable history for Lake Lahontan and other smaller lakes.

Catastrophic Deluges

The uniformitarian doctrine, as developed by Hutton and Lyell, required that the Earth's past history be explained in terms of present-day, observable geologic processes. No "unnatural" or extraordinary events had created the Earth's rocks and surface features. Today uniformitarianism remains a basic geologic concept, but in a less extreme version than that of Hutton and Lyell.

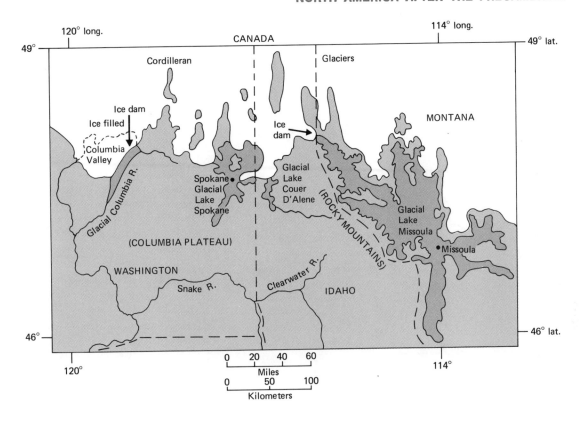

Fig. 15.48 (a) Glacial Lake Missoula.

The Spokane Flood During the Pleistocene, the world's greatest known flood swept across the northern part of the Columbia Plateau. This geologically documented event, called the *Spokane Flood,* left its record in the *Channeled Scablands,* a starkly scenic tract of some 39,000 square kilometers (15,000 square miles) in eastern Washington.

Following the basaltic eruptions, deposition, and deformation in Tertiary time, the surface of the Columbia Plateau in Washington was blanketed by loess. These deposits of windblown silt were derived from lake beds in the western part of the plateau and from ash blown in from volcanos in the adjacent Cascade Mountains. Today loess deposits as much as 60 meters (200 feet) thick form

the rolling hills of the fertile wheat-growing Palouse district in southeastern Washington. Most of the northern Columbia Plateau must have resembled the Palouse district just before the Spokane Flood.

The stage was set when Pleistocene glaciers, nourished by ice fields in the Canadian Rockies, moved down south-trending valleys into the region of the northern plateau. Here, an advancing glacial lobe blocked the Columbia River, diverting it southward through a new channel (subsequently to become the Grand Coulee). Other lobes blocking tributaries to the Columbia created several ice-dammed lakes, including glacial *Lake Missoula* (Fig. 15-48a). This lake originated in the Northern Rockies when ice from breeding grounds in Canada

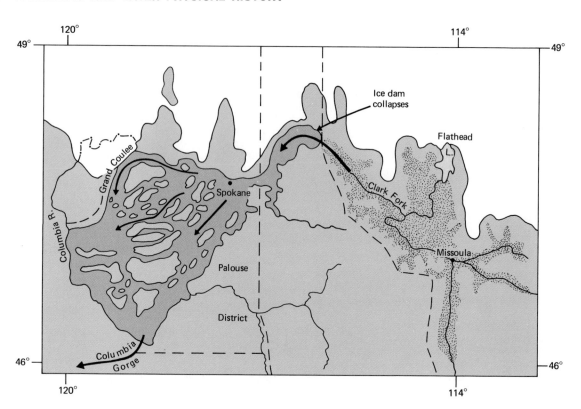

Fig. 15-48 (b) The Spokane Flood. Based on J. H. Bretz, J. T. Pardee, Richmond, Fryxell, Neff, and Weis.

moved down a great valley called the Purcell Trench and blocked the Clarks Fork River. Lake Missoula flooded far back into the deep mountain valleys of western Montana. Based on such evidence as the scars of its former shorelines on valley walls, the lake covered 7800 square kilometers (3000 square miles) and was 290 meters (950 feet) deep at Missoula and nearly 600 meters (2000 feet) deep just behind the ice dam. The lake contained 2000 cubic kilometers (500 cubic miles) of water (half the volume of existing Lake Michigan).

When the ice dam broke, Lake Missoula may have emptied in a day or two. It is estimated that in the narrower parts of the Clarks Fork Valley, the rate of discharge reached an incredible 104 million cubic meters (386 million cubic feet) of water per second.[8] In places the tremendous surge of water heaped gravels into giant ripple marks. These resemble the forms of ordinary current ripples 2.5 to 3 centimeters (1-1.2 inches) high, but the giant ripples of coarse gravels are as much as 9 meters (30 feet) high, 90 meters (300 feet) apart, and 3 kilometers (almost 2 miles) long.

Spilling across the Columbia Plateau, the great

[8] For comparison, the Amazon, the world's largest existing river, discharges some 1.6 million cubic meters (6 million cubic feet) per second.

Fig. 15-49 Retreating falls causing headward growth of a canyon. Water pouring over the falls scours out a plunge pool that undermines the rock in the brink. Its collapse leads to cataract retreat.

Water spilling over the brink of such falls creates plunge pools whose swirling waters constantly undermine rocks at the brink of the upstream channel. The collapse of the undermined rocks, which creates retreating falls at the head of a growing canyon, was a major factor in the origin of Grand Coulee and other scabland canyons (Fig. 15-49). Much of the debris carved from the scablands was deposited in a great mass, flooring a basin in the western part of the Columbia Plateau. Much of the erosional debris went down the Columbia Gorge through the Cascade Mountains and was deposited in western Oregon and Washington in the Columbia River delta.

When the concept of the Spokane Flood was proposed in the 1920s by J. Harlan Bretz of the University of Chicago, many of his geologic confreres summarily rejected his scheme because it seemed a clear reversion to nineteenth-century biblical catastrophism. Several attempts were made to explain the Channeled Scablands in more proper uniformitarian terms, involving "normal" glacial and stream action. Yet over the years, so much geological evidence has been accumulated which supports Bretz's radical, and imaginative, theory that it has become the accepted doctrine for geologists working in the northern Columbia Plateau.

flood created a tremendous braided network of broad channels that stripped the loess cover from the underlying basalt (Fig. 15-48b). Its polygonally jointed blocks were readily quarried by the surging waters to produce a chaotic landscape of scoured basins, mesas, relict patches of loess, and, in places, mazelike networks of steep-walled canyons. Where the flood coursed down the glacially diverted segment of the ancestral Columbia River, it carved out the Grand Coulee, the largest scabland canyon, whose walls are as much as 275 meters (900 feet) high. The cutting of Grand Coulee involved spectacular waterfalls, as indicated by the great transverse cliff now known as Dry Falls, which during the Spokane Flood rivaled the falls at Niagara.

The Bonneville Flood   Geologists have found evidence of yet another deluge in the Columbia Plateau. The Bonneville Flood across the Snake River Plains of southeastern Idaho also resulted from the catastrophic draining of a Pleistocene lake (Fig. 15-50). In this case, however, the sudden release of water did not involve the collapse of an ice dam, but rather the rapid excavation of unconsolidated alluvium when rising Lake Bonneville spilled across a divide, the Red Rocks Pass, and flowed downhill into southeastern Idaho. Here, the waters surged across the basaltic plateau, leaving large rounded boulders on flat surfaces that today stand high above the existing drainage. As in the scabland tracts to the northwest, the flood waters generated great cataracts that rapidly cut back deep canyons at receding falls. Today, in view of the geologic evidence for the Spokane and Bonne- ville floods, it is little wonder that geologists have retreated from the strict version of uniformitarian- ism as propounded by Hutton and Lyell.

Fig. 15-50 Landscape along the Snake River near Twin Falls, Idaho, resulted from the catastrophic draining of Lake Bonneville. View toward the east (the direction from which the waters came) shows a scabland-type topography. Waters flooded across the plateau, covering all the area shown. Giant cataracts cut the short round-headed tributary valleys, scoured out the basin now occupied by a lake, and deepened the main canyon. Courtesy of William B. Hall.

SUGGESTED READINGS

Clark, T. H., and Stearn, C. W., *The Geologic Evolution of North America*, New York, The Ronald Press, 1960.

Dyson, James L., *The World of Ice*, New York, Alfred A. Knopf, 1962.

Flint, R. F., *Glacial and Quaternary Geology*, New York, John Wiley & Sons, 1971.

King, P. B., *The Evolution of North America*, Princeton, N.J., Princeton University Press, 1959.

U.S. Geological Survey, *The Channeled Scablands of Eastern Washington—The Geologic Story of The Spokane Flood*, Washington, D.C., Superintendent of Documents, U.S. Government Printing Office, 1973.

The north end of the Red Sea with the Gulf of Suez (left branch) and the Gulf of Aqaba (right branch). The sea and gulfs are interpreted as axes of divergence where tectonic plates are splitting apart. Egypt is the land on the left, the Sinai Peninsula lies between the branches, and Arabia is at the right. The Mediterranean Sea is visible in the distance. Astronaut view courtesy of N.A.S.A.

16

Fixed Continents to Drifting Plates: Changing Global Concepts

oday much of Earth history is explained as a breakup of supercontinents whose fragments have drifted about as integral parts of spreading and colliding global plates. Since this plate tectonic paradigm provides a dramatic new view of the world and seems confirmed by several lines of evidence, why bother with older theories? The answer is: if we know only the present theory and a smattering of the latest research findings, we miss the excitement of the human endeavor which makes geology a science. The changing and conflicting concepts of geologists concerning the origin of continents and oceans and the evolution of geosynclines and mountains make a nice case history in the nature of scientific operations.

THE PARADIGM OF PERMANENCY

The theory of fixed continents and ocean basins, sometimes called "permanency of continents and ocean basins," dominated geological thinking for well over a hundred years. While geologists were embroiled in the nineteenth-century conflicts that ended in the triumphs of uniformitarianism and organic evolution over catastrophic doctrines, the old paradigm for studies of the Earth's crustal deformation, which Dana had summarized in 1846, remained the accepted doctrine.

The first challenge of any importance appeared around 1910 when Alfred Wegener introduced his concept of horizontally drifting sialic plates. Although Wegener's arguments were persuasively presented, continental drift was strongly rejected[1] by most American geologists (until they were overwhelmed by the physical evidence of sea-floor spreading). European geologists were about evenly divided pro and con as to the validity of Wegener's hypothesis. But for reasons that will become evident, most geologists working in the Earth's Southern

---

1 A professor discussing Wegener's theory once told his class that the other half of a trilobite he found in Newfoundland turned up in Scotland. As biting sarcasm, the joke has now lost its point.

Hemisphere quickly became converts to the concept. Nonetheless, there were long-standing and legitimate geologic arguments against the global migration of continents based on the state of geologic knowledge in the first half of the twentieth century.

In Defense of Fixed Continents

Let's review Dana's paradigm. He assumed that continents formed in the early molten stage of the Earth when granitic material floated to the surface and "froze" into sialic continental masses upon denser underlying sima. The main postulate of the concept was that since Precambrian time the continental blocks and oceanic blocks of the Earth's crust had always had the same relative positions and sizes. Orogeny and shifting seas had modified details of global geography on the high-standing continental blocks, but no major landmass had become deep ocean floor, nor had oceanic floor risen to form a continental mass. The Earth's major crustal blocks had not shifted vertically or horizontally since they originated.

Continental Basements Because Precambrian rocks occur in Paleozoic and later fold belts, the Precambrian shields were extensive by earliest Paleozoic time and unquestionably floored large parts of the geosynclines involved in later orogenies. In North America, for example, Precambrian rocks crop out in the Blue Ridge Mountains far to the southeast of the Canadian shield and far to the southwest in Grand Canyon and the mountains of the Basin and Range province. Moreover, Paleozoic and later belts of mountain-building overlap each other. In Europe rocks contorted in the Hercynian (late Paleozoic) orogeny formed the subsiding floor for the Tethyan geosyncline from which the Alpine-Himalayan mountain chain arose in Cenozoic time.

Sedimentary Rocks Although it might seem strange at first thought, the extensive marine sedimentary rocks now seen on the continents were

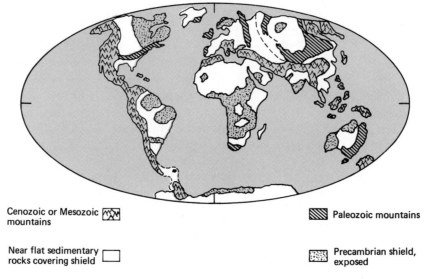

Cenozoic or Mesozoic mountains

Near flat sedimentary rocks covering shield

Paleozoic mountains

Precambrian shield, exposed

Fig. 16-1 Schematic map of the world's younger and older mountain chains as they are today.

good evidence for permanency, as Lyell had suggested. Many of the sediments indicate deposition in shallow seas. The only uncontested deep-water deposits were clearly laid down in mobile geosynclinal belts rather than on uncontorted blocks of the deep ocean basins. Moreover, brown clays, characteristic of abyssal depths, are not found on continental blocks. Also, as previously mentioned, the oceanic depths have no sialic rocks that would indicate foundered continental blocks. Thus the case for permanency seemed strong.

Case History: Eastern North America

The source regions for the sediments laid down in geosynclines, which were later crumpled into mountains, have created a long-standing and intriguing theoretical problem. Many theories were based on studies in eastern North America, where extensive geological investigations had been under way since the mid-1800s.

Archaean Protaxes   In 1890 James Dwight Dana proposed that the Paleozoic conglomerates, sandstones, and shales in North American geosynclines had been derived from the crystalline cores of the Appalachian, Rocky, and Sierra Nevada mountains. These cores were, in Dana's hypothesis, Archaean protaxes that originated far back in Precambrian time and existed throughout the rest of geologic history as a source for the clastic fragments deposited in adjacent geosynclines (Fig. 16-2a). Although a useful working hypothesis in the late nineteenth-century, Dana's early model did not stand the test of subsequent field investigations. The crystalline cores of the Rocky Mountains and Sierra Nevadas were not uplifted and exposed by erosion until Mesozoic and Cenozoic times. In Dana's time it was thought that the crystalline rocks in the Appalachians were all Precambrian, but it has now been demonstrated that they are mainly highly metamorphosed Paleozoic deposits and intrusive plutonic rocks.

Borderlands   Geologists had long recognized that the deposits in New York's Devonian "Delta" and related clastic wedges had been swept westward towards the continental interior. To account for these relations, C. D. Walcott tentatively suggested that the region of the present-day Coastal Plain had contained an uplifted Paleozoic highland, named Appalachia (by H. S. Williams).

The concept of such crystalline borderlands was developed and strongly advocated by Charles Schuchert of Yale University during the first half of the twentieth century.[2] In theory, borderlands were mountainous highlands along the tectonically active continental margins (Fig. 16-2b). They were periodically uplifted and eroded, shedding clastic sediments into subsiding geosynclines encircling the stable interior continental platforms. Three borderlands were proposed: Appalachia, which separated the Appalachian geosyncline from the Atlantic Ocean basin; Llanoria, which extended into the Gulf of Mexico to the south of the Ouachita geosyncline; and Cascadia, which flanked the Cordilleran geosyncline and extended across the region of the present Coast Ranges into the present Pacific Ocean margin. In late Paleozoic or early Mesozoic time, the borderlands disappeared, subsiding beneath the sea in the regions where the present-day coastal plains, continental platforms, and adjacent ocean basins are now located.

The borderland hypothesis never did explain the foundering of lighter sialic masses into the denser simatic rocks of the Earth's underlying crust and mantle. The concept finally had to be abandoned in the 1950s when geophysical investigations showed no evidence of sunken sialic blocks

2 Since Professor Schuchert was coauthor of the leading elementary textbook on historical geology during that period, the borderland concept became doctrine for several generations of American college students.

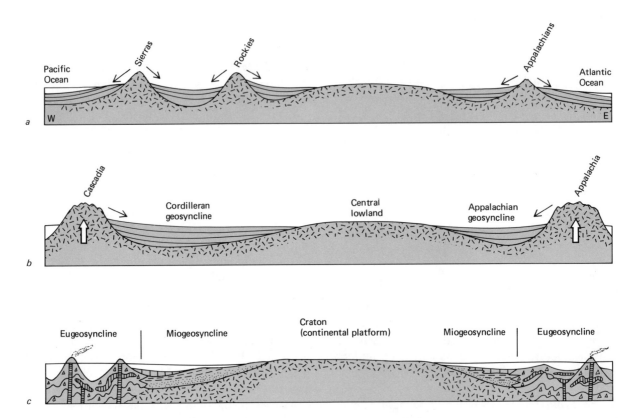

Fig. 16-2 (a) The concept of Archean protaxes. From far back in Precambrian time these crystalline highlands represented by our existing mountains were thought to have shed sediments into adjacent geosynclines. (b) The borderland hypothesis. Tectonically active continental margins periodically rose to create a source region for sediments. Ultimately the borderlands foundered and disappeared. (c) The theory of island arcs takes into account volcanic eugeosynclinal rocks and has existing analogues, as off the east coast of the Asiatic continent.

beneath the continental margins and adjacent floors of the ocean basins.

Island Arcs Starting in the 1940s, Marshall Kay of Columbia University developed the theory of tectonically active island arcs in outer volcanic geosynclines (Fig. 16-2c). Earlier hypotheses had mainly been formulated to explain the limestones, conglomerates, sandstones, and shales in the miogeosynclines and in the clastic wedges, which had

been deposited on continental platforms. Kay stressed the importance of the eugeosynclinal rock assemblages, which are characterized by graywackes, shales, cherts, and volcanic materials. These are the sorts of rocks that would have accumulated in tectonically active regions on sea floors beyond the continental margins where deeply subsiding troughs developed between actively rising chains of volcanic islands. The Japanese and Marianas archipelagos, off the east coast of the Asiatic continent, are present-day examples of such past environments.

Lessons from the Appalachians The concepts stressing protaxes, borderlands, and island arcs as the sources of Paleozoic clastic sediments deposited in the Appalachian region were all developed by geologists who accepted the ruling theory of the fixed positions of continents and ocean basins. Today, of course, geologists are busily fitting Appalachian history into a fundamentally different paradigm—the lateral drifting of global plates. Yet in their time, geologists trained in and guided by the old concepts made many valuable observations about the rock assemblages in the Appalachians. In fitting their new-found facts to the ruling theories —the normal method of scientific investigation— they did discover exceptions. As a result, the primitive working hypothesis of Archaean protaxes was abandoned, and the more dynamic model of continental borderlands adopted; it, in turn, was replaced by the more uniformitarian concept of offshore island arcs. Although protaxes and borderlands may seem mere historical items, best left entombed in dusty technical publications and outdated textbooks, both concepts contributed to the progress of geology.

Dana was right—at least for clastic wedges—in thinking that present-day crystalline highlands had been the source regions for Paleozoic (and many Mesozoic and Cenozoic) sedimentary rocks. His mistake, in thinking that early Precambrian mountains could have survived the ravages of erosion through to the present day, is understandable in light of the knowledge of his time. Nineteenth-century geologists simply had no way of knowing the immense duration of geologic history. And not until the 1930s, when Marland Billings of Harvard University chanced to find a recognizable impression of a Devonian brachiopod in highly metamorphosed schist of New Hampshire's White Mountains, did geologists suspect that some of the crystalline rocks in the Appalachians were younger

than similar-looking gneisses, schists, and granites in the Precambrian shields.

The Paleozoic borderlands along the eastern margin of North America did exist, but in the modern view they were parts of the continents of Europe and Africa. That possibility was, however, never considered by Professor Schuchert, a lifelong vehement opponent of Alfred Wegener's scheme. Marshall Kay, in contrast, became a leading convert to the paradigm of drifting continents. His conception of tectonically active island arcs and deeply subsiding basins survived the plate tectonic revolution. Admittedly, much eugeosynclinal graywacke and shale is now interpreted as continental-slope and deep-sea deposits that originally accumulated along the passive trailing edges of drifting continental platforms—as along the present-day Atlantic margin of North America. Where volcanics are abundant in eugeosynclinal rock sequences, the assemblages are now generally considered to be deep basin deposits associated with island arcs arising in the vicinity of colliding tectonic plate margins.

Since the Appalachian ranges and quietly subsiding Coastal Plain geosyncline lie on the passive east side of the drifting North American platform, the orogenic history of these mountains cannot be related to the existing motions of tectonic plates. Appalachian orogenies resulted from a different and ancient system of spreading axes and subduction zones. Their operation can only be inferred from the record of the rocks preserved in both the North American and European continents (p. 498).

THE CASE FOR A SOUTHERN SUPERCONTINENT

Alfred Wegener's vision of the fragmentation of a late Paleozoic supercontinent, *Pangea*, is now taken for granted in plate tectonic reconstructions of the history of the present-day continents. Why then, for some fifty years, was Pangea considered a "never-never land" by many geologists, particularly Americans?

Some geologists who strongly supported Wegener's view of drifting continents, rather than accepting the ruling theory of fixed continents and ocean basins, doubted the existence of Pangea as a single global continent. Alexander Du Toit, of the University of Johannesburg in South Africa, envisioned two late Paleozoic continents. North America, Europe, and Asia comprised a landmass which he named *Laurasia*. It was separated by the elongate *Tethys* seaway from a great southern continent known as *Gondwanaland*. Although modern paleomagnetic evidence confirms the existence of a late Paleozoic Pangea surrounded by a world ocean, called Panthalassa, the paleogeography of other times is indeed dominated by the two supercontinents of Laurasia and Gondwanaland.

The Necessity for Gondwanaland

In 1858 to explain the similarity of fossilized land plants entombed in the Coal Measures of North America and Europe, Antonio Snyder had drawn his map of Carboniferous paleogeography showing the Americas, Europe, and Africa united in a single landmass. Most of the subsequent evidence, however, that led to the concept of supercontinents and to Wegener's vision of drifting continents came from lands south of the equator.

The notion of a vast southern continent has persisted since antiquity. Ptolemy of Alexandria drew a world map in the second century B.C. showing the Indian Ocean surrounded to the south by *Terra Australis Incognita,* a vast landmass that joined eastern Africa to China. Although nobody had ever seen it, the land made a reasonable analogy with the lands surrounding the Mediterranean and Caspian seas, which were well known then. Ptolemy probably got the idea from Hipparcus (a predecessor in the second century B.C.) who, on noticing the tidal differences between the Atlantic and Indian oceans, theorized that they were separated by a great land barrier. Ptolemy's works disappeared with the fall of the Roman Empire. Their rediscovery in the mid-fifteenth century revived the legendary continent; but not until the late 1700s was the phantom apparently laid to rest when Captain James Cook, under sealed orders from the British Admiralty (to confirm or deny), found no such major continent short of the Antarctic. But perhaps Cook was 200 million years late.

Rocks Geologists working in the Southern Hemisphere long favored some sort of supercontinent in late Paleozoic to mid-Mesozoic time. Rocks of this time interval in Africa and Madagascar, India, Australia, South America, and Antarctica are strikingly similar. Strictly speaking, these Gondwana strata are in India (where the name was given); elsewhere strikingly similar rocks are called by such names as Karroo in Africa, Santa Catherina and São Benito in Brazil, and the Beacon Series in Antarctica. All such strata we shall call Gondwana rocks (Fig. 16-3).

The basal strata, usually lying on a Precambrian basement and sometimes lying on striated glaciated floors, are tillites (lithified glacial deposits) and varvelike claystones. Above them are terrestrial sandstones and shales containing coal seams. Still higher, the rocks are desert-type sandstones, either beneath or interbedded with great basaltic lava flows and often intruded by basaltic dikes and sills. These total sequences are impressively thick, reaching 6000 meters (20,000 feet) in India, as much as 10,500 meters (35,000 feet) in Africa, and not much less in Brazil. The Gondwana rocks indicate a general emergence of the continents. Being terrestrial, they were difficult to date because the standard rock column is based on marine fossils. Fortunately, the Gondwana rocks interfinger with marine rocks in enough places to be correlated with the interval from the late Paleozoic to mid-Mesozoic (Devonian or Carboniferous to Jurassic).

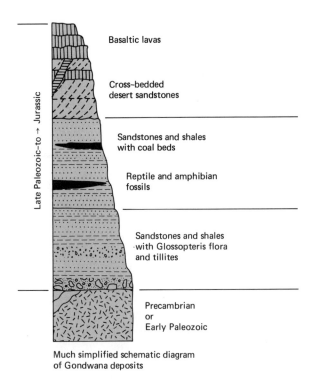

Basaltic lavas

Cross-bedded
desert sandstones

Sandstones and shales
with coal beds

Reptile and amphibian
fossils

Sandstones and shales
with Glossopteris flora
and tillites

Precambrian
or
Early Paleozoic

Late Paleozoic-to → Jurassic

Much simplified schematic diagram
of Gondwana deposits

Fig. 16-3 Much simplified schematic diagram of Gondwana deposits. Although differing in certain details, the similarity of the general sequence of all southern continents made most geologists familiar with these rocks quick converts to the "driftist" concept.

Vertebrate Fossils The nature of Gondwana fossils has long intrigued paleontologists. The late Paleozoic vertebrate animals were apparently quite similar throughout the whole world. Certainly the reptiles were remarkably similar on all the continental masses in middle and late Triassic time. So how did land-dwelling reptiles, such as cotylosaurs, mammallike reptiles, thecodonts, and dinosaurs, migrate from Africa to South America? The South Atlantic Ocean should have been an impassable barrier; moreover, until Cenozoic time the northern and southern continents in the Old World were apparently separated by a major geosynclinal belt occupied by the Tethys Sea. Of particular interest

is a small fish-eating fossil reptile called *Mesosaurus*, found in Triassic rocks of South America and South Africa but nowhere else. Though it lived in water, nobody believes it swam the Atlantic Ocean from one continent to the other.

Plant Fossils Although the vertebrate faunas of Gondwana time were similar on all continents, the plant populations were not. A very distinctive assemblage of seed ferns, the *Glossopteris flora*, characterized the Southern Hemisphere but was notably absent from the northern continents (Fig. 16-4). Glossopteris was apparently adapted to the cool climate indicated by the glacial deposits in the Southern Hemisphere, while tropical and semitropical coal swamps existed across North America, Europe, and Asia. It seemed to many geologists that, if the continents and ocean basins had the same relations in Gondwana time as they do today, the spreading of Glossopteris plants across the South Atlantic and other ocean barriers would have been an insoluble problem. Thus many believed

Fig. 16-4 Glossopteris leaves from the Buckeye Range, Antarctica. Courtesy of Larry Lackey.

that the distribution of both plants and land animals in Gondwana time demanded land connections between continents where the present-day ocean basins exist.

Gondwanaland Interpretations

The southern supercontinent was named Gondwanaland by Eduard Suess, the great Austrian geologist. He also named the Tethys, the elongate ancient seaway separating Gondwanaland from the northern continents.

Sunken Continental Blocks As originally conceived by Suess, Gondwanaland was a continuous Paleozoic land across the Southern Hemisphere from South America on the west to Australia on the east. Later, sometime in the Mesozoic, two great fragments of the continent subsided to create the South Atlantic and Indian Ocean basins, leaving fragments standing as the present southern continents (Fig. 16-5).

The idea neatly solves the distribution of Paleozoic to mid-Mesozoic plants and animals in the Southern Hemisphere. They had simply spread across Gondwanaland before it broke up. It also accounts for truncated continental structures, such as the Cape fold belt at the tip of South Africa; for downfaulted lavas beneath the Arabian Sea off the west coast of India; and for the observation that some ancient sediments, including glacial deposits, apparently came from lands that once existed off the present-day shores.

Despite this, the idea that vertical movements have sunk great continental blocks under the oceans was eventually rejected. Theoretically, it was always hard to explain why great blocks that had long been parts of high-standing continents should ever founder—especially since sialic continental blocks are lighter than the denser sima beneath them. Thus the sial should "float" isostatically on the sima. Most devastating was the eventual geophysical evidence that no granitic, continental-type rocks are in the floors of deep ocean basins.

Land Bridges To geologists who rejected any schemes of fractured supercontinents, land bridges

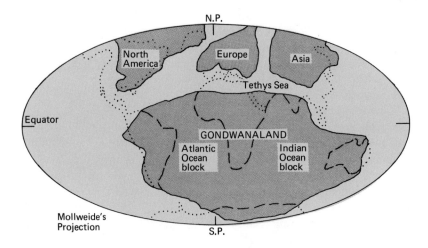

Fig. 16-5 Diagrammatic world map of the late Paleozoic showing the hypothetical Gondwanaland before the fragments subsided to create the South Atlantic and the Indian Oceans.

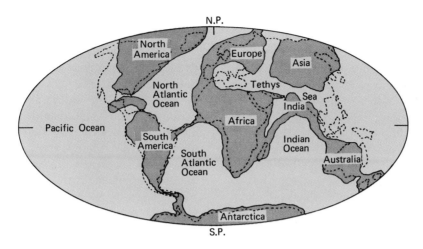

Fig. 16-6 Land bridges as a suggested solution to the late Paleozoic Gondwana problem. Based on Schuchert and others.

seemed a reasonable explanation for the widespread migration of plants and animals in Gondwana time. Today the Isthmus of Panama forms a land bridge between North and South America. During Pleistocene times of lowered sea level, animals must have migrated between North America and Asia across a bridge which was later submerged, along the Bering Straits. Thus some students of the Gondwana problem, notably Charles Schuchert, proposed that similar bridges—which had later sunk into the depths —had formerly provided migration routes across some parts of the ocean basins now separating the Earth's southern continents. The present and recently accepted bridges provided a uniformitarian basis for assuming similar features in the more remote geologic past, and the concept fitted the doctrine of the permanency of continents and ocean basins (Fig. 16-6).

By the 1950s, however, the presence of purely hypothetical land bridges across the oceans was being questioned. After studying the dispersal and evolution of the mammals on the different continents, George Simpson of Harvard strongly opposed any Cenozoic land bridges—except for the well-established routes, such as the Bering Straits

and the Isthmus of Panama. These are either continental shelves where the water is now shallow or zones of active mountain-building. From a study of Mesozoic reptiles, Edwin Colbert, then at the American Museum of Natural History, concluded that there might have been land bridges in the early Mesozoic times.

But as with the theories of foundered continental blocks and foundered borderlands, there were no known physical explanations for sinking land bridges into the depths. Finally, in the recent era of intensive ocean-floor studies, no geophysical evidence was found of sunken bridges in the deep oceanic basins. Today in the plate tectonic era— with the benefit of hindsight—land bridges seem a theoretical contrivance to fit the paradigm of fixed continents and ocean basins.

The Driftists' Case for Gondwanaland Aside from the jigsaw continental fit, Wegener and his followers had assembled a vast amount of interrelated circumstantial evidence for the splitting and drifting of supercontinents. They stressed the *similarity of Gondwana rocks* on all southern continents, and the similarity of Gondwana plants and animals as compared with diversification in later times. The biological problem was neatly solved if plants and animals never had to cross major ocean basins because they lived together on a single landmass before the breakup of Gondwanaland.

Moreover, if the southern continents are properly "slid back together," their *broad structural elements* show striking coincidences. The South American and African Precambrian shields fit neatly; and (with a bit of twisting) the Indian, Antarctic, and Australian shields can also be reassembled as one body. Paleozoic folded mountain belts can be connected: the east-trending ranges through Buenos Aires in South America connect with the Cape folds of South Africa, and these in turn can be extended through Antarctica to the Australian Cordillera. Alexander Du Toit proposed that these ranges rose from a once-continuous geosyncline (the "Samfrau"). In the Northern Hemisphere, the Paleozoic mountain trends, including the crystalline and folded Appalachians in eastern

North America, can be connected to those in northern Europe (Fig. 16-7).

Great lava plateaus in South Africa (on the Drakensberg scarp) and some in southern Brazil, both of early Jurassic age, along with the extensive Deccan "traps" in India—of late Cretaceous and early Cenozoic age—could represent eruptions along fractures accompanying the breakup of Gondwanaland. Also, the apparent offshore source (from areas now occupied by oceans) of thick sedimentary sequences that lie on basement rocks in such places as Southwest Africa, Antarctica, and southeastern South America is easily explained if these continents were once part of a continuous landmass. However, the strongest geologic evidence for drift was the distribution of the late Paleozoic glacial deposits found only in the Southern Hemisphere.

Today Gondwana *tillites* are found mainly in tropic or near-tropic regions. Assuming the doctrine of permanency, late Paleozoic continental glaciers must have covered the equatorial regions at the same time that mild climates prevailed in the higher latitudes. Conceivably the continents were in their same relative positions and the terrestrial poles have migrated—thereby shifting the Earth's broad climatic belts. Put the past poles where you like. If the continents had the same relative positions during late Paleozoic time as now (as permanency demands), the Gondwana tillites would have been so far apart that all the southern continents could not have had separate ice caps. Some must have been in tropical latitudes—unless, that is, the continents were clustered in a single Gondwanaland (Figs. 16-8, 16-9).

Thus continental drift neatly solved the glacial problem, gave land connections for plant and animal migrations, connected shields and mountain trends, and avoided the foundering of continental blocks or land bridges. It satisfied the geophysical model of less dense sialic continents floating isostatically on the denser sima of the oceanic blocks.

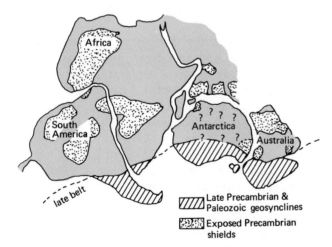

Fig. 16-7 Gondwanaland jigsaw puzzles. The African–South American Precambrian arrangement seems substantiated by recent studies, after Hurley. The continuous geosynclinal belt was first proposed by Du Toit.

Fig. 16-8 Evidence of late Paleozoic Gondwana glaciation at Hallet Cove, south of Adelaide, Australia. Glacially striated pavement is overlain by Permian-age glacial till containing boulders derived from Precambrian bedrock. Photo by the author.

How could such an all-encompassing theory have been doubted?

Objections to Drift The opposition to continental drift was not simply a matter of ingrained thinking and lack of imagination. In the first place, the visual matching of continental outlines is not everywhere as perfect as between Africa and South America, and even here the lengths of the corresponding coasts do not match. In other places, a reasonable matching of margins requires much twisting about of the continents. And it was argued that even a perfect fit of margins did not, in itself, prove drift. Puddles in the low spots of an uneven brick sidewalk would not prove that the bricks had drifted apart; similarly the parallelism of shores might result from a global fracture system producing higher (continental) and lower (oceanic) blocks.

Moreover, the matching of similar rock types and folded mountain belts could be explained by

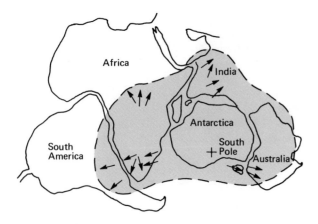

Fig. 16-9 Diagrammatic sketch of A. L. Du Toit's arrangement of the continents into Gondwanaland and a solution of the late Paleozoic glacial problem. Arrows show inferred directions of ice movement; dashed lines represent his inferred ice-cap boundary. After Du Toit.

assuming the permanence of continents and ocean basins. Correspondences in themselves are not proof of drift, for Precambrian shields and later mountain belts have had comparable histories of geosynclinal deposition and orogenic folding, even where there is no possible connection. With suitable twistings, structural elements in the continents could be matched even if they had never been connected, and among "driftists" themselves, different arrangements of the continents in late Paleozoic time were proposed.

The striking similarity in Gondwana rocks could also be explained in terms of permanency. Today similar rocks and soils are developing in such widely separated tropical rainforests as those in Africa and South America; modern glacial deposits are alike in Greenland and Antarctica (which are most obviously not connected); and deserts in different parts of the world are depositional sites of similar sand dunes. Thus it was logically argued that similar rocks reflect similar environments and hence are no proof of former connection.

Nor does the far-flung distribution of Gondwana plants and animals necessarily require a su-

percontinent (or even land bridges). Seeds of the Glossopteris flora might have blown great distances over water, and intervening islands could have provided way-stations in their dispersal; moreover, ocean currents can carry reproductive structures of plants great distances before casting them upon the shore.

Evidence for the worldwide cosmopolitan fauna of land reptiles is not uniformly abundant in all Gondwana rocks—none have been found in Australia. And even granting a worldwide distribution, it was argued that the animals could have migrated across the Northern Hemisphere and then found routes to the south on an arrangement of the continents similar to the present. The Tethys Sea, a seeming barrier, was probably broken by late Paleozoic (Hercynian) mountain-building in various places, which could have given routes across it for land animals.

The strongest opposition to drift always centered on mechanics. Wegener's idea of sialic continental rafts floating across denser sima was never acceptable because it was well established that both were a part of the Earth's solid lithosphere. Even accepting such slippage, the driving forces were a mystery. Two were commonly invoked. The centrifugal force of the Earth's rotation was said to cause the Gondwana continents to drift northward over the bulge of the equator, the mechanism called "Polflucht" by Wegener. The tidal attraction of the Sun and Moon was proposed as dragging the continents westward at the same time. These forces exist, but theoretical calculations by Sir Harold Jeffreys, a leading geophysicist, showed them grossly inadequate, by several million times.

Thus, in the conservative view of permanency of continents and ocean basins, neither the correspondences of continental outlines and structural trends nor the Gondwana rocks and fossils proved continental drift—but they certainly did not dispel the possibility either.

Technical articles and symposia, pro and con, proliferated. Books included Wegener's *Origin of*

Continents and Ocean Basins (1915) and Alexander Du Toit's *Our Wandering Continents* (1927) in support of continental drift; and Walter Bucher's *Crust of the Earth* (1933) in opposition. Lester C. King, like Du Toit a South African, wrote a paper, "The Necessity for Continental Drift," in 1952; but most Americans saw no such necessity then. In general the game was stalemated until the plate tectonic revolution of the 1960s.

The Debate in Perspective

Today, in the plate tectonic era, the evidence for drifting continents seems overwhelming.

New Findings The paleomagnetic evidence of wandering magnetic poles allows us to approximately locate the poles of rotation in geologic time. These locations are different for the separate continents and can only be reconciled if the continents have indeed drifted (Fig. 8-34). The paleomagnetic stripes of normal and reversed polarity in sea-floor plates provide clear evidence for horizontally drifting plates; and so do the ages and magnetic reversals in ocean-bottom sediments. Radiometric dating of rocks with similar structural features in now-separated Precambrian shields, such as Africa and South America, strongly suggest that these masses were once joined together. The jigsaw fit of continents is excellent—if the 1000-meter (3300-foot) depth contours of the continental platforms are matched, using computer techniques, rather than the shorelines.

Global paleoclimatic belts, which the driftists used in establishing the former poles, seem confirmed. Aside from the late Paleozoic patterns when the southern continents were glaciated and northern continents were tropical, the recent discovery of an Ordovician glaciation marked by striated pavements in northwest Africa now confirms the former polar position of Gondwanaland in early Paleozoic time.

New discoveries of vertebrate fossils, notably in Antarctica, confirm the deductions of the driftists about Mesosaurus. In the early 1970s, a number of skeletons of *Thrinaxodon*, a primitive mammal-like reptile, and of *Lystrosaurus*, a primitive reptile that lived in fresh-water swamps and ponds, were found in Antarctica. These early Triassic reptiles —which are identical to species long known from Africa—could never have swum across major oceans; they must have migrated overland when the continental masses were connected.

Reflections on the Nature of Scientific Investigations With the benefit of hindsight—and all the new evidence—one might wonder why Wegener's scheme provoked such strong opposition. To be fair, there were valid arguments against drift (previously mentioned), considering the state of geologic knowledge prior to the plate tectonic revolution. The driftists' evidence, although strong, was purely circumstantial. No physical measurements demonstrating movements of major landmasses were possible in Wegener's time. Although he did attempt to show a shift in the position of Greenland by careful surveying using triangulation, the results were inconclusive. The first successful measurements indicating continental motion stemmed from the pioneering work in the 1950s by P. M. S. Blackett of England and his followers, including Keith Runcorn, who demonstrated paleomagnetic polar wandering of different continents. Direct evidence of continental motion became available when the significance of the paleomagnetic reversals in sea-floor stripes was appreciated.

The main objection was to Wegener's proposed mechanism of sialic continents drifting across simatic ocean floors, which was clearly inadequate (although Wegener recognized that more geophysical studies were needed). Moreover, in the skeptical approach and critical spirit that characterizes the testing of new concepts, scientists tend to put the burden of proof—rightfully—on innovators who challenge an accepted frame of reference.

And there were human factors which are part and parcel of the scientific endeavor. If we accept the puzzle-solving theory for the nature of science, continental drift challenged the long-useful and successful paradigm of the fixity of continents and ocean basins. This frame of reference had provided the basis for the life-work of many geologists who had taken an authoritative stance. And despite his disciples, many scientists looked on Wegener as an "outsider." His original training had been in planetary astronomy and meteorology—not in geology or geophysics. Moreover, as he stressed in his book, *The Origin of Continents and Oceans*, the past patterns of lands and seas could only be determined by the combined work of geodesists, geophysicists, geologists, paleontologists, zoogeographers, plant-geographers, and paleobotanists. Thus he was a "generalist" at a time when "proper" science was dominated by specialists. For this reason, Wegener had difficulty in obtaining an appointment in the German academic community. Although respected as a meteorologist, he was considered by many of his contemporary scientists a fuzzy-minded dilettante because he dealt with so many different fields of science. Yet his approach was truly interdisciplinary in the modern sense, and in a large measure led to his vindication during the plate tectonic revolution.

PLATE HISTORY: LAURASIA AND THE NORTHERN CONTINENTS

The changing paleogeography of the Paleozoic world, when Gondwanaland was the southern supercontinent, was marked by four continents north of Tethys seaway. The hearts of the northern continents were dominated by the present-day Precambrian shields in North America, Europe, and Siberia as well as several smaller Precambrian masses in China. Following active plate histories during the Paleozoic, the ancestral northern continents became united into the supercontinent of Laurasia[3] (which in the late Paleozoic merged with Gondwanaland to form Pangea).

The Classic European Record

Since north Europe—where Hutton, Lyell, Werner, Desmarest, and others founded modern geology—has a strikingly similar history to the classic ground of eastern North America, these two regions serve as a model for the Laurasian continents.

Early Paleozoic Developments Throughout the Cambrian, Ordovician, and Silurian periods, a geosyncline extended through Wales, Scotland, and western Norway to the eastern coast of Greenland, and thence into the region of the present arctic islands. As in the Appalachian region, great thicknesses of graywackes, graptolitic shales, and volcanic materials accumulated in this eugeosyncline. Eastward, thinner shelley facies of sandstone and limestone marked a miogeosyncline extending onto the Baltic-Russian Precambrian platform (Fig. 16-10a). Early spasms of the *Caledonian orogeny* crumpled the eugeosyncline in Ordovician time. Mountain-building reached a climax at the close of the Silurian and continued into the Devonian. During the Devonian, a great clastic wedge was swept eastward from the high Caledonide ranges to form the *Old Red Sandstone* that extended well onto the stable Precambrian continental platforms (Fig. 16-10b). The Old Red deposits are quite similar to strata in the Catskill Delta which developed at the same time in the Appalachian region.

Late Paleozoic Events During late Devonian and early Carboniferous time, the Caledonide ranges were worn down by erosion. South of the remaining

3 Named for the Laurentian region in Canada, Europe, and Asia.

Old Red highlands the crust subsided and most of Europe became an island-dotted sea. Deposits of this episode, corresponding to the Mississippian in the southern Appalachians, include: sandstones and shales of transitional environments; some coal-bearing strata indicating emergent terrestrial regions; and limestones characteristic of shallow seas.

During the *Hercynian orogeny* of upper Carboniferous (Pennsylvanian) and Permian time, Eu-

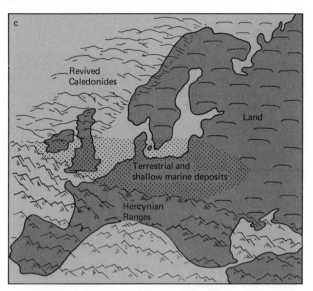

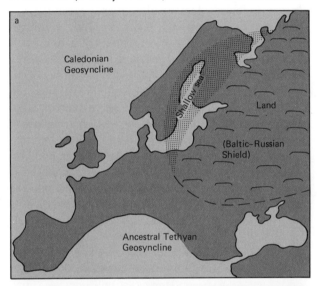

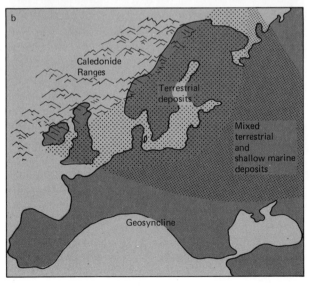

Fig. 16-10 (a) Paleogeography of Europe. Latest Precambrian and earliest Paleozoic (b) Caledonide and Old Red paleogeography. (c) Late Paleozoic-Hercynian time.

rope became a broad basin bordered on the north by the relief of the revived Caledonide highland and on the south by newly risen Hercynian mountains (Fig. 16-10c). Streams from the highlands gradually filled the lowlands with deltaic sands and muds that replaced the shallow seas, creating a broad low-level plain. The Coal Measures, trademark of the Carboniferous, developed across the plain. Forested swamps were drowned repeatedly by lakes or shallow seas, then reappeared as the seas withdrew in a rhythm comparable to that in the interior coal basins of North America. As the Hercynian orogeny intensified, the Coal Measures were themselves folded and in places overridden by fault plates of older rock.

As Hercynian mountain-building waned in Permian and Triassic times, the Carboniferous rocks disappeared beneath a desert waste of shifting dunes, ephemeral lakes, intermittent streams, and landlocked seas in which salt layers were deposited. The record of this paleogeography is the *New Red Sandstone*. These rocks are also comparable to their red equivalents in North America.

The Appalachian-Caledonide Belt

In 1966, several years before plate tectonics became the established doctrine, J. T. Wilson of the University of Toronto proposed that an ancestral Atlantic Ocean basin had opened and then closed during the Paleozoic to create the universal continent of Pangea. This imaginative suggestion of laterally drifting global plates has become the standard interpretation since the plate tectonic revolution. The hypothesis accounts for the presence of Appalachian mountain ranges on the east side of the North American continent, the similarity of west European and North American rock assemblages, and the several phases of geosynclinal and orogenic development.

A Rifted Precambrian Continent It is now thought that in Precambrian time the present European and North American continental masses were parts of a supercontinent. Sometime before the end of the Precambrian era, the ancient supercontinent split and the segments drifted apart—much as they seem to be doing now—to create an ancestral Atlantic Ocean basin. Perhaps this continental fracturing was associated with a paleo-feature like the existing Mid-Atlantic Ridge. But since the ocean-floor plates of the present plate tectonic system are constantly being destroyed and regenerated, any evidence of a very ancient mid-ocean ridge was long ago obliterated. Its existence can only be conjectured from mélange and ophiolite rocks preserved in the continental blocks.

In any case, from late Precambrian to mid-Ordovician time, a situation much like that of the present day existed in which the ancestral Atlantic Ocean was widening. During this episode, miogeosynclinal clastics and limestones were deposited on a subsiding continental platform, while eugeosynclinal sediments accumulated on the passive continental slope and adjacent deep ocean floor. Thus the shelley and graptolitic facies of the early Paleozoic probably reflect a geologic setting that was rather similar to the existing Atlantic coastal plain and adjacent continental margins. Deposition was occurring on a rifted and passive continental margin that was then unrelated to a subduction zone.

Converging Continents The pattern changed markedly in mid-Ordovician when the ancient Atlantic Ocean basin ceased expanding and instead began to close. The resulting compression broke (technically, "decoupled") the formerly outward-moving plate. The oceanic part was then depressed beneath the continental margins as part of a newly formed subduction zone. With the development of active continental margins along an axis of convergence, the early Paleozoic eugeosynclinal deposits on the continental slopes and adjacent ocean bottom were contorted into rising island arcs and locally subsiding basins. More graywackes and volcanic rocks accumulated in these tectonically active belts and were eventually compressed into the complexly folded and faulted continuous Appalachian-Caledonide mountain chain.

Mountain-building was essentially completed in the northern part of the chain during the Taconian and Acadian orogenies of late Ordovician and Devonian time. Here the collision of North America and Europe is recorded in the early Paleozoic mountains extending through New England, eastern Canada and Newfoundland, the British Isles, western Norway, and Greenland to the arctic island of Spitzbergen. Farther south, where the northwest part of Africa collided with North America, mountain-building continued into late Paleozoic or early Triassic time when the Allegheny-Hercynian orogeny created the folded sedimentary Appalachian belt (Fig. 16-11).

This late Paleozoic orogeny was, apparently, associated with the creation of Pangea, Wegener's world-continent in which Laurasia on the north became merged with the southern supercontinent of Gondwana. The single world continent was surrounded by Panthalassa, a universal global ocean.

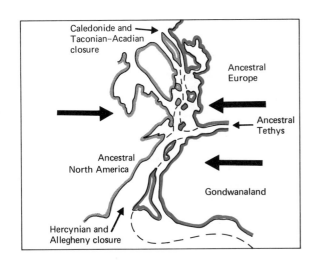

Fig. 16-11 Collision of ancestral continental margins. After drifting apart in the late Precambrian and early Paleozoic, North America, Europe, and Africa began to come together in the mid-Ordovician. Toward the north, geosynclinal deposits were compressed into Caledonide, Taconian, and Acadian ranges from the mid-Ordovician to Devonian times. To the south, Africa closed with North America to produce Allegheny and Hercynian mountains in latest Paleozoic to earliest Triassic times. After J. T. Wilson, Dewey, and others.

The Opening of the Modern Atlantic Ocean Basin
The existing plate tectonic pattern evolved around the end of the Paleozoic. Perhaps the Palisades disturbance, marked by tensional normal faulting and basaltic volcanism in eastern North America as well as mafic dikes in northwest Africa, reflects the initial splitting of Pangea. Why worldwide continental convergence was replaced by the breakup of Pangea is unknown (Fig. 16-12a–d). In any case, the record of the rocks suggests that the fracturing of Laurasia—by the initiation of a new spreading axis, now marked by the Mid-Atlantic Ridge—did not exactly correspond with the suture (welding together) line of the early Paleozoic ancestral continents. Thus some geologists think, based on their studies and speculations, that the southeastern Appalachian Piedmont region was part of Paleozoic ancestral Africa. Parts of eastern New England,

maritime Canada, and southern Newfoundland are relicts of ancestral Europe. On the other hand, northwest Ireland, the Scottish Highlands, and coastal Norway could be slivers of the Paleozoic North American continent (Fig. 16-13).

The Creation of Laurasia

The creation of Laurasia during the Paleozoic involved the collision of four ancestral continents. The interpretation is supported by the rather similar geologic record of geosynclinal deposition and mountain-building found in all landmasses north of the ancestral Tethys.

The Geosynclinal History of the Eurasian Region
In Cambrian time, when the proto-Atlantic separated North America from Europe, a seaway in a subsiding geosyncline lay between Europe and Asia, east of the present-day Ural Mountains. The region which later became modern Asia included a Siberian continent (called *Angara*) that was separated by a broad seaway in which some 7500 meters (25,000 feet) of sediments were eventually deposited from an ancestral continent (called *Cathasia*) preserved in modern China. These ancestral continents seem to have been subdued landmasses, developed on the Precambrian shields, that were surrounded by subsiding geosynclines (Fig. 16-14). The Paleozoic evolution of eastern Europe and the Asiatic proto-continents resembles the history of western Europe and the Appalachian region. Although the geologic record over such a vast region as modern Eurasia is complicated, and the recorded events not always synchronous, a familiar pattern seems evident.

During early Paleozoic time, shallow seas flooded onto the separate proto-continents and deposited a relatively thin section of sandstones, shales, and limestones. Even today, these deposits largely mask the Precambrian continental platforms which, unlike the broad Canadian shield,

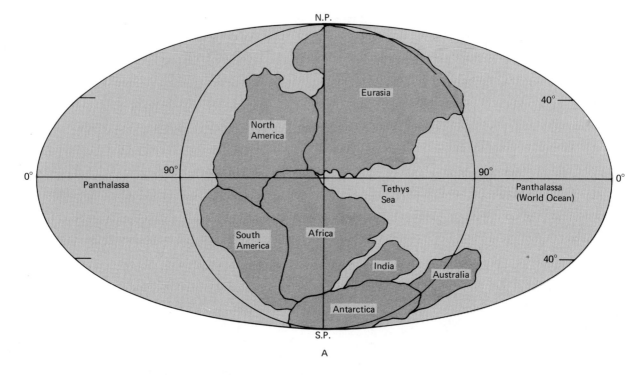

Fig. 16-12 Opening of Atlantic with the Breakup of Pangea.
(a) Permian. A modern reconstruction of the late Paleozoic world 200 million years ago when all the land was concentrated in the single world-continent of Pangea, which was surrounded by the single world-ocean Panthalassa. Adapted from "The Break-up of Pangea" by R. S. Dietz and J. C. Holden, copyright © 1970 by Scientific American, Inc. All rights reserved.

are exposed only as patches in widely separate areas (three in Siberia and six or more in China). In the surrounding, slowly subsiding geosynclines, miogeosynclinal sandstones, shales, and much limestone, as well as the graywackes, shales, and volcanics of eugeosynclinal assemblages, were deposited during Cambrian and Ordovician time.

Orogenic Events In late Silurian and Ordovician times, at different times in different places, orogenies began to deform the deposits. The Uralian geosyncline, which can be used as an example, was mildly affected by Caledonide orogeny (elsewhere in Asia the orogenic pulses may have been stronger). Thereafter the Uralian geosyncline remained a site of marine deposition until the late Paleozoic. Hercynian mountain-building, culminating in late Permian time, buckled the trough into the north-trending Ural mountain range. Its structure, topography, and history are strikingly similar to those of the Appalachians. The rising Urals shed a clastic wedge westward. For at the same time that the New Red Sandstones and other Permo-Triassic rocks were deposited in northwestern Europe, similar deposits containing interbedded salt and gypsum were laid down in eastern Russia.

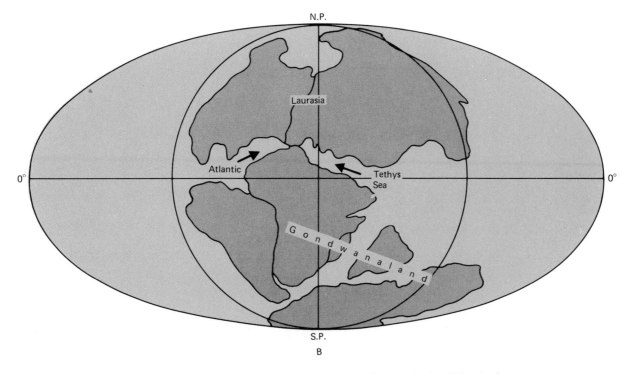

Fig. 16-12 (b) Triassic. The breakup of Pangea was under way by late Triassic time, some 180 million years ago. North America was moving away from Africa as the modern Atlantic Ocean basin began to open on the south. To the north, America and Europe were still joined in Laurasia. Adapted from "The Break-up of Pangea" by R. S. Dietz and J. C. Holden, copyright © 1970 by Scientific American, Inc. All rights reserved.

The Paleozoic and Mesozoic history is comparable throughout Asia. From Devonian into early Carboniferous times, geosynclines continued subsiding and trapping marine sediments in the broad seaway between the ancestral Siberian and Chinese continents, both of which had been enlarged by Caledonide orogenies. In the late Paleozoic, widespread orogenic pulses deformed all the Asiatic geosynclines. These episodes of mountain-building resembled the Allegheny and Hercynian orogenies in eastern North America and western Europe. They were, however, longer lasting, since strong pulses affected the eastern parts of Asia throughout Permian time. As in the rest of Laurasia (North America and Europe), the rise of the late Paleozoic mountains created adjacent terrestrial environments in which the Carboniferous coal swamps flourished.

Thus during the eventful Paleozoic era, the separate proto-continents of Europe, Siberia, China, and also North America were welded by mountain belts into the supercontinent of Laurasia. And Laurasia merged with Gondwanaland to form Pangea (Fig. 16-15). Whether the process resulted from the destruction of geosynclines between fixed continents and ocean basins, or between drifting

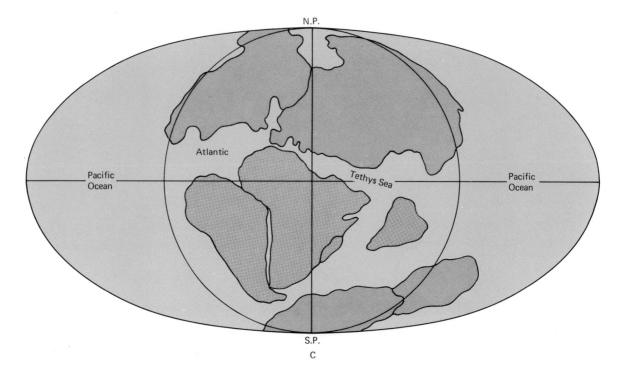

Fig. 16-12 (c) Jurassic. By late Jurassic time, 135 million years ago, Europe was starting to move away from North America as the Atlantic basin grew northward. The northern continents forming Laurasia were separated from the southern by the Atlantic and the great east–west seaway of the Tethys south of Eurasia. The fragmentation of Gondwanaland was well under way. Adapted from "The Break-up of Pangea" by R. S. Dietz and J. C. Holden, copyright © 1970 by Scientific American, Inc. All rights reserved.

plates with subduction zones, makes little difference in the reading of the rocks of Europe and Asia, which are now a combined landmass. Any remaining debate involves the Mesozoic separation of the North American continent and the origin of the North Atlantic Ocean basin. Here, the existence of the volcanic island of Iceland in the middle of the Atlantic was long considered evidence against the perfect fit of continental margins. In the plate tectonic era, however, the interpretation of Iceland as an emergent part of the spreading axis marked by the Mid-Atlantic Ridge seems to confirm the drifting of continents and laterally shifting plates.

THE TETHYAN BELT: SITE OF OUR HIGHEST MOUNTAINS

The great Alpine-Himalayan mountain ranges resulted from the orogenic destruction of the complex Tethys geosyncline during Tertiary time. This event can be interpreted as resulting from the collision of Eurasia with drifting fragments of the original Gondwanaland.

The Dividing Seaway

From sometime in the Precambrian until near the end of Paleozoic time, Gondwanaland was

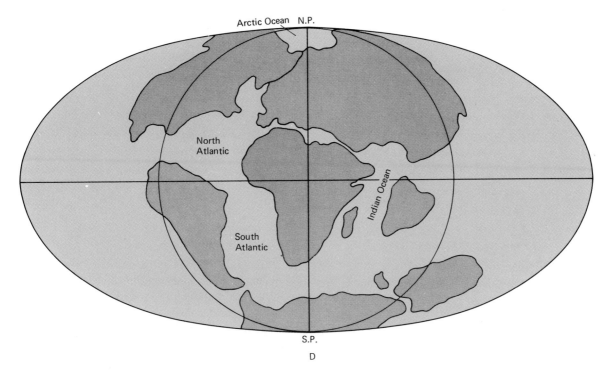

Fig. 16-12 (d) Cretaceous. Except for the connection between Europe and North America across Greenland, the pattern of the modern drifting continents was well established in late Cretaceous time, 65 million years ago. The final splitting of Laurasia, marked by the opening of the northernmost Atlantic, was a Cenozoic event. By Cretaceous time the South Atlantic Ocean was open, and Africa and Europe were converging, closing the Tethys seaway. Adapted from "The Break-up of Pangea" by R. S. Dietz and J. C. Holden, copyright © 1970 by Scientific American, Inc. All rights reserved.

separated from the Laurasian landmasses by an ancestral Tethys seaway girdling almost half the globe.

The Paleozoic Palimpsest[4]   Although much of the geologic record of the early Tethys was destroyed by later events, the region seems to have been a complex of geosynclines adjacent to the continents

and of ocean floors farther offshore. As one trough filled with sediments from island arcs, another would begin to subside in the restive belt. The merging of Gondwanaland and Laurasia into the world continent of Pangea, at the time of Hercynian orogeny, closed the west end of the Tethys near the present-day Straits of Gibraltar. East of this hinge point, the Tethys persisted.

The Later Record   The history of the Tethys is well established from the late Paleozoic onwards. Permo-Triassic deposits in the geosyncline to the south of the mountainous Hercynian belt were

[4] A palimpsest is an ancient parchment which has been used several times (a common practice before paper became cheap and abundant). Because of poor erasing, vestiges of the older writings can still be seen.

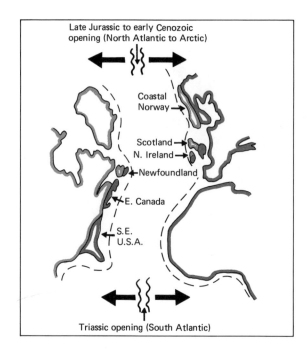

Late Jurassic to early Cenozoic
opening (North Atlantic to Arctic)

Coastal
Norway

Scotland
N. Ireland

Newfoundland

E. Canada

S.E.
U.S.A.

Triassic opening (South Atlantic)

Fig. 16-13 Opening of the present Atlantic Ocean basin. When North America split away from Europe and Africa mainly in Mesozoic time, the line of separation did not exactly follow the line where the old Paleozoic continental margins had joined together. Parts of Paleozoic Africa remained "stuck" to North America, and parts of North America were stuck to Europe. After J. T. Wilson, Dewey, and others.

The Alpine-Himalayan Mountain System

The spectacular Alpine-Himalayan mountain system winds across northwest Africa, southern Europe, and southern Asia in two sinuous belts (Fig. 16-16).

The Double Mountain Belt In places, such as the Alps, the mountain system is compressed into a single highly contorted mass. Throughout most of its length, however, the ranges of the Alpine system loop around *median masses*,[6] which are less disturbed structural blocks within the mountain chain. The orogenic structures around the blocks are thrust outwards and overfolded away from the median masses and onto the flanking continental platforms. The stable median masses in the Alpine chain consist of separate blocks of crystalline basement, metamorphosed and intruded during the Hercynian orogeny. Some of the crystalline blocks have subsided and are deeply buried beneath sedimentary rocks; in others the basement rocks are thinly covered or exposed at the Earth's surface to form crystalline massifs. The median masses have been interpreted as micro-plates, lesser fragments of continental crust. During the general convergence of the main Eurasian and African plates that destroyed the Tethyan geosyncline and created the Alpine-Himalayan chain, the micro-plates had active histories involving geosynclinal deposition and deformation, subduction zones, and transform faults.

Although some geologists have interpreted the Mediterranean region as a complex of deeply foundered blocks, the eastern and western Mediterranean Sea may contain relics of Tethyan ocean floor. In eastern Europe, the Hungarian plains seem developed on a median mass, deeply buried by sediments, that separates the Dinarus and

mainly limestones.[5] At the same time, deposits on the Eurasian platform were largely terrestrial red beds containing some salt and gypsum deposits laid down in land-locked seas. As the late Paleozoic mountains were eroded, their margins subsided and were covered by geosynclinal deposits. Today some of the contorted Hercynian basement rock is exposed in the cores of mountains—such as Mont Blanc in the Alps—which were uplifted in Cenozoic time during the Great Alpine-Himalayan orogenic event.

[5] The highly prized sculptural marble of Carrara in Italy is an example.

[6] Given the descriptive name of *Zwischengebirge* ("between the mountains") by the German geologist Leopold Kober.

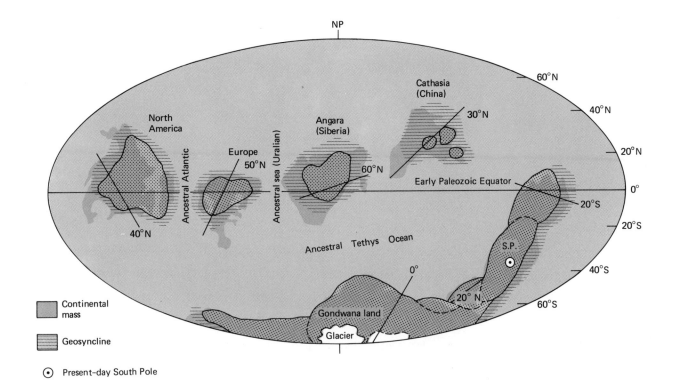

Fig. 16-14 Early Paleozoic paleogeography. A highly speculative reconstruction of how the world might have looked in Cambrian time with Gondwanaland to the south, a broad intervening ancestral Tethys ocean, and four separate northern continents. Since longitude cannot be determined from paleomagnetic data, the east–west distribution of the northern continents is purely speculative, based on geologic evidence of intervening geosynclines and ancestral seas. The ancient positions of the poles and equator can be inferred from paleomagnetic evidence. Dot patterns are Paleozoic heartlands underlain by Precambrian rocks and parallel lines are flanking geosynclines. Outlines of today's continents are given for comparison, with diagonal lines showing present-day latitudes. Compiled from many sources including Dietz and Holden, Dott and Batten, Hintze, and others.

Pindus ranges along the Adriatic Sea from the Carpathian Mountains of east central Europe.

The great Alpine-Himalayan belt is broken between Europe and Asia by the Black Sea. Some geologists attribute the Black Sea to the process of "oceanization," wherein sialic continental crust is converted to mafic ocean floor, but other geologists consider the Black Sea's floor a relic of Tethyan ocean bottom.

South of the Black Sea the double chain continues. In Turkey the Pontic Mountains on the north are separated by the Anatolian Plateau of central Turkey from the Taurus Mountains, which rise precipitously above the Mediterranean Sea to the south. Eastward, the Pontic and Tarus mountains converge in the aptly named Armenian Mountain knot, which is separated from the Caucasus Mountains of Russia only by the Kura River

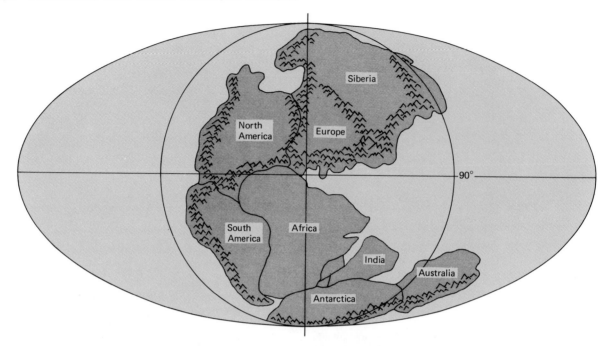

Fig. 16-15 Pangea mountain ranges. The merging of the separate ancestral continents accompanied by the orogenic destruction of geosynclinal belts may have produced a worldwide system of mountain ranges by the end of the Paleozoic. Compiled from Dewey; Dietz and Holden; Umbgrove; and Weeks.

Valley. These mountains are the sources of the Tigris and Euphrates rivers, along which some of mankind's earliest civilizations developed.

Farther eastward, the central Iranian Plateau —a complexly faulted massif like those in Turkey and Hungary—separates the lofty Kopet Dagh and Elburz mountains from the Kurdistan-Zagros ranges to the south.[7] These ranges then converge in the Pamir knot—the roof of the world—whose mountains merge into the Himalayas, the greatest of mountains (Fig. 16-17). The Himalayas seem a combined kink of ranges where the subcontinent of India has driven beneath the main plate of Asia. East of the Himalayas, the great mountain belt veers southward through the highlands of Burma and Thailand into the active volcanic-tectonic belts of the Indonesian archipelagos. These, in turn, merge with the tectonically active "Circle of Fire" which surrounds the Pacific.

The great Alpine-Himalayan mountain ranges were created in Cenozoic time when elements of the supercontinent of Gondwanaland collided with the Eurasian landmass. Different regions along this great mountain trend have had somewhat different detailed histories; yet the Alps of France, Switzerland, and Austria—although more intensely deformed than many mountains—are probably representative of the orogenic destruction of the Tethyan geosyncline.

The Alps

No other mountains are as well known geologically as the Alps, especially the section in

[7] The highly productive Iranian oil fields lie along the western margin of the Kurdistan-Zagros mountain ranges next to the Persian Gulf.

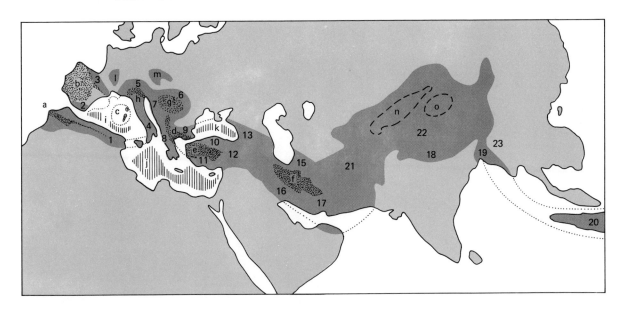

Fig. 16-16 The Alpine-Himalayan mountain system. The great east–west belt of Cenozoic mountains across north Africa and southern Eurasia rose from the Tethyan geosyncline in Mesozoic time. Knots and swirls in the grand pattern developed around median masses. In plate-tectonic interpretations, the median blocks formed micro-continents with associated geosynclines, subduction zones, and transverse faults. Compiled from Dewey, Pittman, and Ryan; Holmes; Lobeck; and Umbgrove.

I. *Median Masses of the Alpine-Himalayan Mountain Chain*

Hercynian basement at or near the surface:
a. Moroccan Meseta
b. Spanish Meseta
c. Sardinian Massif
d. Rhodope Massif
e. Anatolian Plateau
f. Iranian Plateau

Basement deeply buried by sedimentary rocks:
g. Hungarian Plain
h. Po Basin

Tethyan ocean-floor relics:
i. Western Mediterranean
j. Eastern Mediterranean
k. Black Sea

II. *Cenozoic Mountains*
1. Atlas
2. Betic
3. Pyrenees
4. Apennines
5. Alps
6. Carpathians
7. Dinaric
8. Pindus
9. Balkan
10. Pontic
11. Taurus
12. Armenian Knot
13. Caucasus
14. Elburz
15. Kopet Dag
16. Kurdistan
17. Zagros
18. Himalayas
19. Burmese
20. Island Arcs

III. *Uplifted Paleozoic Mountains and Massifs*
Massifs:
l. Massif Central
m. Bohemian Massif
Basins:
n. Tarim Basin
o. Tsaidam Basin

High mountains and plateaus:
21. Pamir Knot (Hindu Kush, Pamir, Alai mountains)
22. Plateau of Tibet

IV. *Mesozoic Mountain Belt*
23. Shan Plateau and Malaysia

Fig. 16-17 The world's greatest mountains, the Himalayas, rise where the tectonic plate of the Indian subcontinent is thrusting under the margin of the plate of Asia. In this astronaut photo from the Apollo 7 spacecraft, the snow line at about 5330 meters (about 17,500 feet), on the left, outlines the main Himalayas. Many of their peaks, including Mount Everest at lower center, rise above 8 kilometers (5 miles) above sea level. The lake-studded high plateau of Tibet is at center and right; the Himalayan "foothills" form the belt just left of the snow-capped mountains; part of the plain of India's Ganges River is at the extreme lower left. Courtesy of N.A.S.A.

Switzerland. Generations of excellent geologists have literally crawled along cliffs and swung from ropes to plot the geologic details exposed in deep chasms and jagged peaks of this scenic region (Fig. 16-18). Yet so intricate are the structures and so complicated by facies changes, both sedimentary and metamorphic, that unraveling the history and

nature of the deformation in the Alps has demanded an extraordinary amount of meticulous field observation blended with grand theory.

The Restless Geosyncline In the Jurassic, there were premonitions of things to come in the Tethys. The geosyncline began slowly buckling into arcuate ridges and intervening troughs as the embryonic Alpine folds developed. During the Cretaceous and early Tertiary, the higher ridges rose above sea level to form festoons of islands. Limestones accumulated as reefs and shell banks in shallow waters of the geosyncline and adjacent shelves. In deep waters of the subsiding troughs, a variety of clastic eugeosynclinal rocks, collectively known as *flysch*, were deposited.

Flysch consists of a thick succession of thin-bedded mudstone layers that repeatedly and rhythmically alternate with layers of coarser clastics, conglomerates, sandstones, and graywackes. The fine muds are apparently carried into the basins by the quiet action of waves and ocean currents. The coarser layers reflect turbidity currents in which clouds of material cascade down the basin slopes and settle into graded bedding. The

Fig. 16-18 Alpine peaks carved on steeply dipping limestones originally deposited in the Tethys Sea. View from Hafelkar above Innsbruck, Austria. Photo by the author.

basins of flysch accumulation must have been in-hospitable to life, since the only fossils found in them are sand-sized micro-fossils and occasional fragments of larger forms that were apparently swept down from surrounding shallow waters. Wherever flysch-type deposits are found (as for example in the Paleozoic Appalachian rocks or the Cretaceous of California), they are interpreted as evidence of a tectonically active marine environment that precedes the main phase of orogeny.

The Climax of Orogeny The Alps emerged in a paroxysm of mid-Tertiary mountain-building. During the main episode of Miocene orogeny, shelf, geosynclinal, and ocean-floor rocks were jammed northward between and over the tops of rising Hercynian basement blocks. Flysch was displaced from its individual depositional troughs, and shallow-water sedimentary rocks were squeezed out of synclines developing between rising buttresses of Hercynian basement rocks. In places sediments of rocks lying upon or forced over rising crystalline massif blocks slid down steep mountain fronts. In this incredibly complex style of Alpine deformation, the dominant major structures are large gravity-emplaced masses involving thrust faults and the great recumbent folds called *nappes*.

The Nappes In the 1800s, most geologists thought that the Alps were relatively simple folds of compression. But just before the end of the century, Maurice Lugeon proved conclusively that the nappes were originally continuous masses that were transported horizontally great distances, one over the other—mainly towards the north from roots far to the south. More than a century of careful observation, detailed mapping, and considerable theorizing culminated in the imaginative theory of Emile Argand (1916).

He proposed that the continental block of Africa had overridden the block of Europe, squeezing the contents of the Tethyan trough into the great Alpine folds that, in his words, "spilled like

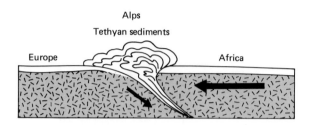

Fig. 16-19 Argand's concept.

breakers on a coast." According to this long-accepted view, the geosynclinal rocks "squirted like toothpaste from a tube" into a tremendous pile of folds, flopped over towards the north (Fig. 16-19). Groups of folds assumed to be rooted in the south traveled the farthest and completely overrode folds whose roots lay farther north. Thus the pre-Alps on the northern margin of the mountains, near Lake Geneva, were considered an erosionally isolated mass, or *klippe*, that had come from far to the south. In Argand's view, the pre-Alps represented the leading edge of a higher fold that once arched completely across Switzerland from roots on the south side of the great pile of Alpine folds.

The nappes have behaved in a very plastic manner. And it was clear to Lugeon and Argand that there had been a tremendous shortening of the Earth's crust, in which a geosynclinal belt estimated to have been some 640 kilometers (400 miles) wide had been compressed into mountains 160 kilometers (100 miles) across. The Alps are indeed a great pile of Tethyan sediments squeezed out of troughs by the closing of a gigantic vise whose jaws were continental plates. But today, Argand's extreme version of plastic deformation has lost favor on theoretical-mechanical grounds and in light of later studies in the French Alps where deformation is less extreme.

The newer view emphasizes the role of local downwarped basins and of resistant crustal blocks called *massifs*. Sediments are now thought to have been squeezed out from separate troughs within the

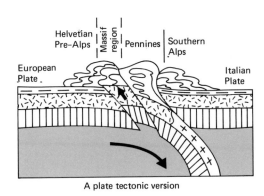

Fig. 16-20 Modern concepts assume less great plastic deformation in the Alps than did Argand. The local blocks and basins involved in deformation have been fitted to plate-tectonic theory. After Dewey and Bird.

broad Tethyan geosyncline, or from synclines developing between massifs rising on the European continental shelf as the orogeny progressed (Fig. 16-20).

Erosion and Post-Orogenic Uplift As the Alps rose in late Oligocene and Miocene times they were eroded and shed a great wedge of debris northward to the flank of the Jura Mountains and into southern Germany. These thick deposits beneath the Swiss Plain are called the *molasse*. They contain soft conglomerates, sandstones, and calcareous mudstones. In contrast to the flysch, the molasse deposits lack graded bedding but are characterized by ripple marks, cross-bedding, and other primary sedimentary structures, as well as fossils. They record partly shallow marine and partly transitional and terrestrial environments whose shifting patterns reflect orogenic pulses. The oscillations of sea and land produced lagoonal and shallow marine deposits, bouldery deltaic fans, sandy alluvial plains, and muddy-bottomed fresh-water lake deposits.

In late Tertiary time, erosion had reduced the Alps to low rolling uplands. Then in Pliocene time,

epeirogenic uplift rejuvenated the mountains as they rose to their present elevations. The former upland surface was deeply dissected by youthful mountain gorges. Relics of the erosion surface are now found in high-country meadows, the Alps that give these mountains their name. During the Pleistocene, extensive valley glaciers carved out the U-shaped troughs, jagged peaks, and sharp ridges that characterize Alpine scenery.

Other mountains of the Alpine-Himalayan belt usually have less intensely deformed structures and different directions of thrust faulting; they expose more crystalline basement; they have greater masses of emplaced granite; and they may have had somewhat different timing of their orogenic pulses. The well-studied Alps, however, give a frame of reference for interpreting flysch and molasse, nappes, and massifs that characterize other major Tertiary ranges (Fig. 16-21).

The Roof of the World

Where the mountain belts that rose from the Tethyan geosyncline pass into Afghanistan, Pakistan, and adjacent regions north of the Indian subcontinent, they create the Earth's greatest mass of extremely high land. Here Tibet, whose average elevations are about 4900 meters (16,000 feet) above sea level, forms the world's highest plateau. The Pamir Mountain knot, whose flat-topped fault-block ranges average about 6100 meters (20,000 feet), includes the Alai Mountains, having 7600-meter (25,000-foot) peaks that are the highest in the Soviet Union, as well as the equally high ranges of the adjacent Hindu Kush Mountains. These folded mountains are the result of Paleozoic orogenies; however, their existing elevations reflect late geologic events, because the southern part of the Tibetan Plateau contains marine rocks whose fossils indicate deposition of strata in the Tethyan Sea during early Cenozoic time. The epeirogenic rise of this "roof of the world," which reflects an un-

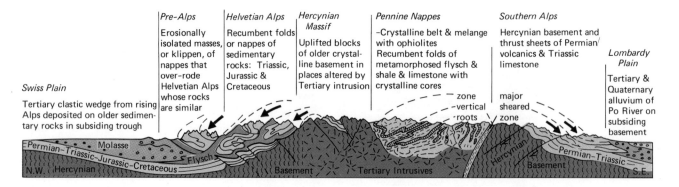

Fig. 16-21 The exceedingly complex structure of the Swiss Alps has been described as a case of "pathologic" orogenic deformation. Yet in its broad elements we can recognize a central region of uplifted massifs and a crystalline and mélange belt. The great folds and thrusts involving sedimentary rocks on the northern and southern Alps have been interpreted by some workers as having slid down from the central regions under the force of gravity. Compiled from Argand, Bailey, Dewey, and Bird, De Sitter, Heim, Holmes, Kober, and Umbgrove.

usually thick sialic crust, accompanied the creation of the Himalaya Mountains (Fig. 16-22).

Tethyan Marine Beginnings Himalayan geologic and structural history is generally similar to that of the Alps, in the order of events. When Asia, to the north, was expanded by the late Paleozoic Hercynian orogeny, the Tethys geosyncline became concentrated along an axis later to become the Himalayan region. Marine deposits in parts of this region were affected by the late Paleozoic folding, but geosynclinal deposition in the area continued until late Cretaceous time. Then seas partially withdrew from the region during a deformational phase associated with submarine volcanism, thrust faulting in the geosyncline, and flysch deposition in actively subsiding troughs. In early Cenozoic time, the sea flooded across the region as far north as southern Tibet.

Himalayan Orogeny The Himalaya Mountains arose during a series of deformational pulses, beginning in the late Eocene and lasting through Tertiary time, that involved intense folding and faulting

along with the deep-seated emplacement of now-exposed granite batholiths. Unlike the minutely studied Swiss Alps, the Himalayas are known only through scattered geologic traverses and limited mapping in small areas; so it is little wonder that interpretations of their basic structure differ. The Swiss geologist Augusto Gansser believes that the Himalayas contain complicated nappe structures, like the Swiss Alps, but with more extensive granitic intrusion. Others envision a simpler structure, folded and faulted but less complicated than the Alps.

In any case, as the Himalayas rose, they were extensively eroded and their debris was shed into a subsiding trough adjacent to the southern mountain border. The deposits created the *Siwalik Series*, a clastic wedge of molasselike deposits some 6½ kilometers (4 miles) thick. A late Tertiary spasm of southward thrust-faulting folded adjacent parts of the Siwalik deposits into impressive ranges known as the Himalayan foothills (Fig. 16-23).

Culminating Events During Quaternary time, the Himalayas rose to their present elevations during a

Fig. 16-22 The top of the world. High ranges to right of center include Mount Everest (highest) 8848 meters (29,029 feet) high. Peak to right of Everest is Mount Lhotse, 8501 meters (27,890 feet) above sea level. Ridge on the left of Everest is the Nuptse, 7906 meters (25,850 feet) high. Photo by the author.

final 1800-meter (6000-foot) epeirogenic uplift (judging from the flat-lying Tertiary marine rocks in the Tibetan Plateau to the north), and streams carved great erosional gorges across the main Himalayan ranges. Thus Triassic limestones, laid down in the Tethys Sea, came to occupy the highest point on Earth, the top of Mount Everest. South of the foothills belt, Siwalik deposits subsided and were buried under several hundreds of meters of alluvium in the modern geosyncline represented by the vast flood plain of the Indus and Ganges rivers.

The upper reaches of these streams are spectacular examples of *antecedent* rivers. That is, they are thought to have maintained their preexisting courses as the Himalayas rose across their paths. These south-flowing streams do not head along a major drainage divide in the highest mountains, but rather to the north. From their sources in the Plateau of Tibet, the streams flow across the axes of highest ranges in tremendous gorges—the gorge of the Indus River is 5100 meters (17,000 feet) deep where it passes through Gilgit in the Kashmir of northern India. The final episode enhancing the spectacular topography of the Himalayas was Quaternary glaciation. Today these ranges, only 28° latitude north of the equator, still

Fig. 16-23 The "foothills" of the Himalayas rival the North American Rockies. Slopes terraced for agriculture near Katmandu, Nepal. Photo by the author.

support large glaciers and permanent snow fields above the 4800-meter (16,000-foot) level.

In the new world view of plate tectonics (and as suggested by the advocates of continental drift in earlier decades), the roof of the world results from the collision of two light sialic continental masses. Where the subcontinent of India is jammed under the Asiatic plate, a deep mountain root results whose isostatic rise has uplifted the Himalayas and formed the Earth's largest mass of high terrain.

THE NORTH AMERICAN CORDILLERA: CONCEPTS AND PROBLEMS

In light of Uniformitarianism, geologists are now fitting the history of the North American Cordillera to plate tectonics. Yet valid as the doctrine seems, it does not guarantee quick answers to all interesting geologic questions. Rather, the paradigm provides the rules for geologic puzzle-solving and determines the nature of currently worthwhile problems to be solved.

The Record from the Rocks

When geologists can relate their hypotheses and theories to the existing regime of drifting plates, their attempts to unravel the geologic history of the complex Cordillera seem fairly successful. But interpretations of ancient Cordilleran plate tectonics, when the global system was different, are highly speculative, since models can only be inferred from the nature of the ancient rocks.

Late Precambrian and Early Paleozoic Speculations
Unlike the Appalachian region, whose Paleozoic history can be well explained by the alternate opening and closing of the Atlantic Ocean, there is no generally accepted plate tectonic scheme for the ancestral interactions of North America and the Pacific Ocean basin. Quiet geosynclinal deposition prevailed on both sides of North America during the late Precambrian, the Cambrian, and into the Ordovician (Fig. 16-24). Since the miogeosynclinal-eugeosynclinal relations in the east are attributed to shelf and slope deposition on the passive trailing edge of a rifted continent, it is tempting to use the same explanation for the comparable western geosynclinal deposits.

But unless the continent was somehow static, a passive eastern continental margin (using the present-day plate tectonic pattern as a model) should have been accompanied by a tectonically active leading edge on a westward-drifting plate. And, if the ancestral Cordilleran geosynclines reflect the previous splitting of a large continent—where is the other continental fragment? There is no evidence of a sialic fragment that drifted out into the

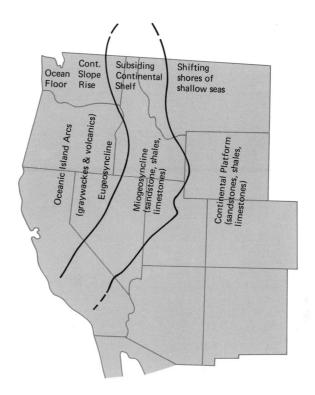

Fig. 16-24 The Cordilleran region from early Paleozoic to Devonian time. The Cordilleran geosyncline with mio-geosynclinal and eugeosynclinal sites of deposition lay west of the continental platform. Compiled from Birchfield and Davis, Fenneman, Hamilton, Haun and Kent, King, Maxwell, and Poole.

Pacific. Perhaps,[8] if there were ancient faults with pronounced lateral displacements as on the existing San Andreas fault, we should look for such a fragment in Alaska.

An alternative interpretation—which avoids the problems of continental rifting—is the suggestion that the early Paleozoic Pacific margin was a region of island arcs. In this hypothesis, the relict volcanics, graywackes, and associated deposits would mark the site of a tectonically active continental margin.

8 Brainstorming is a legitimate preliminary to plate tectonic reconstructions.

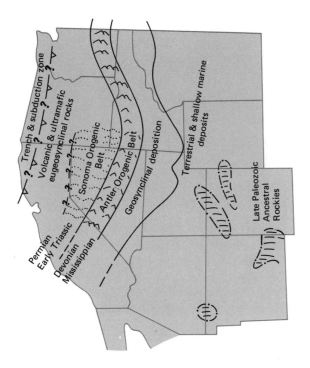

Fig. 16-25 The Cordilleran region from Devonian to early Triassic time. After the development of the Antler and Sonoma belt, the site of the Paleozoic Cordilleran geosyncline was divided into an eastern geosynclinal belt of deposition and a western eugeosynclinal belt along the Pacific border. Compiled from Birchfield and Davis, Fenneman, Hamilton, Haun and Kent, King, Maxwell, and Poole.

Devonian to Triassic Orogeny The Antler orogeny of Devonian and Mississippian times marks a change in Cordilleran evolution. In the Antler event —roughly corresponding in time to the Acadian episode in Appalachian orogenies—field evidence demonstrates that eugeosynclinal deposits were deformed and thrust eastward over miogeosynclinal deposits in central Nevada (Fig. 16-25). Thus rocks of the continental slope and ocean floor were plastered onto the western continental margin. Although its existence can only be inferred from the record of the continental rocks, it is reasonable to assume that a subduction zone had developed along

the colliding boundary of the Paleozoic American and Pacific tectonic plates. Thereafter orogenic deformation—such as the Sonoma event, corresponding in time to the Allegheny-Hercynian orogeny along the Atlantic—can be somewhat confidently related to a subduction zone between colliding plates. In the process, which continues to the present, the Cordilleran continental plate has grown westward into the Pacific Ocean basin.

Eastward Waves of Mesozoic and Early Cenozoic Deformation The Nevadan, Sevier, and Laramide orogenic developments seem to represent a progressive eastward-moving wave of mountain-building onto the sialic continental platform (Fig. 16-26).

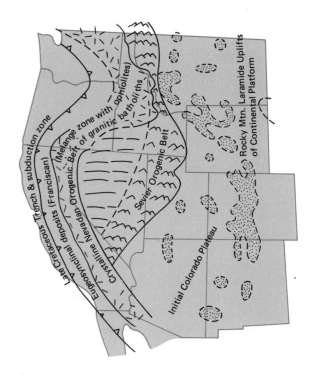

Fig. 16-26 The Cordilleran region in Mesozoic time. By latest Cretaceous time, Nevadan Sevier and Laramide orogenies had created a wide mountainous belt across the Cordillera. Compiled from Birchfield and Davis, Fenneman, Hamilton, Haun and Kent, King, Maxwell, and Poole.

The initial phase of Jurassic and Cretaceous Nevadan orogeny nicely fits the plate tectonic scheme of mélanges and intrusive granitic masses. Eastward across Utah into westernmost Wyoming, the Sevier mountain structures involve low-angle thrusting of contorted masses of miogeosynclinal rocks. During the last phases of orogeny, the thick sialic continental platform was deformed to produce the vertical uplifts in the Colorado Plateau and frontal Rockies of the Wyoming and Colorado regions. Some geologists have suggested that the Sevier-type structures slid, under the force of gravity, off elevated Nevadan welts. However, most now theorize that Sevier structures were pushed eastward by deep-seated compression, somehow related to subducting plates and forces from the Nevadan belts. The cause of the flat-topped Laramide uplifts remains an intriguing problem.

The Present Plate System

Historical interpretations of the complicated Mesozoic and Cenozoic geology of California and the adjacent Pacific basin can be related to the presently active system of continental and oceanic plates.

California and the San Andreas Fault Through Mesozoic time and into the Tertiary, we can envision ongoing orogeny as the Pacific sea-floor plate collided with and descended beneath the North American continental plate (Fig. 16-27). Inland, above the deeper parts of the subducting plate, andesitic and rhyolitic magmas rose into the crust and formed batholiths in a crystalline belt along an axis now exposed along the Sierra Nevadas. Seaward, the Franciscan Series and related mélanges were deposited, crumpled, and plastered against the continental plate.

The next phase of history involves the development of the San Andreas fault. It is theorized that as the North American plate drifted westward, it

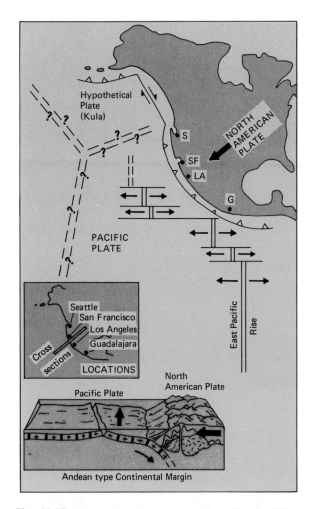

Fig. 16-27 Mesozoic plate relations along the Cordilleran margin. Prior to the development of the San Andreas fault, plates of the Pacific Ocean were probably descending beneath the North American plate and creating a subduction zone like the present one along the west coast of South America. Compiled from Atwater, Anderson, Hintze, and others.

overrode a spreading axis in the oceanic plate which is marked by the East Pacific Rise. The easternmost segment of the rise (which was offset between transform faults) converged with the trench of the subduction zone along southernmost California and

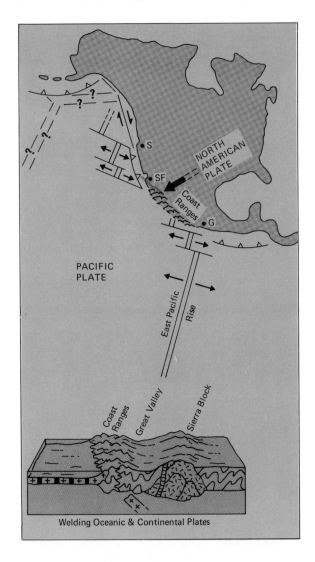

Fig. 16-28 Cenozoic events leading to development of the San Andreas fault. The divergent axis of sea-floor spreading in the East Pacific Rise was over-ridden by the North American plate. Here the subduction zone and the spreading axis became inactive, and oceanic and continental plates welded together. Compiled from Atwater, Anderson, Hintz, and others.

northern Mexico. Here, the merging of rise and trench neutralized each other. The subduction zone was destroyed, oceanic and continental plates were welded together, and crumpled sea-floor deposits emerged as part of the Coast Ranges (Fig. 16-28).

But since the margin of the vast Pacific ocean-floor plate was slowly rotating northward, while the continental plate drifted westward, the newly joined plates sheared, forming the San Andreas strike-slip fault system. On their west side, the Coast Ranges (attached to the oceanic plate) have moved northward relative to rocks east of the fault (attached to the continental plate).

Farther south a segment of the East Pacific Rise slid beneath the continental plate. Here, for some reason, the spreading axis remained active and stretched the overriding continental plate to slowly open up the Gulf of California between the mainland of Mexico and a fracture splinter represented by the peninsula of Baja California. Thereafter the southern part of the San Andreas fault came into existence as Baja California commenced to drift northward along with the rotating Pacific Ocean plate (Fig. 16-29). All in all, this broad scheme is a plausible hypothesis that ties together many scattered observations from a complex geologic region, but it is not the only possible interpretation.

Breakup of the Cordillera That the horsts and grabens of the Basin and Range Province result from extension (stretching) of a broad segment of the American continental plate seems evident to most geologists, but the cause of the extension is a subject of debate. The Basin and Range lies north (and also east) of the Gulf of California (Fig. 16-30). Thus some geologists suggested that the northward extension of the East Pacific Rise remained an active spreading axis after it was overridden by the North American plate. The theory would explain high heat flows recorded in wells drilled into the province, and the broad upwarping and extension producing the horsts and grabens. Another theory,

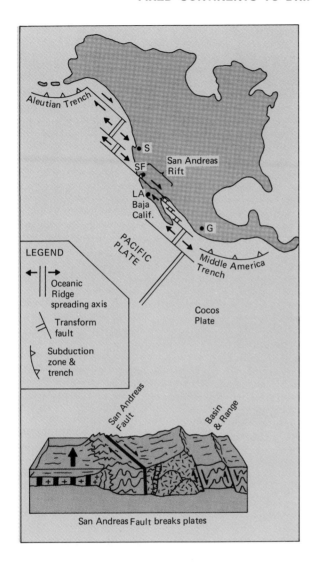

Fig. 16-29 The origin of the San Andreas fault. The northward motion of the Pacific sea-floor plate creates a transform fault along which the southern California Coast Ranges and adjacent Mexico are now being displaced northward. Compiled from Atwater, Anderson, Hintz, and others.

more generally accepted today, relates the origin of the province to the collision of the East Pacific Rise with the North American lithospheric plate and the creation of the San Andreas fault system. The Basin and Range Province is attributed to a twisting and extension of a broad continental region resulting from the northward drift of the adjoining oceanic plate. Yet another theory relates the province to stretching of the continental plate over a broad upwelling of plastic mantle into the overlying crust. The rising mantle material that penetrates and domes up the crust (technically a *diapir*) is thought to have originated after the destruction of the subducting sea-floor plate.

The eastward continuation of the Columbia Plateau volcanics into the Snake River Plains of southern Idaho, thence into the Yellowstone vol-

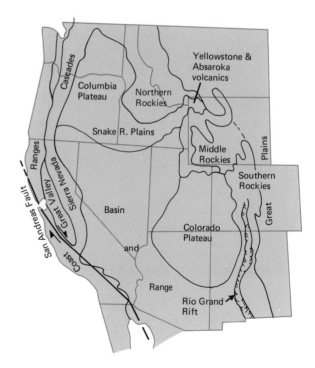

Fig. 16-30 Present-day Cordilleran Region.

Fig. 16-31 Snow on the adjacent country accentuates the trend of the Rio Grande Depression (the dark band extending diagonally from lower left through center). A region of high heat flow, the depression could be over a mantle plume and perhaps a potential axis of spreading. The dark ring (left center) is the Turkey Mountain Dome. The Great Plains are on the right. The Rio Grande Depression is flanked by Basin and Range fault blocks (lower left) and the Southern Rocky Mountains (from center northward). The Colorado Plateau is visible on the left. This astronaut's view extends across New Mexico and Colorado to Wyoming and Nebraska on the horizon; parts of Arizona and Utah are on the left. Courtesy of NASA.

canic plateau of Wyoming, seems significant. The general eastward elongation might reflect a tensional zone developed over mantle plume. This development introduces an interesting possibility, because the Rio Grande depression extending through New Mexico into southern Colorado also seems a tensional zone with associated volcanics that could also be an incipient axis of spreading (Fig. 16-31). One fine day—geologically speaking —these east-trending and north-trending axes of spreading might intersect and send the southwest part of North America drifting off to form a new subcontinent. Thus in light of plate tectonics and our present inadequate knowledge of the North American Cordillera, imaginative geologists have proposed a number of speculative but testable hypotheses whose final acceptance or rejection hinges on many more field investigations combined with physical measurements from geophysical instruments.

SUGGESTED READINGS

Bucher, W. H., *The Deformation of the Earth's Crust*, New York, Hafner Publishing Co., 1957.

Burk, C. A., and Drake, C. L., *The Geology of Continental Margins*, New York, Springer-Verlag Co., 1974.

Du Toit, A. L., *Our Wandering Continents: An Hypothesis of Continental Drifting*, Edinburgh, Oliver & Boyd Co., 1937.

Hintze, L. F., *Geologic History of Utah*, Provo, Utah, Brigham University Geology Studies, Vol. 20, Part 3, 1973 (paperback).

Kay, Marshall, *North American Geosynclines*, Boulder, Colorado, The Geological Society of America, Memoir 48, 1951.

Kummel, Bernhard, *History of the Earth*, San Francisco, W. H. Freeman and Co., 1961.

Sullivan, Walter, *Continents in Motion: The New Earth Debate*, New York, McGraw-Hill Book Co., 1974.

Umbgrove, J. H. F., *Symphony of the Earth*, The Hague, Netherlands, Martinus Nijhoff Publishers, 1950.

Van Waterschoot van der Gracht, W. A. J. M., and others, *Theory of Continental Drift: A Symposium on the Origin and Movement of Land Masses both Intercontinental and Intra-continental, as Proposed by Alfred Wegener*, Tulsa. Okla., American Association of Petroleum Geologists, 1928.

Wilson, J. T., *Did the Atlantic Close and Then Reopen?*, in *Nature*, Vol. 211, pp. 676–681, 1966.

Wilson, J. T., and others, *Continents Adrift*, Readings from *Scientific American*, San Francisco, W. H. Freeman and Co., 1970 (paperback).

Geologic Perspectives and Some Human Concerns

Geology is neither one of the humanities nor a social science, since it is directed outwards towards a better understanding of the Earth. Yet an appreciation of geologic principles is clearly relevant for dealing with increasingly pressing environmental problems.

The Column of Life by Norwegian sculptor, Adolf Gustav Vigeland. Photo courtesy of Annie Marie Walthall.

17

Geology, Environment and the Energy Crises

We live in a time of growing environmental crisis whose proper understanding requires an appreciation of geology. In a sense, geology has always been an environmental science (since it deals with the Earth); yet only since the 1960s has professional interest focused on the field called *environmental geology*. It deals specifically with human interaction with the Earth, both the impact of geology upon mankind and also the human impact on geologic processes.

PEOPLE, FOOD, AND GEOLOGY

At the heart of the environmental crises is the human "population explosion." It stems from the industrial-medical revolution of the last two centuries that eliminated the need for slave labor and dramatically reduced deaths from childbirth and disease. There were earlier times of population expansion, however, accompanying other human breakthroughs in exploiting the Earth environment.

Growth of the Human Population

Based on anthropological evidence, an early population expansion occurred when true men evolved from the primates a million or more years ago. Thereafter the number of people may have risen to somewhat under 10 million as they improved their methods for coping with the environment. In this first cultural revolution, humans developed a social organization based on hunting and gathering of food supplies. They developed efficient stone tools, learned to use fire, and began to migrate from the tropics into the temperate climatic zones. Another population increase began some 10,000 years ago in the Near East. It stemmed from the invention of agriculture, the discovery that food could be produced by cultivating crops, and the domestication of animals. With these developments, human civilizations became possible as permanent villages and cities with complex systems of government emerged. Extensive use was made of geologic resources:

water, soil, building materials, and metals. It is estimated that world population in this phase of human history had reached some 250 million by the time of the birth of Christ.

The Population Explosion By 1850 A.D. the world population reached one billion; the effects of the industrial-medical revolution were becoming evident. Then, in less than a century, world population doubled to some two billion by 1930. In the next 45 years, the population again doubled, to about four billion in 1975. Should the accelerating growth rate continue, the population could again double to some eight billion people by the year 2010 (Fig. 17-1).

The Ghost of Thomas R. Malthus In 1798 the Reverend Thomas Malthus predicted that unrestrained human reproduction would soon far sur-

pass the Earth's power to produce the food needed for mankind. Does the Malthusian principle—which Darwin adopted as a mechanism for animal evolution—apply to people? The failure of Malthus's grim prediction certainly did not result from population control, but rather from a great expansion in food production which he had not foreseen. The nineteenth century was marked by great improvements in agricultural techniques, improvements in food distribution, and the opening of vast new regions to farming, notably in North America. In fact by the 1930s, a glut of farm products in the advanced countries seemed the major problem, since it depressed prices paid to farmers.

Yet in the 1950s, the Malthusian prediction had surfaced again in alarming reports by the United Nations Food and Agriculture Organization. These reports included statistics from the Third World of less-developed countries. However, not all agricultural experts agreed with the statistics that suggested an imminent world food shortage. And in the late 1960s, the introduction of high-yielding varieties of wheat and rice into Asia, during agriculture's "Green Revolution," led to new optimism. Still, although the expansion of populations in the industrialized countries has tended to level off, that in the less developed nations has continued its rapid rise.

The Exploitation of Soils and Agricultural Land

Should the world population continue to increase at the present accelerating rate, the food situation could indeed become critical. The amount of new land that could be brought into cultivation is limited, since most is already needed to support the Earth's present four billion people. The World Food Organization estimated in 1972 that some 11 percent (about 1.5 billion hectares or 3.7 billion acres) of the Earth's surface is suitable for cultivation and crop production; roughly 22 percent (about 3 billion hectares or 7.4 billion acres) serves

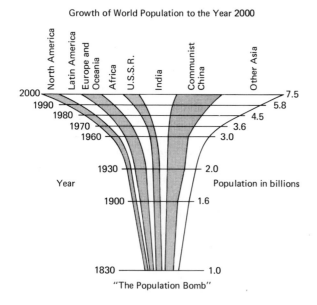

Fig. 17-1 World population. *Environmental Quality*, The First Annual Report of the Council on Environmental Quality, August 1970.

as range, pasture, and meadow for cattle and other livestock; about 30 percent (four billion hectares or ten billion acres) is forest. The rest of the Earth's land surface consists of deserts, arctic tundras, or steep mountains that are unsuitable for agriculture. Some new lands could be opened for farming, but probably not enough to solve the long-term problem.

Modern Agricultural Technology   Increased yields are possible in presently cultivated regions of the less-developed countries through the introduction of mechanized farming and Green Revolution technology. But after its initial successes, the revolution has seemingly passed its climax. The new high-yield crop varieties based on "magic seeds," require much water, fertilizers, and pesticides which are often in short supply in the less-developed countries. Some plant biologists warn that the threat of devastation by pests and plant diseases is increased by concentrating crop production in a few uniform plant strains. A greater genetic pool, in contrast, increases the chances of plant varieties that resist a particular epidemic.

Introduction of highly efficient mechanized farming methods into the less-developed countries of Africa and Asia might seem a simple solution to the food problem. But such techniques require large capital outlays for tractors and equipment, fertilizers, and insecticides; and they consume large amounts of gasoline and other petroleum products whose prices are rising because of the impending energy crisis. Moreover, techniques suitable to the American farm belt cannot be blindly applied to less-developed countries in tropical climatic regions. Aside from the inevitable dislocation of well-entrenched social patterns and tampering with agricultural practices that may be well-adjusted to the regions, there are geological and climatological considerations.

Tropical Soils   Aside from the lack of a significant change from summer to winter temperatures, tropi-

cal soil formation results from factors similar to those operating in temperate zones. Thus there is no single "tropical soil," but rather various kinds reflecting different parent materials, topography, and age, as well as climatological and organic factors. Asia's "rice bowl" in the flood plain and delta of the Mekong River is an example of a highly productive agricultural region where water is abundant and soil is renewed each year by floods which lay down a new layer of silt. Volcanic regions such as Java, Costa Rica, and parts of Africa have highly fertile soils resulting from the weathering of volcanic ash, rich in plant nutrients of calcium, magnesium, and potassium. Some 50 percent of tropical soils, however, as in much of central Africa, South America, and southeastern Asia, are highly leached of plant nutrients. To increase their productivity would require large applications of fertilizer.

It might seem that large tracts of tropical rainforests could be cleared and made into highly productive farmlands. But the lush vegetation reflects high rainfall—not great soil fertility. The soils protected by abundant vegetation are impoverished. They are low in humus, since rapid bacterial action in the warm wet environment rapidly destroys plant litter. They are low in plant nutrients, because high rainfall leaches and washes nutrients out of the porous soil structures.

In some wet tropical regions, weathering produces *laterite* (Latin for "brick"). It results where acid groundwaters remove silica and bases from the soil, leaving residual materials that are mainly oxides of iron and aluminum. When the materials are stripped of their natural cover and exposed to air, they harden into a natural brick. The materials are cut into blocks and dried to form durable building stones in such regions as Thailand. Extreme lateritic weathering has produced valuable ore deposits of aluminum (bauxite) and iron (hematite and limonite). Agriculturally, however, laterization is a serious problem in some tropical regions.

The "lost" Khmer civilization of Cambodia may provide an instructive example. During their height of vitality from the ninth to the sixteenth centuries A.D., the Khmers built great temple-tombs, such as the Angkor Wat near their capital and the splendid walled city of Angkor Thom. Besides their magnificent sculptured architecture built from stone and durable laterite blocks, they laid out carefully planned roads, canals, and irrigated rice fields. Then the high culture declined and their cities were abandoned, to be overwhelmed by dense jungle which hid them until their rediscovery in the nineteenth century. The mysterious collapse of the Khmer civilization has long puzzled archeologists. It has been attributed to various social factors such as constant wars, the craze for elaborate construction, and conversion to Buddhism. Perhaps the basic factor was agricultural failure as cultivated plots deteriorated into laterite.

A modern example may be the Brazilian government's Iata agricultural colony, established in the Amazon Basin of South America. Here, the jungle was cleared using modern heavy equipment, and crops were planted. Five years later, the new fields were virtual pavements of laterite rock. Dahomey in tropical West Africa had a similar experience, although the process there took some 60 years.

Irrigated Lands The modern Israelis have made parts of the Negev Desert flower. Agricultural production might be increased in the Ganges plain of India by the diversion of surface waters and the tapping of underground sources. Irrigation is an ancient agricultural technology, that was in extensive use in Mesopotamia 3000 years before the birth of Christ. Here, along the Tigris and Euphrates rivers the Babylonians created one of the first great civilizations.

Where surface streams are available—and their waters not already totally allocated for use—more land in arid regions could be brought into cultivation. Underground water is another great source for irrigation, although its use is costly because it usually requires pumps and energy sources to drive

them. Unless carefully controlled, however, irrigation can cause *salinization*. Excess irrigation and improper drainage in areas of high evaporation may cause deterioration of cultivated lands by bringing salts to the surface. When salts leached from irrigated lands by percolating waters enter a river, they increase the salinity of the stream for downstream users. In recent times irrigated farms in the Imperial Valley of California and, especially, Mexico have been plagued with 100 percent or more increase in salt content of the waters obtained from the lower reaches of the Colorado River. Costly desalinization facilities were the only solution to the problem. Salinization from extensive irrigation may have been a major cause for the downfall of the Babylonian civilization—a conclusion based on records (inscribed in clay tablets) of ancient temple surveyors and records of deteriorating grain production.

The American Farm Belt With productive soils, a generally beneficent climate, and an advanced agricultural technology, the middle-latitude farm belt of the United States and Canada produces a food surplus. The ability to export food products is a major asset in maintaining a favorable international trade balance, which is needed for a healthy economy. Yet we face problems of another sort, brought on by population expansion and affluence.

In the last 20 years, some 10.9 million hectares (27 million acres) of agricultural land (an area greater than that of Ohio) have been lost in the United States to real-estate development of expanding cities and suburbs and highway and airport construction. And even with efficient modern farming techniques, the United States loses some 3.6 million metric tons (4 million tons) of topsoil—the essential element of agriculture—each year through erosion by wind and water (Fig. 17-2). Put another way, rough estimates of the thickness of topsoil in the United States before intensive agriculture began are on the order of 23 centimeters (9 inches); now the overall thickness is estimated at 15 centi-

Fig. 17-2 Loss of invaluable topsoil from a New Jersey farm. Photo by A. S. Barnhart. U.S. Soil Conservation Service.

meters (6 inches). Even if such estimates are too high, the loss of soil is appreciable. Although the rate of new soil formation varies considerably in response to geological, climatological, and biological factors, the creation of new soil is a slow process requiring, at best, a few decades and perhaps as much as several centuries.

Thus in terms of the critical present-day environmental problems of world food supply and the population explosion, soil is an essentially nonrenewable geologic resource whose availability for agriculture has a definite limit. Moreover, complicating any long-range solutions to these pressing problems is the specter of added pressures brought on by climatic changes of the sort revealed in the geologic record of the last two million years of Earth history.

The "Postglacial" Climate

In the late nineteenth century, G. K. Gilbert, one of America's most outstanding geologists, wrote: "When the work of the geologist is finished and his final comprehensive report is written, the longest and most important chapter will be upon the latest and shortest of the geological periods."

Geologists made impressive progress in the nineteenth and the first half of the twentieth centuries in establishing the concepts of an Ice Age and multiple glaciations, as well as unraveling the general climatic implications of Pleistocene glacial and interglacial history—by diligently mapping moraines, piecing together the sequence of tills and other glacially related deposits, and correctly interpreting the significance of surface and buried soils. Their absolute age determinations, however, were mainly crude estimates of years based on such things as assumed rates of valley cutting (as along the gorge of the Niagara River) and of rock weathering and on the times required for soil development. Some absolute dates had been obtained from annual glacial varves, but study of the Quaternary period was hampered by a general lack of reliable time indicators.

The prevailing geologic opinion was that glaciations were relatively short intervals, interglacial times long lasting, and changes from one climatic regime to another were gradual. Overshadowed by the grander images of the Pleistocene were studies of the time since the great continental glaciers had disappeared from north Europe, Siberia, and North America.

Postglacial time was designated the *Recent* epoch; modern technical publications commonly use the name *Holocene*. Since the "Recent" was generally considered as the beginning of an interglacial episode, it was dramatically predicted that the vast ice sheets would return again, but the onset seemed remote. Significantly, modern soils on Wisconsin deposits are not as well developed as the older interglacial soils on Illinoian and older deposits. Because time and climate are important soil-forming factors, it was concluded that we are in the early stages of an interglacial interval. Based on rough estimates of the rates of weathering and soil formation, some geologists suggested that the Recent epoch began about 25,000 years ago and would probably continue for another 75,000. Over-

all, it seemed that the Earth had entered a long epoch of "normal" and benevolent climate.

Today this comforting view is being challenged. Successive crop failures have plagued Russian agriculture; people by the tens of thousands have starved to death in the African region just south of the Sahara. Are these random disasters, or are they results of continuing climatic deterioration from increasingly glacial conditions? Although competent scientists hold markedly different opinions, a knowledge of the Earth's late Quaternary glacial and climatic history is necessary for attempted predictions.

The European Pollen Record The changing nature of Holocene climates had actually been recognized in the early twentieth century from pollen studies in north Europe. Pollen, the male reproductive dust, is shed in great quantities by seed plants. Under a microscope, the pollen grains of particular kinds of plants can be recognized from the details of their shapes (which are different for different plants). When the tough-shelled wind-blown grains fall into swamps, bogs, or lakes and settle to the bottom—where they are often protected from destruction by oxidation—they provide a record of the plant life in the surrounding region. Thus as sediments gradually build up in a swamp or lake, their pollen content, from bottom to top, can indicate changes in regional vegetation through time. And vegetation reflects climatic zones.

From the study of cores from many sites, the north Europeans established a series of pollen zones that have become a standard for Holocene interpretations. Although they were given Roman numerals and broken down in considerable detail, four general climatic episodes can be interpreted from the pollen zones. The *Boreal stage*, as the glaciers retreated, is characterized by pollen indicating cold climates, such as that of shrubs and plants now found in treeless arctic tundras, or that of fir trees and other evergreens of present-day subarctic forests. The *Atlantic stage*, which followed, was a warmer time, as indicated by the predominant pollens of oaks, beeches, elms, and other deciduous (broad-leaf) trees in the forests of mild temperate climates. The ensuing *Subboreal stage* is marked by a decline of elm trees and the presence of pines, oaks, and other vegetation that tolerates cooler conditions. The pollen record of the *Subatlantic stage* is complicated by the abundance of field and pasture plants reflecting the extensive clearing of forests for agriculture. However, its record of climatic fluctuations is known in considerably more detail than the earlier stages because it corresponds to the time of written historic records and, recently, has received intensive scientific investigation.

New Developments   It now seems clear that the Holocene has been a time of ever-changing climate and that climatic fluctuations may occur quite rapidly. The growth of the great Pleistocene continental glaciers to their maximum extents and their eventual somewhat more rapid retreats and disappearances may be episodes that take from 10 to 20 thousand years because of the large masses of ice involved. But the climatic changes that trigger these events, based on recent studies of the late Pleistocene and Holocene, may occur in a century, more or less. Although they do not rival the major changes from full glacial to interglacial climates, the Holocene climatic fluctuations are instructive, and they have had a significant impact on geological processes, plants and animals, and mankind.

In large measure, the dynamic view of past world climates stems from breakthroughs around 1950 based on new laboratory techniques. Most important was the radiocarbon method, which is widely applicable in determining the absolute ages of late Pleistocene and Holocene events. Among other important developments, the oxygen isotope technique—coupled with absolute dating—gave an invaluable method for determining the relative temperatures of ancient sea water and of glacial ice.

During the late 1940s, Harold Urey determined that the ratio of ordinary oxygen (O_{16}) to the heavier isotope oxygen 18 is related to water temperature (there is slightly more O_{18} in warmer waters). Thus when the *oxygen isotopes* are incorporated into the calcium carbonate shells of marine animals they preserve a record of the relative warmth of the water while the animal lived that can be determined millions of years later by analyzing the shell. The technique became practical about 1950 after instruments (mass spectrometers) were improved to detect the very slight differences in ratios. Thus it became possible to detect relative differences in water temperatures of ancient seas hundreds of millions of years ago.

In Quaternary studies, the oxygen-isotope technique has been used on shells of floating-type foraminifera and other forms that have settled and accumulated on the ocean bottom. Since, as in the law of superposition, older specimens are at the bottom of a core with progressively younger specimens in increasingly higher parts, the changes in water temperatures, which should reflect world climates through the time interval represented in the core, can be determined. Dating the segments of the core in years, however, requires the analysis of radiocarbon or some other radiometric element.

In the 1960s, the oxygen-isotope technique was first applied to ice layers in a long core taken from the Greenland glacier. The oxygen-isotope ratios of various layers are determined by the temperature (below-freezing dew point) at which the snow crystals nourishing the glacier formed. The ratio varies with the seasons, and also with longer-term temperature variations reflecting climatic changes. In the Greenland core, for want of adequate radiometric material, the absolute age determinations had to be based on theoretical assumptions of the accumulation and flow of glacial ice. Nonetheless, a record for almost 1400 meters (4667 feet) of core gave a record going back about 100,000 years, a good record for the last 12,000 years, and a remarkable record for the last 800 years.

Of the recent technological developments, the introduction of *radiocarbon dating* was of the most general importance in environmental studies of late Cenozoic times. Wherever charcoal, wood, bone, shell, and other organic carbon-bearing materials are available, they allow finite dating of late Pleistocene and Holocene materials. Glacial and interglacial deposits, deep sea cores and glacial ice, archeological materials, soils, animal skeletons, and other relics allow important inferences about former environmental conditions. As a result, there has been a new surge of scientific interest in studies of late Pleistocene and Holocene times, the last 40 to 60 thousand years of Earth history. Now, geologists and physical geographers, oceanographers, anthropologists, botanists and zoologists, meteorologists, and other scientists have come to realize that only by cooperative efforts can they tackle the very complicated—and humanly relevant—problems of past, present, and future climatic change.[1] In short, studies of the geologic interval called the Quaternary have become truly interdisciplinary in scope.

In the last few decades, a flood of radiocarbon dates on various materials has been obtained by scientists in various disciplines. Radiocarbon dates —like any procedures involving laboratory techniques—must be properly evaluated. Usually the flush of scientific enthusiasm following the development of a new technique—which at first seems the answer to all previous problems—is tempered by later and conflicting findings. All methods have limitations and inherent difficulties that were not originally appreciated. Nonetheless, by careful cross-checking using separate techniques—as when tree-ring years were checked against "radiocarbon years"—increasingly reliable, although never absolutely accurate, results are obtained.

The present lack of agreement between some Quaternary events that seem reliably dated merely indicates our present inadequate knowledge and the need for better theoretical concepts. For example, local environmental conditions might explain why some alpine glaciers are now melting back while most are stable or advancing. In a grander perspective, the current observation that climatic changes in Europe and North America did not exactly correspond to those in Africa could indicate that there is a pronounced lag between warmings and coolings of the Earth's atmosphere between regions in low and high latitudes. The matter needs further investigation. Yet despite such unanswered questions (which are the lifeblood of the ongoing enterprise called science) a new, and more alarming, view of global climates can be sorted out from the welter of available evidence obtained from studies of changing Pleistocene and Holocene climates.

The Present Concept of Holocene Climates   Three general climatic trends and their approximate dates can be sorted out from a welter of sometimes conflicting Holocene radiocarbon dates. The beginning of the Holocene is commonly placed at about 10,000 years before the present.[2] Some scientists prefer a date of 12,000 B.P. and some suggest dates as recent as 5000 B.P; however, the 10,000-year figure is probably most generally accepted.

The first Holocene climatic interval, which is reflected in the Boreal pollen stage of Europe, represents a global warming trend accompanied by the retreat and disappearance of the Pleistocene continental glaciers from north Europe, North America, and Siberia. Alpine glaciers in the American Cordillera and other mountain belts melted far back up their valleys, and some disappeared completely. Arctic tundra zones retreated as forest

---

1 Samuel Clemens's perceptive comment that "everybody talks about the weather but nobody does anything about it" may now be outdated.

2 Since the time of an event in years before the present (B.P.) would be changing every year, it has been agreed that for purposes of radiocarbon dating B.P. will be considered to start at 1950 A.D.

margins moved poleward and higher into mountains, as the global climatic belts generally expanded into higher latitudes.

During the following climatic interval, beginning some 8000 years ago and ending about 5000 years ago, average temperatures in the middle latitudes were warmer than those of the present day (as much as 3.5° C, or 6.3° F). This interval, represented in the Atlantic pollen stage, has been called the *climatic optimum* (literally, "favorable climate"). It was a time of warm and generally benevolent climates in some places, such as north Europe. In semiarid regions, as in the tropics and western North America, however, the climatic optimum was marked by greatly increased dryness and desert conditions. Thus where the interval was both warm and inhospitably arid, it is often called the *altithermal* (meaning "high temperature").

Roughly 5000 years ago, a general climatic cooling set in that has persisted to the present day. This episode, reflected in the Subboreal and Subatlantic pollen sequences, is referred to as the *Neoglaciation*. It is characterized by several pulses of growth and recession of valley glaciers in the American Cordillera, the Alps, and other mountains. Since Neoglaciation includes the time of written history, its fluctuations have been reconstructed in more detail than the earlier Holocene intervals. An early advance of valley glaciers marks an early phase of Neoglaciation which lasted from about 5000 years to roughly 3000 years B.P. In the next 1000 years a general warming is indicated by glacial recession. It was followed by a resurgence from roughly 2000 to 1000 years ago. After a warm interval and glacial retreat, the glaciers once again made fluctuating advances between 300 and 100 years ago. Thereafter they generally melted back until sometime in the 1940s, when a new global cooling began to reverse the trend. These Neoglacial fluctuations of middle-latitude valley glaciers—although only one of several lines of available evidence—have provided some of the most obvious and accessible evidence for late Holocene climatic changes.

Evidence from Glaciers   The Alps, Rockies, and Sierra Nevada are typical examples of middle-latitude mountains where massive looping moraines lie on adjacent plains beyond the mouths of impressive U-shaped canyons (Fig. 17-4). These moraines, far removed from existing small glaciers, mark the former ends of the long Pleistocene valley glaciers. Their melting and retreat resulting from the general warming in late Pleistocene and early Holocene times is indicated by the long stretches of their abandoned U-shaped canyons.

It was long assumed that the much smaller existing glaciers, high up in the mountains, were the shrunken relics of the Wisconsin-age glaciers. That the modern glaciers represent still another glacial episode was first suggested by François Matthes of the U.S. Geological Survey in the early 1940s. From his studies in and around the Sierra Nevada range, he noted that sediment cores taken from Owens Lake, at the foot of the mountains, contained a salt layer indicating that the lake had dried up at the end of the Pleistocene. The dessication had been followed by a renewal forming the existing lake. Thus Matthes suggested that the existing Sierra glaciers were not Pleistocene relics, but rather new glaciers. They were reborn in what he called a "Little Ice Age," a cooler climatic episode that followed the warmth of the altithermal which had destroyed the Pleistocene glaciers in middle-latitude mountains. Aside from the geologic evidence, Matthes's concept was subsequently supported by pollen studies in the region. Although valley glaciers at higher latitudes, as in the mountains of Canada and Alaska, may well have survived the altithermal before their Little Ice Age expansions, the rebirth of glaciers is recognized as a feature of the Rockies and other ranges.

Evidence that the Little Ice Age, now technically called the Neoglaciation, was marked by several cooler and warmer climatic episodes is furnished by a series of small moraines. They lie far

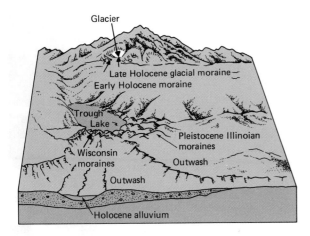

Fig. 17-4 Pleistocene and Holocene moraines in the Rocky Mountains.

back up the glaciated valleys in the "high country" a few kilometers, at most, beyond living glaciers or snow banks. Although all have a fresh appearance, moraines of the oldest of the radiocarbon-dated early Neoglacial advance have stable slopes and weak soil development, whereas those of the latest, in contrast, have unstable slopes and lack any soils. That the ends of many living glaciers have melted back from the youngest moraines reflects the general climatic warming of the nineteenth and twentieth centuries.

Neoglaciation and Some Human Affairs   The Indus (Harapan) civilization—a contemporary of ancient Egypt and Sumeria—flourished for a thousand years, in what is now northwest India and Pakistan, then rapidly declined and disappeared about 3800 years ago (1900 B.C.). Since 1922, when a few bricks and small distinctive soapstone seals of this "lost civilization" were first turned up, over 50 village sites, a major seaport, and two major cities have been discovered. Archeological evidence indicates that this civilization,

which had its cultural center along the Indus River, actually spread across several hundred thousand square kilometers. Much of the region is now the Thar Desert and the inhospitable dry plains of Rajasthan. Along the Indus River, two large cities whose buildings were largely constructed with fire-baked brick show a high degree of organized planning, yield evidence of a highly developed commercial society, and contain large graineries. The Indus culture seems to have been a well-organized, urban-dominated society with a broad agricultural base.

The relatively rapid collapse of this prosperous civilization has never been adequately explained. Hypotheses include: "cultural stagnation"; mismanagement of the land by overcutting of trees to supply wood for firing brick; devastating floods of the Indus River; and invasion by Aryans from the north. Perhaps Holocene climatic change was involved. Recently a pollen-bearing core whose layers were radiocarbon dated was obtained from an ancient lake in Rajasthan. The core indicates several thousand years of favorable climate when grain was extensively cultivated; then, about 4000 B.P., the pollen indicates increasingly arid conditions; around 3500 B.P. the lake completely dried up. Moreover, in parts of the now-dry region, relics of towns have been discovered that are buried by sand dunes. Thus could the end of the Indus civilization reflect successive failures of the grain crop because of global climatic change associated with the onset of Neoglaciation?

Late Holocene climatic events are recorded in the written Norse Sagas and other historic documents that preceded the modern era of scientific weather observations by several hundred years. For about a thousand years, from approximately 400 B.C. to 1300 A.D. (roughly 2350 to 650 B.P.), mild to warm climates dominated Europe. The span encompasses the classical times of the Greeks and

Fig. 17-3 Mammoth Glacier, a Holocene glacier in the Wind River Mountains of Wyoming. Courtesy of Austin Post, University of Washington.

Romans as well as the medieval period. During the latter part of this climatic interval, when the polar ice packs had retreated, Norsemen in open Viking boats explored the North Atlantic, colonized Iceland in the ninth century A.D., and established colonies on the west coast of Greenland where some 10,000 people lived in the eleventh century.

In the 1300s, the climate of Europe and the North Atlantic deteriorated. In northerly regions, cold summers caused an increasing number of crop failures and famines. By the late 1600s, expanding glaciers had buried some Iceland farms that had been used for centuries. As the climate worsened, the North Atlantic became increasingly stormy, and polar pack ice expanded and shed increasing numbers of icebergs that forced navigation routes far southward. Supply trips to the Greenland colonies stopped in the early 1400s. A hundred years later, when ships again reached the island, the settlements were empty of living inhabitants.

A Norse cemetery in Greenland was excavated by archeologists in the 1920s. Skeletons, cloths, and wooden objects in graves of some of the last inhabitants were excellently preserved in permafrost (solidly frozen ground that does not thaw out in summer). Since graves would not have been dug in cementlike permafrost, the cemetery gives further evidence that the condition developed during the climatic deterioration from the fourteenth to nineteenth centuries—the time of the late Neoglacial expansions.

Valley glaciers in the Alps retreated and melted completely or became stagnant relics in high mountain cirques during the mild climatic interval that lasted into the Middle Ages. Farms and villages became established on the floors of scenic U-shaped valleys, not far from the high-mountain ice remnants. Then in the late 1500s and early 1600s the glaciers in the Alps made major advances. Routes traveled for centuries were blocked by ice flowing into high mountain passes. Written records, as in the archives at Chamonix in France, describe lost farmlands and crushed villages overwhelmed by

ice. Thereafter minor glacial retreats alternated with readvances, the last of which, in the 1850s, almost equaled the great glacial advance of the late sixteenth century.

Climatic changes, which are reflected in glacial advances and retreats, are not restricted to middle and high latitudes. They affect the world pattern of climatic belts. From about 700 A.D. until 1600, three great empires, the Ghana, Mali, and Songhai, successively dominated the Sudan of West Africa. This vast region, now known as the *Sahalian zone*, lies in the open plains—the savannas—south of the Sahara Desert and north of the tropical rainforests. The black empires there grew rich and powerful by trading in gold, salt, and ivory. Among their major cities, the legendary Timbuktu (today an undistinguished town) was a center of government, trade, and learning in the sixteenth century. The age of the Sahalian empires ended when Songhai was invaded by the Moroccans in 1591. The military defeat, however, might only represent the final blow to a culture already devastated by climatic events. The fall of the black empires occurred during the Neoglacial episode, whose global cooling could have caused a series of droughts and agricultural disasters in the Sahalian climatic belt, resulting from a weakening of the rainy season brought by the African monsoons. Today this is the region where half a million people starved to death in the great drought of the 1970s.

A Possible Crisis in Climate Glaciations and climatic changes long seemed academic matters—intriguing scientific concepts, but far removed from practical human concerns. During the two hundred years from about the mid-eighteenth to the mid-twentieth centuries—when the world's human population almost quadrupled to about four billion souls—the mild climate favored agricultural production. This time interval was generally considered as "normal" for long-lasting interglacial episodes such as the Holocene. Now, since about 1950, a wealth of new scientific data (from studies of glaciers, pollen, deep-sea cores, and archeology, as well as examination of written historic records) has led to a new and more alarming view of our present situation.

In the grand geologic perspective, climatic changes (which trigger the more gradual expansions and contractions of glaciers as well as major shifts of vegetational patterns) are rapid events that occur in a century or two. After a change, the warm or cold climatic conditions can persist from a few centuries to a few thousands of years. The cold episodes are also times of greater climatic variability. Although cold prevails, warm spells are interspersed, and droughts and floods are more frequent. Since the average world temperature differences between full-glacial and interglacial conditions range from 4° to 6°C (about 7° to 12°F), the fluctuations of 1° to 2°C (from 1.8° to 3.6°F) during the Holocene may seem small. Nonetheless, it is clear that the climatic changes associated with the altithermal, Neoglaciation, and late warm interval had a marked impact on floras and faunas and on human affairs.

Reid Bryson, one of the scientists who are studying climatic variations, points out that in the Northern Hemisphere the present cooling trend (almost 2°C in Iceland, 1° in North Atlantic waters, and half a degree for the hemisphere as a whole) has reduced temperatures to the level of the mid-nineteenth century—a time when Little Ice Age glaciers in the Alps and other middle-latitude mountains made long advances. Today the snow and ice cover in the far north has greatly increased. Growing seasons are as much as two weeks shorter and summer frosts more common in northern agricultural districts. And far to the south, the monsoons have increasingly failed to bring rains to such drought-ridden regions as the Sahalia—already an overgrazed region and now being encroached upon by the Sahara.

To predict future climatic trends—whether there will soon be a general warming, or a leveling off during a cold interval, or even a return to full-glacial conditions—we must learn more about the mechanisms of climatic variations. It seems clear that cooling of the Earth's higher latitudes causes expansion of the cold air in the circumpolar vortices, the great atmospheric swirls covering the Earth's middle and high latitudes. The expansion displaces the global climatic belts towards the equator. Aside from changing precipitation patterns and causing increased climatic variability in middle and high latitudes, the subtropical high-pressure belts creating the world's tropical deserts are forced towards the equator. In the same tropical regions, the cooling at higher latitudes also weakens the summer monsoon effect, creating increased years of drought in the regions where much of the world population is concentrated. From high-level observations, images from orbiting weather satellites, and other scientific and technological advances, the understanding of the mechanics of atmospheric circulation and the ability to predict weather (the short-term condition of the atmosphere) for several weeks in advance have been greatly improved.

Long-range predictions of atmospheric variations are not yet possible. They hinge on the intriguing, humanly relevant, and basic scientific question: what causes climatic changes and Ice Ages? As yet there is no scientifically accepted theory for climatic changes. Solar variations, planetary motions, changes in atmospheric components, and geologic factors are all under consideration. Each working hypothesis has its stout defenders, and all except their own are rejected by other scientists on "firm" evidence. In the end, a solution to the major problem will probably incorporate elements of all these hypotheses. When a theory with predictive value is reached, however, it will certainly be based in large measure on the Holocene record.

In the last few decades, great emphasis has been placed on human intervention with climatic processes. The warming trend following the mid-eighteenth century and the cooling since the mid-twentieth century, both recorded in glaciers, follow the start of the Industrial Revolution (beginning in Great Britain in the 1700s and accelerating elsewhere to the present). Some scientists, impressed by the warming trend that culminated during the first half of this century, argued that carbon dioxide gas released in industrial burning was warming the atmosphere by increasing the greenhouse effect. Thus they predicted long-continued warming that would melt continental glaciers, causing rising sea levels and the drowning of coastal cities, as well as the return of desert altithermal conditions to much of western North America. The cooling trend since about 1950 led other scientists to conclude that industrial smog and other sources of pollution are cooling the atmosphere. Smog contains countless minute solid particles, called *aerosols*, which are responsible for poor visibility. They increase the Earth's *albedo* (reflectivity) so that solar energy is reflected back into space before it can heat the ground and hence the troposphere. With prolonged cooling, the great continental glaciers would again form in North America, Europe, and Siberia and advance over now-populated regions. Sea levels would fall, and large parts of the middle and high latitudes would again become arctic tundras.

The CO_2 and smog mechanisms have a bearing on climatic changes (and they certainly contribute to atmospheric pollution). Yet they alone cannot explain the major climatic changes that occurred long before man created the Industrial Revolution. They could, however, accentuate changes caused by major global or solar mechanisms and they certainly relate to other human environmental problems.

THE GEOLOGIC ENVIRONMENT IN PRACTICAL AFFAIRS

The high standard of living in our affluent modern industrial society depends on good agricultural land, a variety of mineral resources, and trained people to exploit them efficiently. As a pure science, geology's goal is constructing and improving general concepts of the Earth. In the applied science, these concepts are put to practical uses for the benefit of mankind, as in the development of mineral resources. Today at least three-quarters of all geologists work on such economic matters.

Environmental geology, a relatively new development, reflects human concerns about our use of the land and materials of the Earth in practical affairs. In this somewhat restricted sense, environmental geology is an offshoot of economic geology, which deals with mineral resources, and of engineering geology.

Hazards and Planning

Since civil engineering and geology clearly overlap, an appropriate founder for modern geology was William Smith (canal engineer). Today geology is applied increasingly in engineering projects. Preliminary investigations can lead to the cheapest and best sources for building stone, concrete aggregate, and other construction materials of geologic origin. And since all manmade structures rest on rock, preliminary geologic studies can greatly reduce the chance of collapsing dams, leaky reservoirs, highway failures, and settling and cracking of buildings; and it can help locate the safest places for man-made structures in active earthquake regions.

A Recent Case History The impact of Earth processes on man and his works is a major concern of environmental geologic studies. Textbooks often present this aspect in case histories, past lessons which, hopefully, will lead to improved prediction and better future planning.

Often, case histories can be presented as "horror stories" for dramatic effect. A case in point is the 1976 collapse of the Teton earth dam in eastern Idaho. The failure released some 303 billion liters (80 billion gallons) of water that destroyed nearby towns and caused at least nine deaths and a billion dollars in property damage. A commission appointed to study the failure concluded, a few months later, that *piping* was the cause. Technically, piping is subterranean erosion by percolating waters which remove particles from clastic rock materials. The phenomenon has been studied since 1898, when Colonel Clibborn, a civil engineer, predicted the subsequent collapse of the Narora Dam on the Ganges River in India. Three years before the Idaho dam collapsed, a report by the U.S. Geological Survey had warned of potential danger.

The Appreciation of Geologic Processes   Geologic case histories are certainly not lacking (examples presented in previous chapters date back to "Atlantis" and the Minoans). The need has long been to apply our considerable knowledge of geologic processes to future construction and land planning.

Short of a thick textbook, it is impossible to treat the topic of geologic hazards adequately, but certain generalizations are possible. Each of the internal and external geologic processes present problems. People will continue to live near active volcanos, since they provide some of the best soil-forming materials in the wet tropical and subtropical regions. Here, better scientific techniques for predicting violent eruptions are needed. Encouraging progress has been made in predicting the time and severity of earthquakes, and shock-resistant architecture can now be designed. Unfortunately, in the normal course of human affairs, past experiences are often forgotten. Despite the 1906 earthquake, for example, the greatly expanding city of San Francisco has some architecture which is substandard for a seismic region, industries have expanded onto poorly consolidated land reclaimed from the bay, and housing developments have sprung up across the trace of the San Andreas fault.

Less dramatic, but of considerable consequence, are the Earth's slow-acting external landscape processes. When dams built in limestone areas fail to fill reservoirs (because of leaky karst topography), the consequences have been merely financial. The same is true for houses and roads built on slopes undergoing gradual mass movements, and for areas of subsidence over abandoned mines or areas where oil or water has been pumped out. Of more long-term consequence is the destruction of agricultural soils, as in the case of salinization by improper irrigation, or the plowing up for farming of a semi-arid prairie's natural grass cover, which led to the deflation of topsoil in the American "dust bowl" during the drought years of the 1930s.

Although devastating landslides and floods are often in the news, they may seem remote and isolated events (except to the people involved). Nonetheless, many potential areas for such events are geologically evident. A grassed and forested slope may seem a safe and scenic spot for a house or road, but a hummocky topography underlain by angular debris should warn of former slow mass movements or rapid mudflows and landslides. The bottoms of mountain valleys are dangerous construction sites, even if there is no historic record of their flooding. Flood plains, by their geologic nature, are potential disaster sites (Fig. 17-6). Although flat, minimizing construction costs, they are best reserved for agriculture and parks rather than housing developments.

The list of sites with potential hazards is extensive. But in a large measure, future problems can be avoided by prior geologic evaluations followed by sound engineering practices.

Our Limited Geologic Resources

The Babylonians and Egyptians, Greeks and Romans, and other civilizations of the past learned to exploit the Earth environment for agriculture, building materials, and metals. Their energy sources for building their impressive architectural structures were mainly the muscle power of animals and hu-

Fig. 17-5 The impact of geology on mankind. Casts made by pouring plaster into cavities in volcanic ash formerly occupied by the decomposed bodies at Pompeii. Courtesy the *National Geographic* magazine, photo by Lee E. Battaglio, Copyright National Geographic Society.

Fig. 17-6 The obvious dangers of building on a flood plain. Frisky Ranch house, San Isabel National Forest, Colorado. Photo by Gordon Van Buren, U.S. Forest Service.

mans, largely slaves. Wood and charcoal provided the energy for smelting *ores* (rocks from which metals are produced). With a comparatively low world population and limited technology at their disposal, the ancient people's exploitation of geologic resources was minimal.

Since the beginnings of the Industrial Revolution, however, the extraction of the Earth's mineral wealth has accelerated at an alarming rate—until we can foresee the marked depletion and possible future exhaustion of many industrial raw materials and natural fuels, as well as a shortage of fresh water.

Water   Our most essential natural resource is fresh water. Even though water supplies are replenished by the hydrologic cycle, expanding cities, irrigation projects, and industries make the availability of fresh water a matter of universal interest. The problem is no longer restricted to desert and semiarid regions; it now concerns such humid regions as the densely populated eastern United States. An increasing number of geologists are applying their special talents to the better use of all our water resources, underground supplies as well as surface waters in lakes and streams.

Economic Minerals and the Metallic Ores   Sulfur, halite, potassium salts, gypsum, nitrates, and phosphates are among the many other geologic raw materials consumed in large amounts by modern industry. Agricultural use of chemical fertilizers, which are derived from geologic materials, has become increasingly important. In the face of the rapidly expanding world population, more efficient food production is essential.

From antiquity the higher civilizations have used metals. The bronze and iron ages are now replaced by the "age of steel," which consumes iron ore and many other metals at an extravagant rate. Copper, lead, tin, aluminum, manganese, zinc, silver, platinum, and gold are only a partial list of other metals now in great demand. Although metallic minerals occur in the three great groups of rocks, their concentration into ore bodies (metallic mineral accumulations which can be worked at a profit) is exceedingly rare. Improved technology has made lower-grade mineral deposits into ore bodies as richer sources become depleted, but ores clearly present an economic problem. No mineral deposits are renewed: they are gone once a mine is exhausted.

Some scientists take an optimistic view of our metallic resources, because many such elements are fairly widely distributed in some of the Earth's crustal rocks in minute quantities. They believe that improved technology for processing very low-grade metal-bearing rocks will solve the problem. Unfortunately, to extract usable metals from great volumes of valueless rock minerals requires enormously increased use of energy. And the imminent and predictable depletion of our most-used energy-producing materials has precipitated the current energy crisis.

The Fossil Fuels   Oil and coal are fossil fuels in which solar energy of the geologic past has been stored by the incomplete decay of plant and animal materials. Since it takes very special conditions of burial and millions of years of geologic time to create appreciable deposits of coal and oil, these energy-producing materials are, in the human time-perspective, essentially nonrenewable resources.

The Industrial Revolution was built on coal. This burnable rock is largely the remains of woody plants that accumulated in swamps whose toxic waters of low oxygen content inhibited bacterial action and thus prevented decay, or rotting. After the swamp deposits were buried under later sedi-

ments, pressure and, in some cases, heat increased the ratio of elemental, or free, carbon by driving off water and volatile substances. The density and hardness of the coal also increased. Coals range from little-altered lignites whose woody character is apparent, through bituminous coals, to anthracite (hard coal) of glassy luster and the highest carbon content. World reserves of coal are tremendous—seven trillion tons by some estimates—so that generally only the thickest, most accessible, and best grades are being extensively worked. After World War II, the use of coal slowed because oil and natural gas are more easily obtained (by drilling a hole), their extraction scars the landscape less, and they are more convenient to handle and use. Now, in the energy crisis, coal production has regained its importance. The American coal reserves far exceed our dwindling supplies of petroleum. Western coals, as in Wyoming and Montana, will be extensively developed. Although these coals are less efficient heat producers, they are low in sulphur that creates polluting gases during combustion, they can be used in manufacturing many chemical products, including gasoline, and the reserves are enormous. Nonetheless, the most convenient fossil fuels, in the short term, will still be natural gas and petroleum.

With the tremendous and ever-increasing consumption of oil—for heating, for gasoline, diesel, and jet engines, and for such industrial products as plastics and dyes—it is not surprising that over 40 percent of all geologists in the United States are in the petroleum industry. Although the origin of oil and gas is still not completely known, most probably originated in relatively stagnant basins on the sea floor from plant and animal residues in highly organic oozes. Conversion of these oozes to oil seems to require increased temperatures and pressures resulting from burial under later sediments.

Commercial "pools" of oil require special geologic conditions. These include: a *source bed*, often a dark shale; a *reservoir bed*, such as a permeable sandstone through which fluids can move; an im-

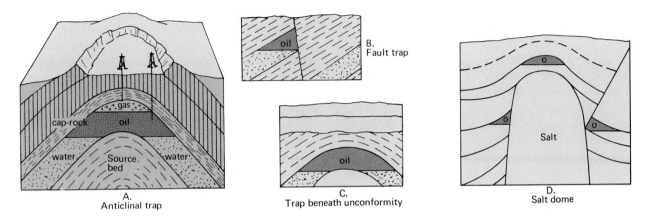

Fig. 17-7 Typical oil traps which have been commercially successful.

permeable *cap rock* to prevent escape of the oil; and some sort of *structure* to trap the oil. The first, and most obvious, traps to be investigated extensively were anticlines. Now the search is for less easily located traps which generally require geophysical and other special techniques. For example, traps concealed by unconformities, and therefore not indicated by surface mapping, can be located by portable seismographs. Gravity surveys to detect light masses at depth are used in the Gulf Coast to locate salt domes—great plugs of salt that rise from depth to intrude overlying sediments and thereby create a variety of traps. In general, the search for oil is limited to regions of thick sediments containing marine rocks (Fig. 17-7).

The Energy Crisis In late 1973 and early 1974, the American public discovered the energy crisis during the Arab oil producers' embargo. The gasoline shortage led the public to search for a scapegoat. The oil industry was accused of withholding limitless supplies to increase prices, government was accused of bungling and hampering the industry, environmentalists were accused of blocking needed energy developments, and the Arabs were accused of greed and politics. If there were elements of truth in the complaints, the fact is that our profligate consump-

tion of oil and fossil fuels has brought on a real energy crisis.

Petroleum products from oil were used to tar boats in ancient Babylon, and oil was used by American Indians for its "medicinal" properties. Commercial exploitation, however, began in 1859 when a Colonel Drake drilled the first oil well near Titusville, Pennsylvania. Most of this oil was converted to kerosene for lamps, before the development of the gasoline engine. Since 1900 world

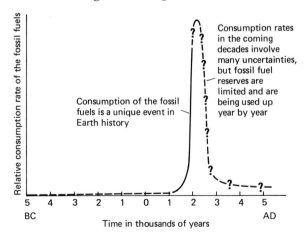

Fig. 17-8 The "Age of Fossil Fuels" illustrated schematically. From R. J. Ordway, *Earth Science and the Environment*, D. Van Nostrand Company.

production of crude oil, which reflects demand and use, has doubled about every ten years—an escalating rate of increase that cannot go on indefinitely.

In the mid-1970s, the United States—having some 6 percent of the world's population—used about one-third of the total world energy production. A number of estimates have been made in the last few decades of how much oil is to be found in the United States and the ultimate production that is possible. These estimates are based on the areas of sedimentary rock that might contain oil, past records of exploration for oil, the amount of drilling needed to produce a successful well, the actual production of wells, improvements in drilling technology, and other important factors. Even though the predictions of future oil production make use of mathematical and statistical treatments, the end results are always based on certain initial assumptions (informed guesses) that control the final results. Thus different experts have come up with different estimates.

In the early 1960s, a very encouraging study by the U.S. Geological Survey predicted that the ultimate oil production of the United States would be some 93,810 billion liters (590 billion barrels).[3] Since less than 20 percent had been discovered by 1959, the report concluded that some 73,140 billion liters (460 billion barrels), 80 percent, remained to be discovered. A more sobering estimate of U.S. oil production was made in the late 1960s by M. King Hubbert, an eminent geologist and geophysicist. The earlier estimate was based on the assumption that in the future the amount of drilling needed to produce a given quantity of oil would remain the same. Hubbert took into account the fact that as time goes by, oil is becoming harder and harder to find. In 1937, for example, the overall rate of oil discovery was 26,553 liters (167 barrels) for each foot of drilling. Thereafter the success ratio has drastically declined to the present level of about

3 One U.S. barrel of oil contains 159 liters (42 gallons).

5565 liters (35 barrels) for each foot of drilling.[4]

In the 1970s, estimates of oil production still vary but they are on the conservative side. In round figures, about 5900 billion liters (100 billion barrels) of oil have been produced and consumed. About 6042 billion liters (38 billion barrels) are calculated to be in proven reserves (in known oil fields). The U.S. Geological Survey estimated in 1975 that undiscovered and recoverable oil was some 16,695 billion liters (105 billion barrels). The rate of consumption, however, continues to rise, and most studies indicate that in the mid-1970s United

4 "Dry holes" may produce water, but not oil. Even if a well strikes oil, it is not considered profitable unless it produces a million or more barrels. By 1961 only one well in 70 was profitable in the "wildcat" search for new oil fields.

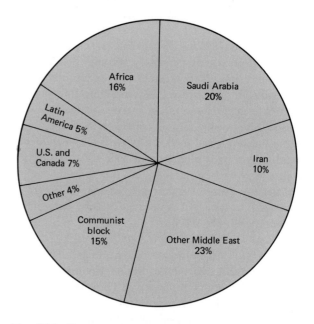

Fig. 17-9 Known world resources total 667 billion barrels (Saudi Arabian reserves may be understated by a large amount). To put these figures in perspective, the United States in 1973 consumed more than 17 million barrels each day, whereas it produced 11 million barrels. Moreover, world petroleum demand is presently rising at an annual rate of 7 percent, which means a doubling every 10 years. *Oil and Gas Journal*, 12 December 1972.

States oil production had reached its peak and would now steadily decline. Over half of the U.S. oil production has been since 1950, and shortly after the year 2000 only about 10 percent would remain. To maintain anything like the present increase in the rate of consumption, vast quantities of oil will have to be imported.

Estimates of the world's total final production of petroleum, including oil-rich states like those in the Middle East, vary from 333,900 billion to 214,-650 billion liters (2100 billion to 1350 billion barrels). The lower estimate would lead to a peak production in the 1980s and the higher to a peak around the year 2000—after which world productions will markedly decline. Tremendous as the production figures may seem, however, 90 percent of the world's petroleum would be exhausted before 2025 by the former estimate and slightly after that date by the latter—if the present rate of consumption were to continue. The consumption and potential exhaustion of natural gas, a closely related energy-producer, follows a similar pattern in existing estimates. Thus the phasing out of oil and gas as prime energy sources is measured in decades—a half-century or less. Clearly they will have to be supplemented and eventually replaced by other energy sources, of which the most immediately available is coal.

That coal was a burnable rock was known in Great Britain almost a thousand years ago, but its use was negligible before 1800. From about 1860 (when useful records begin) the world production of coal can be calculated. The rate of use doubled about every 16 years, tapered off between World Wars I and II, then doubled about every 20 years thereafter. Coal is dirty to handle, expensive to transport, and dangerous to mine underground where gas explosions can occur. Surface strip mining scars the landscape, and coal-burning—if unregulated—pollutes the atmosphere for miles around. But the tremendous reserves in the United States may make it our fossil fuel of the future. Estimates of usable reserves vary, but 1500 billion metric tons,

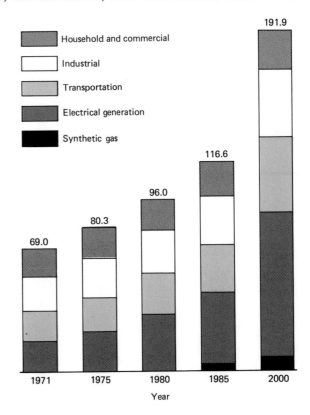

Fig. 17-10 United States energy consumption by sector. 1971–2000. W. G. Dupree, Jr., and J. A. West, *United States Energy Through the Year 2000,* U.S. Department of the Interior, December 1972.

about a quarter of the world supply, might be minable in the United States. The Soviet Union might produce 4300 billion metric tons, almost two-thirds of the world coal supply. Thus coal, which was the major fossil fuel of the nineteenth century, may return to the dominant position in the twenty-first. It could be a major energy source for 300 to 400 years, but would be so for only half that time if other energy sources are not extensively developed (Fig. 17-12).

Some Other Energy Sources Supplementary energy sources can help in the growing crisis, but all

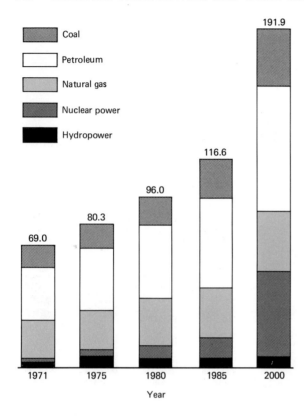

Fig. 17-11 United States energy consumption by source, 1971–2000. W. G. Dupree, Jr., and J. A. West, *United States Energy Through the Year 2000,* U.S. Department of the Interior, December 1972.

have limitations or need developmental work. The tar sands of Alberta and oil shales of Colorado, Utah, and Wyoming could each produce several hundred billion barrels of petroleum products. Developmental studies have been made, but there has been no commercial production because these areas cannot compete with oil from wells. In the present state of technology, they would require strip mining and processing. The oil shales, especially, would require the disposal of large amounts of waste, and their processing would demand a considerable amount of the water that is already allocated in the dry lands of the west.

Hydroelectric power might be further developed, mainly in the energy-deficient countries of the tropics. Elsewhere, many of the good sites are already in use, and future expansion requires the building of more large dams. Aside from esthetic considerations, dams upset the ecological balance of a stream, which may adversely affect food production by farming and fishing—as happened when the Nile River was blocked by the great Aswan Dam. The useful life of any dam is limited because its reservoir forms a settling basin that gradually fills with the load brought in by streams.

In places where the rise and fall of ocean tides is pronounced, energy-producing installations could be constructed. Geothermal power, already exploited in a few regions, could be somewhat increased in certain areas of volcanic activity or high heat flow. Direct sunlight (solar energy) and wind, long used in a small way, can help reduce our dependence on energy from the diminishing fossil fuels. All these sources will probably play an increased role in the future; but barring some unforeseen technological breakthrough, they can provide only a small percentage of the total energy needs. None seem likely sources for the large amounts of fuel used in the internal combustion engines that provide most of the transportation in our present mobile and affluent society.

Nuclear Power The most likely major sources for additional energy requirements of the late twentieth century involve *nuclear-fission* power. Such sources worry many people because of the possibilities—however remote—of accidents releasing dangerous materials, the environmental problems of disposing of long-lived radioactive wastes, and the chance that terrorists might seize plants for their political purposes. With awareness and the state of modern technology, however, the risks should be reducible to the point where they are far outweighed by the benefits of the energy produced.

Operating nuclear power plants now use fission (atom-splitting) of the rare isotope uranium-235.

Fig. 17-12   Coal, a fossil fuel for the next few centuries. Strip mining in north-central Wyoming. Photo by Gary B. Glass, courtesy of the Wyoming State Geological Survey.

In the process, the isotope is used up, and useless radioactive "garbage" results. Current estimates indicate that ore reserves for producing U-235 are only adequate to refuel existing plants and supply about half the needs of proposed new installations. It is hoped that *breeder reactors*, now in the developmental stage, will be operative in power plants by the late 1980s. Breeder reactors convert the relatively abundant but slowly decaying isotopes of U-238 and thorium-232 into products capable of sustaining the chain reactions needed for power generation. In principle, they should create more new nuclear fuel than they "burn," thus solving the raw materials supply problem.

Nuclear fusion is the great hope for solving twenty-first-century energy problems. The theory of nuclear fusion, the energy source of the Sun and stars, has been known for some 40 years. The process involves the combining of heavy isotopes of hydrogen (called deuterium and tritium) to create the more complex atom of helium, with the release of considerable energy. Abundant supplies of "heavy" hydrogen are available in the oceans. The theory of fusion was successfully applied to the release of untamed energy in the explosion of hydrogen bombs. Ever since, the possibilities for harnessing the energy for useful purposes have been investigated. The technical problems of releasing such energy in controlled increments for power generation are formidable. Tremendous energy is needed to start the reaction (atom bombs trigger the H-bombs) and the heats involved, which reach hundreds of millions of degrees centigrade, are hotter than the surface of the Sun. No known materials can withstand such temperatures; however, tests show that streams of the superhot ionized gases can be held and contained in powerful magnetic fields.

If fusion energy is ever harnessed for practical purposes, the heat could drive steam-power plants that generate electricity. Part of the electricity could be used to break water down into its two components, oxygen and hydrogen. Hydrogen, which can be pumped long distances through pipelines, could replace natural gas for industrial purposes and heating, and it could replace gasoline as the fuel for aircraft, trucks, and automobiles. Unfortunately, it is not certain that nuclear fusion can be tamed and put to use in power plants, because major developmental problems remain unsolved.

EPILOGUE

The present concepts of geology rest upon a long history of careful study of details and development of techniques that extend our range of observation; of cautious reasoning blended with a flair for imaginative speculation; and of debates and arguments between geologists holding different opinions. Today the plate tectonic paradigm has become the accepted frame of reference for studies of the Earth's architecture and internal dynamics. There are, however, some very competent geologists who do not accept the scheme and still defend the older paradigm of fixed continents and ocean basins. Although plate tectonics seems well established from several separate lines of evidence, it will certainly be modified and improved by future generations of geologists (perhaps even rejected in some future time). For geology is not mystical, divine, or ultimate knowledge; rather it represents a collection of mental images of an ever-changing Earth sifted from the dedicated work of highly argumentative men and women.

Now that the excitement of the plate tectonic revolution is over, and operations have settled into a more normal scientific routine, perhaps the next major breakthrough in geology and related sciences will relate to an acceptable general mechanism for the causes of Ice Ages and climatic changes. This would have an important bearing on the interrelated problems of the population explosion, world food supplies, and our diminishing geologic resources. Perhaps we have presented an overly pessimistic view of the global environmental crises. But at the least, we should be aware of the problems, because the human impact upon the Earth, and vice versa, is a matter for concern. Scientists studying population, food, climate, and resources have differing opinions ranging from guarded optimism to Malthusian pessimism. Wisely used, science and technology can alleviate and perhaps solve some of the problems, but there is no guarantee that they will provide all the answers, especially where complex social and political matters are involved. One way or another, however, the problems will be solved: perhaps by the harsh realities of natural processes; more hopefully by cooperative and intelligent human planning based in part on appreciation of the Earth and its processes.

Fig. 17-13 "The world-wide, highly radioactive marker bed at this locality one meter thick, conformably overlays a beer-can conglomerate." Cartoon by Gunther Von Gotsche, from Wyoming Geological Association Newsletter.

SUGGESTED READINGS

Bryson, R. A., *A Perspective on Climatic Change*, in *Science*, Vol. 184, pp. 733–760, 1974.

Committee on Resources and Man, *Resources and Man*, San Francisco, W. H. Freeman and Co., National Academy of Sciences–National Research Council, 1969.

Denton, G. H., and Porter, S. C., *Neoglaciation*, in *Scientific American*, Vol. 222, No. 6, pp. 100–110, 1970.

Dyson, J. L., *The World of Ice*, New York, Alfred A. Knopf Co., 1962.

Energy Issue of *Science*, Vol. 184, No. 4134, pp. 212–402, 1974.

Food Issue of *Science*, Vol. 188, No. 4188, pp. 502–662, 1975.

National Academy of Sciences, *The Earth in Human Affairs*, San Francisco, Canfield Press of Harper & Row Publishers, 1972.

Park, C. F., Jr., *Earthbound—Minerals, Energy and Man's Future*, San Francisco, Freeman Cooper & Co., 1975.

Appendix A: Elementary Aspects of Maps

Since maps are an essential geologic tool, their general characteristics and features should be understood. A map is a two-dimensional diagram of an area of the Earth's surface and is usually to an exact scale. That is, it is a drawing on flat paper of the ground and associated features, and usually shows things at a small fraction of their true size.

Maps give two general types of information: *planimetric information*, which gives the place or location of things, and *topographic information*, which shows relief, or landscape features. Planimetric information includes such things as the location of roads, cities, and towns (as on a highway map), land ownership, the location of rivers, lake and sea shores, and the location of rock types or units (on geologic maps). Topographic information indicates slopes, the unevenness of the ground associated with mountains and valleys, plains, and other relief features.

Map Orientation and Declination

With a few exceptions, maps are constructed so that the top of the map is north, the bottom south, the right side east, and the left side west. To read a map in the field, it is best to orient the map with the ground. To do this, turn the top of the map towards north on the ground; then points on the map will correspond in direction to points on the ground surface.

Declination is the difference between true north (the direction to the North Geographic Pole) and magnetic north (the direction in which a compass needle points). It is given in degrees of angular measure and shown by a V-shaped symbol in which a T or star indicates True North, and a half arrow, sometimes with an MN, shows Magnetic North. (Some declination symbols show still another north, marked by a Y or GN, which is Grid North of the military grid system.)

True North of the Geographic Grid (the system of latitude and longitude) coincides with the Earth's axis of rotation, but the Earth's magnetic poles do not coincide with the axis of rotation. The Magnetic North Pole is in the Arctic Islands of Canada; and to complicate matters further, the magnetic pole slowly migrates with time, thus changing declinations. The change is slight, however, for recently made maps. Charts and tables are available to determine necessary corrections.

One should know the declination in an area for the following reasons: declinations are different on different maps; most directions in the field are taken with magnetic compasses (some compasses can be adjusted to read true directions, but some cannot); and maps are constructed in reference to true directions (not magnetic).

Map Scales

The scale of a map is the ratio between the distance separating two points as measured on the map and the actual distance separating the same two points on the ground. For instance, on a map the distance between the centers of city A and city B, as measured with a ruler, might be 1 inch[1] (map distance); but the actual distance between the two cities is 50 miles (ground distance). The general scale relation can be given as a simple formula:

$$\text{map scale} = \frac{\text{map distance}}{\text{ground distance}}.$$

Map scale can be presented in three different ways; however, all give the ratio of map to ground distance. The *Fractional Scale* (sometimes called the *R.F.*, meaning representative fraction) is the most important and widely used. For example, $\frac{1}{10,000}$ (sometimes written as 1:10,000) is a fractional scale indicating that one unit measured on the map is equal to 10,000 of the same units on the ground. No particular unit of measure is implied

[1] Metric conversions will not be used in this appendix because almost all available American maps use English units.

or required for use with such a scale. Thus any unit can be substituted into the scale. As examples, 1:10,000 can be taken as 1 inch (on the map) equals 10,000 inches (on the ground), or 1 foot (map) equals 10,000 feet (ground) or 1 centimeter equals 10,000 centimeters. So long as the same unit of measure is assumed for the top and bottom of the fraction, any units may be used. Thus this is a truly international scale wherever arabic numerals are used.

The *Word Scale* makes it easier to visualize scale relations. In it, the map distance is stated as some convenient small unit, and the ground distance is given in a common larger unit. Examples would be: "1 inch equals 2 miles," or "1 centimeter equals 1 kilometer," or "1″ = 1000 ft." Although convenient, such scales are not international since you could read them only if you knew the language and units of measure used.

The *Graphical Scale* is most useful for measuring distances on maps. In these scales an actual line is drawn on the map (usually at the bottom) and divided into segments which are labeled with corresponding ground distances. These scales are constructed to any convenient ground units; but, where more than one certain unit is desired, a separate scale must be drawn for each.

THE UNITED STATES LAND SURVEY

The United States Land Survey system, invented by Thomas Jefferson, was adopted by Congress in 1785 for the survey of the Northwest Territories. It is now used for location and land designation in the United States, except for the original thirteen colonies, some other eastern states, and Texas.

The system is based on squares, 6 miles on a side, which are called Congressional Townships. These townships are located in reference to the intersection of a true north-south line, or Principal Meridian (Fig. A-1a), and a true east-west line, a

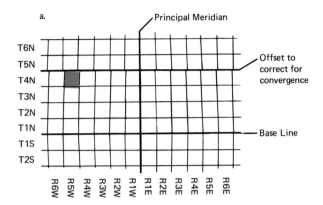

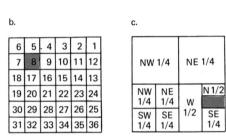

U.S. Land Survey Diagrams

Fig. A-1 U.S. land survey diagrams: (a) Division into townships. (b) Divisions within a township. (c) Divisions within a section.

geographic parallel called a Base Line. Some 32 different systems of townships are controlled by intersecting Principal Meridians and Base Lines which are given either numbers or names.

Land in the first east-west row of townships north of a given Base Line is designated as being in Township One North (T.1N.); land in the second row is designated as being in T.2N., etc. South of the Base Line, land is designated as T.1S., T.2S., etc. The north-south columns of townships are designated as Ranges. Range One West (R.1W.), and Range One East (R.1E.) lie on their respective sides of a Principal Meridian. R.2E. and R.2W. are the next columns out from the Principal Meridian, R.3W. and R.3E. next beyond, and so on, to the limits of the area controlled by a particular Meridian and Base Line. Thus the Congressional Township in the accompanying illustration would be designated as T.4N., R.5W.

Since the townships are bounded on east and west by true north-south lines, or geographic meridians, the townships would become markedly narrower progressively farther away from the Base Line because meridians converge towards the poles. Thus a correction is made by setting back the townships to a 6 mile width at the north side of each four sets of townships (i.e., at T.4N., T.8N., T.12N., etc.).

Each township is divided into 36-mile squares, called Sections. The sections are numbered as shown in Fig. A-1b. Thus land shown by the shaded square would be designated as S 8 T.4N., R.5W.

The sections are divided according to aliquot parts; for example, into the northeast, southeast, northwest, and southwest quarters; or northern and southern, and eastern and western halves. In turn, a unit such as a quarter-section can be further divided into parts using the same system (Fig. A-1c), and these smaller parts can be further subdivided by the same method. Thus in the accompanying diagram the area described (the shaded rectangle) is the S½ NE¼ SE¼ sec. 8 T.4N., R.5W. of the 6th Principal Meridian. The smaller land designations of the U.S. Land Survey may seem complicated, but always remember that land designations start with the smallest unit and progress to the larger ones.

Introduction to Topographic Maps

Four main methods are used to represent relief, or topography, on maps, and some maps combine several methods. In the method of *altitude tints*, different colors are used for specified ranges of elevations. For instance, green might represent elevations from zero feet (sea level) to 1000 feet above sea level; yellow might indicate elevations from 1000 to 2000 feet; brown, 2000 to 3000 feet,

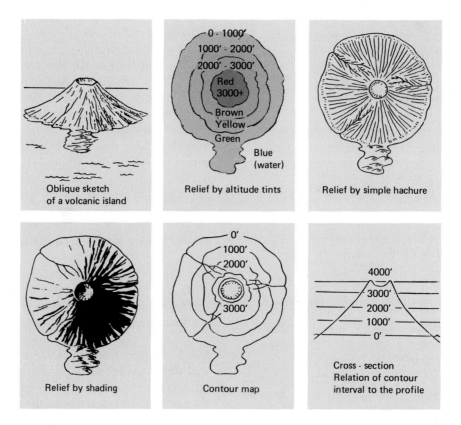

Fig. A-2 Methods of showing relief on maps. The same volcanic island is illustrated by various methods.

red, over 3000 feet (Fig. A-2). However, check each map used, for the colors for various ranges of elevations are not always the same. This system of altitude tints is commonly used on small-scale maps (which cover large areas), air navigation maps, and some large-scale (small area) foreign maps.

The method of *hachures* makes use of many short lines, each aligned with the direction of steepest slope. Although complicated methods of hachuring were constructed for use on detailed maps of small areas in the past, simple hachures are used today mostly for small-scale maps requiring no great detail.

Relief is also shown by *shading*—most commonly assuming a northwest source of illumination (representing the Sun's position), so that east-to-south slopes are shown by shadows and west-to-north slopes appear illuminated. Shaded relief maps give a strong visual impression of topography, resembling the ground as it would be seen from an airplane. Although shading is sometimes used alone as on maps issued by some airlines, it is often combined with other methods.

Although each of the methods just described for showing relief serves certain purposes, the most precise information as to elevations is provided by the system of *contour lines*. A contour is a line on a map connecting points of equal elevation. On any particular map, the contour lines have a constant vertical spacing called the contour interval (a few maps use two different intervals). To help visualize contours, consider that a river has been dammed and the water rises in the resulting lake. If air photos are taken at each 5-foot vertical rise of the lake surface, the shorelines represent contours having 5-foot contour interval (Fig. A-3). Where slopes around the lake are gentle, these successive "contours" will be widely spaced because the water had spread far *laterally* during its rise. If slopes are steep, the "contours," or lake shores, will be closer together, since the water is restricted as the lake deepens. The same idea applies to the reading of contour maps; closely spaced contour lines mean steeper slopes, and widely spaced ones indicate gentler slopes.

The contour interval selected for a particular map depends on the scale of the map and the relief of the area represented. Small-scale maps (covering large areas) usually require large contour intervals of 50, 100, 200 feet, or more. On the other hand, maps of smaller areas (large scale) tend to have intervals of 25 feet or less. Relief also determines the C.I. (contour interval) used. Mountainous regions (high relief) require large contour intervals

so that the lines will not merge into a solid and unreadable mass. Nearly flat areas (low relief) require small contour intervals to bring out topographic detail. Some maps of very flat areas use contour intervals as small as a foot.

On the commonly used maps of the U.S. Geological Survey, the contours (shown as brown lines) require close observation because they are often closely spaced (Fig. A-4). Because of this, not every contour is labeled as to elevation. By convention, every fifth contour is made heavier than the others and labeled. To get the elevation of an unlabeled contour, find the nearest labeled one and—using the contour interval—count and calculate the elevation of the unmarked line. If a point lies between contours, its elevation can be approximated. If the point is between a 20-foot and 30-foot contour, one can safely say its elevation is more than 20 feet and less than 30 feet. However, assuming the ground slope is constant between the contours, a closer approximation is possible. If the point is half-way between the contours horizontally on the map, you can assume it is also half-way vertically, and therefore has an elevation of 25 feet. A point one-quarter of the distance from a 50-foot contour towards a 60-foot contour could be assumed to have an elevation of about 52.5 feet, and so on.

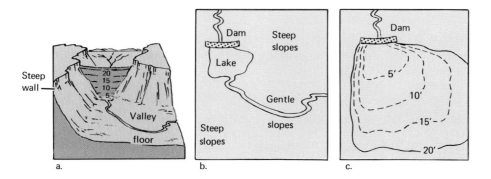

Fig. A-3 (a) Oblique view of a lake starting to fill behind a dam. (b) Map view showing the lake shore after a five foot vertical rise in water level. (c) Lake shore shown at successive rises in water level. The dotted lines approximate contours.

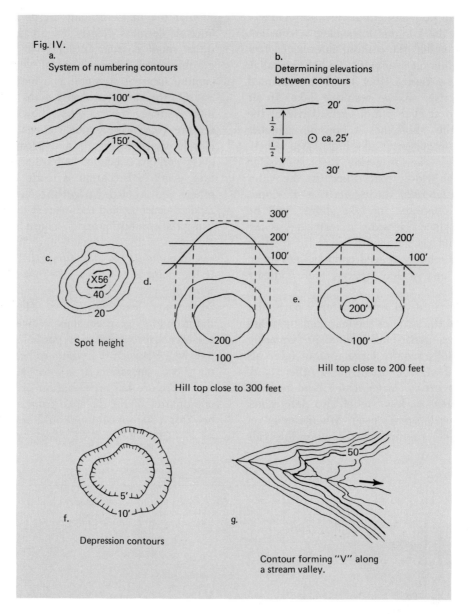

Fig. IV.

a.
System of numbering contours

b.
Determining elevations
between contours

c.

Spot height

d.

Hill top close to 300 feet

e.

Hill top close to 200 feet

f.

Depression contours

g.

Contour forming "V" along
a stream valley.

Fig. A-4   Contours as they are indicated and read on maps of the U.S. Geological Survey.

Hills are special problems. Some do have their tops marked by spot heights (brown ×'s) giving the elevation. In other cases, the elevation can be safely stated as higher than the last contour shown on the hill and lower than the next not appearing. Thus the summit of a hill might be described as more than 100 feet, but less than 120 feet. However, a closer approximation is possible assuming the hill is smoothly rounded. Where the diameter of the last closed contour is wide, you can assume the highest point is near the next highest contour that does not show. If the hill were a flat-topped mesa, you could be wrong; but usually, by studying several hilltops on the map, the probability of mesas can be determined. On the other hand, if the last closed contour has a small diameter—compared to the general contour spacing—one can assume that the height is close to the last closed contour.

Because most landscapes are cut by stream valleys, large and small, contours commonly have V-shaped deflections where they cross a drainage line. A simple but useful rule for reading or sketch-ing contour lines is that the point of the V is directed upstream and the widening of the V is downstream.

Special contours show closed depressions such as volcanic craters or hollows containing lakes. Such depression contours are marked by hachures pointing towards the center of the depression. These contours are read in the same way as regular contours in terms of contour interval and elevations.

This discussion points out what contours are. However, the real art of reading contour maps is the ability to interpret geologic landforms, structures, and rock types indicated by the contour patterns. This demands some training and the ability to visualize the Earth's surface in three dimensions from squiggly brown lines on a flat piece of paper.

Appendix B: Metric Conversion

English to Metric	Metric to English

English to Metric

1 inch = 2.540 centimeters
1 foot = 0.305 meter
1 yard = 0.914 meter
1 mile = 1.609 kilometers

1 square inch = 6.452 square centimeters
1 square foot = 0.093 square meter
1 square yard = 0.836 square meter
1 square mile = 2.590 square kilometers

1 cubic inch = 16.387 cubic centimeters
1 cubic foot = 0.028 cubic meter
1 cubic yard = 0.765 cubic meter
1 cubic mile = 4.168 cubic kilometers

1 quart = 0.946 liter
1 gallon = 3.785 liters

1 ounce = 28.350 grams
1 pound = 0.454 kilogram
1 short ton (2000 pounds) = 907.185 kilograms
1 short ton = 0.907 metric ton

Metric to English

1 centimeter = 0.394 inch
1 meter = 3.281 feet
1 meter = 1.094 yards
1 kilometer = 0.621 mile

1 square centimeter = 0.155 square inch
1 square meter = 10.764 square feet
1 square meter = 1.196 square yards
1 square kilometer = 0.386 square mile

1 cubic centimeter = 0.061 cubic inch
1 cubic meter = 35.315 cubic feet
1 cubic meter = 1.308 cubic yards
1 cubic kilometer = 0.240 cubic mile

1 liter = 1.057 quarts
1 liter = 0.264 gallons

1 gram = 0.035 ounces
1 kilogram = 2.205 pounds
1 kilogram = 0.001 ton (short)
1 metric ton = 1.102 tons (short)

Density:

1 pound per cubic foot = 0.016 gram per cubic centimeter

1 gram per cubic centimeter = 62.4 pounds per cubic foot

Velocity:

1 mile per hour = 1.609 kilometers per hour

1 kilometer per hour = 0.621 mile per hour

TEMPERATURE CONVERSION

Water freezes at 32°F and boils at 212°F (at normal sea level pressures); thus the two points are separated by 180°F. 1°F = 5/9th of 1°C

Celsius (Centigrade) Scale:

Water freezes at 0°C and boils at 100°C (normal pressure); thus the two points are separated by 100°C. 1°C = 9/5th of 1°F

To convert Fahrenheit to Celsius: subtract 32 (add 32 if below zero) and multiply by 5/9.

To convert Celsius to Fahrenheit: multiply by 9/5 and add 32.

Kelvin Scale:

Zero is −273°C; water freezes at 273°K and boils (normal pressure) at 373°K. 1°K = 1°C

Glossary

This glossary contains the "less ordinary" geologic terms, some common ones that are usually misused, and some nontechnical words that may be troublesome to the average student. It is designed to help the reader of this particular book; it is no substitute for a standard English dictionary, or the more specialized and extensive *Dictionary of Geologic Terms* or the *Glossary of Geology and Related Sciences* published by the American Geological Institute.

Abrasion As applied to wind action, the natural sand blasting of rock surfaces by windblown material.

Absolute dating Dating of a geologic feature or event in years.

Abyssal Related to the deepest parts of the ocean.

Acraniata A chordate animal with no definite head, brain, or paired appendages.

Adaptive Showing favorable adjustment to a situation or surroundings.

Adventitious Formed in odd places or without any particular order.

Agate A very finely crystalline kind of quartz containing colored layers.

Agglomerate A volcanic rock composed of coarse, often angular, fragments.

Aggrade To build up or raise a stream channel or land surface by deposition.

Algae A large group of nonvascular plants that have chlorophyl, but lack roots, stems, or leaves.

Alluvial fan A fan-shaped landform developed where a stream deposits material when flowing into an area of lesser slopes, as, for example, a mountain stream emerging into a flat-floored major valley.

Alluvium Clastic deposits laid down by the action of modern streams.

Amino acids The building blocks of protein.

Ammonoid An extinct shelled cephalopod having folded or wrinkled cross-partitions in its shell.

Amoeba A very simple, very small, one-celled animal.

Amphibian A vertebrate animal which has gills in the immature swimming stage, can travel on land, but (with a few exceptions) must breed in water.

Analogy A similarity of function or properties without necessarily being identical. Or, reasoning by comparing something to something better known.

Andesite A fine-grained volcanic rock lacking quartz, and composed mainly of feldspar (plagioclase) and dark iron-magnesian silicates. Named after the Andes Mountains whose volcanos contain this rock.

Angiosperm Flowering plant. The dominant vegetation of the present day. Seeds are covered. Many trees, shrubs, vegetables, and flowers are examples.

Anorthosite An igneous plutonic rock largely composed of a single type of plagioclase feldspar (often called labradorite). Anorthosites are only moderately abundant in the Earth's crust, but they seem to be the dominant rock in the Moon's highlands.

Antecedent Refers to a stream that maintains its previous course when mountains or hills are uplifted across it.

Anthracite "Hard" coal. It is black with a glassy luster, and has a high percentage of fixed carbon.

Anthropoid Of a group of higher primates including man, apes, and monkeys.

Anthropomorphism Ascription of human attributes to other natural things.

Antibiotic Harmful to or destroying bacteria.

Anticline A fold in which the center moved upwards relative to the sides when it formed. It is, or was, convex upwards.

Aphanitic Refers to textures of fine-grained igneous rocks which have crystals too small to be readily visible without magnification.

Aquifer A rock layer or zone from which water can be obtained.

Archaic Belonging to an earlier time.

Archipelago A chain or group of islands.

Arête Sharp, knifelike ridge between glaciated valleys.

Arkose A sandstone containing abundant grains of feldspar as well as quartz.

Artesian spring A place where water is rising from depth under pressure.

Artiodactyl A hoofed mammal having an even number of useful toes, such as the pig and cow.

Arthropod An invertebrate animal having jointed legs and a segmented shell of chitin. Insects, spiders, and crabs are examples.

Ash (volcanic) Fine, dust-sized solid material erupted by volcanos.

Astrogeology The study of planets and other solid bodies in the solar system.

Aves The birds.

Badland A region almost totally lacking vegetation and eroded into a rough miniature mountain country of intricately dissected gulleys, ridges, and pinnacles.

Bar A ridge, usually of sand or gravel, in a sea, lake, or stream. Bars may be submerged or rise above the water.

Barrier beach A long sand ridge rising above the sea which is separated from the shore by a lagoon.

Basalt A dark-colored, fine-grained, igneous rock containing dark feldspars and ferromagnesian minerals. The most abundant type of volcanic rock worldwide.

Batholith A very large intrusive body of granite or granitic rocks. They are exposed at the ground surface only after long erosion.

Bathyscaphe A specially designed submarine for very deep diving.

Beach A sand-, pebble-, or cobble-covered slope along the water's edge.

Bed load Stream-carried material that remains in contact with or bounces along the bottom of the stream channel.

Bedrock Solid rock either exposed at the Earth's surface or covered by unconsolidated materials.

Belemnoid An extinct cephalopod commonly leaving its cigar-shaped internal skeleton as a fossil.

Benioff zone A planar zone where earthquake centers are concentrated that slopes at about 45°, more or less, from beneath the oceanic trenches underneath the continental margins.

Biased Slanted. Showing partiality or preference.

Bituminous coal "Soft" coal. It is black, has a blocky fracture, and gives off volatile matter on burning.

Bivalve An animal having two shells, such as a clam.

Bomb (volcanic) Large fragments blown out of a volcano as solid or plastic masses.

Brachiopod A marine animal having two unequal shells, commonly attached to the sea bottom by a fleshy stalk.

Breakers Collapsing waves that have moved into shallow water near shore.

Bryophytes A group of nonvascular plants that may have stems and leaves, but lack true roots.

Bryozoans Small colonial animals living mainly in the sea. Called *moss animals*, they build lacelike, twiglike, or mounded colonies.

Calcite The mineral composed of calcium carbonate ($CaCO_3$).

Caldera A large depression, commonly steep-sided and several times as wide as it is deep, found in volcanic areas, often in the center of a volcano.

Canine Of or related to a dog. Also a stabbing tooth in mammals.

Capacity The ability of a stream to transport in terms of the total amount of debris it can carry past a given point in a given unit of time.

Carapace A bony or chitinous shell covering an animal's back.

Carbohydrate An organic compound containing carbon, hydrogen, and oxygen—such as sugar or starch.

Carbonaceous Coaly or containing carbon.

Carbonation As a form of chemical weathering, refers to reactions involving carbon dioxide and water that create soluble carbonates.

Cartilaginous Composed of gristle, or tough elastic animal substance.

Catastrophism The concept that widespread, or worldwide, often violent events of far greater magnitude than can be observed at the present time have created rocks and topography of the Earth and suddenly exterminated large groups of living things at various times.

Cave A natural cavity, large enough to be entered by man, that extends back into total darkness.

Cephalopods The most advanced group of molluscs. They swim by expelling a water jet. The squid, octopus, and shelled nautilus are examples.

Cetacea Mammals leading a fishlike existence such as the whale.

Chert A very dense sedimentary rock composed largely of very finely crystalline silica.

Chitinous Having an organic composition like fingernails.

Chlorophyl The green substance in plants that is responsible for photosynthesis.

Chordate An animal with a backbone or stiffening internal flexible rod.

Chronology The measuring of time or ordering of events.

Cinders (volcanic) Rough sand to pebble-sized materials erupted by volcanos.

Circumpolar vortex A global swirl of air comprising the Earth's lower atmosphere (troposphere). Two exist from about 30° latitude, north and south, to the respective polar regions.

Cirque The rounded, armchairlike head of a glacial valley.

Clastic Composed of fragments of older rocks or solid organic material.

Clay May refer to certain specific very finely crystalline *minerals*, or to materials consisting of solid particles of the finest *size*, or to materials that are plastic and *moldable* when wet.

Claystone A consolidated sedimentary rock consisting largely of clay particles.

Cleavage (mineral) A tendency to split more readily in certain parallel directions than in others.

Coagulate To form a clot or jellylike mass.

Coal A combustible black (sometimes brown) rock formed by the alteration of plants, mainly woody, so that the carbon content is increased.

Coast A zone of indefinite width from the water's edge inland to a major topographic change.

Coelenterata A group (phylum) of simple animals having a two-layered body and a simple digestive tract with a single opening which may be surrounded by tentacles. Jellyfish are an example.

Colloidal suspension Refers to very fine particles—like those of clay—held up in water by the continuous movement and jostling of water molecules.

Colonial Refers to a group of similar animals living in such close connection that the individuals are not readily recognizable from the mass of the resulting structure.

Competence The ability of a stream to transport in terms of the largest-sized particles or boulders it can move.

Concretions Nodular masses which are more resistant than the sedimentary beds containing them.

Conduction The transfer of heat (or sound or electricity) through material without movement of masses of the material.

Condylarths Primitive hoofed mammals.

Conglomerate A sedimentary rock composed mainly of rounded fragments larger than sand. It is composed of pebbles, cobbles, or larger fragments.

Convection The transfer of heat through a liquid or gas by movements of masses of the material.

Convergent boundary In plate tectonic theory, place where major plates of lithosphere are colliding (as along the west coast of South America) and where oceanic trenches and adjacent deformation and mountain-building are active.

Coral A marine organism that lives attached to the bottom as a solitary individual or together in colonies. Their calcareous external skeletons may form large limestone masses, or reefs. They are coelenterates.

Cordillera A mass of mountain ranges, valleys, and plains which has an overall elongation or trend. More specifically, the north-trending mountainous region of North America extending from the edge of the Great Plains to the Pacific Ocean.

Correlate To determine that rocks in different places are of equivalent age. Or that events in separate areas occurred at the same time.

Corrosion Stream erosion accomplished by solution of rocks.

Cosmic Belonging to the universe.

Cosmic ray Intense, penetrating radiation coming from outer space.

Cosmogony A branch of astronomy dealing with hypotheses for the origin of the universe.

Cotylosaur An extinct ancestral reptile.

Creationist One who believes that each and every type of living thing was specially made by a supreme power or being, and that, once created, plants and animals never change (an antievolutionist).

Crinoid A marine invertebrate animal attached to the sea floor by a flexible jointed stalk, and having a head or cup with many radiating arms. A "sea lily."

Cross-laminations Sloping layers lying at a distinct angle to the larger layers which contain them (cross-bedding).

Crossopterygian A fish with lungs and fins with stalks of bone surrounded by flesh.

Crystal A solid bounded by smooth faces whose arrangement results from an orderly and repeated internal arrangement of atoms.

Crystalline Having a definite and repeated arrangement of atoms (ions).

Crystalline rocks Commonly used loosely to refer to igneous and metamorphic rocks (not sedimentary). More exactly, the term applies to any rock having easily discernable crystals.

Cycad A tree with a stubby trunk and palmlike fronds. A gymnosperm plant.

Daughter product Element produced from another element during radioactive decay.

Deduction Reasoning from a general principle to particular conclusions or consequences.

Deflation The blowing away of material by wind action.

Deformation Change in original shape or volume.

Delta A low plain underlain by deposits laid down where a stream flows into an ocean, lake, or other standing body of water.

Dendritic Refers to a stream pattern that is generally treelike in its branchings.

Density Mass per unit volume.

Deposition (geologic) The settling out of particles or precipitation of materials in solution to form a rock or potential rock.

Desert A region of sparse vegetation that usually does not support much life.

Desilication Includes various chemical weathering reactions that remove silica from rocks and soil.

Dew point The temperature at which an air mass is saturated with water vapor.

Diamond Composed of pure carbon, the hardest mineral known.

Diatom A single-celled microscopic plant that lives in water and secretes a siliceous skeleton.

Differentiate To separate into unlike parts.

Dike A flat, tabular igneous intrusion that cuts across older structures.

Dilitancy The process involved in recent theorizing on the causes of earthquakes. Rocks being deformed along a fault gradually increase in volume because of increased pore spaces resulting from minute fracturing and mineral-grain rotation.

Dip The inclination of a rock layer or surface measured from the horizontal. It is at right angles to the strike.

Discharge (stream) The amount of water flowing through a cross-section of a stream in a given unit of time.

Disconformity An unconformity wherein the older and younger beds are parallel.

Divergent boundary In plate tectonic theory, an axis of spreading, as occurs in mid-oceanic ridges, where major plates of lithosphere are moving apart and new rocks are forming by eruptions from the Earth's mantle.

Dramatis personae The cast of characters in a play.

Drift (glacial) A general term for any deposits associated with glaciation. They could be laid down by ice, meltwater streams, or in glacial lakes.

Dune A mound, ridge, or hill piled up by wind. Usually dunes consist of sand, but silt and clay dunes also exist.

Dunite A crystalline rock largely composed of the mineral olivine. A variety of peridotite.

Echinoderms   A group (phylum) of animals having a five-fold radial symmetry. An example is the starfish.

Echinoid   Free-moving, marine, invertebrate animals having a five-rayed symmetry. They belong to the phylum Echinodermata. Sea urchins are an example.

Eclogite   A granular rock largely composed of garnet and pyroxene.

Ecology   The study of organisms and how they relate to their environment.

Edentate   A mammal that lacks teeth or has simple teeth without enamel.

Effluent   Flowing out of. May refer to a stream receiving some water from groundwater.

Elasticity   The property applied to bodies which return to their original form after a distorting force is removed.

Embryology   The science dealing with the development of plants and animals after fertilization but before they develop their typical form.

Empirical   Based on observation with little or no interpretation.

Enigma   Something puzzling.

Environment   All the conditions in a place which influence communities of life.

Epeirogeny   Broad movements of the Earth's crust. May cause major uplift and downwarping, but the rocks involved are not intensely folded.

Epicenter   The apparent point of maximum intensity of an earthquake at the surface of the ground.

Epoch   A unit of geologic time. A subdivision of a geologic period.

Era   A major division of geologic time. Generally five eras are recognized (see the geologic time scale).

Erosion   The progressive removal, or carving away, of rock and other Earth materials by geologic processes acting on or near the Earth's surface.

Erratic   A glacially transported boulder of different rock type than the rock on which it rests.

Escape velocity   The velocity a particle or object must reach to escape from the gravitational attraction of the Earth (or other massive body).

Esker   A long, sinuous ridge of stratified meltwater material laid down in streams flowing beneath a glacier.

Estuary   The lower part of streams or drainage channels which are affected by the rise and fall of tide from the adjacent sea.

Eugeosyncline (eugeocline)   Either: the outer part of a broad regional downwarp (geosyncline), underlain by oceanic crust, and containing volcanic rocks along with shales and poorly sorted sandstones (graywackes); or shales and graywackes deposited in continental slopes and continental rises.

Eurypterid   An extinct arthropod with poison glands but no antennae.

Eustatic change   A worldwide change in sea level.

Evaporite   A rock made of minerals whose deposition resulted from complete, or extensive, evaporation of water in which they were dissolved.

Evolution   Progressive change, often to more complex forms.

Exhumed   Reexposed by the removal of rock material.

Exoskeleton   A shell or hard protective outer cover of an invertebrate animal.

Extrusive (igneous)   Molten rock material that rises and pours or is ejected onto the Earth's surface.

Facies   Differences from place to place in a rock unit laid down at the same time. (The meaning of the term is complicated by several different interpretations.)

Fathom   1.8 meters (6 feet). A nautical unit for depth measurement.

Fault   A fracture along which rock masses have been offset.

Fauna   A group of animals that lived together in a place or time.

Ferns   Plants with large feathery leaves on whose backs reproductive spores are produced. They are the more primitive, dominant group of living plants (Pteropsida).

Ferromagnesian minerals   Generally dark minerals containing much iron and magnesium as well as silicon and oxygen.

Filamentous   Threadlike.

Finite   Measured in years (of dating).

Fiord   A narrow, steep-walled, glacial trough flooded by the sea.

Fission (nuclear)   An atomic process in which more complex atoms break down into other simpler atoms, with the release of energy.

Fissure   A long, narrow fracture, or crack, in rocks.

Flocculation   The process of aggregating, coming together, in fine clots. Applies to the settling of clay particles in water.

Flora   A group of plants that lived together in a place or time.

Fluid   A liquid or gas. Something that readily changes shape and flows.

Focus (plural: foci)   The true center of an earthquake. The place of maximum intensity at depth.

Foliation   A tendency of metamorphic rocks to part along parallel planes or contorted surfaces.

Foraminifera   One-celled animals living mainly in salt water that secrete calcite or cement foreign grains together to form a protective cover.

Formation   Rock masses forming a unit convenient for description and mapping.

Fossil   Any evidence of ancient life preserved in rock by natural causes.

Fungi   A group of nonvascular plants lacking roots, stems, or leaves. They lack chlorophyl and feed on organic matter.

Fusion (atomic)   An atomic process in which two simpler types of atom merge into a different kind of atom, with the release of energy.

Gametophyte   A plant which bears male and female elements which produce a nonsexual, spore-producing plant.

Gastropod   A snail.

Geanticline   A complex mountainous belt of folded and faulted rocks from which sediments for a geosyncline are derived.

Genealogy   A history of the descent of a person or family from their various ancestors.

Genetic   Relating to origin.

Genus (plural: genera)   A group of related animals. A group of species.

Geochemistry   The science dealing with the abundances and distribution of chemical elements in the Earth's crust, hydrosphere (water), and atmosphere.

Geochronology   The study of time in relation to the history of the Earth.

Geology   The science which deals largely with the composition, structure, and topography of the solid Earth; the internal and external forces acting on it; its history of physical change and past life.

Geometric   Having to do with space relations.

Geophysics   The science which applies the theory, methods, and techniques of experimental physics to the study of the Earth including the hydrosphere (waters) and atmosphere.

Geosyncline   A large, generally linear downwarp filled with sedimentary and, in some cases, extrusive igneous rocks.

Geyser   A hot spring that periodically erupts water and steam into the air.

Glacier   A large natural mass of ice, derived from snow, that originates on land and shows evidence of flowage.

Graben   A structural valley representing a down-dropped block between normal faults.

Graphite   A very soft, black to grey mineral composed of pure carbon.

Graptolites   An extinct group of marine invertebrates whose relation to other major groups is not surely known. Usually preserved in dark shales as small branches having many small cups.

Gravity   The force of attraction between masses in the universe. (Most commonly related to the mass of the Earth).

Graywacke   A sandstone containing angular quartz and feldspar fragments and some rock fragments, with clay-sized filling in between the sand-sized grains. A generally tough, dark-colored rock.

Groundwater   Water underground in the zone where rocks are saturated.

Guyot   A submarine mountain having a conical, volcanic shape but with a flat, truncated top.

Gymnosperms   Plant reproducing by seeds which are usually exposed and held in cones. Pine and fir trees are examples.

Hadley cell   Great atmospheric convection cell, driven by the strong heating of the Earth's equatorial belt, which may be the major mechanism controlling the circulation of air in the Earth's tropical and subtropical regions.

Half-life   The time required for half the atoms of a radioactive element to break down into different products.

Halite   The mineral composed of sodium chloride. Rock salt.

Headland   Generally, a bold projection of land into an ocean, sea, or lake.

Hematite A common iron oxide mineral.

Heterogeneous Containing a mixture of different things.

Hiatus A "gap" in the rock record resulting from erosion or nondeposition.

Hierarchy People or things arranged in a graded system of higher to lower ranks.

Holocene Time since the Pleistocene continental glaciers retreated and disappeared from north Europe, Siberia, and North America. Estimates of its beginning vary, but a date of about 10,000 years ago seems reasonable.

Hornfels A tough, unfoliated metamorphic rock resulting from contact metamorphism of fine-grained rocks.

Horst A mountain uplifted between normal faults.

Hybrid The offspring of two different types of plants or animals.

Hydration The chemical combination of water with a compound to form a new compound.

Hydrosphere All the water on the Earth's rock crust including oceans, lakes, swamps, streams, glaciers, and similar features.

Hypothesis A tentative explanation that needs more testing for certain observations or facts. A possible theory.

Igneous rock Rock formed by solidification of once-melted material, or magma.

Incisor A front or nipping tooth.

Induction Reasoning from particular facts or observations to a general conclusion.

Infiltration In relation to groundwater, refers to the movement of water from the surface into the ground.

Insectivore A mammal that lives on insects and worms.

Intensity The strength of an earthquake measured by readily observable features such as the effects on man-made structures. It differs from magnitude, which is earthquake strength measured by instruments.

Intrusive (igneous) Penetrating into older rocks as a molten mass and later solidified at depth without reaching the Earth's surface.

Invertebrate An animal without a backbone.

Isostasy The concept that large lighter masses of the Earth's crust are floating on denser material beneath.

Joint A fracture which has not offset rock on either side. (If slippage has occurred, a fracture is called a fault.)

°K (Kelvin) A temperature scale like the Centigrade except that zero is −273°C.

Karst Regions in limestone, or similar rocks, that have a topography marked by sink holes, disappearing streams, and other features resulting from extensive solution of the rock.

Kettle A depression in glacial drift left where a block of ice has melted.

Kinetic Resulting from motion.

Labyrinthodont An extinct amphibian with an infolded pattern of tooth enamel like that of a lobe-finned fish.

Laccolith A lens-shaped igneous intrusion that bows up surrounding sediments.

Lagoon A protected body of water between reefs, barrier beaches, or islands and the shore.

Larva The early form of an animal when it does not resemble the parent.

Laterite A red soil in humid tropical regions. It contains much oxidized iron and aluminum and is leached of silica.

Lava The molten rock material flowing onto the Earth's surface from volcanic action. Also, the same material after it becomes rock.

Leach To remove a soluble substance by downward-filtering water.

Lignite A brown to brownish-black coal that still retains texture of the original wood.

Limb One side, or flank, of a fold.

Limestone A sedimentary rock composed mainly of calcium carbonate, the mineral calcite.

Lithification The conversion of loose sediments into consolidated rock.

Lithologic Relating to the overall physical characteristics of a rock unit or specimens. Usually refers to features that can be observed directly or by using at the most a magnifying glass.

Lithosphere The outer solid or rock part of the earth. Essentially the earth's crust and uppermost solid mantle.

Little Ice Age The time marked by minor glacial expansions during the last few thousand years, following the world's warm climatic interval that ended the Pleistocene.

Load The sediment and other materials carried along by a stream.

Loess Deposits of windblown dust which are largely silt-size and generally tan to buff in color. (Sometimes reworked by streams.)

Luster The appearance of a material in reflected light.

L-waves The slowest-moving earthquake waves, which follow the Earth's surface.

Lycopsida Spore-bearing plants with solid stems and minute spirally arranged leaves.

Maar A volcanic explosion crater not associated with lava.

Magma Molten parent material of igneous rocks.

Mammoth A large, extinct, hairy elephant that lived in northerly climatic zones.

Manganese A metallic element used in steel-making.

Mantle A major zone of the Earth between the crust and core. Thought to consist of dense rocks containing mostly iron-rich silicate minerals.

Maria Darker areas on the Moon's surface originally thought to be seas.

Marine Related to the sea.

Marsupial A mammal whose young are born in a very immature stage and cared for in the mother's pouch.

Mastodon An extinct relative of the elephant.

Meander A bend of large arc in a river.

Metabolism The chemical and physical processes in a living organism involving the making and breakdown of protoplasm.

Metamorphic rocks Rocks so changed by heat and pressure that their original characteristics are lost.

Meteorite A particle or mass from outer space forming an incandescent streak as it is burned by friction on entering the Earth's atmosphere.

Meteoroid A solid particle from outer space ranging from sand-size to great blocks.

Microseisms Almost continuous weak tremors in the Earth from winds, storms, and other causes not related to earthquakes.

Mineral A naturally occurring solid having a definite repeated internal structure and a characteristic chemical composition. It is crystalline.

Miogeosyncline (miogeocline) A broad downwarp, occurring on the margins of continental platforms, that contains nonvolcanic rocks dominated by conglomerates, well-sorted sandstones, shales, and limestones.

Miscegenation Interbreeding between different races.

Mohole A project to drill through the Earth's crust into the mantle. Temporarily, at least, it's canceled.

Molar A cheek tooth for grinding or shearing.

Mollusca A numerous and varied group (phylum) of invertebrate animals having well-developed circulatory, digestive, and sensory systems. Clams, snails, and squids are examples.

Monadnock An isolated hill or low mountain standing above a much-eroded, low rolling plain (or peneplain).

Monocline A flexure causing a local steepening in strata between generally horizontal rocks at its top and bottom.

Monotreme An egg-laying mammal.

Moraine Refers to certain landforms composed of glacial till.

Mountains Generally, any high, rugged topography. Geologically, rugged regions of volcanic or markedly deformed rocks.

Nappe Large rock mass that has moved some distance either by low-angle faulting, recumbent folding, or both.

Nautiloid A shelled cephalopod having simple saucer-shaped cross-partitions in its shell.

Nebula A large luminous cloud of gas and dust in space.

Neoglaciation A technical term generally comparable to "Little Ice Age."

Neptunist An eighteenth- or early-nineteenth-century geologist who believed that all the Earth's crustal rocks were precipitated from sea water.

Névé Granular ice derived from snow (also called firn).

Nitrate A compound containing nitrogen and oxygen (NO_3).

Noble gas Gas such as neon and argon whose atoms do not tend to combine into compounds.

Nova A star which suddenly increases greatly in brightness, then fades. A bursting star.

Nucleus A central unit around which other matter is gathered. The central part of an atom. Or a directive body found in some cells.

Orbit The path of a body moving around another body.

Ore An accumulation of minerals or rock from which a metal can be extracted at an economic profit.

Organic Related to life. Or, compound of carbon.

Ornithiscian Refers to a dinosaur group whose pelvic or hip structure was like that of birds.

Orogeny Mountain-building associated with folding and faulting.

Ostracoderm A primitive jawless fish with a head armored by bone.

Outcrop An exposure of bedrock at the ground surface.

Oxidation The combination of a substance with oxygen. Also, the loss of electrons from an atom, or ion.

Paleogeography The physical geographic features of a region at some given time in the geologic past.

Paleontology The science that deals with past life based on the study of fossils.

Pediment A generally gravel-covered surface sloping away from the base of a mountain front. It is cut fairly evenly across rocks, although some rocks may be hard and others poorly consolidated.

Pelecypod A clam.

Peneplain A low rolling surface of broad extent near sea level which results from prolonged erosion of formerly elevated land.

Peridotite A coarse-grained igneous rock containing olivine and sometimes other ferromagnesian minerals. It has no quartz and little, if any, feldspar.

Period A fundamental unit of geological time which has worldwide application.

Perissodactyl A hoofed mammal having an odd number of useful toes. The horse and rhinoceros are examples.

Permeability In relation to groundwater, refers to the ability of rocks or rock material to transmit water or other fluids.

Phosphate A compound containing phosphorous and oxygen (PO_4^{-3}).

Photosynthesis The making of carbohydrates from water and carbon dioxide using the energy of sunlight.

Phylum A major division of the plant or animal kingdoms.

Pisces The fish.

Placental Refers to mammals whose young are born in a relatively mature state.

Placoderm A primitive fish with jaws. Some had bone armor.

Planet A large nonluminous body moving in orbit around a star.

Plankton Floating or drifting life in the sea.

Plant A form of life having no central nervous system, no means of locomotion, that can make use of simple food materials from soil, water, air, or other organisms.

Plastic Having some characteristics of solids and some of liquids.

Plateau Generally, any extensive flat-topped highland. Geologically, a highland underlain by nearly horizontal rock layers which may be flat-topped or highly dissected.

Pleistocene See the geologic time scale. A time division corresponding to the latest major glaciation of the Earth. The term also refers to deposits of that time.

Plume In plate tectonic theory, a rising mass of the Earth's hot mantle that may or may not be part of a great convection cell.

Plunge The dip of a fold axis.

Plutonic Refers to rocks formed at considerable depths beneath the Earth's surface. Includes some igneous and some metamorphic rocks.

Pollen Minute grains carrying the male elements for fertilization in seed plants.

Polymorphic change A change of one mineral to another without leaving the solid state.

Porifera A group (phylum) of multi-celled, simple animals having no separate organs or special tissues. The sponges.

Porosity Refers to the total void, or pore, space in a rock or rock material.

Porphyritic (texture) Refers to igneous rocks in which larger crystals are surrounded by finer-grained or glassy material.

Precipitation Chemically, the separation of substances from solution. Also, the falling of rain or snow from the atmosphere.

Predator One who preys on others.

Preempt To be established beforehand.

Primate A mammal with five digits bearing flat nails on feet, or hands adapted for grasping. Man and monkeys are primates.

Primordial First, or original.

Proboscidian A mammal whose nose and upper lip form a trunk, e.g., the elephant.

Progenitor An ancestor, or parent.

Protoplasm The semiliquid, somewhat granular material of an animal or plant cell.

Prototype A primitive form.

Protozoa The single-celled animals.

Psilopsida Extinct group of the most primitive vascular plants.

Pteropsida The most numerous group of living vascular plants. Includes ferns, gymnosperms, and angiosperms.

Pumice A very lightweight and light-colored volcanic rock resulting from solidification of a very frothy lava. Being noncrystalline, it is structurally a glass.

Pyroclastic rocks Rocks composed of volcanic particles and fragments.

Quadruped A four-footed animal.

Quantitative Relating to amounts which can be stated in numbers.

Radiation (plants and animals) The spreading of organisms into new environments.

Radioactive Refers to elements that break down into other elements by giving off particles from their nuclei.

Radiolaria A group of marine, one-celled animals having complicated siliceous skeletons.

Radiometric Related to the measurement of the breakdown of radioactive elements.

Radium A radioactive element obtained mainly from uranium ores.

Radula A rasplike tongue characteristic of certain invertebrates (molluscs).

Red bed A general name for red sedimentary rocks which are most commonly sandstones or shales.

Regolith The same as waste-mantle.

Rejuvenation Increase of a stream's energy so that it starts deepening its valley. Climatic change or uplift of a region might cause this.

Reptile A cold-blooded, air-breathing vertebrate with a scaly or platy skin. Most lay eggs with shells, although in a few the young are born alive.

Reverse fault A fault whose hanging wall has moved up relative to the footwall.

Rift A valley formed by sinking of a strip between two faults. Usually a graben; sometimes a major strike-slip fault.

Rill The smallest type of distinct stream. A small trickle of water or its dry channel.

Rills (on the Moon) Trenchlike features of uncertain depth, but up to 1.6 kilometers (a mile) wide and 250 kilometers (150 miles) long.

Ripple marks Small ridges and troughs in unconsolidated materials such as sand, caused by moving water or air.

Rock Natural aggregate or mass of mineral matter forming an appreciable part of the Earth's crust.

Rodent A mammal whose teeth continue to grow through life and are designed for gnawing. Rats and squirrels are examples.

Ruminant A hoofed animal that stores food in a complex stomach, then returns the food to the mouth for more chewing before final swallowing and digestion.

Saltation A hopping or skipping motion of particles carried along by wind or streams. The particles periodically drop to the ground or channel bottom, bounce, and become suspended until they again drop.

Sarcodina A group of one-celled animals that extrude movable parts for locomotion.

Saurischian Refers to a dinosaur group whose pelvic or hip structure was like most other reptiles.

Savanna A grassland with sparse clumps of trees and having marked wet and dry seasons.

Scabland A starkly scenic terrain, trenched in places by a network of canyons, characteristic of basaltic areas where soil and covering materials have been largely stripped off.

Scarp A steep slope or clifflike form.

Schist A medium- to coarse-grained metamorphic rock that is mainly composed of flat tabular minerals.

Science The human search for understanding of the physical world during which the accumulation of new observations, with much argument, leads to the gradual—sometimes drastic—changing of general concepts. Also defined as: "controlled imagination" (a concise definition); "doing your damndest to solve your problems" (pungent); "getting government research grants" (realistic); "using my techniques to solve the problems of the physical universe" (pompous); and "knowledge, as of general truths or particular facts, obtained and shown to be correct by accurate observation and thinking" (incorrect).

Scoria Dark-colored volcanic lava having basaltic composition and glassy texture.

Scorpion A small land-dwelling arthropod, looking somewhat like a miniature lobster and having a jointed tail with poisonous stinger.

Sedentary Not given to much movement. Fixed in place.

Sedimentary rocks Rocks formed from materials once carried in water or air that have since settled, been precipitated, or even secreted by plants and animals, to form layers on the Earth's crust. Or, rock formed by deposition at the Earth's surface of material derived from older rocks.

Seed A plant's many-celled reproductive structure having an embryo plant with a food supply and protective cover.

Seismic Related to earthquakes or other earth vibrations.

Seismograph An instrument for recording earthquake waves and other vibrations in the Earth.

Seismology The science dealing with earthquakes and Earth vibrations.

Septum An internal partition.

Series A time-rock unit corresponding to an epoch.

Shark A fish with a cartilaginous skeleton.

Shelf The zone of shallow sea marginal to a continent.

Shield Broad rolling lowlands eroded on largely crystalline rocks forming the core of a continent.

Shrew A small, active, mouselike insectivore.

Sial The granitic zone of the Earth's crust. The term is derived from silicon and aluminum which are abundant in its rocks.

Silica The compound silicon dioxide (SiO_2).

Silicate Compounds containing silicon and oxygen in a crystalline structure based on the silicon tetrahedron.

Sill A flat tabular igneous intrusion that parallels older structures.

Siltstone A consolidated sedimentary rock consisting of fragments between sand and clay in size.

Sima The basaltic lower part of the Earth's crust whose rocks contain silicate minerals rich in iron and magnesium.

Sink A closed depression in the ground resulting from solution or collapse in limestone areas.

Siphuncle An internal calcareous tube in a shelled cephalopod.

Slate A very fine-grained metamorphic rock which tends to split into smooth, flat chips and slabs.

Slump A type of mass movement in which a block of material moves as a unit down and out on a curved plane of slippage.

Soil Unconsolidated Earth material characterized by horizons roughly parallel to the ground which are produced by organic and inorganic processes. In civil engineering, any unconsolidated Earth material is termed soil.

Sounding (nautical) Determination of the depth of a water body.

Species A group of plants or animals of very similar structure and relationship. Living species interbreed and produce fertile offspring.

Spectroscope An optical instrument for separating light into different wave lengths for study.

Speculative Related to theorizing from very little evidence.

Sphenopsida Plants with spore-bearing cones, hollow-jointed stems, and small leaves at the joints.

Spit (coastal) A small land projection into the sea. Many are sandy extensions of a beach into open water.

Spore An asexual, usually single-celled, reproductive structure of plants.

Sporophyte A nonsexual plant which produces spores that develop into sexual plants.

Spring A natural place where water seeps out of the ground for extended periods.

Stack A steep-sided pinnacle isolated by wave attack from a cliff of which it was once a part.

Stalactite An iciclelike projection of limestone projecting down from the ceiling in a cave.

Stalagmite A mound or projection of limestone projecting up from the floor of a cave.

Starch A carbohydrate food substance.

Stratification A layered or bedded structure formed during the deposition of sedimentary rocks.

Stream A mixture of water and rock waste flowing in a definite linear channel.

Striation A fine linear groove or ridge.

Strike The compass direction of an imaginary horizontal line on a sloping surface or rock layer.

Subaerial On land (as opposed to being under water).

Subduction zone In plate tectonic theory, a place where one major plate of lithosphere is descending under another.

Subsidence A mass movement, a dominantly vertical settling of material resulting from loss of mass at depth.

Subterranean Underground.

Superposition The laying down of rock layers in order, one on top of the other. It implies that the older layers are beneath younger layers. Also, downcutting by a stream so that it acquires a course across a formerly buried structure.

S-wave The secondary seismic wave. It passes through the body of the Earth and gives a shaking or up-and-down motion to particles in its path.

Syncline A fold in which the center moved down relative to the sides when it formed. It is, or was, concave downwards.

System A time-rock unit corresponding to a geologic period.

Talc A very soft mineral with a greasy "feel." A magnesian silicate found in metamorphic rocks.

Taxonomic Pertaining to the systematic classification of plants and animals.

Taxonomist A scientist who specializes in the classification of plants or animals.

Technology The application of science to practical or economic uses.

Tectogene A large, deeply downbuckled pod of sediments and volcanic rocks.

Tectonic Relating to structures in the Earth's crust caused by deformation (not of volcanic origin).

Tentacle A fleshy armlike projection from an invertebrate animal.

Terrace A steplike landform having a flat or gently sloping surface that is bounded on one side by a more steeply rising surface and on the other by a descending one.

Test A hard cover or supporting structure of invertebrate animals.

Tetrapod A vertebrate animal with four limbs. Amphibians, reptiles, birds, and mammals are tetrapods.

Thallus A plant body consisting of a mass of tissue with little differentiation into organs typical of higher plants.

Thecodont An extinct reptile that was ancestral to crocodiles, birds, and the extinct flying reptiles.

Theory An explanation of relationships between certain observed phenomena that has withstood considerable testing.

Therapsids An extinct reptile group with some mammallike structures.

Till Deposits laid down by the direct action of glacial ice. Usually a poorly sorted mixture of particles from clay to boulders in size.

Trace The line made by the intersection of a fault plane with the ground surface.

Traction Refers to a method of stream transport where the load carried is in constant contact with the channel bottom. Movement is by rolling, skidding, and so on.

Transmutation The change of one chemical element into another.

Trellis Stream pattern with several long, parallel trunk streams that are joined at right angles by short tributary streams.

Trilobite An extinct arthropod whose shell was divided into three longitudinal sections.

Tuff Volcanic ash consolidated into rock.

Tundra Treeless arctic areas.

Turbid Cloudy, filled with stirred-up fine sediments.

Ultraviolet light Extremely short wave lengths of light.

Unconformity A buried erosion surface with older rocks beneath and markedly younger rocks above.

Ungulate A hoofed mammal, such as a horse.

Uniformitarianism The general geologic rule that "the present is the key to the past," or that the processes operating today have acted the same throughout geologic time.

Unloading A weathering process in which rock breaks into slabs roughly parallel to the ground because of a slight expansion in rock after a thick mass of overlying material is eroded away.

Uranium A heavy radioactive metallic element found only in compounds.

Urea A very soluble crystalline solid found in the urine of mammals. Also produced synthetically for use in making plastics and other products.

Varve A thin sedimentary layer that represents one

Thallophyta A group of nonvascular plants lacking roots, stems, or leaves.

year of deposition. It may consist of a summer and a winter lamination.

Vascular Refers to plants or animals having a system of vessels and ducts for distributing dissolved foods and nutrients.

Ventifacts Pebbles or cobbles with flat faces often intersecting as sharp ridges that have been modified by windblown sand.

Vertebrate Animal with an internal, jointed spinal column of cartilage or bone, a skull, and paired appendages.

Vesicles Small cavities as found in some volcanic rocks.

Vestigial A small or useless organ or limb that was functional in ancestral animals.

Virus A substance that contains the chemical elements of protoplasm and acts as a living disease-producing organism in plants or animals but when isolated forms a crystalline structure like a mineral. It seems to link the mineral and life kingdoms.

Viscous Fluid but resisting flow. Does not change form quickly as more perfect fluids do.

Volatile Easily vaporized.

Volcanism (vulcanism) The processes creating volcanic rocks, volcanos, and related eruptive features.

Volcano A mountain resulting from the ejection, or eruption, of molten rock material and fragments from a vent extending into the Earth.

Vulcanist An eighteenth- or early-nineteenth-century geologist who recognized that basalt and granite were igneous rocks and believed that the Earth's internal heat dominated all major geologic processes.

Waste-mantle (regolith) The unconsolidated rock materials that may cover solid bedrock.

Water table The top of the zone in which rocks are saturated with groundwater.

Water witch A person who claims to locate water underground using forked sticks, rings suspended from threads, and other divining instruments.

Weathering The physical and mechanical breakdown of rock in place at or near the Earth's surface because of exposure to the atmosphere.

Well A man-made excavation to obtain underground water.

Wind gap A notch through a ridge that is no longer occupied by the stream that caused it.

Yardangs Sharp crested ridges separated by troughs resulting from wind erosion in soft rocks.

Index

CENOZOIC ERA,
TERTIARY PERIOD

	PALEOCENE 65 mil. yrs. ago*	EOCENE 54—3 mil. yrs.*	OLIGOCENE 38—7 mil. yrs.*	MIOCENE 26 mil. yrs.*	PLIOCENE 5 mil. yrs.*
MONOTREMES				Platypus	
MARSUPIALS		Opossums			
INSECTIVORES		Shrew		Moles	
BATS					
PRIMATES	Tree Shrew / Prosimians	Lemurs *Notharctus* / Tarsiers		Proto-anthropoids / *Ramapithecus* 14 million yrs. ago	3,000,000 yrs. ago *Australopithecus* "Lucy" *Homo* family (Johanson & Taieb 1974- Hadar, Ethiopia) / 4,000,000 Early hominids (Howell & Coppens 1969 Omo River, Kenya)
EDENTATES			Ancestral armadillos		Anteater
CETACEA			Whale		Porpoises
RODENTS			Squirrel-like mammals	Beaver	
RABBITS			Rabbits	Pikas	
CREODONTS					
CARNIVORES		Ancestral dogs		Hyaenas	Ancestral bears
CONDYLARTHS	*Phenacodus*				
PROBOSCIDIANS				Early mastodonts	
UINTATHERES					
PERISSODACTYLS		Ancestral horse / Titanothere	*Mesohippus*	*Merychippus* / *Moropus*	*Pliohippus*
ARTIODACTYLS		Oreodont	Camels / Entelodonts	Early antelope / *Dinohyus*	*Alticamelus* / *Synthetoceras*

PLACENTALS

UNGULATES

Date of Beginning of Epoch